Compendium of Organic Synthetic Methods

Compendium of Organic Synthetic Methods

IAN T. HARRISON

and

SHUYEN HARRISON

SYNTEX RESEARCH
PALO ALTO, CALIFORNIA

WILEY - INTERSCIENCE
A Division of John Wiley & Sons, Inc.
New York · London · Sydney · Toronto

PREFACE

Compendium of Organic Synthetic Methods is a systematic listing of functional group transformations designed for use by bench chemists, persons planning syntheses, students attending courses on synthetic chemistry, and teachers of these courses. The idea for this compilation came from the observation that organic chemists spend a large proportion of their time searching a formidable original literature for these hard-to-find synthetic methods.

A key feature of this book is the classification of reactions on the basis of the functional group of the starting material and of the product, without reference to the reaction mechanism. We wished to produce as comprehensive a set of reactions as possible, covering all branches of organic chemistry. Reactions giving low yields or requiring exotic reagents are not omitted. Consequently reactions included cover the full range of methods from boiling in oil to treatment with fluorine or orange-peel enzymes.

The presentation of each synthetic method in the form of representative reactions without discussion follows the plan used successfully in *Steroid Reactions* (Djerassi, Holden-Day). The limitations of such compilations containing much information but few words are obvious; there is, however, a great need for a comprehensive one-volume listing of synthetic methods as an intermediary between the chemist and the literature. The reader interested in a detailed discussion of synthetic methods should consult *Reagents for Organic Synthesis* (Fieser and Fieser, Wiley) and *Survey of Organic Syntheses* (Buehler and Pearson, Wiley).

We apologize to authors for the space-saving anonymity of references and for referring in many instances not to papers by the originators of a reaction but rather to subsequent articles by other authors. We make no apology, however, for omitting unnecessary reference punctuation, and avoiding the use of *ibid.* and other sources of confusion.

Ian T. Harrison
Shuyen Harrison

Palo Alto, California
March 1971

CONTENTS

ABBREVIATIONS

Ac	acetyl
Bu	butyl
DCC	dicyclohexylcarbodiimide
DDQ	2,3-dichloro-5,6-dicyanobenzoquinone
DMA	dimethylacetamide
DMF	dimethylformamide
Et	ethyl
HMPA	hexamethylphosphoramide
Me	methyl
Ms	methanesulfonyl
NBA	*N*-bromoacetamide
NBS	*N*-bromosuccinimide
NCS	*N*-chlorosuccinimide
Ni	Raney nickel
Ph	phenyl
Pr	propyl
Pyr	pyridine
THF	tetrahydrofuran
THP	tetrahydropyranyl
Ts	*p*-toluenesulfonyl

INDEX

Sections—heavy type
Pages—light type

PREPARATION OF → (rows)
FROM → (columns)

*Cell format: **section** (heavy type) / page (light type). Blank cells correspond to sections for which no examples were found in the literature.*

PREPARATION OF ↓ \ FROM →	Acetylenes	Carboxylic acids, acid halides, anhydrides	Alcohols, phenols	Aldehydes	Alkyls, methylenes, aryls	Amides	Amines	Esters	Ethers, epoxides	Halides, sulfonates	Hydrides (RH)	Ketones	Nitriles	Olefins
Acetylenes	**1**/1	**16**/16	**31**/75	**46**/132	**61**/177	**76**/203	**91**/230	**106**/271		**136**/329		**166**/379	**181**/457	**196**/479
Carboxylic acids, acid halides, anhydrides	**2**/2	**17**/18	**32**/76	**47**/132	**62**/178	**77**/204	**92**/230	**107**/272	**122**/309	**137**/329	**152**/357	**167**/380	**182**/457	**197**/482
Alcohols, phenols	**3**/3	**18**/26	**33**/78	**48**/137	**63**/180	**78**/208	**93**/232	**108**/280	**123**/310	**138**/331	**153**/359	**168**/386	**183**/459	**198**/484
Aldehydes	**4**/3	**19**/31	**34**/80	**49**/144	**64**/181	**79**/209	**94**/233	**109**/287	**124**/316	**139**/338	**154**/363	**169**/396	**184**/460	**199**/489
Alkyls, methylenes, aryls		**20**/36		**50**/146	**65**/182	**80**/210					**155**/365	**170**/400	**185**/464	**200**/493
Amides		**21**/39	**36**/85	**51**/148	**66**/184	**81**/211	**96**/236	**111**/289		**141**/338	**156**/366	**171**/403	**186**/464	**201**/495
Amines	**7**/5	**22**/41	**37**/86	**52**/150	**67**/184	**82**/213	**97**/240	**112**/290	**127**/318	**142**/339	**157**/367	**172**/404	**187**/465	**202**/496
Esters	**8**/6	**23**/42	**38**/87	**53**/152	**68**/185	**83**/218	**98**/249	**113**/291	**128**/318	**143**/342	**158**/368	**173**/406	**188**/467	**203**/498
Ethers, epoxides		**24**/46	**39**/92	**54**/154	**69**/186	**84**/220	**99**/249	**114**/293	**129**/320	**144**/343	**159**/369	**174**/408	**189**/467	**204**/501
Halides, sulfonates, sulfates	**10**/6	**25**/47	**40**/102	**55**/156	**70**/186	**85**/220	**100**/250	**115**/295	**130**/320	**145**/345	**160**/370	**175**/411	**190**/468	**205**/504
Hydrides (RH)		**26**/53	**41**/107	**56**/162	**71**/191	**86**/222	**101**/255	**116**/299	**131**/322	**146**/349	**161**/375	**176**/417	**191**/471	**206**/512
Ketones	**12**/10	**27**/56	**42**/110	**57**/164	**72**/193	**87**/223	**102**/258	**117**/302	**132**/323	**147**/353	**162**/375	**177**/419	**192**/473	**207**/513
Nitriles		**28**/62		**58**/166	**73**/198	**88**/225	**103**/262	**118**/304			**163**/376	**178**/433	**193**/474	**208**/520
Olefins	**14**/13	**29**/64	**44**/119	**59**/168	**74**/198	**89**/227	**104**/264	**119**/305	**134**/325	**149**/354	**164**/377	**179**/435	**194**/475	**209**/520
Miscellaneous compounds	**15**/13	**30**/68	**45**/122	**60**/172	**75**/202	**90**/229	**105**/266	**120**/307	**135**/328	**150**/356	**165**/377	**180**/442	**195**/476	**210**/526

PROTECTION

	Sect.	Pg.
Carboxylic acids	30A	71
Alcohols, phenols	45A	124
Aldehydes	60A	174
Amines	105A	266
Ketones	180A	449

Blanks in the table correspond to sections for which no examples were found in the literature

xi

INTRODUCTION

Classification and Organization. *Compendium of Organic Synthetic Methods* contains approximately 3000 examples of published chemical transformations classified according to the reacting functional group of the starting material and the functional group formed. Those reactions that give products with the same functional group form a chapter. The reactions in each chapter are further classified into sections on the basis of the functional group of the starting material. Within each section reactions are listed in a somewhat arbitrary order although an effort has been made to put chain lengthening processes before degradations.

The classification is unaffected by allylic, vinylic, or acetylenic unsaturation, which appears in both starting material and product, or by increases or decreases in the length of carbon chains. For example, the reactions t-BuOH $\rightarrow$ t-BuCOOH, PhCH$_2$OH $\rightarrow$ PhCOOH, and PhCH $=$ CHCH$_2$OH $\rightarrow$ PhCH $=$ CHCOOH are all found in Section 18 on carboxylic acids from alcohols.

The terms hydrides, alkyls, and aryls classify compounds containing reacting hydrogens, alkyl groups, and aryl groups, respectively; for example, RCH$_2$-H $\rightarrow$ RCH$_2$COOH (carboxylic acids from hydrides), RMe $\rightarrow$ RCOOH (carboxylic acids from alkyls), RPh $\rightarrow$ RCOOH (carboxylic acids from aryls). Note the distinction between R$_2$CO $\rightarrow$ R$_2$CH$_2$ (methylenes from ketones) and RCOR' $\rightarrow$ RH (hydrides from ketones).

The following examples illustrate the application of the classification scheme to some potentially confusing cases:

RCH$=$CHCOOH $\rightarrow$ RCH$=$CH$_2$	(hydrides from carboxylic acids)
RCH$=$CH$_2$ $\rightarrow$ RCH$=$CHCOOH	(carboxylic acids from hydrides)
ArH $\rightarrow$ ArCOOH	(carboxylic acids from hydrides)
ArH $\rightarrow$ ArOAc	(esters from hydrides)
RCHO $\rightarrow$ RH	(hydrides from aldehydes)
RCH$=$CHCHO $\rightarrow$ RCH$=$CH$_2$	(hydrides from aldehydes)
RCHO $\rightarrow$ RCH$_3$	(alkyls from aldehydes)
R$_2$CH$_2$ $\rightarrow$ R$_2$CO	(ketones from methylenes)
RCH$_2$COR $\rightarrow$ R$_2$CHCOR	(ketones from ketones)
RCH$=$CH$_2$ $\rightarrow$ RCH$_2$CH$_3$	(alkyls from olefins)
RBr $+$ RC$\equiv$CH $\rightarrow$ RC$\equiv$CR	(acetylenes from halides, also acetylenes from acetylenes)

ROH + RCOOH → RCOOR   (esters from alcohols, also esters from carboxylic acids)

Sulfonic esters are grouped with halides. Hydrazines are listed with amines and hydrazides with amides.

Yields quoted are overall with allowance for percentage of conversion, impurities, etc. and therefore often differ from the values given in the original paper.

Trivial reactions not described in the given references but required to complete a sequence are indicated by a dashed arrow.

How to Use the Book. Examples of the preparation of one functional group from another are located via the index on p. xi, which gives the corresponding section and page. Thus Section 1 contains examples of the preparation of acetylenes from other acetylenes; Section 2, acetylenes from carboxylic acids; Section 3, acetylenes from alcohols; etc.

Sections giving examples of the reactions of a functional group are found in horizontal rows of the index. Thus Section 1 gives examples of the reactions of acetylenes forming other acetylenes; Section 16, reactions of acetylenes forming carboxylic acids; Section 31, reactions of acetylenes forming alcohols; etc.

Examples of alkylation, dealkylation, homologation, isomerization, transposition, etc. are found in Sections 1, 17, 33, and so forth, which lie close to a diagonal of the index. These sections correspond to the preparation of acetylenes from acetylenes, carboxylic acids from carboxylic acids, alcohols and phenols from alcohols and phenols, etc.

Examples of the protection of carboxylic acids, alcohols, phenols, aldehydes, amines, and ketones are also indexed on page xi.

Examples of name reactions can be found by first considering the nature of the starting material and product. The Wittig reaction, for example, is to be found in Section 199 on olefins from aldehydes and Section 207 on olefins from ketones.

The pairs of functional groups, alcohol–ester, carboxylic acid–ester, amine–amide, carboxylic acid–amide, can be interconverted by quite trivial reactions. When a member of these groups is the desired product or starting material, the other member should of course also be consulted in the text.

The original literature must be used to determine the generality of reactions. A reaction given in this book for a primary aliphatic substrate may in fact also be applicable to tertiary or aromatic compounds.

The references given usually yield a further crop of references to previous work. Subsequent publications can be found through Science Citation Index.

Reactions Included in the Book. Interconversions of monofunctional compounds form the major part of this compilation. Reactions of bifunctional compounds in which the two functions are identical and which give monofunctional products are also included; for example, $R_2C(COOH)_2 \rightarrow R_2CO$ (ketones from carboxylic acids).

Examples of the removal of one functional group from bifunctional compounds and the preparation of functional groups from groups not listed on the index are included in the miscellaneous sections; for example, $RCH(Br)COR \rightarrow RCH_2COR$, $RCH(OH)COR \rightarrow RCH_2COR$ and $RCH=CHCOR \rightarrow RCH_2CH_2COR$ (ketones from miscellaneous compounds), $RCH_2NO_2 \rightarrow RCHO$ (aldehydes from miscellaneous compounds).

Reactions are included even when full experimental details are lacking from the given reference. In some cases the quoted reaction is a minor part of a paper or may have been investigated from a purely mechanistic aspect. When several references are given, the first refers to the reaction illustrated; others give further examples, related reactions, or reviews.

Reactions Not Included in the Book. Reactions forming bifunctional products are not included, for example, $RCH=CH_2 \rightarrow RCH(OH)CH_2OH$. Chain lengthening processes via unsaturated intermediates, for instance, $RCHO \rightarrow [RCH= CHCOOEt] \rightarrow RCH_2CH_2COOEt$, are only partially covered. Ring forming reactions and reactions that involve several functional centers (e.g., the Diels-Alder reaction) are represented by very few examples. These gaps will be filled by a second volume, presently under consideration, which will include reactions forming unsaturated and other bifunctional products.

Reactions published after early 1971 are not included.

Compendium of Organic Synthetic Methods

Chapter 1 PREPARATION OF ACETYLENES

Section 1 <u>Acetylenes from Acetylenes</u>

Review: The Synthesis of Acetylenes Org React (1949) 5 1

$C_5H_{11}C{\equiv}CH \xrightarrow[\text{2 BuI HMPA}]{\text{1 i-PrMgCl } Et_2O} C_5H_{11}C{\equiv}CBu$ 66%

Bull Soc Chim Fr (1964) 2000

$BuC{\equiv}CH \xrightarrow[\text{2 } Et_2SO_4 \text{ } Et_2O]{\text{1 EtMgBr } Et_2O} BuC{\equiv}CEt$ 70%

JACS (1936) 58 796
JOC (1959) 24 840

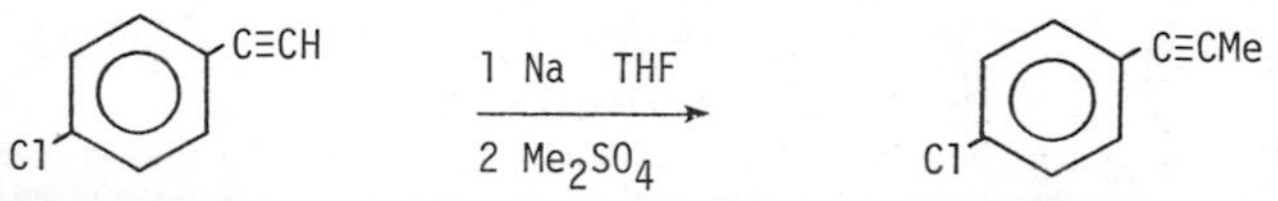
C≡CH
Cl
1 Na THF
2 Me2SO4
C≡CMe
Cl

36%

Bull Soc Chim Fr (1965) 1525

Further examples of the reaction   RC≡CH + R'X ⟶ RC≡CR'   are included in
section 10 (Acetylenes from Halides, Sulfonates and Sulfates)

PhC≡CH $\xrightarrow{\text{CH}_2\text{I}_2\quad \text{Zn-Cu}\quad \text{Et}_2\text{O}}$ PhC≡CMe   37%

Bull Soc Chim Fr (1965) 1525
Tetrahedron (1958) 3 197

EtCH$_2$C≡CH $\xrightarrow{\text{KOH}\quad \text{EtOH}\quad 170\text{-}180°}$ EtC≡CMe   ~70%

JACS (1951) 73 1273
Quart Rev (1970) 24 585

C$_5$H$_{11}$C≡CCH$_3$ $\xrightarrow[150°]{\text{NaNH}_2\quad 1,2,4\text{-trimethylbenzene}}$ C$_5$H$_{11}$CH$_2$C≡CH   80%

Org React (1949) 5 1

BuC≡C(CH$_2$)$_3$CH$_3$ $\xrightarrow{\text{NaNH}_2\quad \text{mineral oil}}$ Bu(CH$_2$)$_4$C≡CH

Org React (1949) 5 1

Section 2   <u>Acetylenes from Carboxylic Acids and Acid Halides</u>
oooooooooooooooooooooooooooooooooooooooooooooooooooooo

BuC≡CCOCl $\xrightarrow{\substack{1\ \text{Ph}_3\text{P=CHPh}\quad \text{Et}_3\text{N}\quad \text{C}_6\text{H}_6 \\ 2\ \sim 280°}}$ BuC≡C-C≡CPh

JCS (1964) 543

$$PhCH_2COOH \xrightarrow[\text{2 PhMgBr-CdCl}_2]{\text{1 SOCl}_2} PhCH_2COPh \xrightarrow[\text{2 NaNH}_2]{\text{1 PCl}_5} PhC{\equiv}CPh \qquad 34\%$$

Helv (1938) <u>21</u> 1356

Section 3   Acetylenes from Alcohols

$$BuOH \xrightarrow{\text{Na TsCl}} BuOTs \xrightarrow{\text{NaC}{\equiv}\text{CH NH}_3} BuC{\equiv}CH \qquad 37\text{-}47\%$$

JACS (1937) <u>59</u> 1490
Org React (1949) <u>5</u> 1

Section 4   Acetylenes from Aldehydes

$$PhCHO \xrightarrow[\text{Et}_2O]{\text{Ph}_3\text{P=CHO-}\bigcirc\text{-Me}} PhCH{=}CHO\text{-}\bigcirc\text{-Me} \xrightarrow{\text{PhLi}} PhC{\equiv}CH \qquad 29\%$$

Ber (1962) <u>95</u> 2514

$$AcOCH_2CH{=}\underset{\underset{Me}{|}}{C}(CH_2)_2CHO \xrightarrow[\substack{\text{lithium piperidide} \\ \text{2 MeONa MeOH} \\ \text{3 BuLi Et}_2O}]{\text{1 Ph}_3\overset{+}{P}CH_2Cl \; \overset{-}{Cl} \; Et_2O} HOCH_2CH{=}\underset{\underset{Me}{|}}{C}(CH_2)_2C{\equiv}CH \qquad 55\%$$

JACS (1969) <u>91</u> 4318

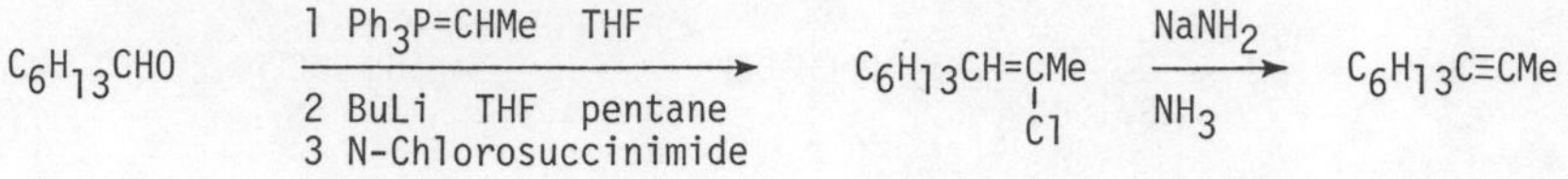
C6H13CHO
1 Ph3P=CHMe THF
2 BuLi THF pentane
3 N-Chlorosuccinimide
C6H13CH=CMe
Cl
NaNH2
NH3
C6H13C≡CMe

Tetr Lett (1970) 447

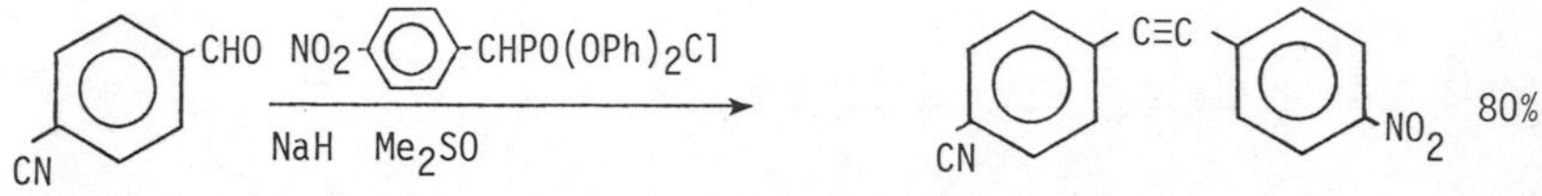
CHO
CN
NO2
CHPO(OPh)2Cl
NaH Me2SO
C≡C
NO2
80%

JACS (1965) 87 2777

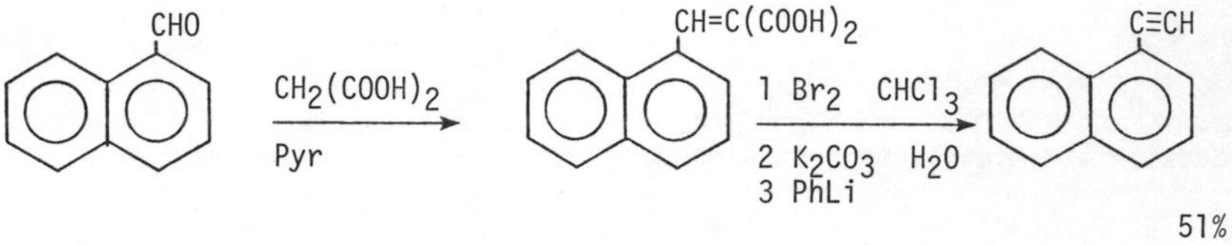
CHO
CH2(COOH)2
Pyr
CH=C(COOH)2
1 Br2 CHCl3
2 K2CO3 H2O
3 PhLi
C≡CH
51%

Compt Rend (1949) 229 660

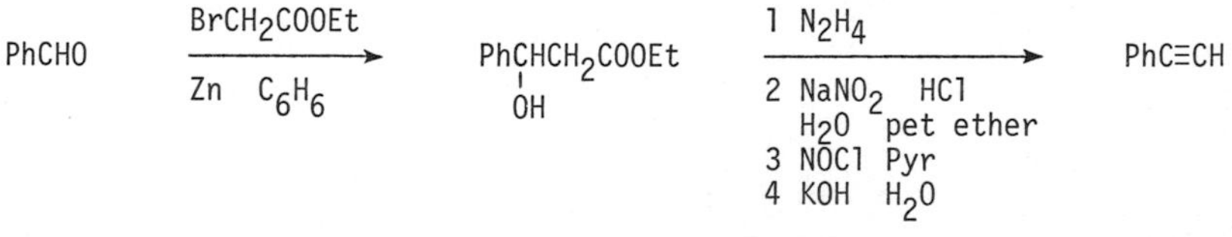
PhCHO
BrCH2COOEt
Zn C6H6
PhCHCH2COOEt
OH
1 N2H4
2 NaNO2 HCl
H2O pet ether
3 NOCl Pyr
4 KOH H2O
PhC≡CH

JACS (1951) 73 4199

Section 5   Acetylenes from Alkyls, Methylenes and Aryls

No examples

Section 6   Acetylenes from Amides

No examples

Section 7   Acetylenes from Amines

$$
\begin{array}{c}
\overset{+}{} \\
EtCHCH_2NMe_3 \\
\underset{\underset{+}{NMe_3}}{|} \quad 2Br^-
\end{array}
\quad
\xrightarrow[\text{2 100-250°}]{\text{1 Ag}_2\text{O   H}_2\text{O}}
\quad
EtC{\equiv}CH
$$

JACS (1939) <u>61</u> 1943

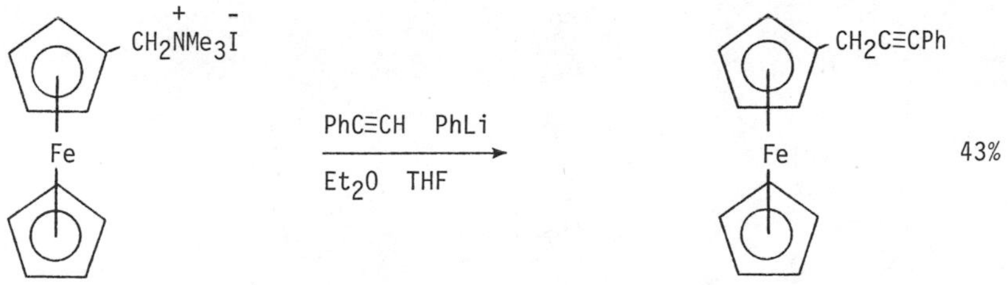

$$
\xrightarrow[\text{Et}_2\text{O   THF}]{\text{PhC}{\equiv}\text{CH   PhLi}}
$$

43%

JCS (1963) 2990

Section 8   Acetylenes from Esters
             °°°°°°°°°°°°°°°°°°°°°°°°°°°

$$\text{MeCHCH}_2\text{OAc} \xrightarrow{\ 450°\ } \text{MeC≡CH}$$
$$\quad\ \ \big|$$
$$\quad\ \ \text{OAc}$$

Izv (1959) 43
(Chem Abs 54 7547)

Section 9   Acetylenes from Ethers
             °°°°°°°°°°°°°°°°°°°°°°°°°°°

No examples

Section 10   Acetylenes from Halides, Sulfonates and Sulfates
             °°°

Review: The Synthesis of Acetylenes   Org React (1949) 5 1

$$\text{i-PrBr} \xrightarrow[\ 150°\ ]{\text{PhC≡CLi}\quad\text{dioxane}} \text{i-PrC≡CPh} \qquad\qquad 65\%$$

Annalen (1958) 614 37

$$\xrightarrow[\text{Me}_2\text{SO}]{\text{LiC≡CH·H}_2\text{NCH}_2\text{CH}_2\text{NH}_2}$$

53%

JACS (1969) 91 4771
Angew (1959) 71 245

$Me_2C=CH(CH_2)_2\underset{\underset{Me}{|}}{C}=CHCH_2Br$ $\xrightarrow[\substack{2\ AgNO_3 \\ 3\ NaCN}]{1\ LiCH_2C\equiv CSiMe_3}$ $Me_2C=CH(CH_2)_2\underset{\underset{Me}{|}}{C}=CHCH_2CH_2C\equiv CH$

Tetr Lett (1968) 5041
(1970) 2249

BuOTs $\xrightarrow[\text{Toluene}]{PhC\equiv CNa}$ $PhC\equiv CBu$ 65-70%

Org React (1949) 5 1
JACS (1937) 59 1490

Et_2SO_4 $\xrightarrow{NaC\equiv CH\ \ NH_3}$ $EtC\equiv CH$ 60%

JACS (1931) 53 289

$i\text{-}Pr_2SO_4$ $\xrightarrow{NaC\equiv CH\ \ NH_3}$ $i\text{-}PrC\equiv CH$ 29-50%

JACS (1937) 59 1490

EtBr $\xrightarrow[\substack{2\ BrCH_2C\equiv CCH_2Br}]{1\ Mg\ \ Et_2O}}$ $EtCH_2C\equiv CCH_2Et$

JCS (1946) 1009

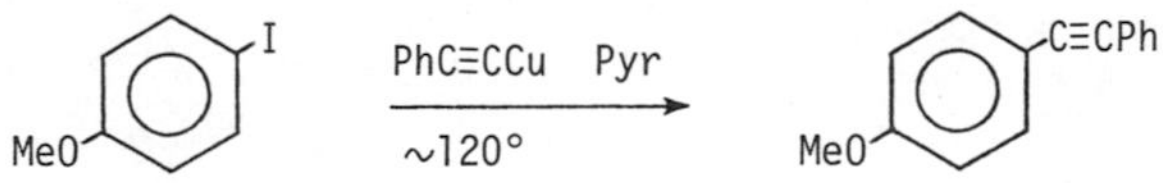

JOC (1963) 28 3313
JACS (1964) 86 4358

PhCH=CHBr $\xrightarrow{\text{PhC}\equiv\text{CCu} \quad \text{DMF}}$ PhCH=CHC≡CPh   75%

Chem Comm (1967) 1259

PhI $\xrightarrow{\begin{array}{l}\text{1 CuC}\equiv\text{CCH(OEt)}_2 \quad \text{pyr}\\ \text{2 Acid hydrolysis}\\ \text{3 NaOH \ MeOH \ H}_2\text{O}\end{array}}$ PhC≡CH

JCS <u>C</u> (1969) 2173

Further examples of the reaction RC≡CH + R'X → RC≡CR' are included in section 1 (Acetylenes from Acetylenes)

BuBr $\xrightarrow{\begin{array}{l}\text{1 Mg \ Et}_2\text{0}\\ \text{2 PhC}\equiv\text{CBr \ CoCl}_2\end{array}}$ BuC≡CPh   32%

JCS (1954) 1704

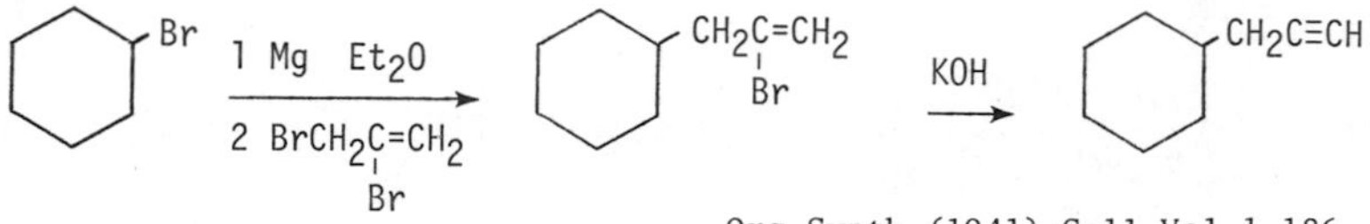

Org Synth (1941) Coll Vol 1 186
Org React (1949) <u>5</u> 1

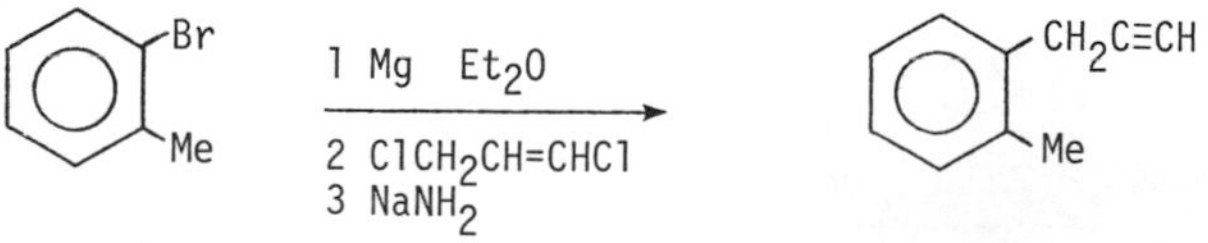

Compt Rend (1925) <u>181</u> 555

$$\text{[cyclohexyl-Cl]} \xrightarrow[\text{2 NaNH}_2 \quad \text{NH}_3]{\text{1 CH}_2\text{=CHCl} \quad \text{AlCl}_3} \text{[cyclohexyl-C≡CH]} \qquad 36\%$$

Rec Trav Chim (1965) <u>84</u> 31

$$\begin{array}{c} \text{EtCHCH}_2\text{Br} \\ | \\ \text{Br} \end{array} \xrightarrow[\begin{array}{c} \text{2 Ag}_2\text{O} \quad \text{H}_2\text{O} \\ \text{3 100-250°} \end{array}]{\text{1 Me}_3\text{N}} \text{EtC≡CH} \qquad 9\%$$

JACS (1939) <u>61</u> 1943

$$\begin{array}{c} \text{C}_8\text{H}_{17}\text{CH-CH}_2 \\ | \quad | \\ \text{Br} \quad \text{Br} \end{array} \xrightarrow{\text{NaNH}_2 \quad \text{Me}_2\text{SO}} \text{C}_8\text{H}_{17}\text{C≡CH} \qquad 54\%$$

Tetrahedron (1970) <u>26</u> 2127
JACS (1934) <u>56</u> 2120

$$\begin{array}{c} \text{EtCH}_2\text{CH-CH}_2 \\ | \quad | \\ \text{Br} \quad \text{Br} \end{array} \xrightarrow{\text{KOH} \quad \text{EtOH}} \text{EtCH}_2\text{C≡CH}$$

Org React (1949) <u>5</u> 1

Further examples of the conversion of dibromides into acetylenes are
included in section 14 (Acetylenes from Olefins)

Section 11   Acetylenes from Hydrides (RH)
　　　　　　ooooooooooooooooooooooooooooooooo

No examples

Section 12   Acetylenes from Ketones
oooooooooooooooooooooooo

Review: The Synthesis of Acetylenes   Org React (1949) 5 1

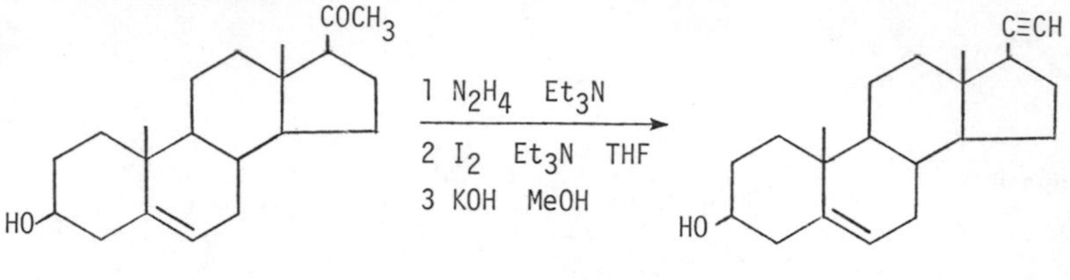

1 N_2H_4 Et_3N
2 I_2 Et_3N THF
3 KOH   MeOH

JOC (1969) 34 3502

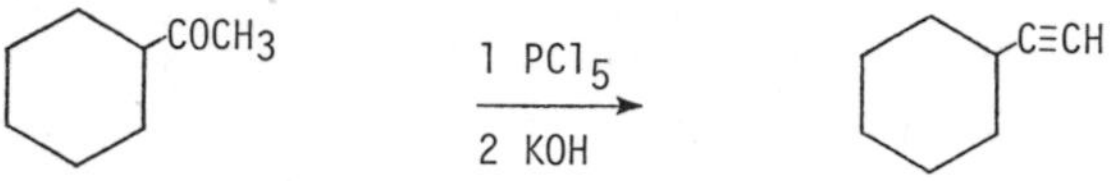

1 PCl_5
2 KOH

~35%

JACS (1932) 54 1184

1 PCl_5 C_6H_6
2 $NaNH_2$ NH_3

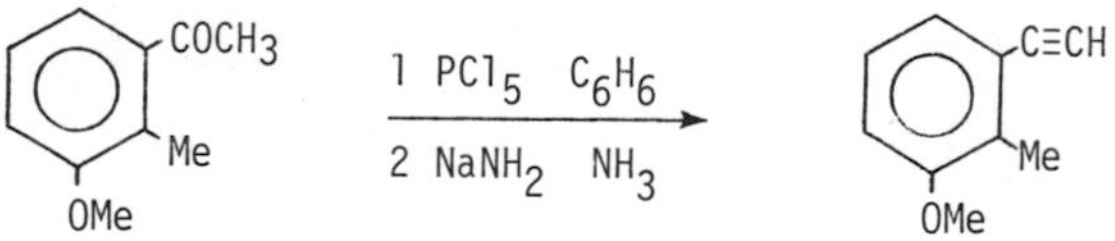

73%

Tetrahedron (1969) 25 4249

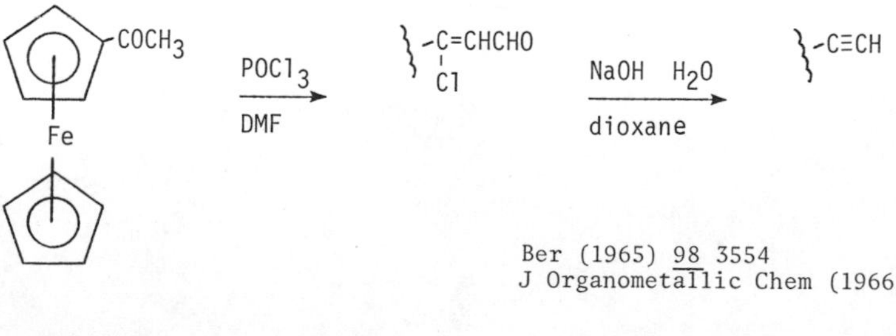

$POCl_3$
DMF

$NaOH$ H_2O
dioxane

77%

Ber (1965) 98 3554
J Organometallic Chem (1966) 6 173

PhCH$_2$COMe $\xrightarrow[\text{2 (CF}_3\text{COO)}_2\text{Hg}_2 \quad \text{Et}_2\text{O}]{\text{1 N}_2\text{H}_4 \cdot \text{H}_2\text{O}}$ PhC≡CMe   48%

JOC (1966) 31 624

PhCOCH$_3$ --→ PhCOCH$_2$Br $\xrightarrow{\text{(PhO)}_3\text{P}}$ PhC≡CH

Dokl (1965) 163 656
(Chem Abs 63 11338)

PhCOCH$_3$ --→ PhCOCH$_2$Cl $\xrightarrow[\text{2 NaNH}_2 \quad \text{Et}_2\text{O} \quad \text{NH}_3]{\text{1 (EtO)}_3\text{P}}$ PhC≡CH   ~75%

JCS (1963) 3712

PhCO
$\underset{\text{Et}}{|}$ $\xrightarrow[\text{Zn} \quad \text{C}_6\text{H}_6]{\text{BrCH}_2\text{COOEt}}$ $\overset{\text{Et}}{\underset{\text{OH}}{\text{PhCCH}_2\text{COOEt}}}$ $\xrightarrow{\begin{array}{l}\text{1 N}_2\text{H}_4 \\ \text{2 NaNO}_2 \quad \text{HCl} \\ \quad \text{H}_2\text{O} \quad \text{pet ether} \\ \text{3 KOH} \quad \text{EtOH} \quad \text{H}_2\text{O}\end{array}}$ PhC≡CEt   80%

JACS (1951) 73 4199

$\xrightarrow{\begin{array}{l}\text{1 N}_2\text{H}_4 \\ \text{2 MnO}_2 \\ \text{3 CHCl}_3 \quad \text{t-BuOK} \\ \text{4 MeLi}\end{array}}$

Chem Ind (1969) 1306

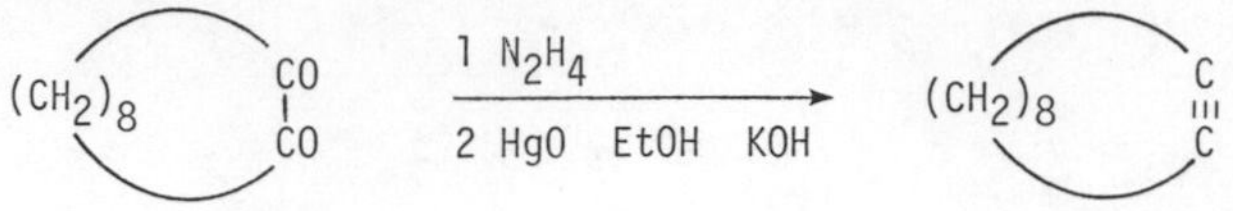

$(CH_2)_8 \begin{matrix} CO \\ | \\ CO \end{matrix}$ $\xrightarrow[\text{2 HgO EtOH KOH}]{\text{1 } N_2H_4}$ $(CH_2)_8 \begin{matrix} C \\ ||| \\ C \end{matrix}$

JACS (1952) <u>74</u> 3636 3643

PhCOCOPh $\xrightarrow[\text{HgO } C_6H_6]{N_2H_4 \cdot H_2O \quad \text{PrOH}}$ PhC≡CPh   67-73%

Org Synth (1963) Coll Vol 4 377

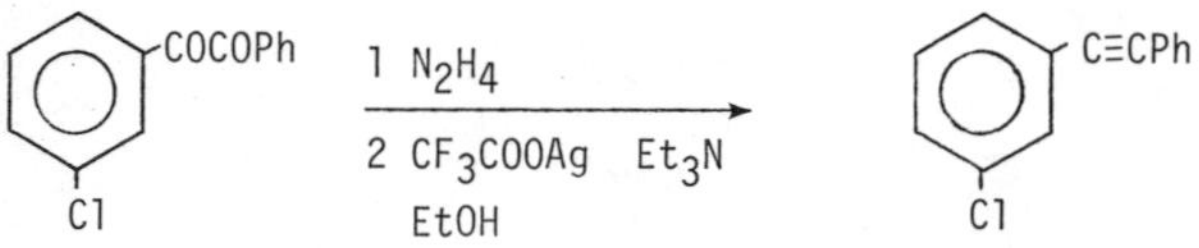

$\xrightarrow[\substack{\text{2 } CF_3COOAg \quad Et_3N \\ \text{EtOH}}]{\text{1 } N_2H_4}$ 80%

JOC (1958) <u>23</u> 665

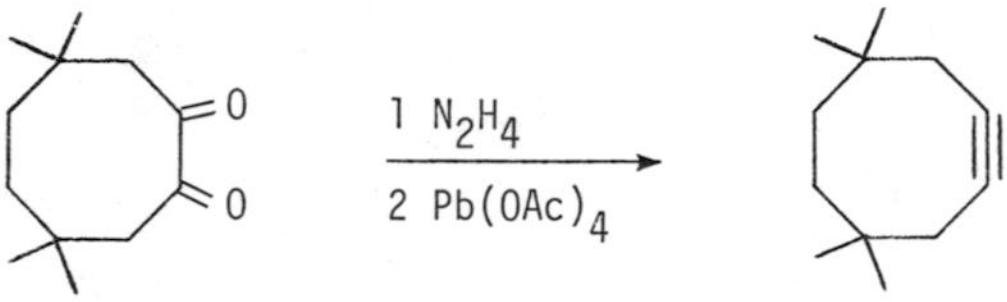

$\xrightarrow[\text{2 Pb(OAc)}_4]{\text{1 } N_2H_4}$ 26%

Tetr Lett (1968) 4511

PhCOCOPh $\xrightarrow{\text{(EtO)}_3P \quad 215°}$ PhC≡CPh   60%

JOC (1964) <u>29</u> 2243

Section 13   Acetylenes from Nitriles
             °°°°°°°°°°°°°°°°°°°°°°°°°°°°°°°

No examples

Section 14   Acetylenes from Olefins
             °°°°°°°°°°°°°°°°°°°°°°°°°°°°°°°°

$BuCH=CH_2$ $\xrightarrow{\text{Li} \quad \text{THF}}$ $BuC{\equiv}CLi$ $\xdashrightarrow{H_2O}$ $BuC{\equiv}CH$

~65%

JOC (1967) <u>32</u> 105

$\dfrac{1 \ Br_2 \quad CCl_4}{2 \ NaNH_2 \quad HMPA}$

48%

Tetr Lett (1970) 41
Tetrahedron (1970) <u>26</u> 2127

$C_8H_{17}CH{=}CH(CH_2)_7COOH$ $\xrightarrow[2 \ NaNH_2 \quad NH_3 \quad Et_2O]{1 \ Br_2}$ $C_8H_{17}C{\equiv}C(CH_2)_7COOH$ 68%

J Am Oil Chem Soc (1951) <u>28</u> 27
(Chem Abs <u>45</u> 8449)
Org Synth (1947) <u>27</u> 76

Section 15   Acetylenes from Miscellaneous Compounds
             °°

Review: The Synthesis of Acetylenes   Org React (1949) <u>5</u> 1

$$PhCH=CHBr \xrightarrow{\text{BuLi} \quad Et_2O} PhC\equiv CLi \xrightarrow{\text{H}_2\text{O}} PhC\equiv CH$$

JACS (1940) <u>62</u> 2327

$$PhCH=CHCl \xrightarrow{\text{PhLi} \quad Et_2O} PhC\equiv CH \qquad\qquad 100\%$$

Ber (1941) <u>74</u>B 1474

$$PhCH=CHBr \xrightarrow{\text{NaH} \quad HMPA} PhC\equiv CH \qquad\qquad 78\%$$

Bull Soc Chim Fr (1966) 1293

$$C_6H_{13}\underset{Br}{C}=CH_2 \xrightarrow{\text{NaH} \quad HMPA} C_6H_{13}C\equiv CH \qquad\qquad 70\%$$

Bull Soc Chim Fr (1966) 1293

$$PhC=CHBr \atop Me \xrightarrow{\text{NaNH}_2 \quad HMPA \quad C_6H_6} PhC\equiv CMe \qquad\qquad 40\%$$

Bull Soc Chim Fr (1966) 1293

$$CF_3\underset{Cl}{C}=\underset{Cl}{C}CF_3 \xrightarrow{\text{Zn} \quad EtOH} CF_3C\equiv CCF_3 \qquad\qquad 54\%$$

JACS (1949) <u>71</u> 298

$PhCH=CClF$ $\xrightarrow{\text{BuLi}}$ $PhC\equiv CBu$ ~20%

Angew (1963) $\underline{75}$ 638
(Internat Ed $\underline{2}$ 477)

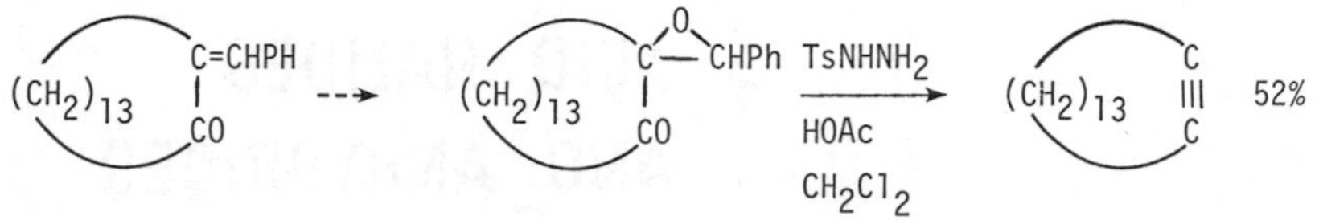

Helv (1967) $\underline{50}$ 2101

$\underset{\overset{|}{OH}}{PhCOCHPh}$ $\xrightarrow[\text{2 } C_6H_{13}COOH \quad 500°]{\text{1 } HC(OEt)_3 \quad HOAc}$ $PhC\equiv CPh$ 40%

Chem Comm (1970) 206

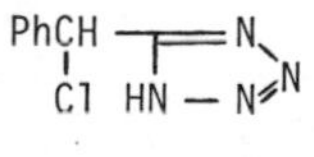

$\xrightarrow[\overline{176°}]{\text{Mesitylene}}$ $PhC\equiv CH$ 59%

Tetr Lett (1966) 1663

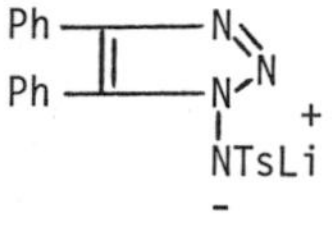

$\xrightarrow{h\nu \quad \text{dioxane}}$ $PhC\equiv CPh$ 85%

Angew (1964) $\underline{76}$ 144
(Internat Ed $\underline{3}$ 138)

$CH_2=CHCH_2\overset{+}{N}Me_3\overset{-}{Br}$ $\xrightarrow[\text{2 } 310-325°]{\text{1 } Ag_2O \quad H_2O}$ $HC\equiv CMe$ <34%

JOC (1949) $\underline{14}$ 1

Chapter 2   PREPARATION
OF
CARBOXYLIC ACIDS
ACID HALIDES
AND ANHYDRIDES

Section 16   Carboxylic Acids from Acetylenes
oooooooooooooooooooooooooooooooooooo

$BuC\equiv CH$ $\xrightarrow[\begin{array}{l}\text{2 BuLi   THF}\\\text{3 } CO_2\end{array}]{\text{1 (i-Bu)}_2\text{AlH}}$ $BuCH_2CH(COOH)_2$ $\xrightarrow{\triangle}$ $BuCH_2CH_2COOH$

62%

Tetr Lett (1966) 6021

$$\begin{array}{c}CH\\ \vertdots\\C\\ (CH_2)_8\\COOH\end{array} \xrightarrow[\text{2 } CO_2 \text{ Et}_2O]{\text{1 NaNH}_2 \text{ NH}_3} \begin{array}{c}CCOOH\\ \vertdots\\C\\ (CH_2)_8\\COOH\end{array} \xrightarrow{H_2 \text{ Ni } H_2O} \begin{array}{c}CH_2COOH\\ CH_2\\ (CH_2)_8\\COOH\end{array}$$ 33%

JACS (1945) 67 1171

$C_6H_{13}C\equiv CH$ $\xrightarrow[\text{2 TsCl}]{\text{1 Na}}$ $C_6H_{13}C\equiv CCl$ $\xrightarrow{\text{KOH   EtOH}}$ $C_6H_{13}CH_2COOH$ 42%

Annales de Chimie (1931) 16 309

BuC≡CH $\dfrac{\text{1 Dicyclohexylborane   THF}}{\text{2 m-Chloroperbenzoic acid   THF}}$ → BuCH$_2$COOH

JACS (1967) <u>89</u> 291

PhC≡CH $\dfrac{\text{CF}_3\text{COO}_2\text{H   Na}_2\text{HPO}_4}{\text{CH}_2\text{Cl}_2}$ → PhCOOH + PhCH$_2$COOH

                    25%   38%

JACS (1964) <u>86</u> 4866

C$_8$H$_{17}$C≡C(CH$_2$)$_7$COOH $\dfrac{\text{KMnO}_4\ \ \text{H}_2\text{O}}{\text{pH 12}}$ → HOOC(CH$_2$)$_7$COOH   80%

JOC (1952) <u>17</u> 1063

BuC≡CBu $\dfrac{\text{1 O}_3\ \ \text{CCl}_4\ \ \text{HOAc}}{\text{2 NaI}}$ → BuCOOH   35%

Annalen (1953) <u>583</u> 29

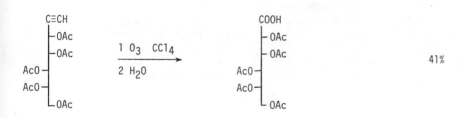

Carbohydrate Res (1966) <u>2</u> 315

Carboxylic acids may also be prepared by conversion of acetylenes into esters or amides, followed by hydrolysis. See section 106 (Esters from Acetylenes) and section 76 (Amides from Acetylenes)

Section 17   Carboxylic Acids, Acid Halides and Anhydrides
              from Carboxylic Acids and Acid Halides

$$C_5H_{11}COCl \xrightarrow[\substack{2\ NaOH\ EtOH \\ 3\ N_2H_4\ KOH\ triethanolamine}]{1\ Cyclododecanone\ morpholine\ enamine\ Et_3N} C_5H_{11}(CH_2)_{12}COOH$$

~70%

Ber (1967) 100 4010

$$\begin{array}{c} COCl \\ | \\ (CH_2)_8 \\ | \\ CH \\ \| \\ CH_2 \end{array} \xrightarrow[\substack{2\ MeONa\ MeOH \\ 3\ I(CH_2)_{10}COOEt\ K_2CO_3 \\ 4\ KOH\ MeOH}]{\substack{Na \\ 1\ MeCOCHCOOEt\ C_6H_6}} \begin{array}{c} COOH \\ | \\ (CH_2)_{11} \\ | \\ CO \\ | \\ (CH_2)_8 \\ | \\ CH \\ \| \\ CH_2 \end{array} \xrightarrow[\substack{KOH \\ triethylene \\ glycol}]{N_2H_4} \begin{array}{c} COOH \\ | \\ (CH_2)_{11} \\ | \\ CH_2 \\ | \\ (CH_2)_8 \\ | \\ CH \\ \| \\ CH_2 \end{array}$$

Arkiv Kemi (1949) 1 99
(Chem Abs 43 7414)

$$C_5H_{11}COCl \xrightarrow[\substack{2\ H_2\ Pd-C \\ 3\ 78°\ (decarbox) \\ 4\ H_2SO_4}]{\substack{(CH_2)_6COOCH_2Ph \\ | \\ 1\ NaC(COOCH_2Ph)_2}} C_5H_{11}CO(CH_2)_7COOH \xrightarrow[NaOH]{N_2H_4} C_5H_{11}(CH_2)_8COOH$$

56%

JCS (1950) 174

$$\begin{array}{c} COCl \\ | \\ (CH_2)_8 \\ | \\ CH \\ \| \\ CH_2 \end{array} \xrightarrow[2\ C_6H_6\ reflux]{\substack{(CH_2)_{10}COO-\bigcirc_O \\ | \\ 1\ NaC(COO-\bigcirc^O)_2}} \begin{array}{c} COOH \\ | \\ (CH_2)_{11} \\ | \\ CO \\ | \\ (CH_2)_8 \\ | \\ CH \\ \| \\ CH_2 \end{array} \xrightarrow[\substack{diethylene \\ glycol}]{N_2H_4\ KOH} \begin{array}{c} COOH \\ | \\ (CH_2)_{11} \\ | \\ CH_2 \\ | \\ (CH_2)_8 \\ | \\ CH \\ \| \\ CH_2 \end{array}$$

JCS (1952) 3945

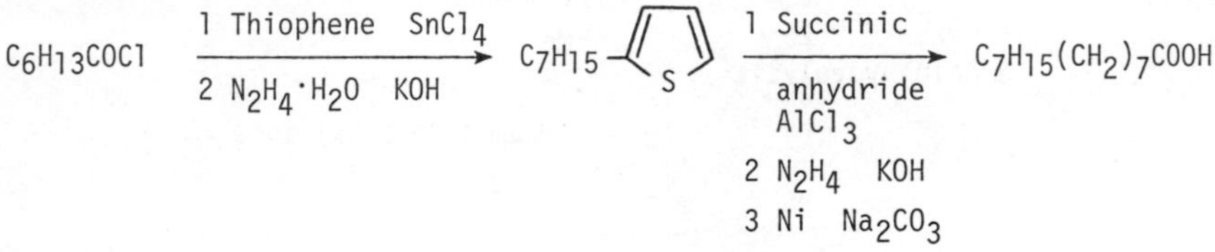
C6H13COCl
1 Thiophene SnCl4
2 N2H4·H2O KOH
C7H15
S
1 Succinic anhydride AlCl3
2 N2H4 KOH
3 Ni Na2CO3
C7H15(CH2)7COOH

JCS (1954) 4162
(1962) 350

$EtCOOH \xrightarrow{}$ $Et(CH_2)_3COOH$ 28%

1 $SOCl_2$
2 CH_2N_2 Et_2O
3 $CH_2{=}CO$ toluene
4 N_2H_4 NaOH diethylene glycol

Annalen (1964) 678 113

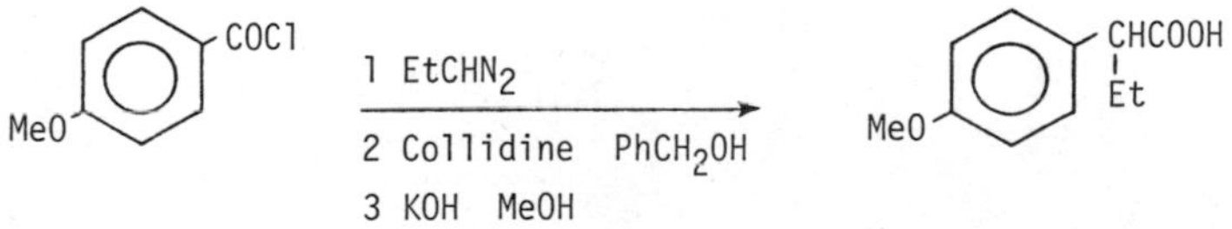
COCl
MeO
1 EtCHN2
2 Collidine PhCH2OH
3 KOH MeOH
CHCOOH
Et
MeO

37%

JOC (1948) 13 763

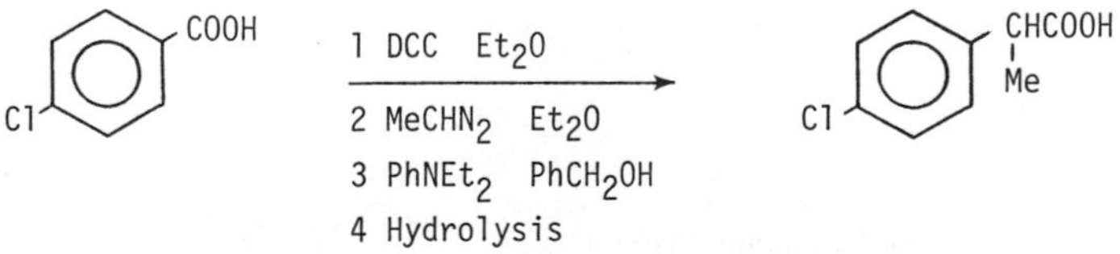
COOH
Cl
1 DCC Et2O
2 MeCHN2 Et2O
3 PhNEt2 PhCH2OH
4 Hydrolysis
CHCOOH
Me
Cl

17%

JCS C (1970) 971

PhCOCl $\xrightarrow{\begin{array}{l}\text{1 MeCHN}_2 \ \ \text{Et}_2\text{O}\\ \text{2 Ag}_2\text{O} \ \ \text{PhNH}_2\\ \text{3 Acid hydrolysis}\end{array}}$ PhCHCOOH
                                              |
                                              Me

Chem Ind (1955) 1673

$C_{14}H_{29}CH_2COCl$ $\xrightarrow{\begin{array}{l}\text{1 CH}_2\text{N}_2 \ \ \text{Et}_2\text{O}\\ \text{2 HCl} \ \ \text{Et}_2\text{O}\end{array}}$ $C_{14}H_{29}CH_2COCH_2Cl$ $\xrightarrow{\text{KOH}}$ $C_{14}H_{29}CHCOOH$ 70%
                                                                                                                    |
                                                                                                                    Me

Chem Phys Lipids (1968) 2 213

$\xrightarrow{\begin{array}{l}\text{1 ClCOOEt} \ \ \text{Et}_3\text{N} \ \ \text{CH}_2\text{Cl}_2\\ \text{2 CH}_2\text{N}_2 \ \ \text{Et}_2\text{O}\\ \text{3 h}\nu \ \ \text{dioane} \ \ \text{H}_2\text{O}\end{array}}$ $\}$–CH$_2$COOH   ~34%

JCS C (1969) 1319
Tetr Lett (1969) 4517

$\xrightarrow{\begin{array}{l}\text{1 CH}_2\text{N}_2 \ \ \text{Et}_2\text{O}\\ \text{2 Ag}_2\text{O} \ \ \text{Na}_2\text{CO}_3 \ \ \text{Na}_2\text{S}_2\text{O}_3\\ \text{dioxane} \ \ \text{H}_2\text{O}\end{array}}$ ~74%

Org React (1942) 1 38
JCS (1950) 926

$\xrightarrow{\begin{array}{l}\text{1 Me}_2\overset{+}{\text{S}}\overset{-}{\text{OCH}}_2 \ \ \text{THF}\\ \text{2 h}\nu \ \ \text{H}_2\text{O}\end{array}}$ ~80%

JACS (1964) 86 1640

$BuCH_2COOH$ $\quad$ $\xrightarrow[\begin{array}{l}2\ BuLi \\ 3\ C_6H_{13}Br\end{array}]{1\ NaH\ \ (i\text{-}Pr)_2NH\ \ THF}$ $\quad$ $\underset{C_6H_{13}}{BuCHCOOH}$ $\qquad$ 86%

JACS (1970) <u>92</u> 1397

$C_5H_{11}CH_2COOH$ $\quad$ $\xrightarrow[\begin{array}{l}2\ HMPA \\ 3\ BuBr\end{array}]{1\ BuLi\ \ (i\text{-}Pr)_2NH\ \ hexane\ \ THF}$ $\quad$ $\underset{Bu}{C_5H_{11}CHCOOH}$ $\qquad$ 93%

JOC (1970) <u>35</u> 262

$\underset{Bu}{C_5H_{11}CHCOOH}$ $\quad$ $\xrightarrow[2\ BuBr]{1\ BuLi\ \ (i\text{-}Pr)_2NH\ \ hexane\ \ THF}$ $\quad$ $\overset{Bu}{\underset{Bu}{C_5H_{11}CCOOH}}$ $\qquad$ 50%

JOC (1970) <u>35</u> 262

$PhCH_2COOH$ $\quad$ $\xrightarrow[2\ i\text{-}PrBr]{1\ Sodium\text{-}naphthalene\ \ THF}$ $\quad$ $\underset{i\text{-}Pr}{PhCHCOOH}$ $\qquad$ 88%

JOC (1967) <u>32</u> 2797

$PhCH_2COOH$ $\quad$ $\xrightarrow[2\ EtI]{1\ i\text{-}PrMgCl\ \ HMPA\ \ Et_2O}$ $\quad$ $\underset{Et}{PhCHCOOH}$

Bull Soc Chim Fr (1964) 2000

$$\text{(o-CH}_3\text{C}_6\text{H}_4\text{COOH)} \xrightarrow[\text{2 BuBr}]{\text{1 (i-Pr)}_2\text{NLi   heptane   THF}} \text{(o-CH}_2\text{Bu C}_6\text{H}_4\text{COOH)} \quad 69\text{-}73\%$$

JACS (1970) 92 1396

Alkylation of acids may also be accomplished via ester or amide intermediates. See section 113 (Esters from Esters) and section 81 (Amides from Amides)

COOH		COOMe		COOH
CHMe		CMe		CHMe
CH$_2$	1 Br$_2$ P$_4$	CH	KMnO$_4$ Me$_2$CO	CH$_2$
CHMe	2 MeOH	CHMe		CHMe
CH$_2$	3 Pyr	CH$_2$		CH$_2$R
CHMe		CHMe		
CH$_2$R		CH$_2$R		

JCS (1963) 3081

$$\text{PrCH}_2\text{CH}_2\text{COCl} \xrightarrow[\substack{\text{2 i-PrCH}_2\text{OH}\\\text{3 PhNEt}_2}]{\text{1 Br}_2} \text{PrCH=CHCOOCH}_2\text{Pr-i} \xrightarrow[\text{300-340}°]{\text{KOH}} \text{PrCOOH}$$

Biochem J (1951) 50 163

$$\text{CCl}_3\text{COOH} \xrightarrow[\text{DMF}]{\text{SOCl}_2} \text{CCl}_3\text{COCl} \quad 89\%$$

Helv (1959) 42 1653
Org Synth (1932) Coll Vol 1 12

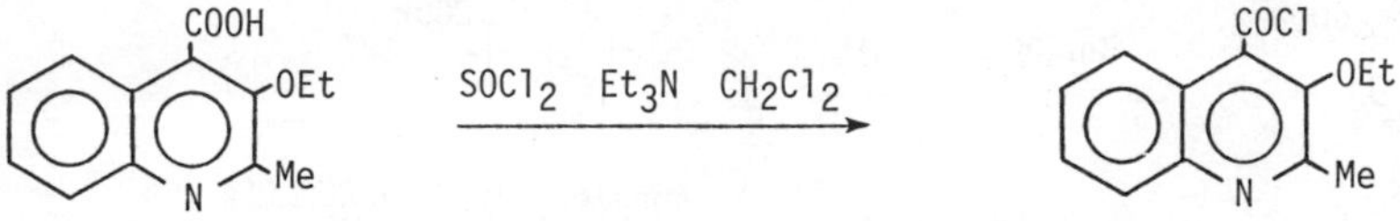

$$SOCl_2 \quad Et_3N \quad CH_2Cl_2$$

JCS (1963) 491

$$PhCH_2CH_2COOH \xrightarrow{(COCl)_2 \quad C_6H_6} PhCH_2CH_2COCl \qquad 98\%$$

JACS (1920) 42 599
Can J Chem (1955) 33 1515

$$t\text{-BuCOOH} \xrightarrow{PhCOCl} t\text{-BuCOCl} \qquad 92\%$$

JACS (1938) 60 1325

$$PhCH_2COOH \xrightarrow{PCl_3} PhCH_2COCl$$

Org Synth (1943) Coll Vol 2 156

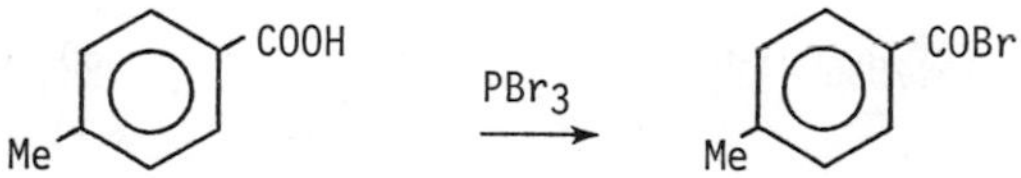

$$PBr_3$$

65%

JCS (1934) 1406

$$Ph_3P \quad CCl_4$$

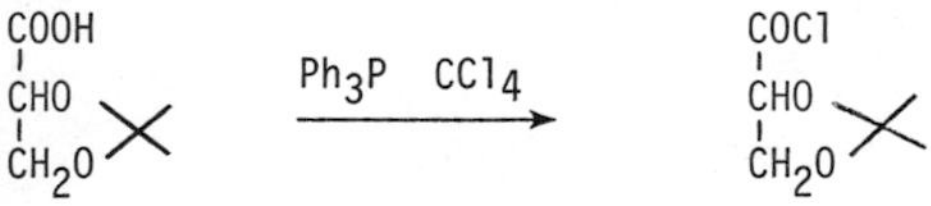

JACS (1966) 88 3440

$$\text{Ph}_3\text{PBr}_2 \quad \text{PhCl}$$

Annalen (1966) <u>693</u> 132

$$\text{C}_{15}\text{H}_{31}\text{COOH} \xrightarrow[\text{HCl} \quad \text{CHCl}_3]{\text{NN'-Carbonyldiimidazole}} \text{C}_{15}\text{H}_{31}\text{COCl} \qquad 68\%$$

Annalen (1966) <u>694</u> 78

$$\text{PhCOOH} \longrightarrow \text{PhCOCl} \qquad 77\%$$

Ber (1963) <u>96</u> 1387

$$\text{MeCH=CHCOOH} \xrightarrow{\text{MeCCl}_2\text{OEt}} \text{MeCH=CHCOCl} \qquad 79\%$$

Rec Trav Chim (1957) <u>76</u> 969

$$\text{C}_{17}\text{H}_{35}\text{COOH} \xrightarrow{\text{Ac}_2\text{O}} (\text{C}_{17}\text{H}_{35}\text{CO})_2\text{O} \qquad 50\text{-}80\%$$

JACS (1941) <u>63</u> 699

$$\text{PhCOOH} \xrightarrow{\text{TsCl} \quad \text{pyr}} (\text{PhCO})_2\text{O} \qquad 97\%$$

JACS (1955) <u>77</u> 6214

$C_6H_{13}COONa$ $\xrightarrow{\quad C_6H_{13}COCl \quad H_2O \quad}$ $(C_6H_{13}CO)_2O$ 60%

JCS (1964) 755

$C_{15}H_{31}COOH$ $\xrightarrow{\quad DCC \quad CCl_4 \quad}$ $(C_{15}H_{31}CO)_2O$ 85%

J Lipid Res (1966) 7 174

$Me[CH_2]_7CH-CH[CH_2]_7COOH$ $\xrightarrow[\quad Et_3N \quad THF \quad]{ClCOOEt}$ $(Me[CH_2]_7CH-CH[CH_2]_7CO)_2O$
$\quad\quad\quad OH \; OH$ $\quad\quad\quad\quad\quad OH \; OH$

JOC (1963) 28 1905

$C_{15}H_{31}COOH$ $\xrightarrow{\quad N\text{-Trifluoroacetyl imidazole} \quad C_6H_6 \quad}$ $(C_{15}H_{31}CO)_2O$ 54%

Ber (1962) 95 2073

$PhCOOAg$ $\xrightarrow{\quad CS_2 \quad}$ $(PhCO)_2O$ 98%

Proc Chem Soc (1957) 20

$C_6H_{13}COOH$ $\xrightarrow{\quad HC{\equiv}COMe \quad}$ $(C_6H_{13}CO)_2O$ 67%

JCS (1954) 1860

PhCOOH $\xrightarrow{\text{1 CH}_2\text{=CO}}$ (PhCO)$_2$O   96%
         2 Distillation

                                    JACS (1932) 54 3427

PrCOOH $\xrightarrow{\text{PhCOCH=CHCOPh}\quad \text{Bu}_3\text{P}\quad \text{C}_6\text{H}_6}$ (PrCO)$_2$O   77%

                                    JOC (1964) 29 1385

Section 18   Carboxylic Acids from Alcohols and Phenols
             °°

                                    Ber (1967) 100 978

PhCHOH $\xrightarrow{\text{CH}_2\text{=CCl}_2\quad \text{H}_2\text{SO}_4\quad \text{BF}_3}$ PhCHCH$_2$COOH   50%
 | |
 Me   Me

                                    Angew (1965) 77 967
                                    (Internat Ed 4 956)

t-BuOH $\xrightarrow{\text{HCOOH}\quad \text{H}_2\text{SO}_4}$ t-BuCOOH   75%

                                    Org Synth (1966) 46 72
                                    JACS (1961) 83 3980

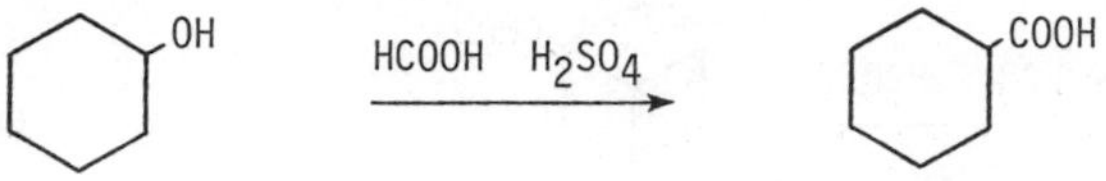

$$\text{cyclohexanol} \xrightarrow{\text{HCOOH} \quad H_2SO_4} \text{cyclohexanecarboxylic acid}$$

75%

Ber (1966) 99 1149

$$\text{MeOH} \xrightarrow[\text{HI} \quad H_2O \quad C_6H_6]{\text{CO (400 psi)} \quad RhCl_3 \cdot 3H_2O} \text{MeCOOH}$$

99%

Chem Comm (1968) 1578

$$FCH_2(CH_2)_8CH_2OH \xrightarrow{CrO_3 \quad HOAc \quad H_2O} FCH_2(CH_2)_8COOH$$

93%

JACS (1956) 78 2255
      (1960) 82 6147

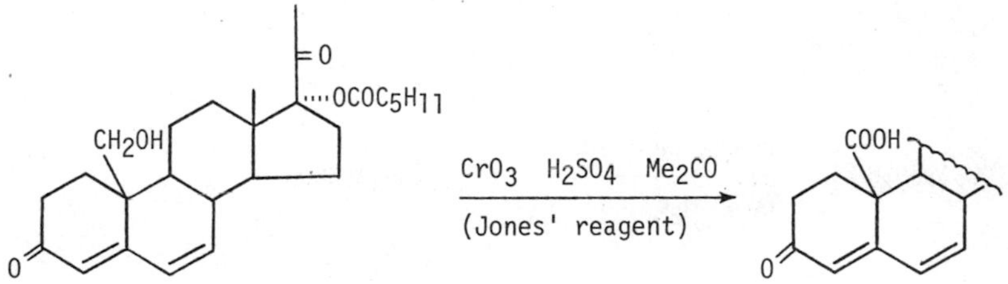

$$\xrightarrow[\text{(Jones' reagent)}]{CrO_3 \quad H_2SO_4 \quad Me_2CO}$$

77%

Helv (1967) 50 269
J Med Chem (1970) 13 926

$$Cl_3C(CH_2)_3CH_2OH \xrightarrow{KMnO_4 \quad H_2O} Cl_3C(CH_2)_3COOH$$

92%

Bull Chem Soc Jap (1963) 36 1264

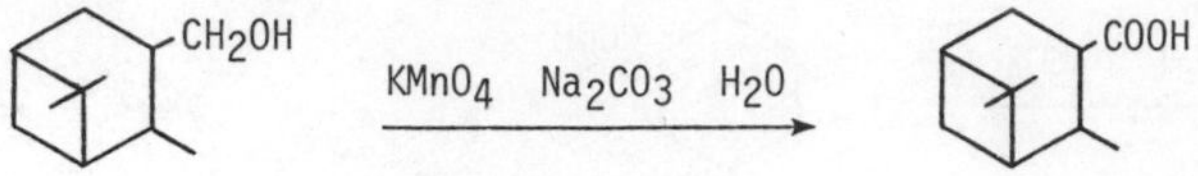
CH2OH
KMnO4 Na2CO3 H2O
COOH

66%

Bull Acad Polon (1964) 12 15
(Chem Abs 61 1895)
JACS (1950) 72 2953

$Br(CH_2)_5CH_2OH$ —HNO_3→ $Br(CH_2)_5COOH$

JACS (1950) 72 5137

$C_9H_{19}CH_2OH$ —N_2O_4→ $C_9H_{19}COOH$

Ber (1956) 89 202

$C_5H_{11}CH_2OH$ —RuO_4 CCl_4→ $C_5H_{11}COOH$ 10%

JACS (1958) 80 6682

$Ph(CH_2)_2CH_2OH$ —Nickel peroxide / NaOH H_2O→ $Ph(CH_2)_2COOH$ 70%

JOC (1962) 27 1597

$PhCH{=}CHCH_2OH$ —Nickel peroxide / NaOH H_2O→ $PhCH{=}CHCOOH$ 81%

JOC (1962) 27 1597

EtCHCH$_2$CH$_2$OH $\xrightarrow{\text{AgO}}$ EtCHCH$_2$COOH   100%
  | |
 Me   Me

Tetr Lett (1968) 5685

(HOCH$_2$)$_3$CCH$_2$OH $\xrightarrow{\text{O}_2 \quad \text{Pt} \quad \text{NaHCO}_3 \quad \text{H}_2\text{O}}$ (HOCH$_2$)$_3$CCOOH   50%

Ber (1956) <u>89</u> 1648
Angew (1957) <u>69</u> 600

Me(CH$_2$)$_{14}$CH$_2$OH $\xrightarrow{\text{t-Butyl chromate}}$ Me(CH$_2$)$_{14}$COOH   54%

Bull Chem Soc Jap (1965) <u>38</u> 893

PhCH$_2$OH $\xrightarrow{\text{CCl}_4 \quad \text{KOH} \quad \text{t-BuOH} \quad \text{H}_2\text{O}}$ PhCOOH   75%

JACS (1969) <u>91</u> 7510

MeCH$_2$OH $\xrightarrow{\text{Xenic acid} \quad \text{H}_2\text{O}}$ MeCOOH

JACS (1964) <u>86</u> 2078

$\xrightarrow{\text{KOH \quad fusion}}$

47%

Helv (1944) <u>27</u> 1727
JACS (1948) <u>70</u> 3485

Me$_2$COH

COOH

$$\xrightarrow[\text{NH}_4\text{VO}_3]{\text{H}_2\text{O}_2 \quad \text{HBr} \quad \text{H}_2\text{O}}$$

COOH

COOH

COOH

56%

JCS (1961) 4082

$$\xrightarrow[\text{2 H}_2\text{O}_2 \quad \text{H}_2\text{O}]{\text{1 O}_3 \quad \text{EtOAc}}$$

—COOH

55%

Zh Org Khim (1967) <u>3</u> 1636
(Chem Abs <u>68</u> 29874)
Tetr Lett (1967) 4729

$$\xrightarrow[\text{H}_2\text{O}]{\text{CrO}_3 \quad \text{HIO}_4}$$

—OH

—COOH

JCS <u>C</u> (1966) 1918

C$_8$H$_{17}$CHCH$_2$OH
    OH

$$\xrightarrow{\text{O}_2 \quad \text{cobalt laurate} \quad \text{PhCN}}$$

C$_8$H$_{17}$COOH

70%

Tetr Lett (1968) 5689

Section 19   Carboxylic Acids and Acid Halides from Aldehydes

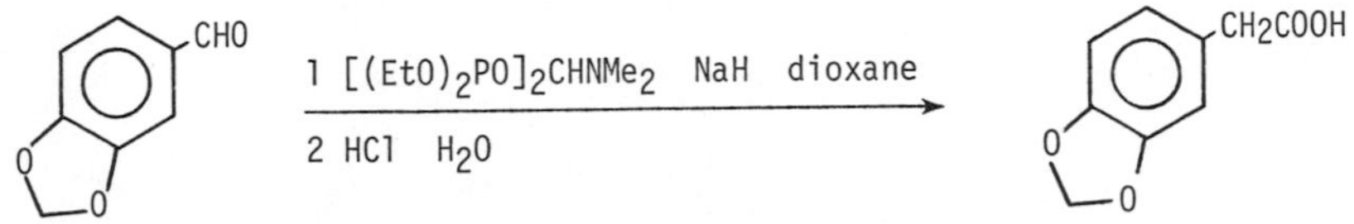

$$1 \; [(EtO)_2PO]_2CHNMe_2 \quad NaH \quad dioxane$$
$$2 \; HCl \quad H_2O$$

Angew (1968) 80 364
(Internat Ed 7 391)

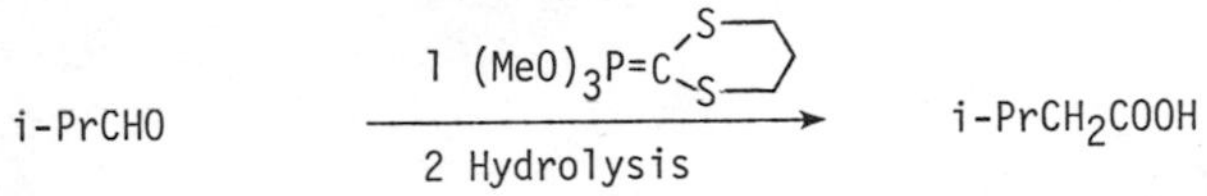

i-PrCHO

$$1 \; (MeO)_3P=C{\overset{S}{\underset{S}{\big\langle}}}$$

2 Hydrolysis

i-PrCH_2COOH

Tetr Lett (1967) 3201

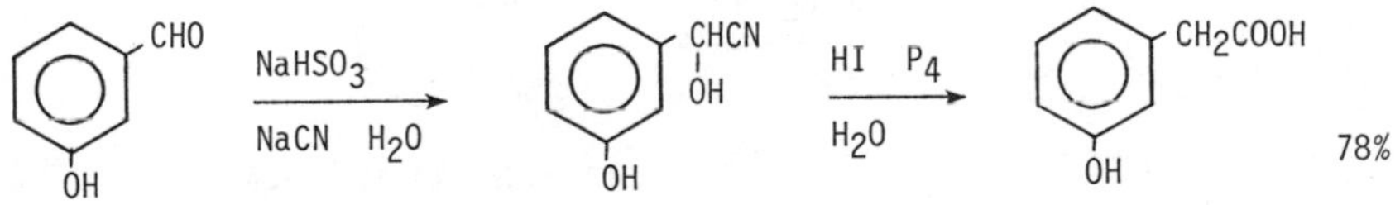

$$NaHSO_3$$
$$NaCN \quad H_2O$$

$$HI \quad P_4$$
$$H_2O$$

78%

JOC (1956) 21 1149

PhCHO

$$1 \; PhCONHCH_2COOH$$
$$2 \; Base \; or \; acid$$

$$PhCH_2COCOOH$$

$$H_2O_2$$

$$PhCH_2COOH$$

Org React (1942) 1 210

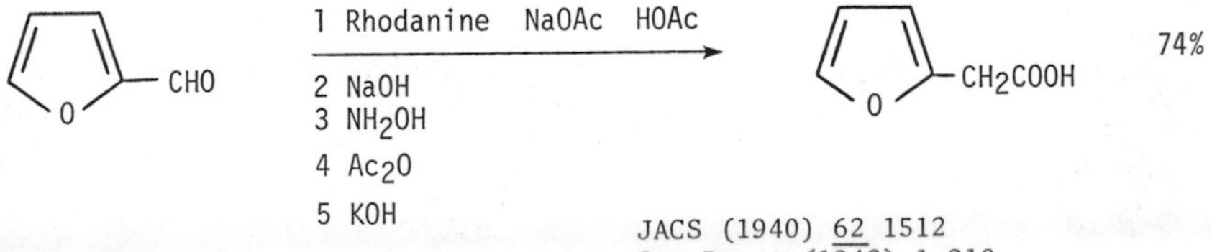

1 Rhodanine   NaOAc   HOAc
2 NaOH
3 NH_2OH
4 Ac_2O
5 KOH

74%

JACS (1940) 62 1512
Org React (1942) 1 210

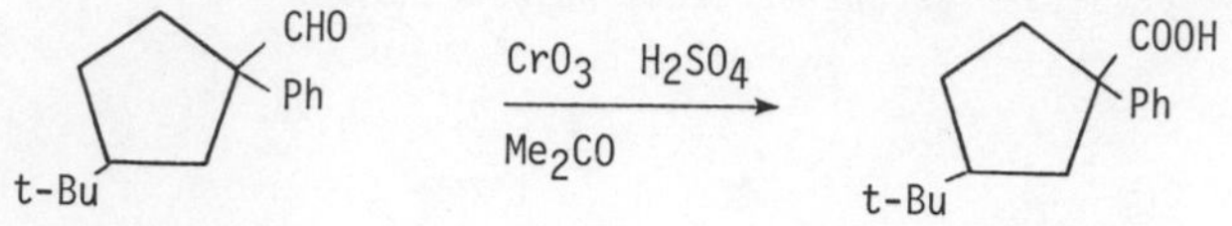

$$\xrightarrow{\text{CrO}_3 \quad \text{H}_2\text{SO}_4}{\text{Me}_2\text{CO}}$$

85%

JCS $\underline{C}$ (1970) 1168

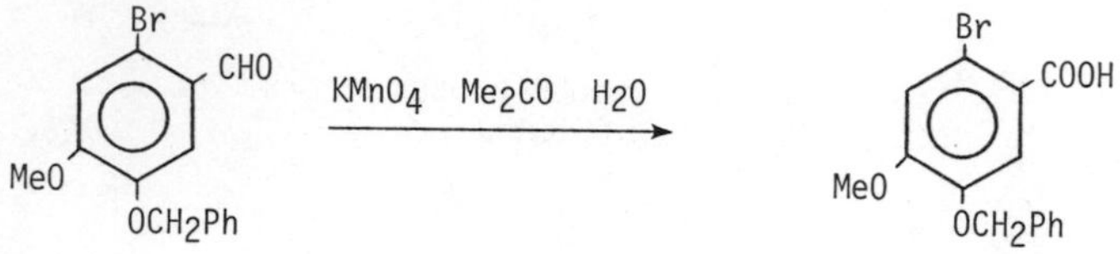

$$\xrightarrow{\text{KMnO}_4 \quad \text{Me}_2\text{CO} \quad \text{H}_2\text{O}}$$

JCS $\underline{C}$ (1970) 1208

$$\xrightarrow{\text{KMnO}_4 \quad \text{Pyr}}$$

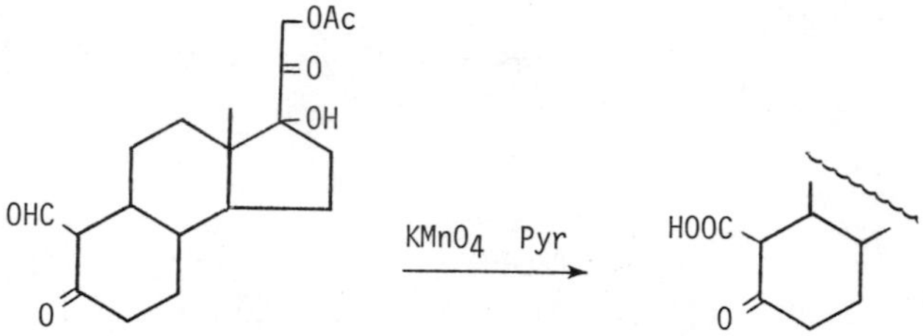

Steroids (1964) $\underline{3}$ 639

$$C_6H_{13}CHO \xrightarrow{\text{KMnO}_4 \quad \text{H}_2\text{SO}_4 \quad \text{H}_2\text{O}} C_6H_{13}COOH \qquad 76\text{-}78\%$$

Org Synth (1943) Coll Vol 2 315

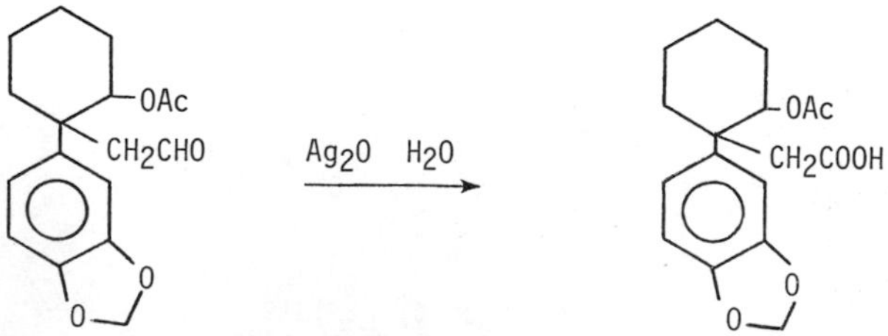

$$\xrightarrow{\text{Ag}_2\text{O} \quad \text{H}_2\text{O}}$$

93%

Tetrahedron (1968) $\underline{24}$ 6583

$Me_2C=CH(CH_2)_2\underset{\underset{Me}{|}}{C}=CH(CH_2)_3CHO$ $\xrightarrow[\text{THF}\ \ H_2O]{\text{AgO}}$ $Me_2C=CH(CH_2)_2\underset{\underset{Me}{|}}{C}=CH(CH_2)_3COOH$ 55%

Tetr Lett (1969) 1837
JACS (1968) 90 5617

PrCHO $\xrightarrow{\text{Argentic picolinate}}$ PrCOOH

Tetr Lett (1967) 415

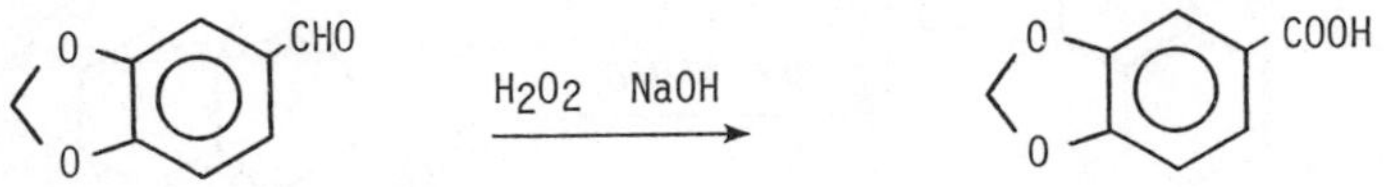

$\xrightarrow{H_2O_2\ \ NaOH}$ 90%

Monatsh (1955) 86 325

$C_6H_{13}CHO$ $\xrightarrow{\text{MeCOO}_2H}$ $C_6H_{13}COOH$ 88%

Org React (1957) 9 73

PrCHO $\xrightarrow[\text{PhCMe}_2\ \ \text{OsO}_4\ \ \text{MeOH}]{\overset{\overset{\text{OOH}}{|}}{}}$ PrCOOH

JCS (1950) 2169

$\xrightarrow{SeO_2\ \ H_2O_2\ \ t\text{-BuOH}}$

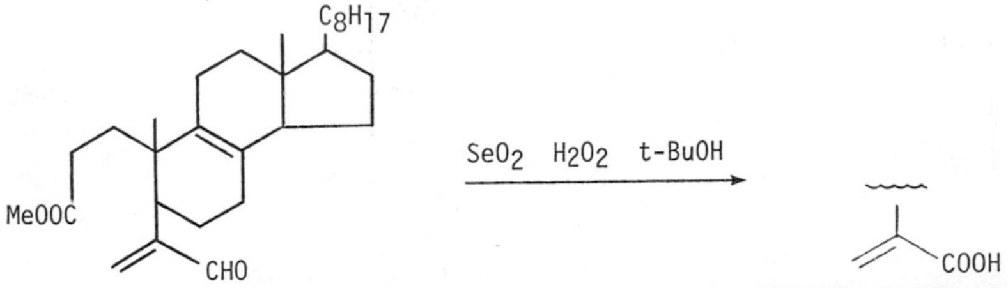

Chem Comm (1969) 945

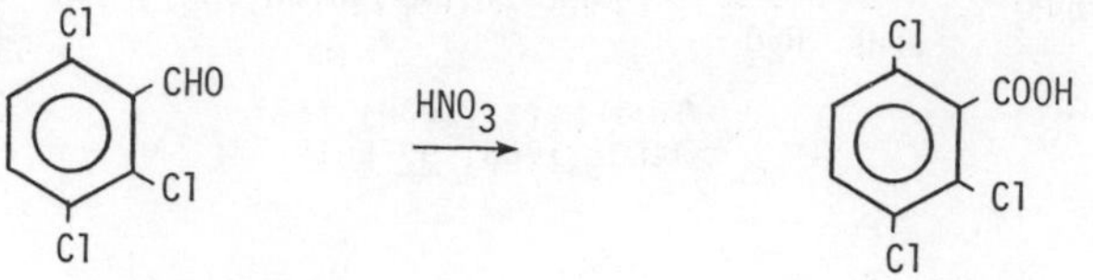

JCS (1951) 1208

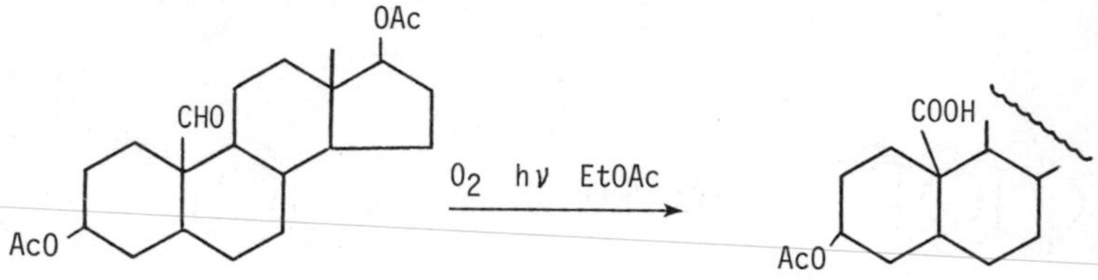

24%

Tetr Lett (1966) 2507

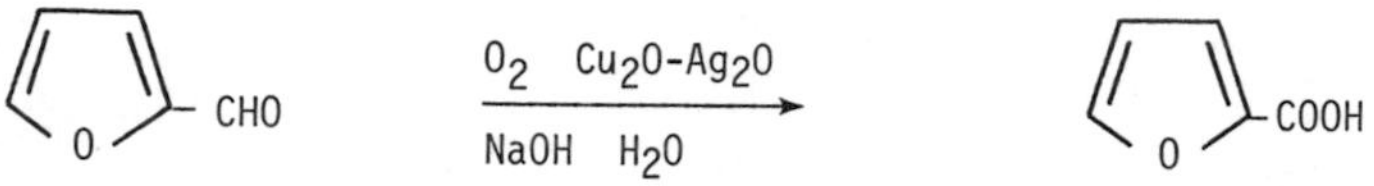

50%

Org Synth (1963) Coll Vol 4 493

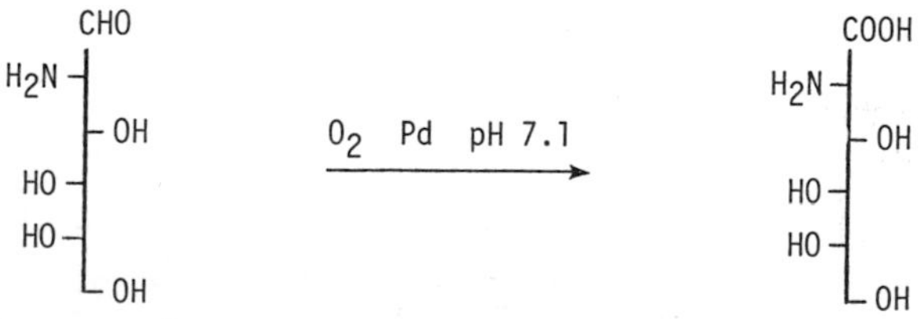

86-90%

Helv (1957) <u>40</u> 2383

54-60%

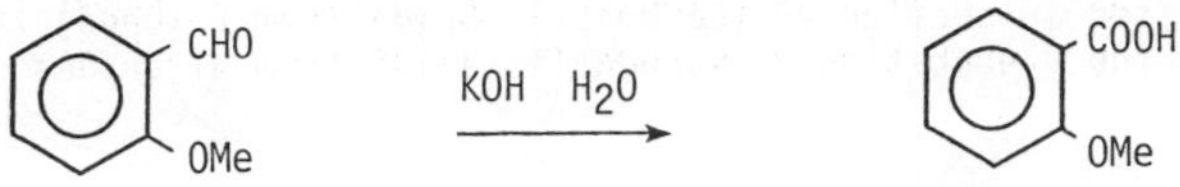

$$\xrightarrow{\text{KOH   H}_2\text{O}}$$

34%

Org React (1944) 2 94
Org Synth (1963) Coll Vol 4 974

BuCHCHO
  |
  Et

$$\xrightarrow{\text{NaOH   H}_2\text{O}}$$

BuCHCOOH
  |
  Et

20%

Helv (1951) 34 1211

i-PrCHO

$$\xrightarrow[\text{2 KOH   diethylene glycol}]{\text{1 NH}_2\text{OH·HCl   NaOAc   EtOH   H}_2\text{O}}$$

i-PrCOOH

88%

JOC (1962) 27 629

MeOCHCHO
  |
  O
  |
MeOCHCHO

$$\xrightarrow[\text{KHCO}_3\text{   H}_2\text{O}]{\text{I}_2\text{   KI   K}_2\text{CO}_3}$$

MeOCHCOOH
  |
  O
  |
MeOCHCOOH

JACS (1954) 76 3188

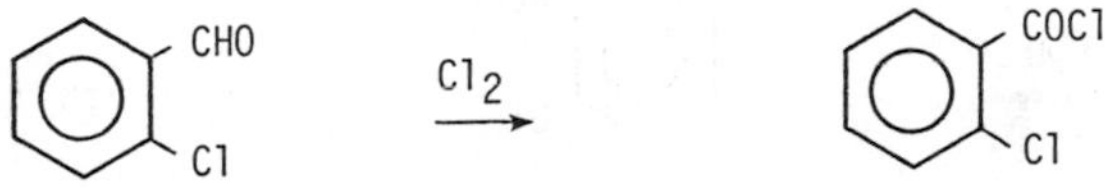

$$\xrightarrow{\text{Cl}_2}$$

70%

Org Synth (1941) Coll Vol 1 155

i-PrCH$_2$CHO

$$\xrightarrow{\text{Benzoyl peroxide   CCl}_4}$$

i-PrCH$_2$COCl

60%

JACS (1947) 69 2916

Carboxylic acids may also be prepared by conversion of aldehydes into esters, followed by hydrolysis. See section 109 (Esters from Aldehydes). Some of the methods listed in section 27 (Carboxylic Acids from Ketones) may also be applied to the preparation of carboxylic acids from aldehydes

Section 20 Carboxylic Acids from Alkyls, Methylenes and Aryls

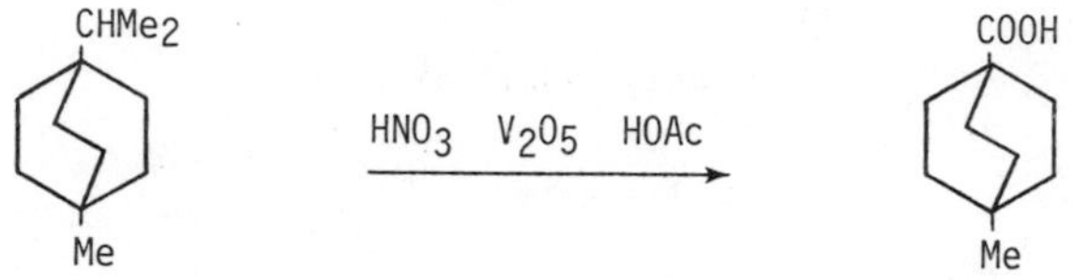
CHMe2
Me
HNO3 V2O5 HOAc
COOH
Me

25%

J Med Chem (1970) 13 254

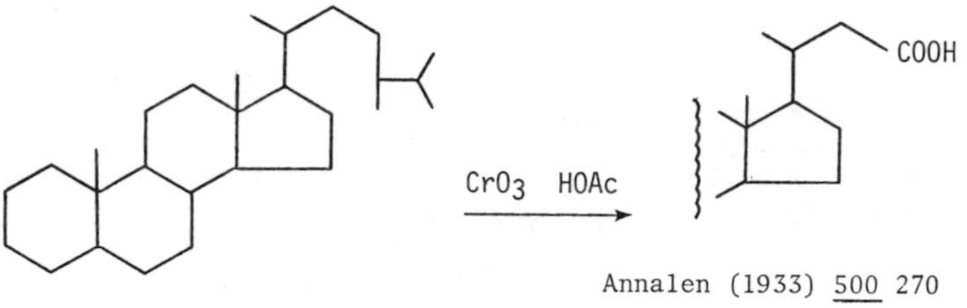
CrO3 HOAc
COOH

0.3%

Annalen (1933) 500 270

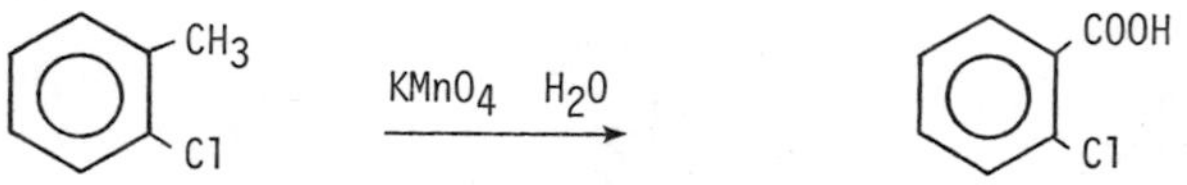
CH3
Cl
KMnO4 H2O
COOH
Cl

76-78%

Org Synth (1943) Coll Vol 2 135

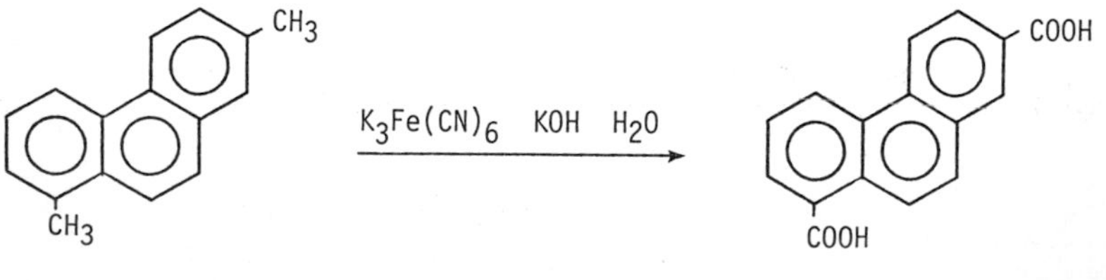
CH3
CH3
K3Fe(CN)6 KOH H2O
COOH
COOH

Helv (1931) 14 233

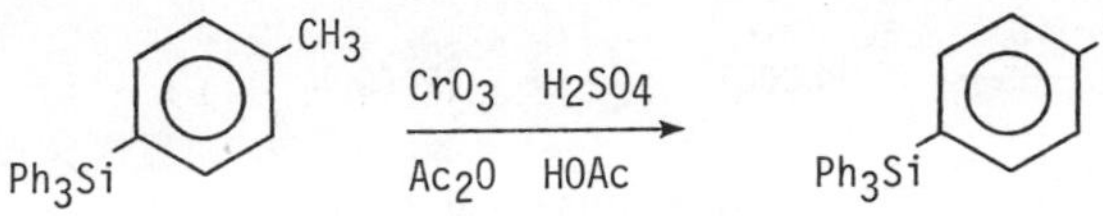
CH3
Ph3Si
CrO3 H2SO4
Ac2O HOAc
COOH
Ph3Si

81%

JACS (1956) 78 1689

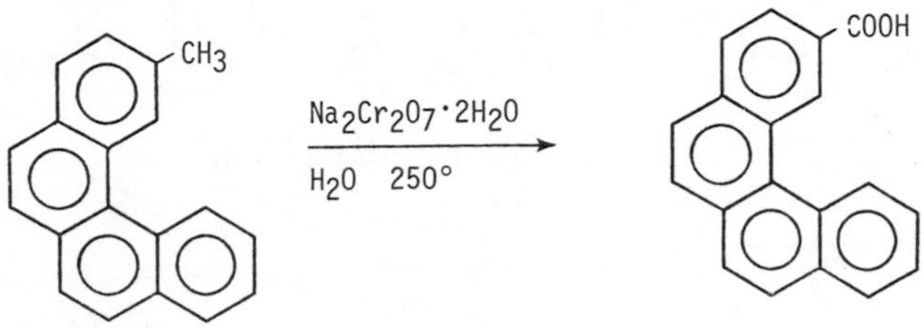
CH3
Na2Cr2O7·2H2O
H2O 250°
COOH

64%

JOC (1961) 26 1759

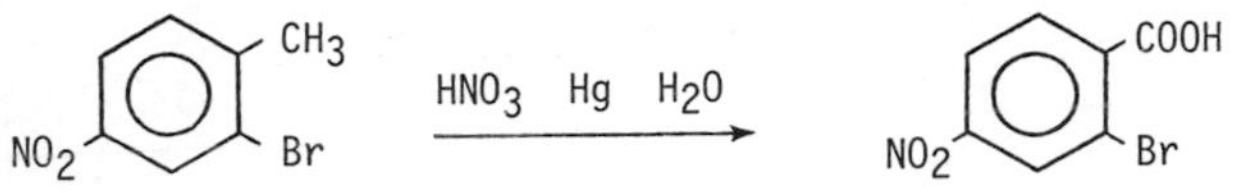
CH3
NO2
Br
HNO3 Hg H2O
COOH
NO2
Br

76%

Ber (1961) 94 834

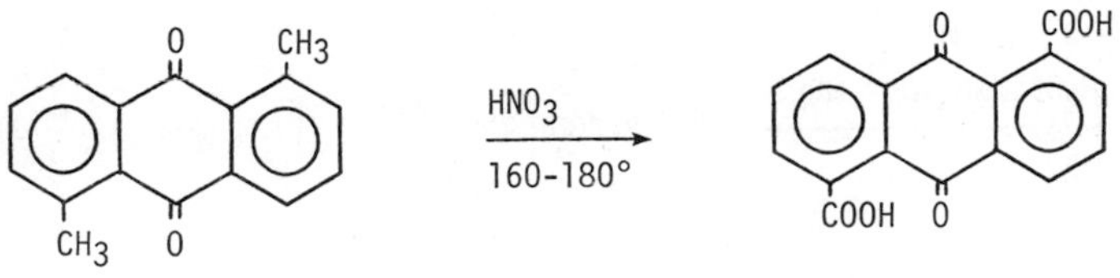
O
CH3
CH3
O
HNO3
160-180°
O
COOH
COOH
O

JCS (1960) 341
JOC (1960) 25 668

$PhCH_3$ —Argentic picolinate→ PhCOOH

Tetr Lett (1967) 415

PhCH$_3$ $\xrightarrow{\text{Cl}_2}$ PhCCl$_3$ $\xrightarrow{\text{TiO}_2}$ PhCOCl $\dashrightarrow$ PhCOOH

JACS (1958) 80 3483

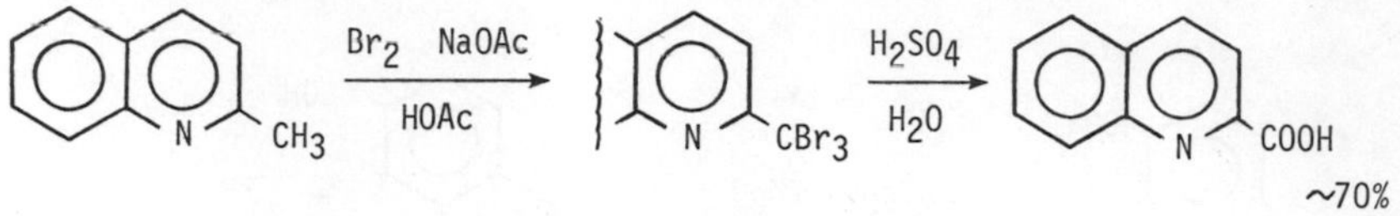

JACS (1946) 68 1840

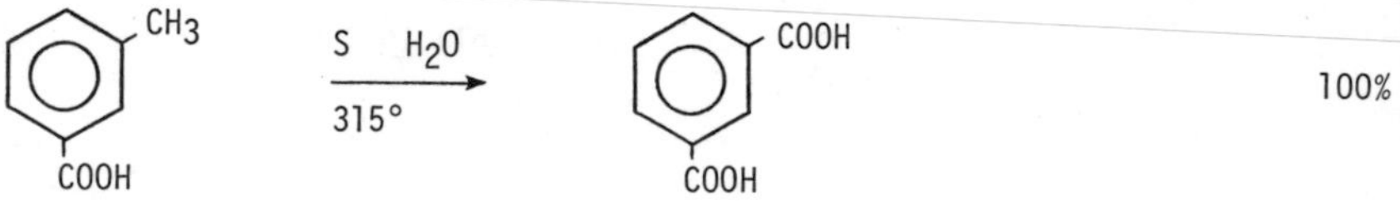

~70%

$\xrightarrow[315°]{\text{S \quad H}_2\text{O}}$

100%

JOC (1961) 26 2929

PhCHMe$_2$ $\xrightarrow[315°]{\text{SO}_2 \quad \text{H}_2\text{O}}$ PhCOOH

24%

JOC (1961) 26 2929

PhCH$_2$Me $\xrightarrow[\text{H}_2\text{O}]{\text{H}_2\text{O}_2 \quad \text{NH}_4\text{VO}_3 \quad \text{HBr}}$ PhCOOH

40%

JCS (1961) 4082

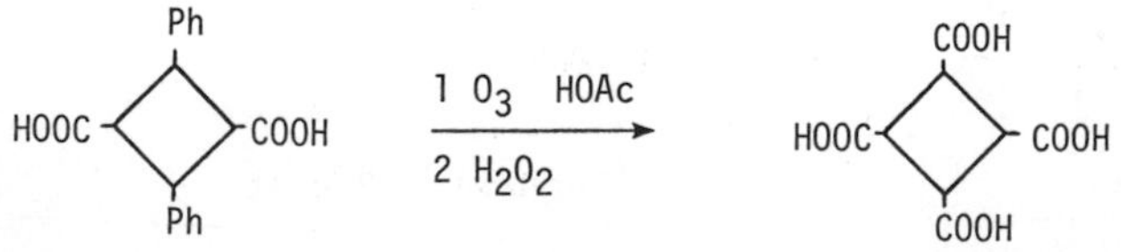

$\xrightarrow[\text{2 H}_2\text{O}_2]{\text{1 O}_3 \quad \text{HOAc}}$

Ber (1960) 93 2521

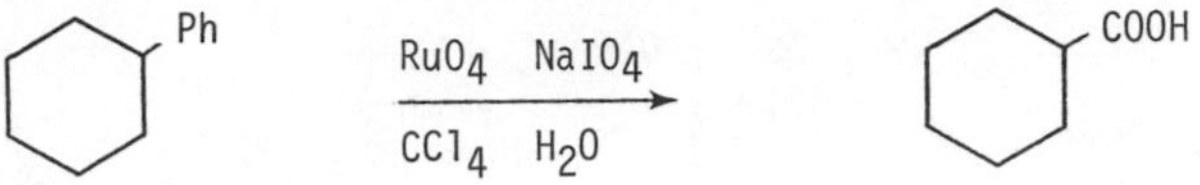
Ph
RuO4 NaIO4
CCl4 H2O
COOH

25%

Tetr Lett (1967) 4729
Chem Comm (1970) 1420

Section 21 Carboxylic Acids from Amides

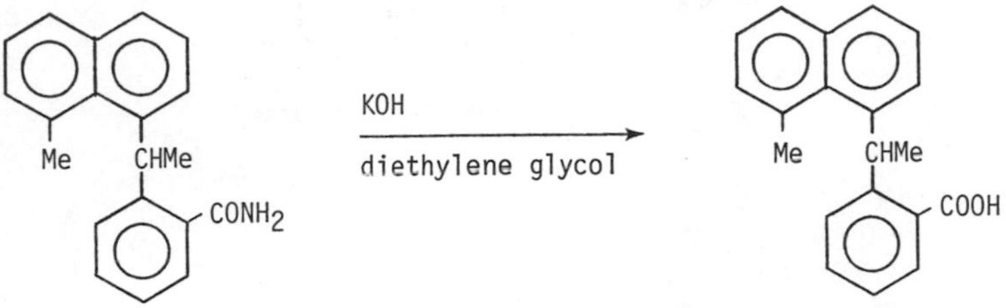
Me
CHMe
CONH2
KOH
diethylene glycol
Me
CHMe
COOH

94%

JOC (1950) 15 617

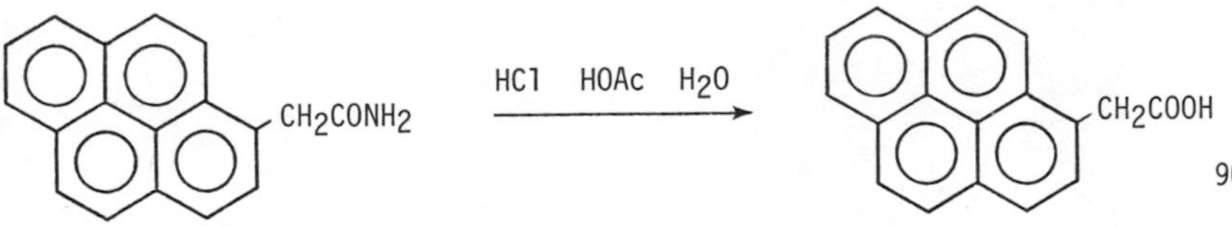
CH2CONH2
HCl HOAc H2O
CH2COOH

90%

JACS (1941) 63 2494

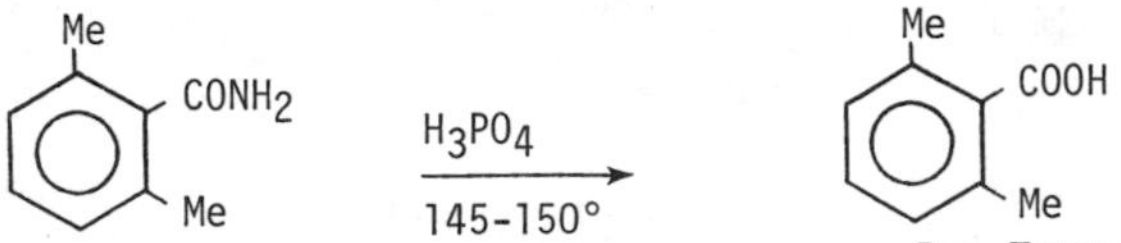
Me
CONH2
Me
H3PO4
145-150°
Me
COOH
Me

70%

Rec Trav Chim (1927) 46 600

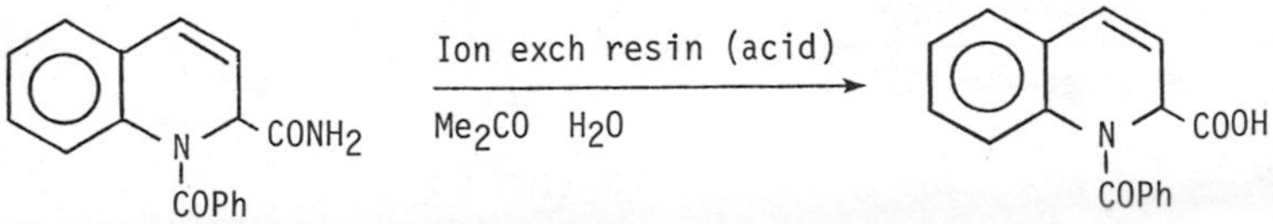
N
CONH2
COPh
Ion exch resin (acid)
Me2CO H2O
N
COOH
COPh

61%

Chem Ind (1957) 736

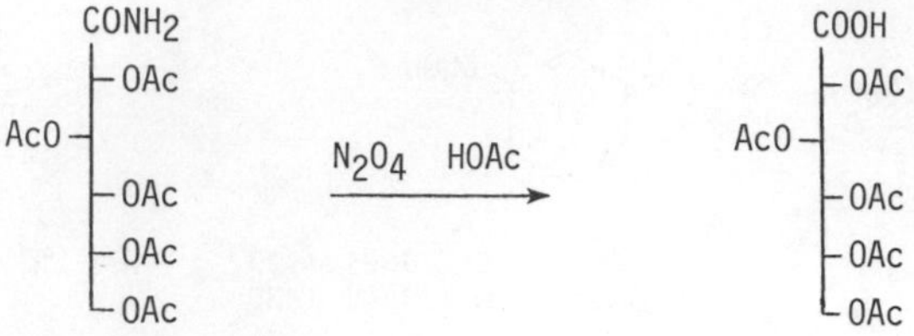

$$\xrightarrow{\quad N_2O_4 \quad HOAc \quad}$$

58%

JACS (1938) <u>60</u> 235

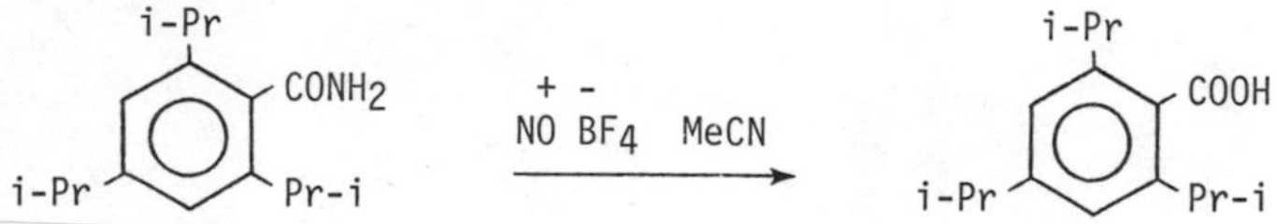

$$\xrightarrow[\text{NO BF}_4 \quad \text{MeCN}]{+ \quad -}$$

70%

JOC (1965) <u>30</u> 2386

Bu$_3$CCONH$_2$ $\xrightarrow{\quad BuONO \quad HCl \quad HOAc \quad}$ Bu$_3$CCOOH   79%

JACS (1948) <u>70</u> 3091
J Med Chem (1966) <u>9</u> 603

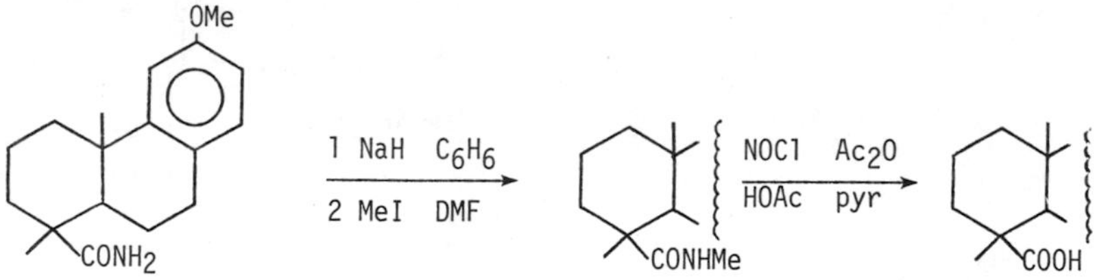

$$\xrightarrow[\text{2 MeI \quad DMF}]{\text{1 NaH \quad C}_6\text{H}_6}$$

$$\xrightarrow[\text{HOAc \quad pyr}]{\text{NOCl \quad Ac}_2\text{O}}$$

66%

JACS (1961) <u>83</u> 1492

PhCONHCHMe $\xrightarrow[\text{2 Pentane \quad 25°}]{\text{1 NaNO}_2 \quad \text{Ac}_2\text{O}}$ PhCOOH   64%
  |
  Et

JACS (1955) <u>77</u> 6011

PhCON⟩ + [catechol]PCl$_3$ ⟶ PhCOOH   67%

Ber (1963) $\underline{96}$ 1387

EtOOCCH$_2$NHCOCH(CH$_2$)$_2$CONHNHPh $\xrightarrow[\text{H}_2\text{O}]{\text{MnO}_2 \quad \text{HOAc}}$ EtOOCCH$_2$NHCOCH(CH$_2$)$_2$COOH   82%
$\qquad\qquad$ | $\qquad\qquad\qquad\qquad\qquad\qquad\qquad\qquad$ |
$\qquad$ NHCOOCH$_2$Ph $\qquad\qquad\qquad\qquad\qquad\qquad\qquad$ NHCOOCH$_2$Ph

JOC (1963) $\underline{28}$ 453

NHR'
|
RCHCONHNHPh $\xrightarrow[\text{cellosolve} \quad \text{H}_2\text{O}]{\text{FeCl}_3 \quad \text{HCl}}$
$\qquad\qquad$ NHR'
$\qquad\qquad$ |
$\qquad\qquad$ RCHCOOH   85%

JACS (1957) $\underline{79}$ 637   645

Carboxylic acids may also be prepared by conversion of amides into esters, followed by hydrolysis. See section 111 (Esters from Amides)

Section 22   Carboxylic Acids from Amines
∘∘∘∘∘∘∘∘∘∘∘∘∘∘∘∘∘∘∘∘∘∘∘∘∘∘∘∘∘∘∘∘∘∘∘∘

CH$_3$CH$_2$NEt$_2$ $\xrightarrow[\text{2 CO}_2]{\text{1 i-PrLi}}$ i-Pr(CH$_2$)$_2$COOH   25%

JACS (1969) $\underline{91}$ 6362

$\overset{+}{N_2} \overset{-}{BF_4}$ $\xrightarrow[]{\text{Ni(CO)}_4 \quad \text{HOAc}}$ ... COOH   52%

JCS (1962) 686

KMnO₄   NaOH   H₂O

27%

JACS (1951) 73 4122

Section 23   Carboxylic Acids and Acid Halides from Esters

Hydrolysis and cleavage of esters to carboxylic acids . . . page 42-45
Degradation of esters to carboxylic acids 45
Acid halides from esters 46

$C_9H_{19}CH=CCOOMe$
             |
             Me

KOH   EtOH
──────────►

$C_9H_{19}CH=CCOOH$
             |
             Me

Org Synth (1963) Coll Vol 4 608

$MeOOC(CH_2)_9COOMe$

Ba(OH)₂   MeOH
──────────────►

$MeOOC(CH_2)_9COOH$ 60%

Org Synth (1963) Coll Vol 4 635

$(i-Pr)_3CCOOMe$

t-BuOK   Me₂SO
──────────────►

$(i-Pr)_3CCOOH$ 100%

Tetr Lett (1964) 2969
JCS (1965) 1290

KOH-dicyclohexyl-
18-crown-6
──────────────►

94%

JACS (1967) 89 7017

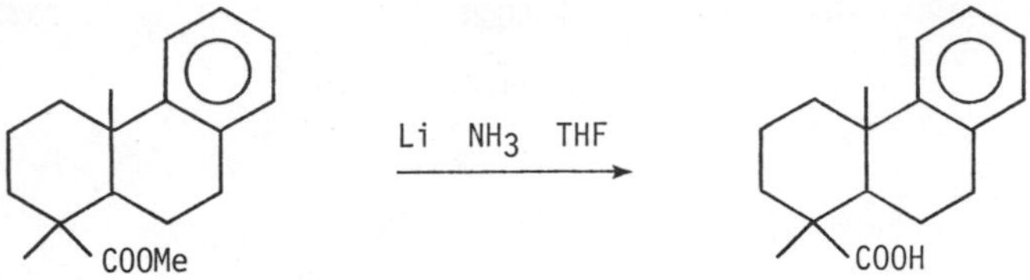

Li NH$_3$ THF

59%

JACS (1958) 80 217

PhCOOMe →(Me$_3$N MeOH) PhCOOH

JACS (1933) 55 4079

MeCOOPh →(Guanidine H$_2$O EtOH) MeCOOH

Bull Chem Soc Jap (1966) 39 852
JACS (1964) 86 837

BrCH$_2$CHCOOMe →(HBr H$_2$O) BrCH$_2$CHCOOH   72%
    Br   Br

JACS (1940) 62 3495

MeCOOC$_5$H$_{11}$ →(Ion exch resin (acid) H$_2$O) MeCOOH

JCS (1952) 1607

Further examples of the reaction RCOOR' → RCOOH + R'OH are included in section 38 (Alcohols from Esters) and section 30A (Protection of Carboxylic Acids)

PhCOOBu-t $\xrightarrow{\text{MeOH \quad reflux \quad 4 days}}$ PhCOOH   23%

JACS (1941) 63 3382

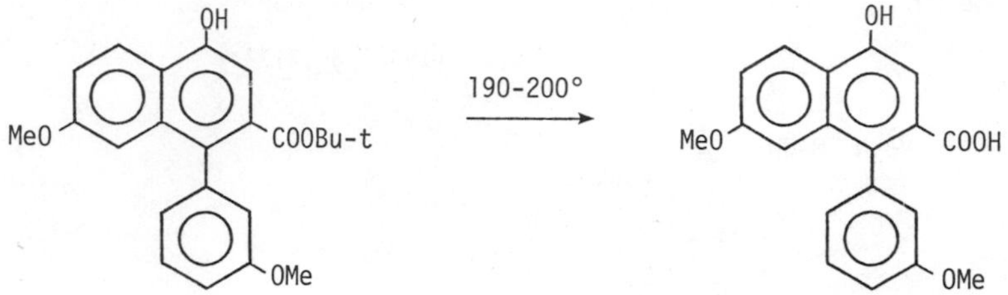

190-200°   100%

JOC (1962) 27 519

LiI   DMF

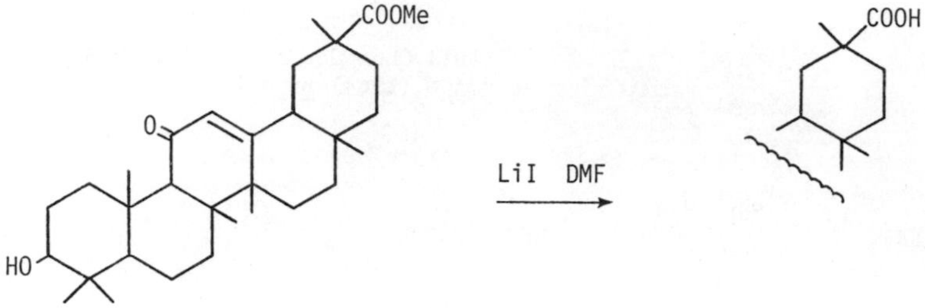

JCS (1965) 6655
For cleavage of ethyl esters see JOC (1963) 28 2184

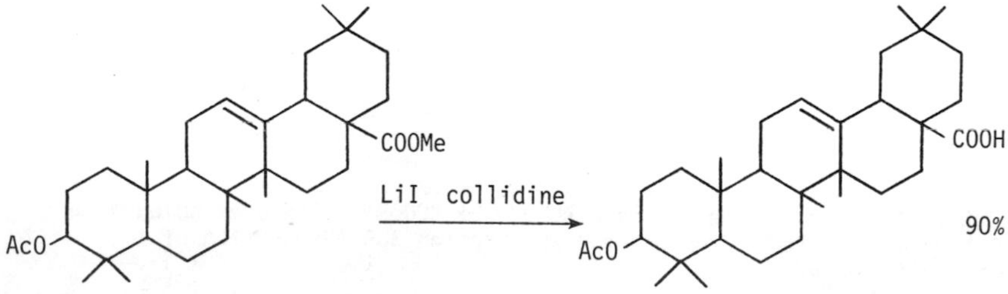

LiI   collidine   90%

Helv (1960) 43 113

$(i-Pr)_3CCOOMe$ $\xrightarrow{\text{PrSLi   HMPA}}$ $(i-Pr)_3CCOOH$ 99%

Tetr Lett (1970) 4459

$PhCOOCH_2Ph$ $\xrightarrow{\text{H}_2 \text{ Pd Et}_2\text{O}}$ $PhCOOH$

JOC (1958) 23 1700
Org React (1953) 7 263

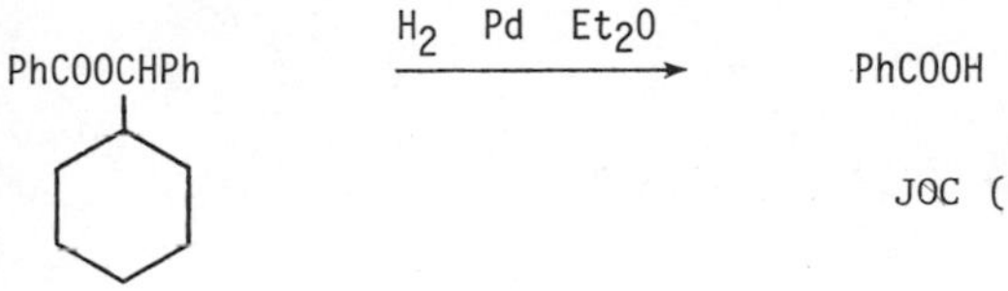

$PhCOOCHPh$ $\xrightarrow{\text{H}_2 \text{ Pd Et}_2\text{O}}$ $PhCOOH$

JOC (1958) 23 1700

Further examples of the cleavage of benzyl esters are included in section
30A (Protection of Carboxylic Acids)

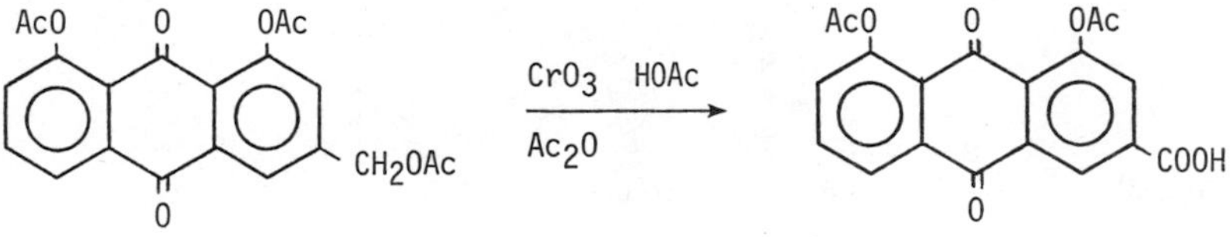

JOC (1961) 26 979

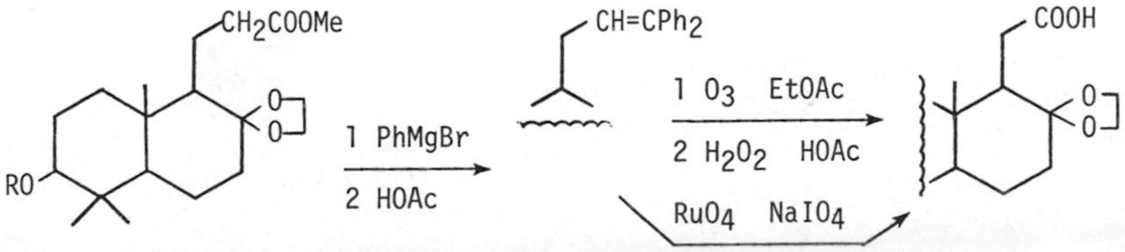

JACS (1963) 85 3419

$$C_7H_{15}COOC=CH_2 \quad \xrightarrow{\text{HF} \quad Et_2O} \quad C_7H_{15}COF \qquad 50\%$$
$$\underset{Me}{|}$$

JOC (1969) <u>34</u> 2486

PhCOOBu $\xrightarrow{}$ PhCOCl   91%

Ber (1963) <u>96</u> 1387

Section 24   Carboxylic Acids from Ethers

$$CH_3CH_2OEt \quad \xrightarrow[\text{2 } CO_2]{\text{1 i-PrLi} \quad Et_2O} \quad i\text{-}PrCH_2CH_2COOH \qquad 5\%$$

JACS (1953) <u>75</u> 1771
(1955) <u>77</u> 2806

$$CH_2=CHCH_2OPh \quad \xrightarrow[\text{2 } CO_2]{\text{1 Mg} \quad BrCH_2CH_2Br \quad THF} \quad CH_2=CHCH_2COOH \qquad 63\%$$

J Organometallic Chem (1969) <u>18</u> 249

$$(EtCH_2)_2O \quad \xrightarrow{Br_2 \quad H_2O} \quad EtCOOH \qquad 100\%$$

JACS (1967) <u>89</u> 3550

$$(PhCH_2)_2O \quad \xrightarrow{Br_2 \quad H_2O} \quad PhCOOH \qquad 55\%$$

JACS (1967) <u>89</u> 3550

$$Me(CH_2)_{14}CH_2OEt \xrightarrow{\quad CrO_3 \quad HOAc \quad CH_2Cl_2 \quad} Me(CH_2)_{14}COOH \qquad 55\%$$

Chem Comm (1966) 752

Carboxylic acids may also be prepared by oxidation of ethers to esters, followed by hydrolysis. See section 114 (Esters from Ethers)

Section 25   Carboxylic Acids and Acid Halides from Halides

Cl
|
$(CH_2)_4$
|
t-Bu
$\xrightarrow[\substack{2\ EtOOC(CH_2)_{14}CHO \\ 3\ KOH\quad EtOH\quad H_2O}]{1\ Mg\quad Et_2O}$
COOH
|
$(CH_2)_{14}$
|
CHOH
|
$(CH_2)_4$
|
t-Bu
$\xrightarrow[\substack{2\ KOH \\ 3\ H_2\quad Pt\quad MeOH}]{1\ PBr_3}$
COOH
|
$(CH_2)_{19}$
|
t-Bu

JACS (1950) 72 5139

Br
|
CH_2
|
CHMe
|
Et
$\xrightarrow[\substack{2\ CdCl_2 \\ 3\ MeOOC(CH_2)_2COCl}]{1\ Mg\quad Et_2O}$
COOMe
|
$(CH_2)_2$
|
CO
|
CH_2
|
CHMe
|
Et
$\xrightarrow[H_2O]{Zn\quad HCl}$
COOH
|
$(CH_2)_4$
|
CHMe
|
Et
$\qquad 46\%$

JACS (1944) 66 46

$PhCH_2Cl \xrightarrow[\substack{2\ N_2H_4\quad NaOH\quad MeOH}]{1 \quad\quad KI\quad KOH\quad H_2O} PhCH_2(CH_2)_5COOH \qquad 69\%$

Ber (1952) 85 61 1061

$C_{16}H_{33}Br$ $\xrightarrow[\text{2 Cyclohexanone}]{\text{1 Mg   Et}_2\text{O}}$ [cyclohexane with OH and $C_{16}H_{33}$] $\xrightarrow[\text{2 N}_2\text{H}_4\quad\text{NaOH}]{\text{1 CrO}_3\quad\text{HOAc}}$ $C_{16}H_{33}(CH_2)_5COOH$

73%

JACS (1948) 70 3352

[o-bromotoluene] $\xrightarrow[\text{2 Cyclopentanone}]{\text{1 Mg   Et}_2\text{O}}$ [cyclopentyl-substituted aryl OH with Me] $\xrightarrow[\substack{\text{2 N}_2\text{H}_4\quad\text{NaOH}\\ \text{diethylene}\\ \text{glycol}}]{\text{1 CrO}_3\quad\text{HOAc}}$ [aryl $(CH_2)_4COOH$ with Me]

JOC (1958) 23 584

$\underset{\overset{|}{Me}}{C_9H_{19}CHCH_2Br}$ $\xrightarrow[\substack{\text{2 } \bigcirc\text{-CHO}\\ \text{3 HCl EtOH}\\ \text{4 N}_2\text{H}_4\quad\text{KOH}}]{\text{1 Mg   Et}_2\text{O}}$ $\underset{\overset{|}{Me}}{C_9H_{19}CHCH_2(CH_2)_4COOH}$

Z Physiol Chem (1951) 287 65

[cyclohexyl-Cl] $\xrightarrow[\substack{\text{2 CH}_2\text{=CH}_2\\ \text{3 CO}_2}]{\text{1 Li   Et}_2\text{O}}$ [cyclohexyl-$(CH_2)_2COOH$]

47%

JACS (1969) 91 6362

$PhCH_2Cl$ $\xrightarrow[\substack{\text{2 CH}_2\text{=CHCN}\\ \text{KOH   t-BuOH}}]{\text{1 NaCH(SO}_2\text{Et)}_2}$ $\underset{\overset{\displaystyle SO_2Et}{\underset{\displaystyle SO_2Et}{|}}}{PhCH_2\overset{|}{C}CH_2CH_2CN}$ $\xrightarrow[\substack{\text{HOAc}\\ \text{2 Ni   NaOH}\\ \text{H}_2\text{O}}]{\text{1 HCl   H}_2\text{O}}$ $PhCH_2(CH_2)_3COOH$

JACS (1952) 74 1225

$$BuBr \xrightarrow[\text{NaOEt}]{\text{MeCOCH}_2\text{COOEt}} \underset{\overset{|}{\text{COMe}}}{\text{BuCHCOOEt}} \xrightarrow{\text{KOH}\quad \text{H}_2\text{O}} BuCH_2COOH$$

Org Synth (1941) Coll Vol 1 248
JACS (1930) 52 5005

1 Mg   THF
2 CdCl$_2$
3 BrCH$_2$COOEt
4 NaOH

62%

JOC (1968) 33 1675

$$C_8H_{17}I \xrightarrow[\text{2 HCl \quad H}_2\text{O \quad dioxane}]{\text{1 CH}_3\text{COOBu-t \quad NaNH}_2 \quad \text{NH}_3} C_8H_{17}CH_2COOH \qquad 70\%$$

JACS (1959) 81 5817

$$PhCH_2Cl \xrightarrow[\text{2 HCl \quad H}_2\text{O \quad dioxane}]{\text{1 EtCH}_2\text{COOBu-t \quad NaNH}_2 \quad \text{NH}_3} \underset{\overset{|}{\text{Et}}}{PhCH_2CHCOOH} \qquad 92\%$$

JACS (1959) 81 5817

Further examples of the alkylation of esters with halides are included
in section 113 (Esters from Esters). Examples of the alkylation of acids
with halides are included in section 17 (Carboxylic Acids from Carboxylic
Acids)

$$PhCH_2I \xrightarrow[\text{2 KOH \quad MeOH}]{\text{1 Ph}_3\text{P=CHCOOMe \quad EtOAc}} PhCH_2CH_2COOH \qquad 75\%$$

Tetr Lett (1960) (4) 5
Ber (1962) 95 2921

$C_{18}H_{37}I$ $\xrightarrow{\begin{array}{l}1 \text{ EtCH(COOMe)}_2 \quad \text{Na} \quad \text{BuOH} \\ \hline 2 \text{ NaOH} \quad H_2O \\ 3 \text{ 160-170°}\end{array}}$ $C_{18}H_{37}\underset{\underset{Et}{|}}{CH}COOH$ 83%

JCS (1953) 3031
Org React (1957) $\underline{9}$ 107

t-BuCl $\xrightarrow{\begin{array}{c}CH_2(COOEt)_2 \\ \hline BF_3 \cdot Et_2O\end{array}}$ $t\text{-BuCH(COOEt)}_2 \xrightarrow{\text{NaOH}} t\text{-BuCH}_2COOH$
                                                          50%

Naturwiss (1964) $\underline{51}$ 288

RBr $\xrightarrow{\quad\quad\quad\quad}$ [oxazoline structure]   BuLi   [oxazoline structure] $\xrightarrow[H_2O]{HBr \quad NaBr}$ RCH_2COOH 85%

JACS (1969) $\underline{91}$ 5886

t-BuCl $\xrightarrow{CH_2=CCl_2 \quad BF_3 \quad H_2SO_4}$ $t\text{-BuCH}_2COOH$ 78%

Angew (1965) $\underline{77}$ 967
(Internat Ed $\underline{4}$ 956)

[naphthalene with Cl substituent] $\xrightarrow{\begin{array}{l}1 \text{ Mg} \quad BrCH_2CH_2Br \quad Et_2O \\ \hline 2 \text{ CO}_2\end{array}}$ [naphthalene with COOH substituent]   56%

(Procedure for unreactive halides)
JOC (1959) $\underline{24}$ 504

$\underset{\underset{Et}{|}}{\overset{\overset{Me}{|}}{PrCCl}}$ $\xrightarrow{\begin{array}{l}1 \text{ Mg} \quad Et_2O \\ \hline 2 \text{ CO}_2 \quad (50 \text{ psi})\end{array}}$ $\underset{\underset{Et}{|}}{\overset{\overset{Me}{|}}{PrCCOOH}}$ 25%

JACS (1949) $\underline{71}$ 1877

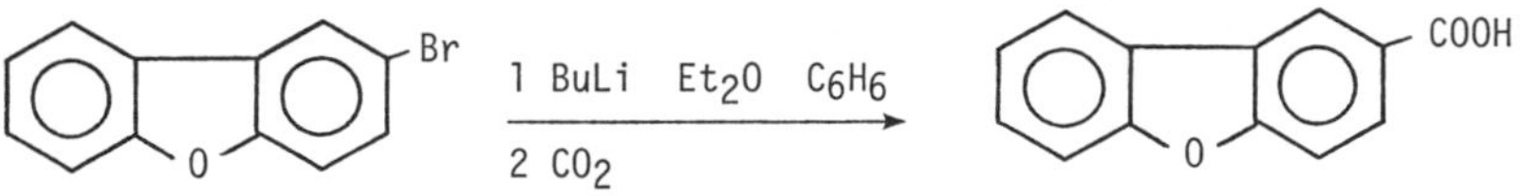

JACS (1939) <u>61</u> 1371

i-PrI $\xrightarrow{\text{LiC(SPr-i)}_3}$ i-PrC(SPr-i)$_3$ $\xrightarrow[\text{H}_2\text{O}]{\text{HgCl}_2 \quad \text{Me}_2\text{CO}}$ i-PrCOOH

Angew (1967) <u>79</u> 468
(Internat Ed <u>6</u> 442)

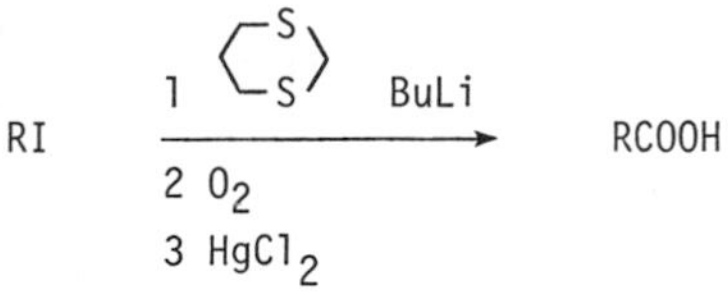

RI $\xrightarrow[\substack{2\ \text{O}_2 \\ 3\ \text{HgCl}_2}]{1\quad\text{BuLi}}$ RCOOH

Angew (1965) <u>77</u> 1134
(Internat Ed <u>4</u> 1075)

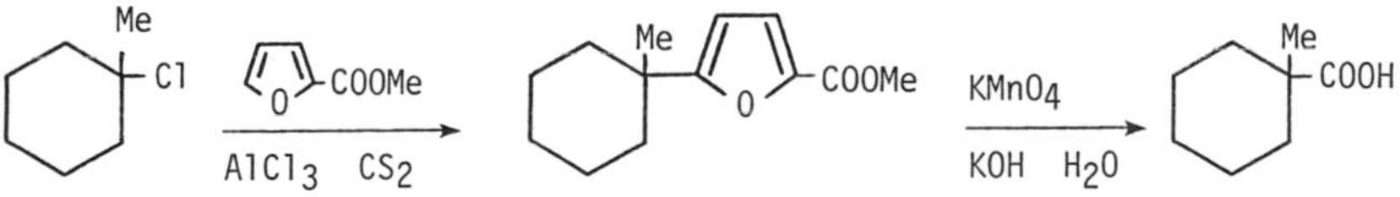

Helv (1935) <u>18</u> 721

PhBr $\xrightarrow[\text{KOAc} \quad \text{H}_2\text{O}]{\text{CO (300 kg/cm}^2\text{)} \quad \text{Ni(OAc)}_2}$ PhCOOH

Bull Chem Soc Jap (1967) <u>40</u> 2203

MeCH=CHCH$_2$Cl $\xrightarrow[(\pi\text{-allyl-PdCl})_2]{\text{CO (500 atmos)}}$ MeCH=CHCH$_2$COCl   81%

JCS (1964) 1588

$$PhCH_2Cl \xrightarrow[\text{(Ph}_3\text{P)}_2\text{Rh(CO)Cl}]{\text{CO (100 atmos)}} PhCH_2COCl$$

Tetr Lett (1966) 4713
JACS (1968) 90 99

$$Me_2\underset{\underset{Et}{|}}{C}Cl \xrightarrow{\text{CO   AgClO}_4 \quad \text{PhNO}_2} Me_2\underset{\underset{Et}{|}}{C}COOH \qquad 100\%$$

JACS (1960) 82 1261

$$BuBr \dashrightarrow (BuS)_2 \xrightarrow[\text{di-t-butyl peroxide}]{\text{CO   (EtO)}_3\text{P}} BuCOSBu \xrightarrow{\text{NaOH}} BuCOOH$$

JACS (1960) 82 2181

$$BuI \xrightarrow{\text{Me}_2\text{NLi   CHBr}_3 \quad \text{HMPA}} BuCBr_3 \xrightarrow{\text{NaOH}} BuCOOH$$

Compt Rend (1967) C 264 1609

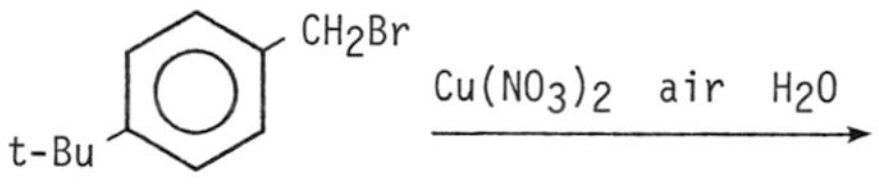

JCS (1935) 1847

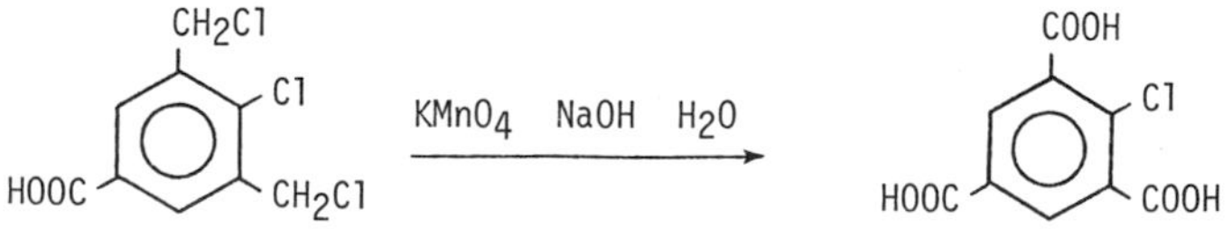

J Pharm Soc Jap (1950) 70 538

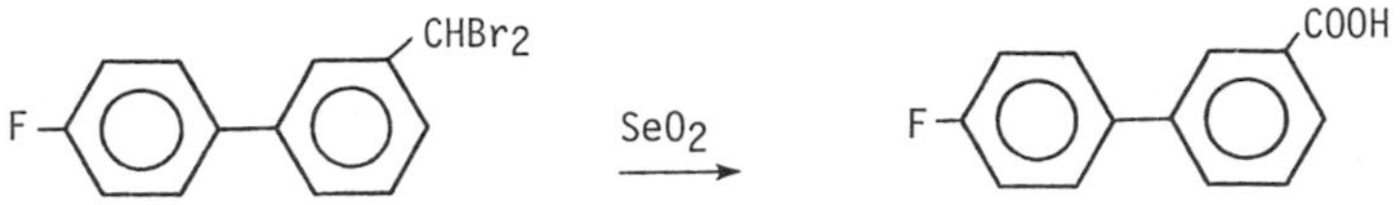

JACS (1966) 88 3318
JCS (1962) 186

Carboxylic acids may also be prepared by conversion of halides into esters or amides followed by hydrolysis. See section 115 (Esters from Halides) and section 85 (Amides from Halides)

Section 26   Carboxylic Acids from Hydrides (RH)
ooooooooooooooooooooooooooooooooooooooo

Reactions in which a hydrogen is replaced by carboxyl or a carboxyl containing chain, e.g. RH → RCOOH or RCH$_2$COOH (R=alkyl, vinyl or aryl), are included in this section. For reactions in which alkyl or aryl groups are oxidized to carboxyl, e.g. RCH$_3$ → RCOOH, see section 20 (Carboxylic Acids from Alkyls, Methylenes and Aryls)

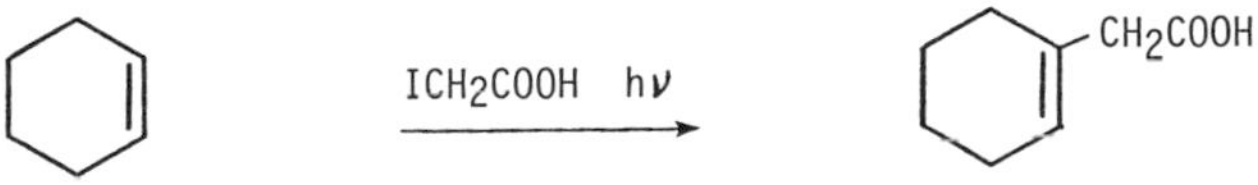

ACS Div Petr Chem (1966) 11 241
(Chem Abs 66 104510)

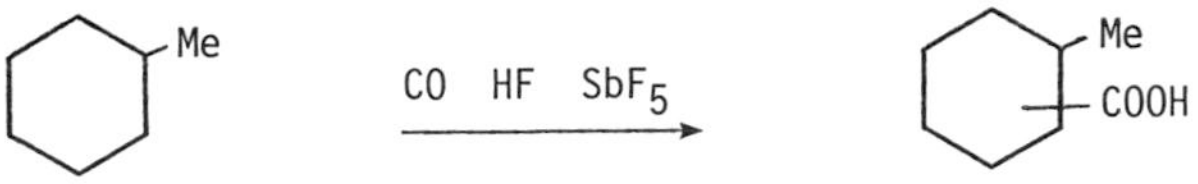

Ber (1967) 100 984

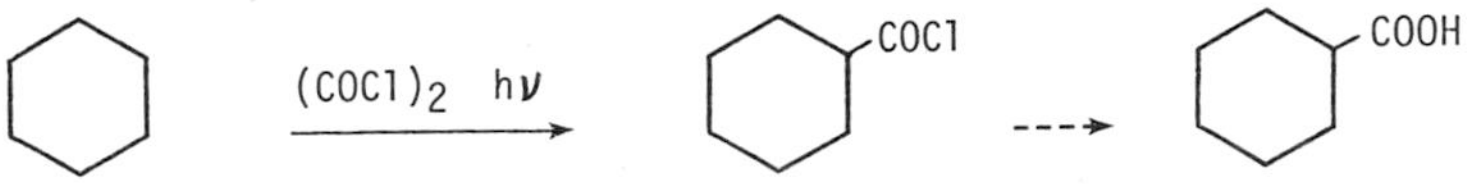

JCS (1963) 3918

$C_8H_{17}H$ $\xrightarrow[\text{2 } CO_2]{\text{1 } (i\text{-Bu})_2Hg\text{-K}}$ $C_8H_{17}COOH$ 24-35%

Monatsh (1967) <u>98</u> 763

$PhCH_3$ $\xrightarrow[\text{2 } CO_2]{\text{1 Na   PhCl}}$ $PhCH_2COOH$ 77%

JACS (1940) <u>62</u> 1514
Chem Rev (1957) <u>57</u> 867

 $\xrightarrow[\text{2 } H_2O \text{ dioxane}]{\text{1 } CO(CN)_2}$

Chem Ind (1961) 1116

PhH $\xrightarrow{\overparen{(CH_2)_nCOO} \quad AlCl_3}$ $Ph(CH_2)_nCOOH$ 44% (n=3)
51% (n=4)

JACS (1952) <u>74</u> 1591

 $\xrightarrow{ClCH_2COOH \quad Fe_2O_3 \quad KBr}$ <45%

JACS (1950) <u>72</u> 4302
Synthesis (1970) 628

$PhCH_2Me$ $\xrightarrow[\text{2 } CO_2]{\text{1 BuLi}\cdot Me_2NCH_2CH_2NMe_2}$ $\underset{\underset{Me}{|}}{PhCHCOOH}$ +

JOC (1970) <u>35</u> 10
Chimia (1970) <u>24</u> 109

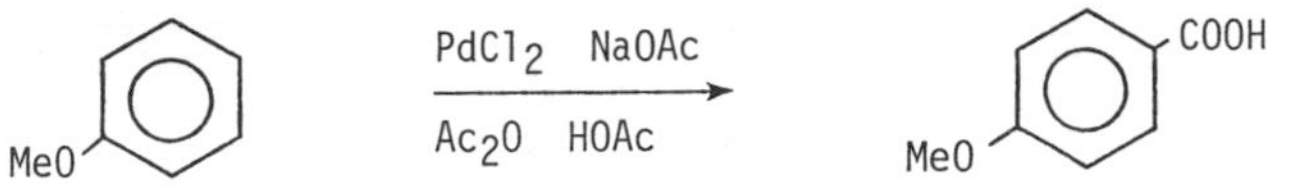

Tetr Lett (1969) 1019

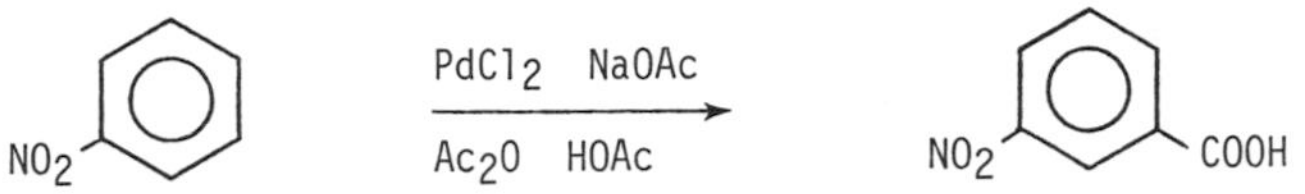

Tetr Lett (1969) 1019

PhH $\xrightarrow{\text{CO}_2 \text{ (or COCl}_2) \quad \text{AlCl}_3}$ PhCOCl ---→ PhCOOH   low yield

Chem Rev (1955) 55 229

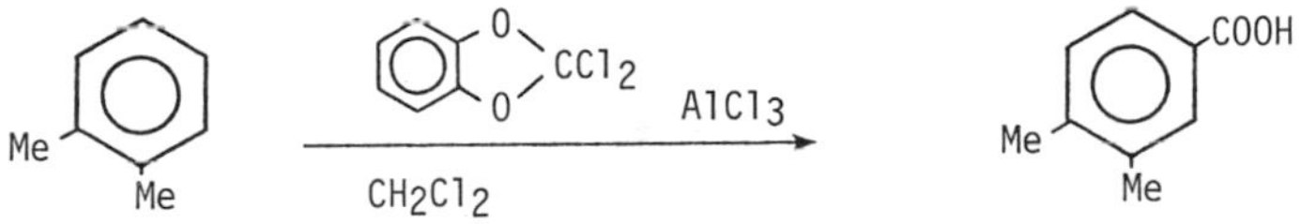

91%

Ber (1963) 96 1382

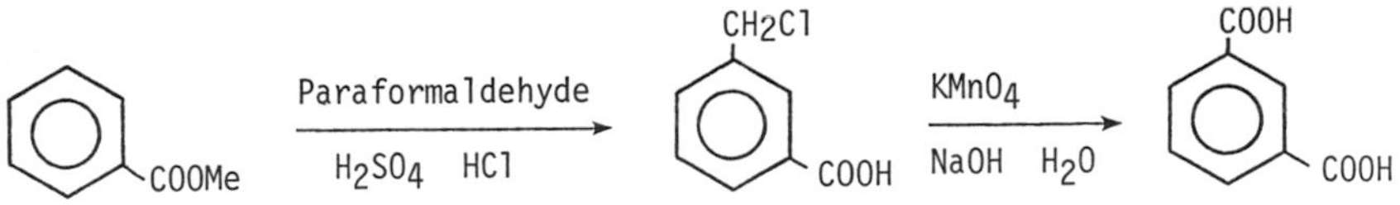

J Pharm Soc Jap (1950) 70 535

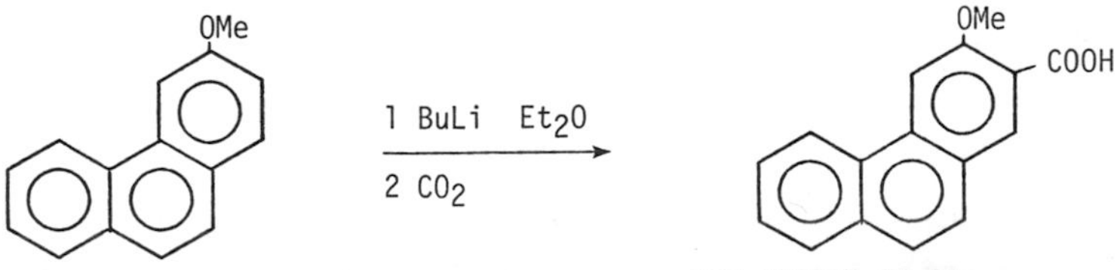

33%

JOC (1970) 35 10
Org React (1954) 8 258

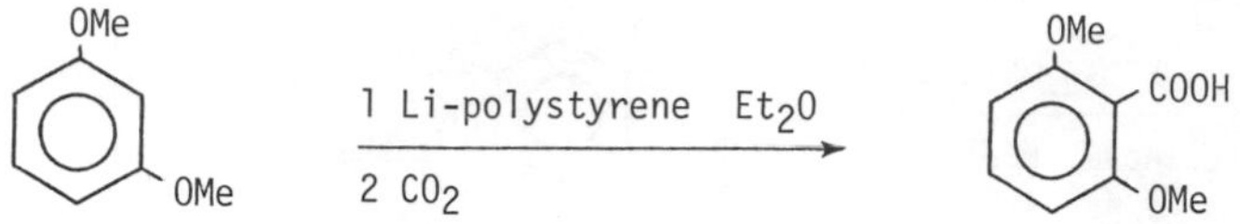
OMe
OMe
1 Li-polystyrene Et2O
2 CO2
OMe
COOH
OMe

60%

Ber (1964) 97 3098

Carboxylic acids may also be prepared by conversion of hydrides (RH) into esters or amides, followed by hydrolysis. See section 116 (Esters from Hydrides) and section 86 (Amides from Hydrides)

Section 27 Carboxylic Acids from Ketones

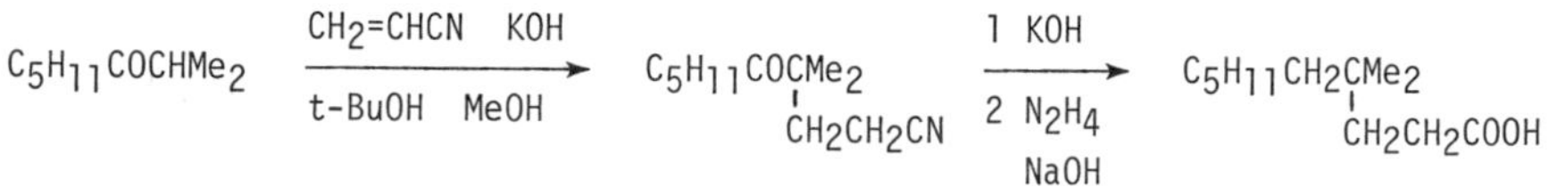
C5H11COCHMe2
CH2=CHCN KOH
t-BuOH MeOH
C5H11COCMe2
CH2CH2CN
1 KOH
2 N2H4
NaOH
C5H11CH2CMe2
CH2CH2COOH

JCS (1948) 1741

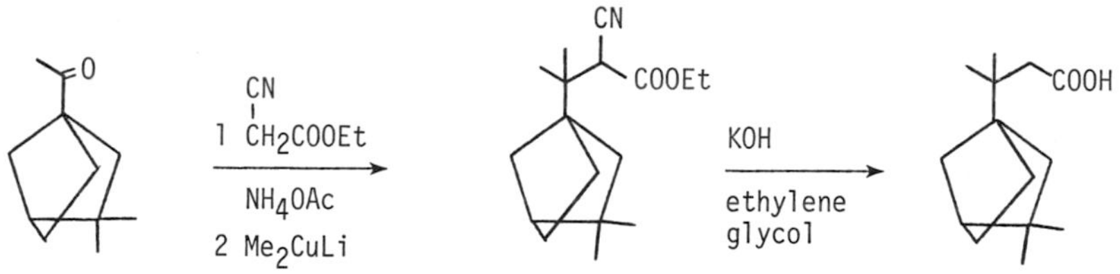
O
CN
1 CH2COOEt
NH4OAc
2 Me2CuLi
CN
COOEt
KOH
ethylene
glycol
COOH

43%

Tetr Lett (1967) 2893

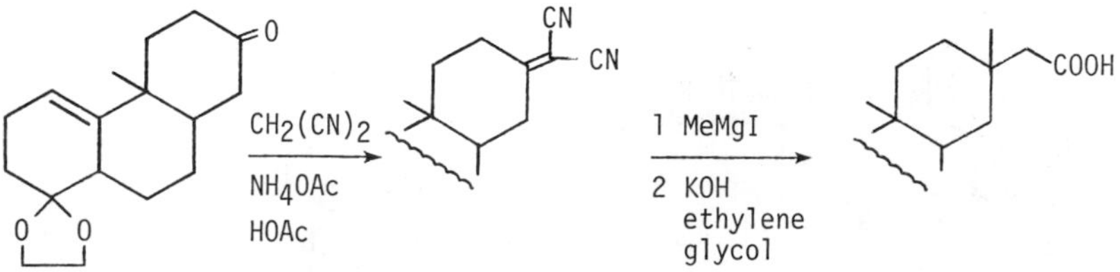
O
O O
CH2(CN)2
NH4OAc
HOAc
CN
CN
1 MeMgI
2 KOH
ethylene
glycol
COOH

Tetr Lett (1964) 1763

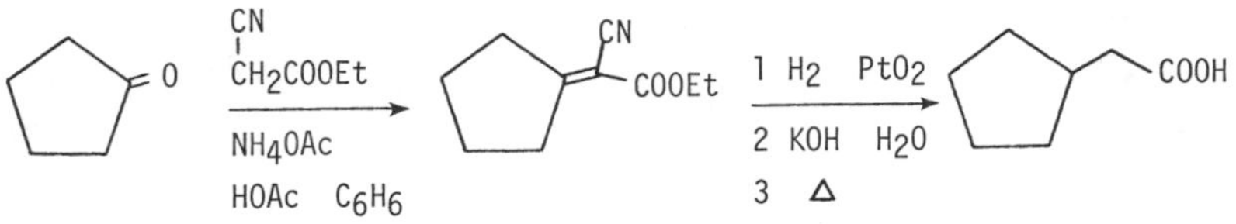

JACS (1948) 70 497
       (1959) 81 5397

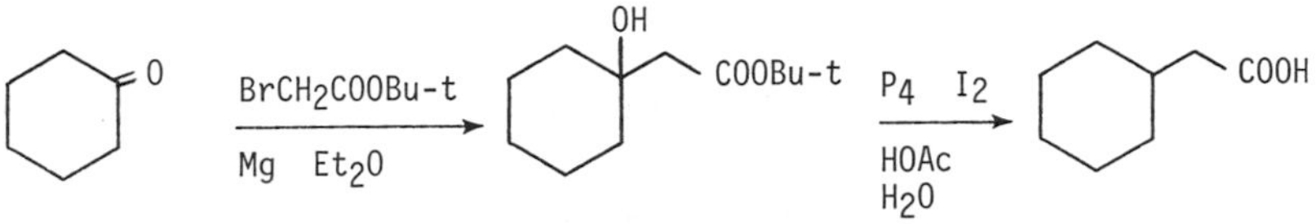

JOC (1966) 31 983
JACS (1946) 68 2339

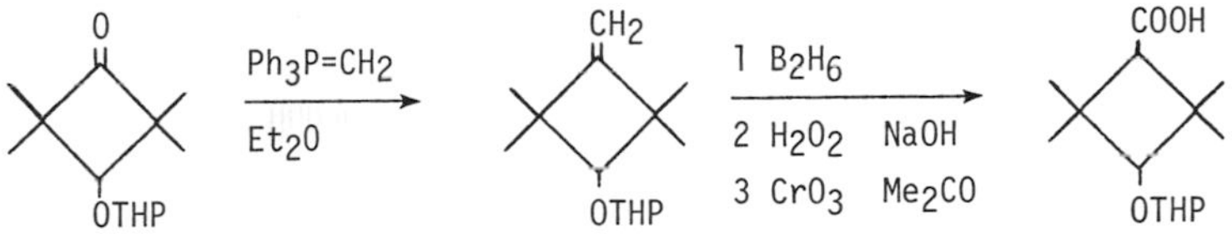

39%

Chem Comm (1968) 206
JACS (1960) 82 2498

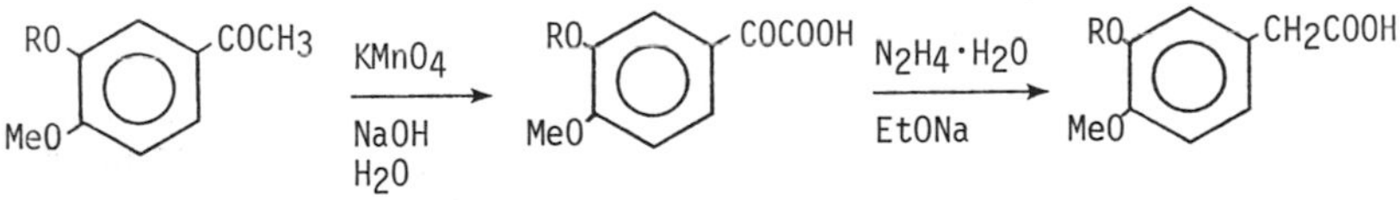

Monatsh (1952) 83 883

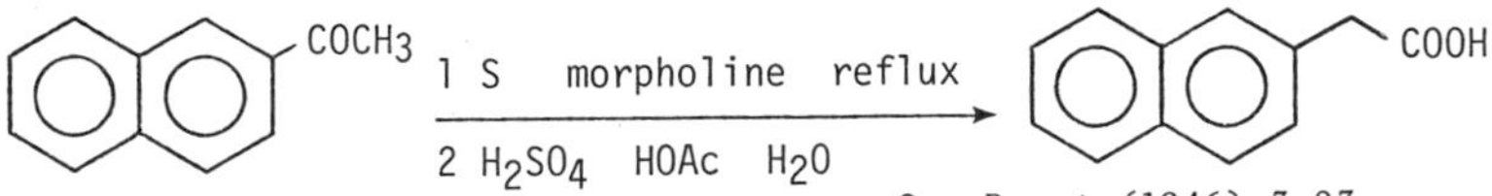

76%

Org React (1946) 3 83

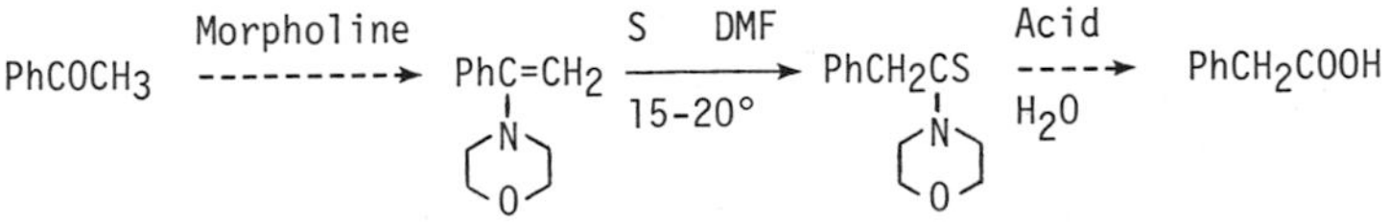

PhCOCH₃ ------→ PhC=CH₂ ──→ PhCH₂CS ----→ PhCH₂COOH

Angew (1964) 76 861
(Internat Ed 3 705)

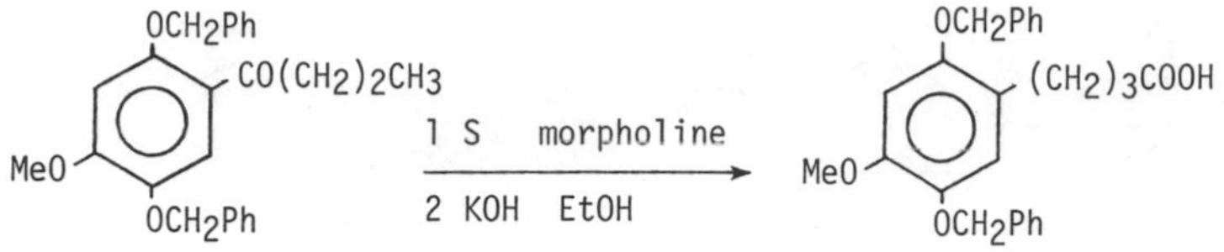

Helv (1964) 47 1996   37%

$Ph_2CHCOCH_3$ $\xrightarrow[\text{t-BuOH} \quad H_2O]{CCl_4 \quad KOH}$ Ph_2CHCH_2COOH 70%

JACS (1969) 91 7510

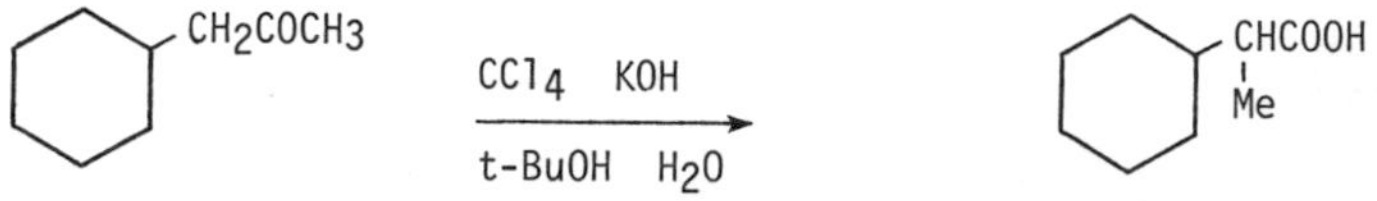

$\xrightarrow[\text{t-BuOH} \quad H_2O]{CCl_4 \quad KOH}$

60%

JACS (1969) 91 7510

$\xrightarrow[\text{Et}_2O]{KOH}$

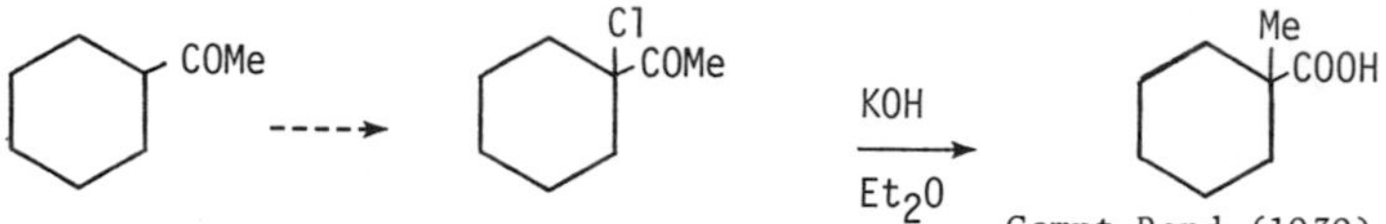

Compt Rend (1939) 208 1020
JACS (1960) 82 4307
Org React (1960) 11 261

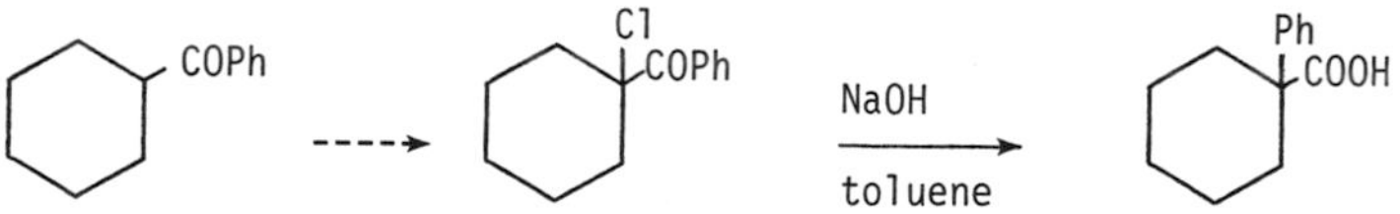

$\xrightarrow[\text{toluene}]{NaOH}$

51%

JACS (1952) 74 5352

H_2O_2 SeO_2 t-BuOH

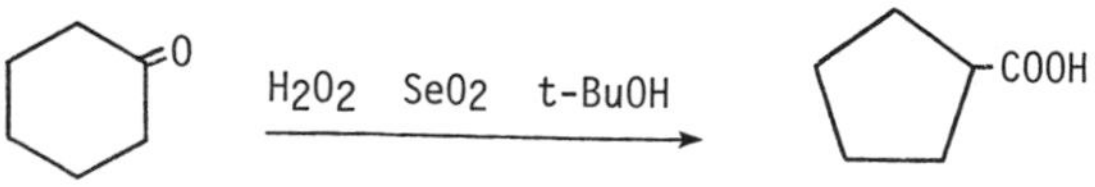

< 27%

JOC (1957) 22 1680
Compt Rend C (1967) 265 578
Annalen (1965) 681 30

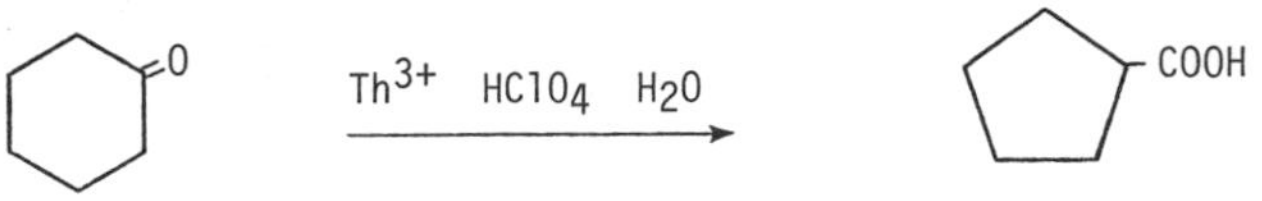
O
Th3+ HClO4 H2O
COOH

Tetr Lett (1966) 1779

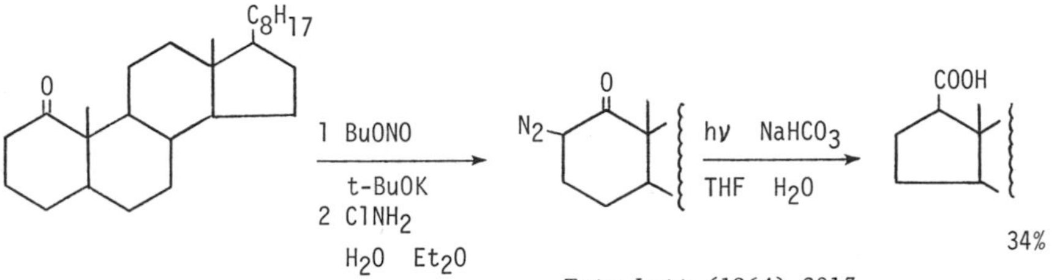
C8H17
O
1 BuONO
t-BuOK
2 ClNH2
H2O Et2O
N2
O
hν NaHCO3
THF H2O
COOH
34%

Tetr Lett (1964) 2813

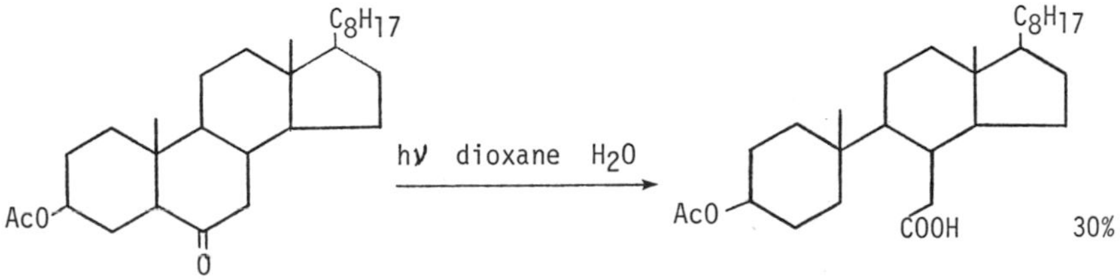
C8H17
AcO
O
hν dioxane H2O
C8H17
AcO
COOH
30%

Ber (1964) 97 958
Angew (1965) 77 229
(Internat Ed 4 211)

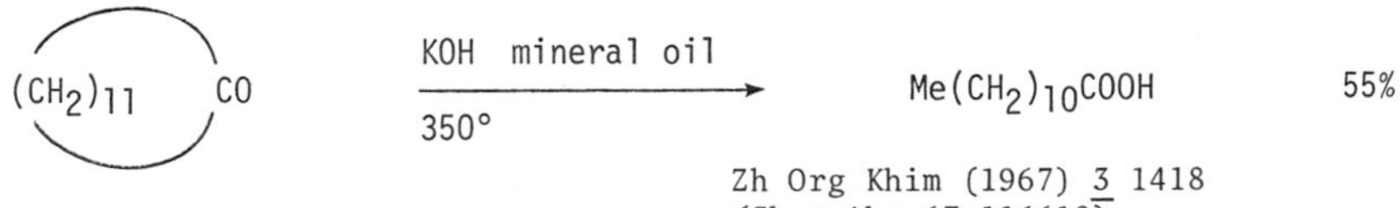
(CH2)11
CO
KOH mineral oil
350°
Me(CH2)10COOH
55%

Zh Org Khim (1967) 3 1418
(Chem Abs 67 116618)

$C_{16}H_{33}CH_2COPh$ —(1 MeI $NaNH_2$; 2 $NaNH_2$)→ $C_{16}H_{33}C(Me)_2CONH_2$ —($NaNO_2$, H_2SO_4, H_2O)→ $C_{16}H_{33}C(Me)_2COOH$

JCS (1942) 488

t-BuOK   t-BuOH

165°

43%

JACS (1969) 91 1009

t-BuOK   Me₂SO   H₂O

room temp

65%

Tetr Lett (1964) 3251

Ph₂CO

t-BuOK   monoglyme   H₂O

30°

PhCOOH   90%

JACS (1967) 89 946

t-BuCOMe

CCl₄   KOH

t-BuOH   H₂O

t-BuCOOH   80%

JACS (1969) 91 7510

1 I₂   Pyr

2 KOH
  diethylene glycol
  H₂O

60-70%

JACS (1951) 73 3803

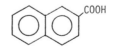

Cl₂   NaOH   H₂O

87%

Org Synth (1943) Coll Vol 2 428

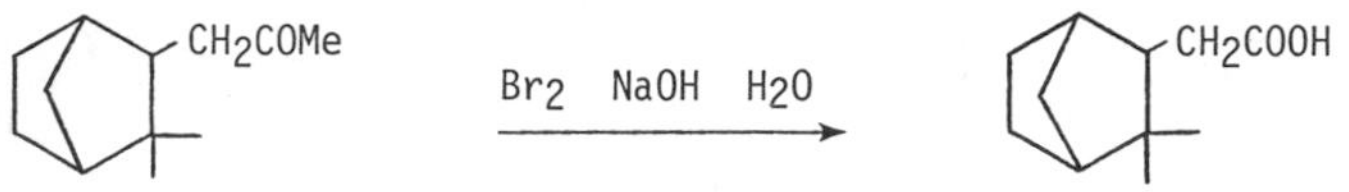

Aust J Chem (1967) 20 2033

PhCOPr $\xrightarrow{\begin{array}{c}\text{1 NaOBr } H_2O\\\hline\text{2 NaOH}\end{array}}$ PhCOOH   48%

JACS (1950) 72 1642

Et$_3$CCOEt $\xrightarrow{\begin{array}{c}\text{HNO}_3\text{ }H_2O\\\hline\end{array}}$ Et$_3$CCOOH   71%

Bull Soc Chim Fr (1967) 2011
JACS (1942) 64 2421

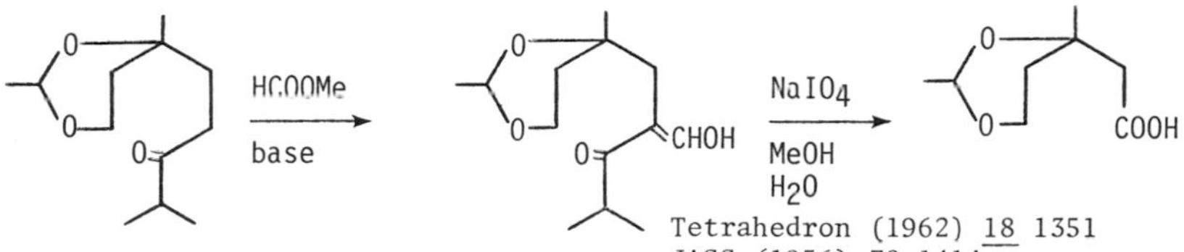

Tetrahedron (1962) 18 1351
JACS (1956) 78 1414

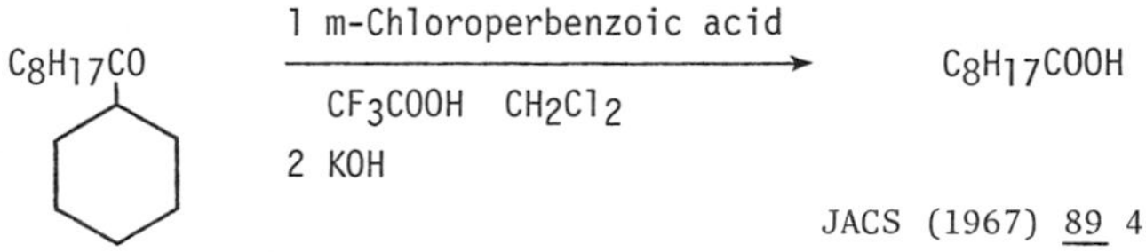

C$_8$H$_{17}$CO $\xrightarrow{\begin{array}{c}\text{1 m-Chloroperbenzoic acid}\\\hline\text{CF}_3\text{COOH   CH}_2\text{Cl}_2\\\text{2 KOH}\end{array}}$ C$_8$H$_{17}$COOH   79%

JACS (1967) 89 4530

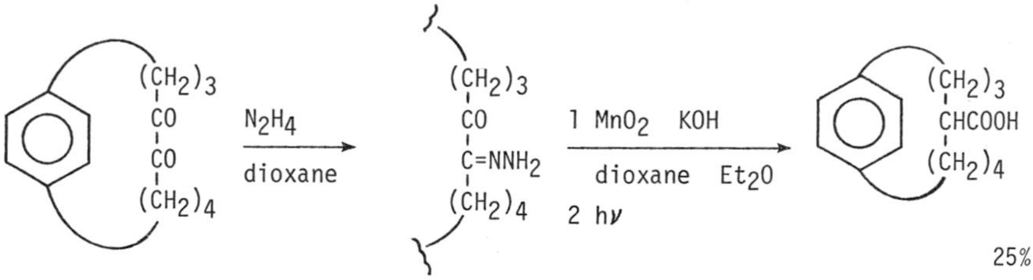

25%

JACS (1963) 85 1171

PhCOCOMe $\xrightarrow{\text{NaOBr H}_2\text{O}}$ PhCOOH   91%

JACS (1950) 72 1642

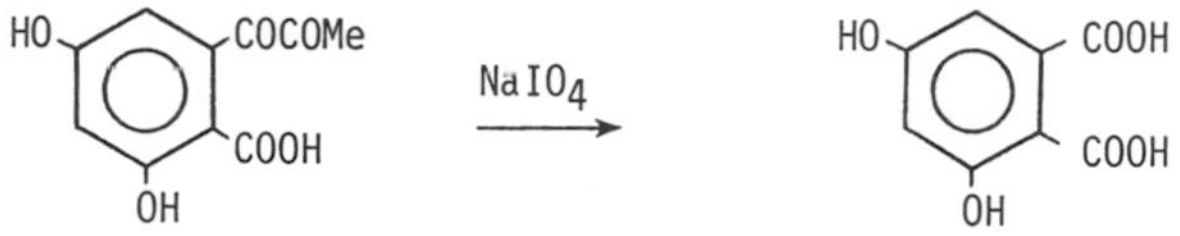

$\xrightarrow{\text{NaIO}_4}$

JCS (1935) 1467
Org React (1944) 2 341

$Me(CH_2)_7COCO(CH_2)_7COOH$ $\xrightarrow[\text{HOAc}]{\text{MeCOO}_2\text{H}}$ $Me(CH_2)_7COOH$ 100%

Org React (1957) 9 73

Carboxylic acids may also be prepared by conversion of ketones into
esters or amides, followed by hydrolysis. See section 117 (Esters
from Ketones) and section 87 (Amides from Ketones)

Section 28   Carboxylic Acids from Nitriles

$\xrightarrow{\text{H}_2\text{SO}_4 \ \text{H}_2\text{O}}$

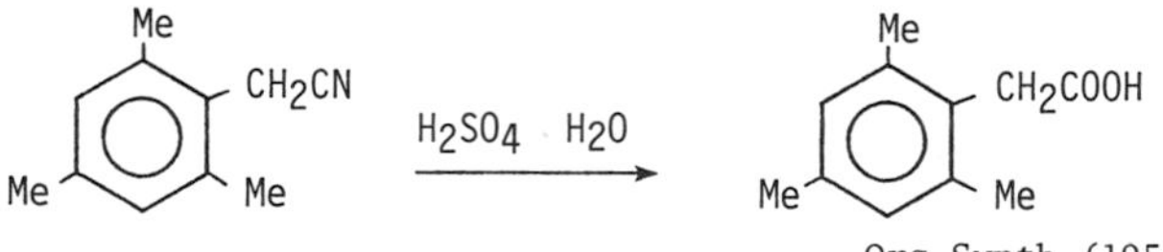

Org Synth (1955) Coll Vol 3 557

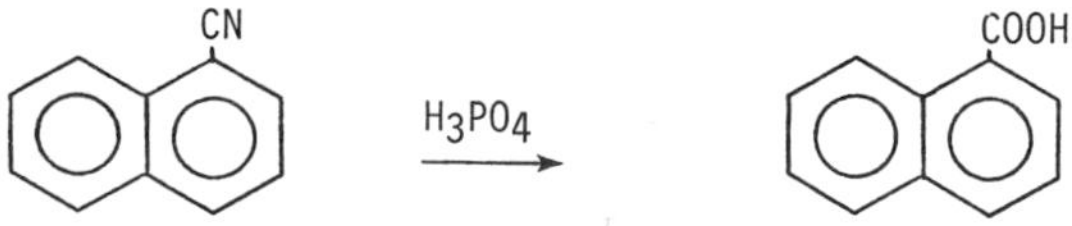

$\xrightarrow{\text{H}_3\text{PO}_4}$

70-90%

Rec Trav Chim (1927) 46 600

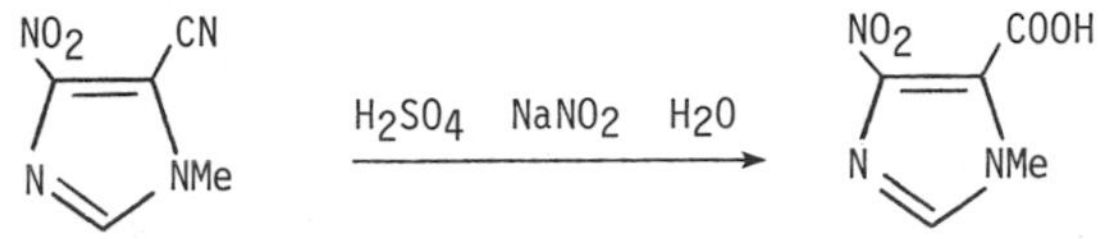

H_2SO_4 $NaNO_2$ H_2O

92%

JCS (1945) 751

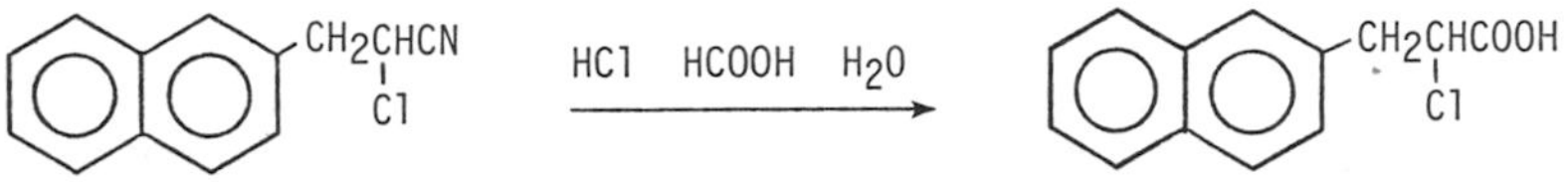

HCl   HCOOH H_2O

80%

Proc Chem Soc (1962) 117

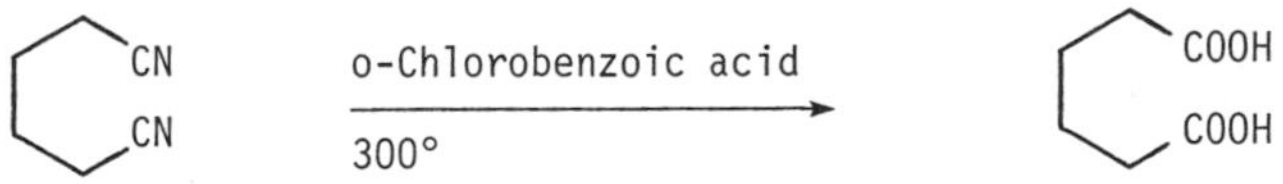

o-Chlorobenzoic acid

300°

97%

Annalen (1968) 716 78

1 H_2O_2 KOH H_2O

2 H_3PO_4

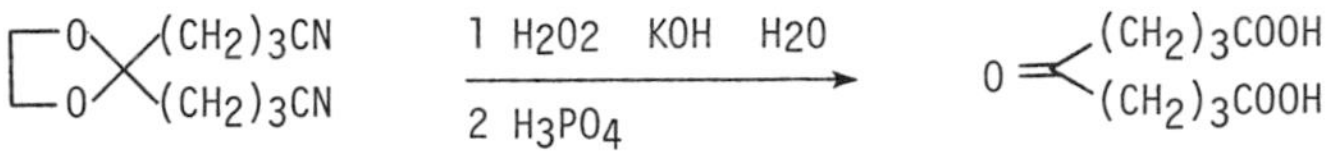

85%

JCS (1962) 4722

PhCN

H_2S NH_3 EtOH

$PhCSNH_2$ ---->   PhCOOH

Ber (1890) 23 158
JOC (1951) 16 131

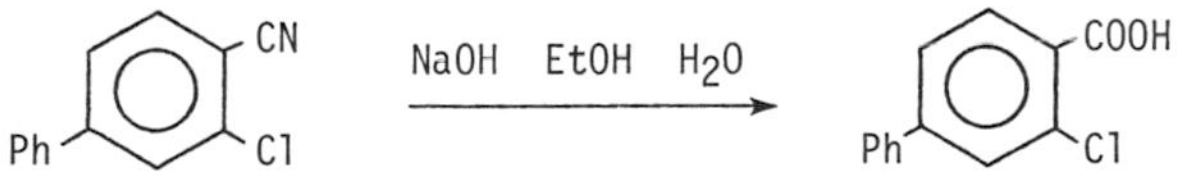

NaOH   EtOH H_2O

81-90%

JCS C (1966) 840

Me — [ring] — CN   →   KOH H₂O / diethylene glycol   →   Me — [ring] — COOH   93%

Chem Pharm Bull (1969) <u>17</u> 1564

Carboxylic acids may also be prepared by conversion of nitriles into esters or amides, followed by hydrolysis. See section 118 (Esters from Nitriles) and section 88 (Amides from Nitriles)

Section 29   Carboxylic Acids from Olefins

1 t-BuLi Et₃N

2 CO₂

→   [bicyclic] — Bu-t / COOH   30-45%

JOC (1965) <u>30</u> 917
JACS (1969) <u>91</u> 6362

$Me_2C=CH_2$ → $CH_2=CCl_2$ BF_3 H_2SO_4 → Me_3CCH_2COOH 75%

Angew (1965) <u>77</u> 967
(Internat Ed <u>4</u> 956)

$C_6H_{13}CH=CH_2$ → CH_3COOH / di-t-butyl peroxide   → $C_6H_{13}(CH_2)_3COOH$ ~70%

JCS (1965) 1918

$BuCH=CH_2$ → $BrCH_2COOH$ hν   → $Bu(CH_2)_3COOH$ 80%

Chem Comm (1967) 435

$C_6H_{13}CH=CH_2$ $\xrightarrow[\substack{\text{di-t-butyl} \\ \text{peroxide}}]{CH_2(COOEt)_2}$ $C_6H_{13}CH_2CH_2CH(COOEt)_2$ $\xrightarrow[\substack{2 \quad \Delta}]{1 \text{ Hydrolysis}}$ $C_6H_{13}(CH_2)_3COOH$

Chem Ind (1961) 830

$BuCH=CH_2$ $\xrightarrow[\substack{\text{di-t-butyl} \\ \text{peroxide}}]{HC(COOEt)_3}$ $BuCH_2CH_2C(COOEt)_3$ $\dashrightarrow[\substack{2 \quad \Delta}]{1 \text{ Hydrolysis}}$ $Bu(CH_2)_3COOH$

Synthesis (1970) 124

$C_6H_{13}CH=CH_2$ $\xrightarrow[\substack{\text{borane} \\ 2 \text{ CO}}]{1 \text{ Di-cyclohexyl}}$ $C_6H_{13}CH_2CH_2CO$ ⬡ $\xrightarrow[\substack{\text{acid} \quad CF_3COOH \\ CH_2Cl_2}]{\text{m-Chloroperbenzoic}}$ $C_6H_{13}CH_2CH_2COOH$

79%

JACS (1967) 89 4530

CO (200 atmos) H_2SO_4

80%

JACS (1960) 82 1261

HCOOH H_2SO_4

9%

Tetrahedron (1965) 21 2641

HCOOH H_2SO_4

Ber (1966) 99 1149

$$\text{CO (300 atmos)} \quad H_2SO_4 \longrightarrow$$

Me Me / Et COOH   60%

Ber (1964) 97 3088

$$\xrightarrow{\;\;\;\;\;} \quad \text{(F, Cl bicyclic)} \quad \xrightarrow{H_2SO_4} \quad \text{(cyclohexane-COOH)} \quad 90\%$$

Ber (1968) 101 1291

$$C_6H_{13}CH=CH_2 \quad \xrightarrow[\text{HCl} \quad Me_2CO \quad H_2O]{Ni(CO)_4 \quad h\nu} \quad C_6H_{13}CHCOOH \quad \text{(+ isomers)}$$

$$\underset{Me}{|}$$

43%

Angew (1965) 77 813
(Internat Ed 4 790)

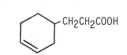

$$\xrightarrow[\text{2 } CO_2]{\text{1 PrMgBr} \quad TiCl_4 \quad THF}$$

49%

JOC (1970) 35 392

Me OH
—CH$_2$CH=CH$_2$
—Me

$$\xrightarrow[\text{2 } CrO_3]{\text{1 } B_2H_6 \quad THF}$$

Me OH
—CH$_2$CH$_2$COOH
—Me

(→ lactone)

Chem Comm (1968) 122

$$\begin{array}{l} CH_2 \\ \| \\ CH \\ | \\ (CH_2)_8 \\ | \\ COOH \end{array} \quad \xrightarrow[\text{2 NaOH}]{\text{1 } (NH_4)_2S \quad (NH_4)_2S_2O_3 \quad H_2O} \quad \begin{array}{l} COOH \\ | \\ CH_2 \\ | \\ (CH_2)_8 \\ | \\ COOH \end{array} \quad 35\%$$

JACS (1946) 68 2033

$Me_2CH(CH_2)_3\underset{\underset{Me}{|}}{C}HCH=CH_2$ $\xrightarrow{\text{KMnO}_4 \quad \text{NaHCO}_3 \quad \text{Me}_2\text{CO}}$ $Me_2CH(CH_2)_3\underset{\underset{Me}{|}}{C}HCOOH$

45%

JACS (1943) 65 745

$\xrightarrow{\begin{array}{c}\text{KMnO}_4 \quad \text{NaIO}_4 \quad \text{K}_2\text{CO}_3 \\ \hline \text{t-BuOH} \quad \text{H}_2\text{O}\end{array}}$

Can J Chem (1967) 45 1439

$\begin{array}{c}COOMe \\ | \\ (CH_2)_7 \\ | \\ CH \\ \| \\ CH \\ | \\ (CH_2)_7 \\ | \\ Me\end{array}$ $\xrightarrow{\begin{array}{c}\text{KMnO}_4 \quad \text{NaIO}_4 \\ \hline \text{t-BuOH} \quad \text{H}_2\text{O}\end{array}}$ $\begin{array}{c}COOMe \\ | \\ (CH_2)_7 \\ | \\ COOH \\ \\ COOH \\ | \\ (CH_2)_7 \\ | \\ Me\end{array}$

98-99%

Can J Chem (1956) 34 1413
JACS (1961) 83 4819

$\xrightarrow{\begin{array}{c}\text{RuO}_4 \quad \text{NaIO}_4 \\ \hline \text{Me}_2\text{CO} \quad \text{H}_2\text{O}\end{array}}$

89%

JACS (1963) 85 3419

$C_{11}H_{23}CH=CH_2$ $\xrightarrow{\begin{array}{c}1 \quad O_3 \\ \hline 2 \quad Ag_2O \quad NaOH \quad H_2O\end{array}}$ $C_{11}H_{23}COOH$

Ber (1942) 75B 656

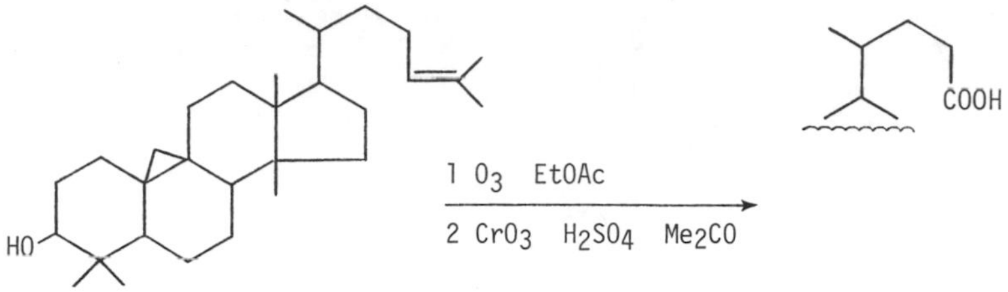

$$\xrightarrow[\text{2 CrO}_3 \ \text{H}_2\text{SO}_4 \ \text{Me}_2\text{CO}]{\text{1 O}_3 \ \text{EtOAc}}$$

80%

Tetr Lett (1969) 1733

Carboxylic acids may also be prepared by conversion of olefins into esters or amides, followed by hydrolysis. See section 119 (Esters from Olefins) and section 89 (Amides from Olefins)

Section 30   Carboxylic Acids from Miscellaneous Compounds

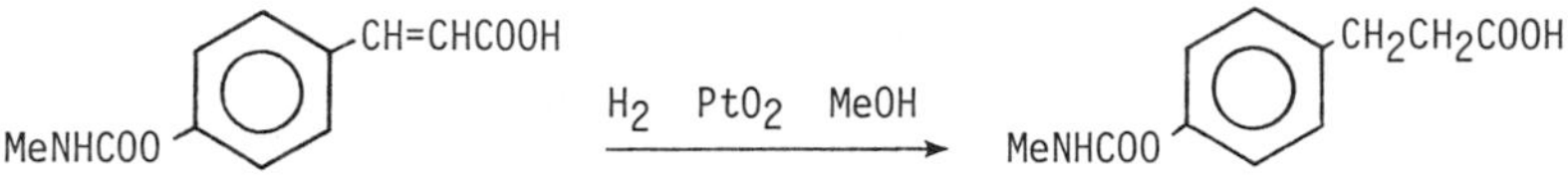

$$\xrightarrow{\text{H}_2 \ \text{PtO}_2 \ \text{MeOH}}$$

70%

Chem Pharm Bull (1970) 18 243

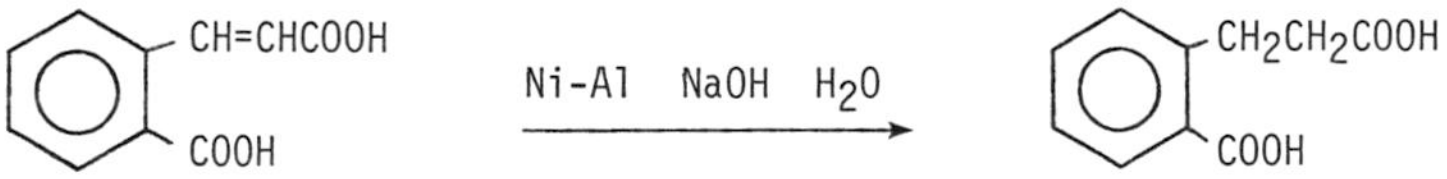

$$\xrightarrow{\text{Ni-Al} \ \text{NaOH} \ \text{H}_2\text{O}}$$

92-95%

Org Synth (1963) Coll Vol 4 136

$$\text{PhCH=CHCOOH} \xrightarrow{\text{N}_2\text{H}_4 \ \text{Ni} \ \text{H}_2\text{O}} \text{PhCH}_2\text{CH}_2\text{COOH}$$

85%

Act Chem Scand (1961) 15 1200

PhCH=CHCOOH $\xrightarrow{\quad H_2 \quad (Ph_3P)_3RhCl \quad}$ PhCH$_2$CH$_2$COOH   75%

Compt Rend C (1966) 263 251
Chem Comm (1969) 1365

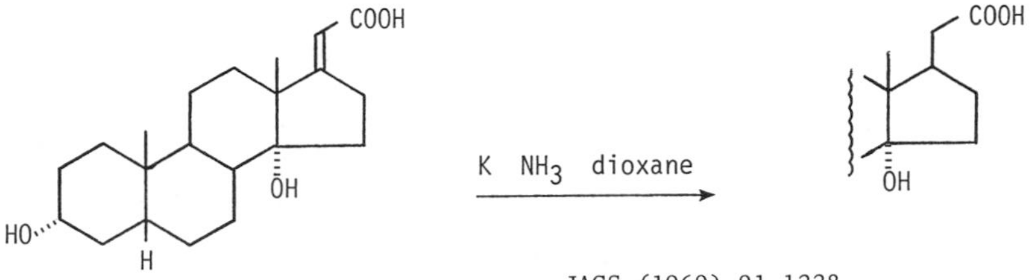

$\xrightarrow{\quad K \quad NH_3 \quad dioxane \quad}$ 95%

JACS (1969) 91 1228
Compt Rend (1969) C 268 640

PhCH=CHCOOH $\xrightarrow{\quad P_4 \quad KI \quad H_3PO_4 \quad}$ PhCH$_2$CH$_2$COOH   80%

Helv (1939) 22 601

PhCHCOOH $\xrightarrow{\quad P_4 \quad KI \quad H_3PO_4 \quad}$ PhCH$_2$COOH   90%
|
OH

Helv (1939) 22 601
Org Synth (1932) Coll Vol 1 224

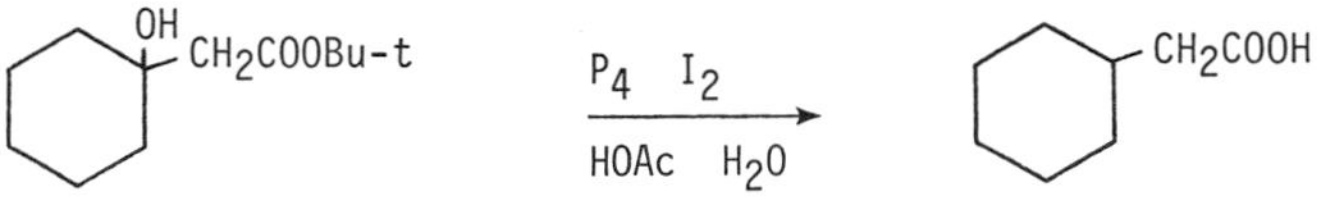

$\xrightarrow[\text{HOAc} \quad H_2O]{\quad P_4 \quad I_2 \quad}$

JOC (1966) 31 983

PhCHCOOH $\xrightarrow[\text{tetralin}]{\quad H_2 \quad Pd\text{-}BaSO_4 \quad}$ PhCH$_2$COOH
|
OAc

Org React (1953) 7 263

$$PrCH_2NO_2 \xrightarrow{\text{H}_2\text{SO}_4 \quad \text{H}_2\text{O}} PrCOOH \qquad \sim 90\%$$

Ind Eng Chem (1939) 31 118

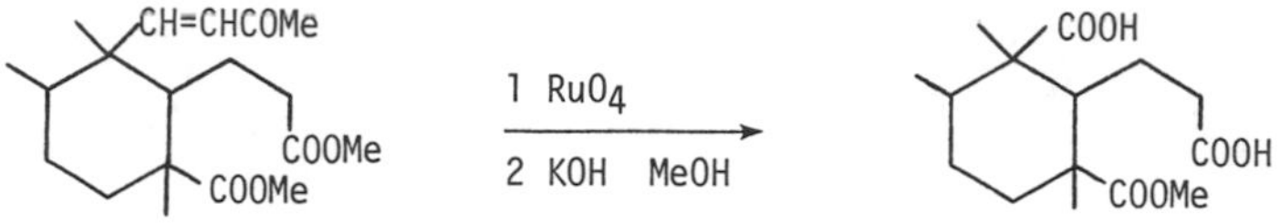

1 RuO$_4$

2 KOH   MeOH

Tetr Lett (1968) 2681

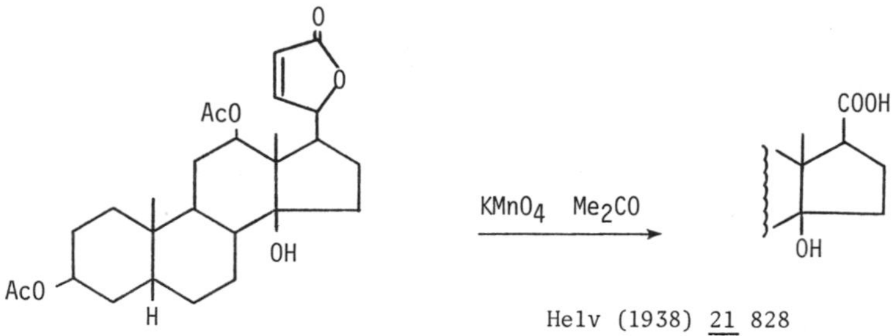

KMnO$_4$ Me$_2$CO

Helv (1938) 21 828

$$C_8H_{17}\underset{OH}{\text{C}}\text{OCH}(CH_2)_7COOH \xrightarrow{\text{KIO}_4 \quad \text{H}_2\text{SO}_4 \quad \text{EtOH}} C_8H_{17}COOH \qquad 92\%$$

JCS (1936) 1788
(1935) 1467

$$Ph\underset{OH}{\text{C}}\text{HCOOH} \xrightarrow[\text{NaOH} \quad \text{H}_2\text{O}]{\text{Nickel peroxide}} PhCOOH \qquad 90\%$$

Chem Pharm Bull (1964) 12 403

Section 30A   Protection of Carboxylic Acids

RCOOH $\xrightarrow{\quad\text{- - - -}\quad}$ RCOOCH$_2$CCl$_3$

$\xleftarrow{\quad\quad}$
Zn   HOAc   H$_2$O

JACS (1966) 88 852

RNHCH$_2$COOH $\xrightarrow{\begin{array}{c}\text{NO}_2\!\!-\!\!\bigcirc\!\!-\!\!\text{SCH}_2\text{CH}_2\text{OH}\\ \hline \text{TsOH}\quad\text{C}_6\text{H}_6\end{array}}$ RNHCH$_2$COOCH$_2$CH$_2$S$-\bigcirc-$NO$_2$

$\xleftarrow{\begin{array}{c}\\ \hline\end{array}}$
1 H$_2$O$_2$ (NH$_4$)$_6$Mo$_7$O$_{24}$
2 pH 10-10.5

JCS C (1969) 2495

RCOOH $\xrightarrow{\quad\text{- - - -}\quad}$ RCOOCH$_2$CH$_2$SMe $\xrightarrow{\quad\text{H}_2\text{O}_2\quad\text{(NH}_4\text{)}_6\text{Mo}_7\text{O}_{24}\quad}$ RCOOCH$_2$CH$_2$SO$_2$Me

$\xleftarrow{\quad\quad}$
1 MeI
2 pH 10-10.5
                        pH 10-11

(stable to acid)

Tetr Lett (1968) 2525

$\xrightarrow{\quad\text{(Me}_3\text{Si)}_2\quad\text{CHCl}_3\quad}$

$\xleftarrow{\quad\text{H}_2\text{O}\quad}$ COOSiMe$_3$

Annalen (1964) 673 166
JACS (1966) 88 3390

$\xrightarrow{\quad\text{Me}_2\text{C=CH}_2\quad\text{H}_2\text{SO}_4\quad}$

$\xleftarrow{\quad\text{CF}_3\text{COOH}\quad}$ COOCMe$_3$

J Med Chem (1966) 9 444

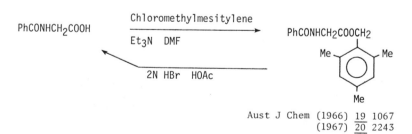

JACS (1957) 79 1995
JCS (1952) 3945

JACS (1969) 91 5674

JCS C (1970) 964

Aust J Chem (1966) 19 1067
(1967) 20 2243

Further examples of the preparation and cleavage of benzyl esters
are included in section 107 (Esters from Carboxylic Acids) and
section 23 (Carboxylic Acids from Esters)

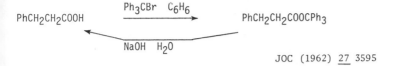

PhCH$_2$CH$_2$COOH $\xrightarrow[\text{NaOH   H}_2\text{O}]{\text{Ph}_3\text{CBr   C}_6\text{H}_6}$ PhCH$_2$CH$_2$COOCPh$_3$

JOC (1962) <u>27</u> 3595

RNHCH$_2$COOH $\xrightarrow[\text{HCl   dioxane   H}_2\text{O}]{\text{Ph}_3\text{CCl   Et}_3\text{N}}$ RNHCH$_2$COOCPh$_3$

JCS <u>C</u> (1966) 1191

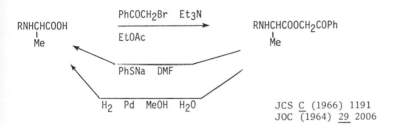

RNHCHCOOH   PhCOCH$_2$Br Et$_3$N   RNHCHCOOCH$_2$COPh
    | $\xrightarrow{\text{EtOAc}}$ |
    Me   Me

PhSNa   DMF

H$_2$ Pd   MeOH   H$_2$O

JCS <u>C</u> (1966) 1191
JOC (1964) <u>29</u> 2006

PhCH=CHCOOH $\xrightarrow[\text{base}]{\text{p-Bromophenacyl bromide}}$ PhCH=CHCOOCH$_2$CO

Zn   HOAc

Br

Tetr Lett (1970) 343

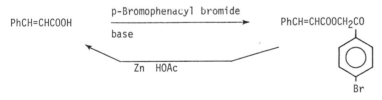

C$_5$H$_{11}$COOH $\dashrightarrow$ C$_5$H$_{11}\overset{\text{O}}{\overset{||}{\text{C}}}$OS

$\xleftarrow{\text{h}\nu\ \text{C}_6\text{H}_6}$

NO$_2$

NO$_2$

JCS (1965) 3571
JACS (1970) <u>92</u> 6333

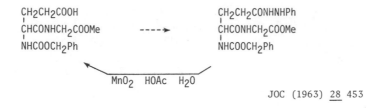

CH₂CH₂COOH
|
CHCONHCH₂COOMe   - - - - ▶
|
NHCOOCH₂Ph

CH₂CH₂CONHNHPh
|
CHCONHCH₂COOMe
|
NHCOOCH₂Ph

MnO₂   HOAc   H₂O

JOC (1963) 28 453

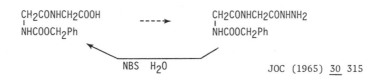

CH₂CONHCH₂COOH
| - - - - ▶
NHCOOCH₂Ph

CH₂CONHCH₂CONHNH₂
|
NHCOOCH₂Ph

NBS   H₂O

JOC (1965) 30 315

RCH₂COOH   - - - ▶   RCH₂COOMe   - - - ▶   RCH₂CONHOH

HIO₄   H₂O

JACS (1960) 82 4903

R⟨benzene ring⟩COOH

NH₂
|
Me₂CCH₂OH
———————————▶

HCl   H₂O   EtOH

R⟨benzene ring⟩C=N with oxazoline ring

(Stable to RMgX
and CrO₃)

JACS (1970) 92 6646

Other reactions useful for the protection of carboxylic acids are included
in section 107 (Esters from Carboxylic Acids), section 23 (Carboxylic Acids
from Esters), section 77 (Amides from Carboxylic Acids) and section 21
(Carboxylic Acids from Amides)

Chapter 3   PREPARATION
OF
ALCOHOLS
AND PHENOLS

$BuC\equiv CH$ $\xrightarrow[\text{2 MeCOEt Et}_2O]{\text{1 EtMgBr Et}_2O}$ $BuC\equiv C\underset{\underset{OH}{|}}{\overset{\overset{Et}{|}}{C}}Me$ $\xrightarrow{H_2 \ Pt}$ $BuCH_2CH_2\underset{\underset{OH}{|}}{\overset{\overset{Et}{|}}{C}}Me$

92%

JACS (1941) <u>63</u> 186
Tetr Lett (1965) 1619

$BuC\equiv CH$ $\xrightarrow[\substack{\text{2 MeLi} \\ \text{3 EtBr} \\ \text{4 NaOH } H_2O}]{\text{1 } B_2H_6 \ \text{THF}}$ $Bu\underset{\underset{OH}{|}}{C}HEt$

JACS (1967) <u>89</u> 291

$BuC\equiv CH$ $\xrightarrow[\substack{\text{2 MeONa} \\ \text{3 MeI} \\ \text{4 } O_2}]{\text{1 (i-Bu)}_2AlH \ \text{THF}}$ $BuCH_2\underset{\underset{Me}{|}}{C}HOH$ 90%

Tetr Lett (1966) 6021

$BuC\equiv CH$ $\xrightarrow[\text{2 } H_2O_2 \ \text{NaOH}]{\text{1 } B_2H_6 \ \text{THF}}$ $BuCH_2CH_2OH$ 80%

JACS (1967) <u>89</u> 291

Section 32 Alcohols and Phenols from Carboxylic Acids, Acid Chlorides and Anhydrides

$$PhCOOH \xrightarrow[H_2O]{H_2\ (205\ atmos)\ \ ReO_3} PhCH_2OH \qquad 43\%$$

JOC (1963) 28 2345

$$C_9H_{19}COOH \xrightarrow{H_2\ (173\ atmos)\ \ Rh_2O_7} C_9H_{19}CH_2OH \qquad 100\%$$

JOC (1959) 24 1847

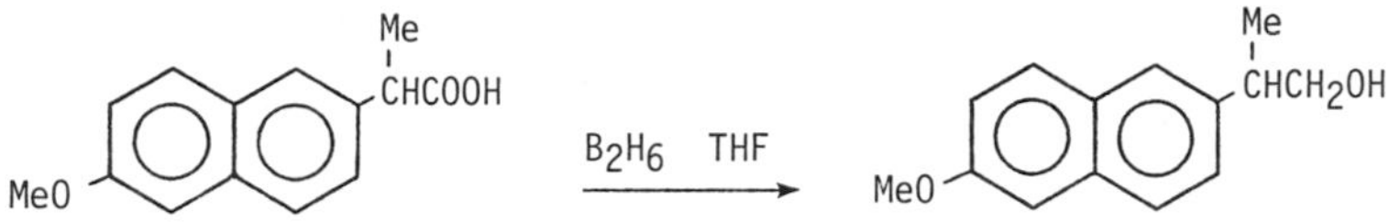
Me
CHCOOH
MeO
B2H6 THF
Me
CHCH2OH
MeO

87%

J Med Chem (1970) 13 203

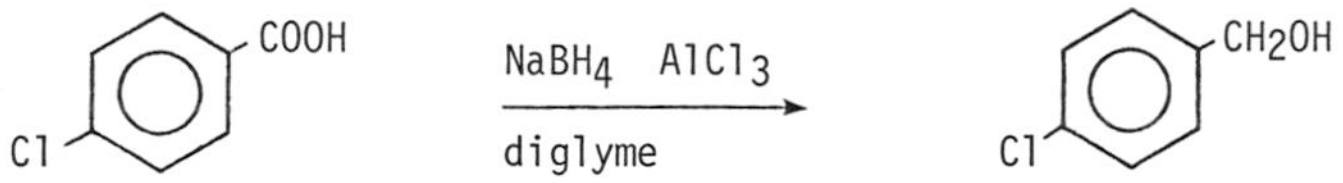
COOH
Cl
NaBH4 AlCl3
diglyme
CH2OH
Cl

64%

JACS (1956) 78 2582

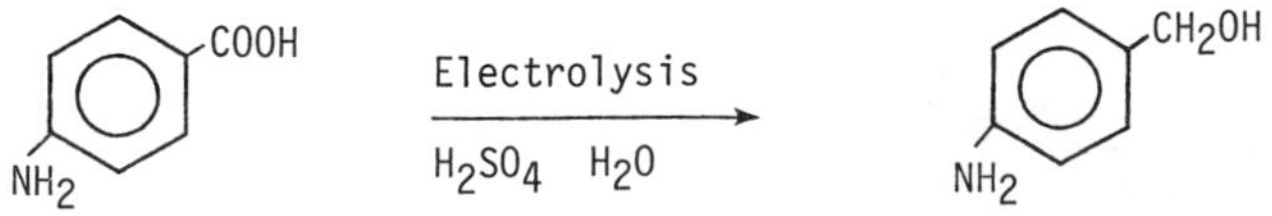
COOH
NH2
Electrolysis
H2SO4 H2O
CH2OH
NH2

56%

JCS (1942) 98

Ph$_3$CCOOH $\xrightarrow{\text{LiAlH}_4 \quad \text{THF}}$ Ph$_3$CCH$_2$OH

$\downarrow$

Ph$_3$CCOCl $\xrightarrow{\text{LiAlH}_4 \quad \text{THF}}$ ↗

Org React (1951) $\underline{6}$ 469

BrCH$_2$CH$_2$COCl $\xrightarrow{\text{LiAlH}_4 \quad \text{AlCl}_3 \quad \text{Et}_2\text{O}}$ BrCH$_2$CH$_2$CH$_2$OH   90%

JACS (1959) $\underline{81}$ 610

C$_{15}$H$_{31}$COCl $\xrightarrow{\text{NaBH}_4 \quad \text{dioxane}}$ C$_{15}$H$_{31}$CH$_2$OH   87%

JACS (1949) $\underline{71}$ 122

C$_5$H$_{11}$COOH $\xrightarrow[\text{THF}]{\text{ClCOOEt} \quad \text{Et}_3\text{N}}$ C$_5$H$_{11}\underset{\text{OCOOEt}}{\overset{\text{CO}}{|}}$ $\xrightarrow[\text{H}_2\text{O}]{\text{NaBH}_4 \quad \text{THF}}$ C$_5$H$_{11}$CH$_2$OH   70%

Chem Pharm Bull (1968) $\underline{16}$ 492

PhCH$_2$CONH ⟨β-lactam/S ring structure⟩ COOH $\xrightarrow{\text{1 ClCOOEt} \atop \text{2 NaN}_3}$ ⟨CON$_3$⟩ $\xrightarrow{\text{NaBH}_4}$ ⟨CH$_2$OH⟩   78%

J Med Chem (1964) $\underline{7}$ 483

C$_{15}$H$_{31}$COCl $\xrightarrow[\text{pyr}]{\text{PhCH}_2\text{SH}}$ C$_{15}$H$_{31}$COSCH$_2$Ph $\xrightarrow{\text{Ni} \quad \text{Et}_2\text{O}}$ C$_{15}$H$_{31}$CH$_2$OH

Helv (1946) $\underline{29}$ 684

C$_6$H$_{13}$COCl $\xrightarrow{\text{1 m-Chloroperbenzoic acid}}$ C$_6$H$_{13}$OH   66%

pyr hexane
2 Hexane   cyclohexane   reflux
3 Base

JOC (1965) $\underline{30}$ 3760

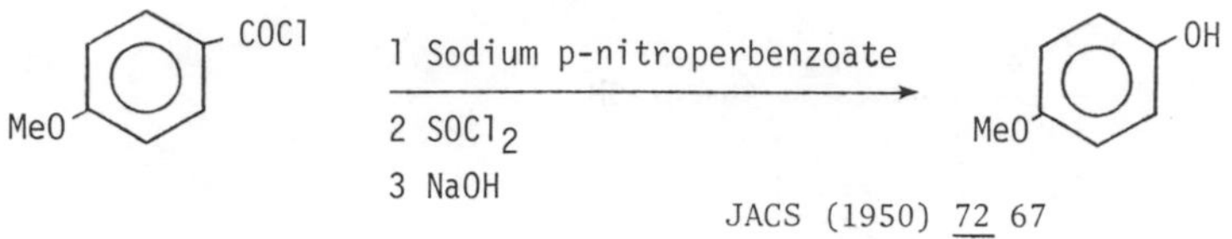
COCl
MeO
1 Sodium p-nitroperbenzoate
2 SOCl2
3 NaOH
OH
MeO

JACS (1950) 72 67

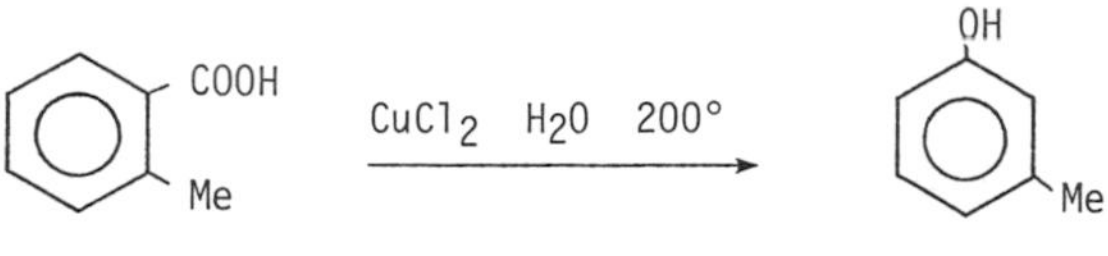
COOH
Me
CuCl2 H2O 200°
OH
Me

JOC (1961) 26 3144

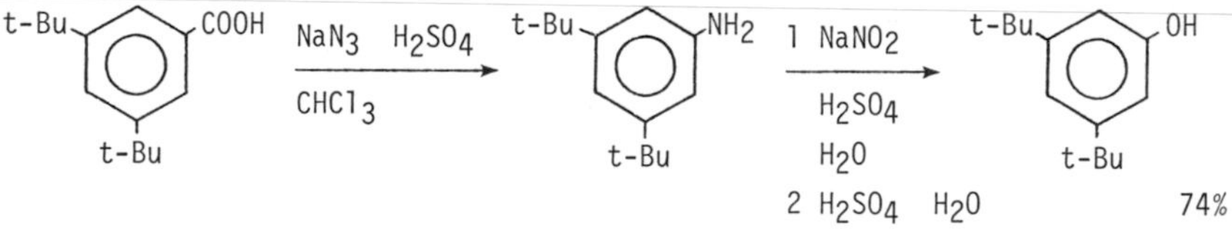
t-Bu
COOH
t-Bu
NaN3 H2SO4
CHCl3
t-Bu
NH2
t-Bu
1 NaNO2
H2SO4
H2O
2 H2SO4 H2O
t-Bu
OH
t-Bu
74%

Can J Chem (1963) 41 1653

Alcohols may also be prepared by esterification of carboxylic acids followed by reduction. See section 38 (Alcohols from Esters)

Section 33 Alcohols from Alcohols and Phenols

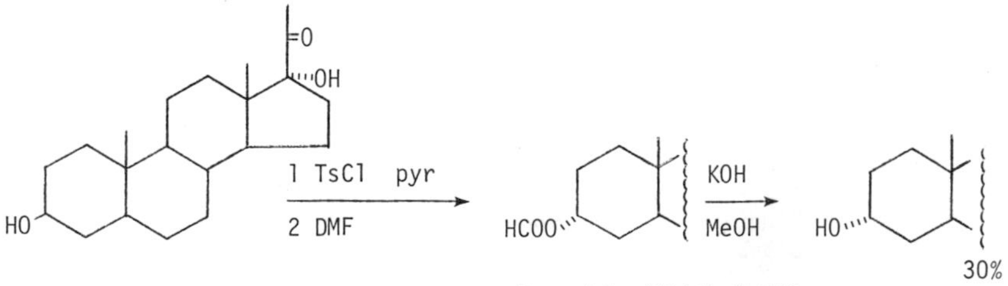
=O
OH
HO
1 TsCl pyr
2 DMF
HCOO
KOH
MeOH
HO
30%

Steroids (1964) 3 359

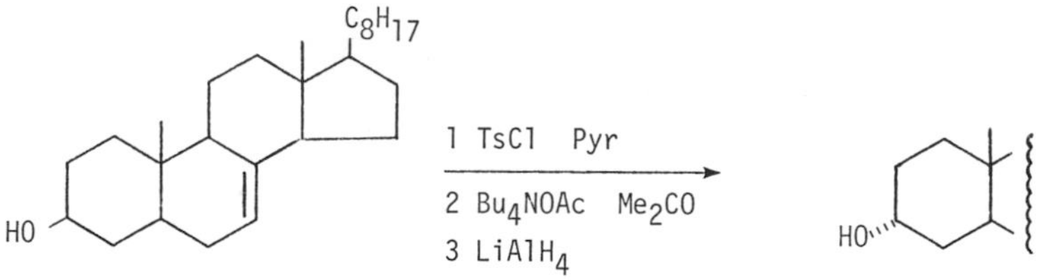
C_8H_{17}
HO
1 TsCl Pyr
2 Bu_4NOAc Me_2CO
3 $LiAlH_4$
HO

92%

JCS C (1969) 1605

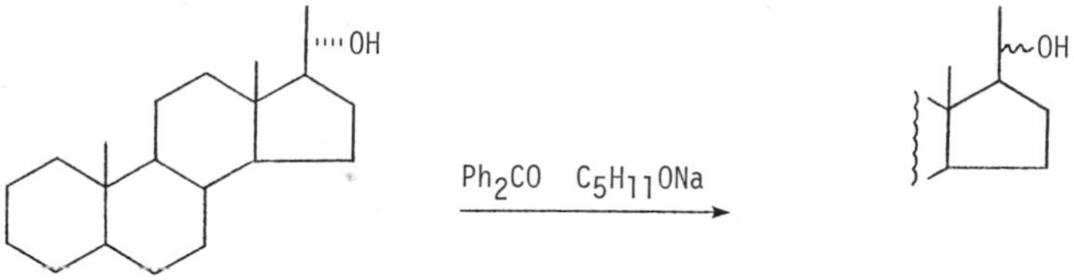
OH
Ph_2CO $C_5H_{11}ONa$
OH

JCS C (1969) 969

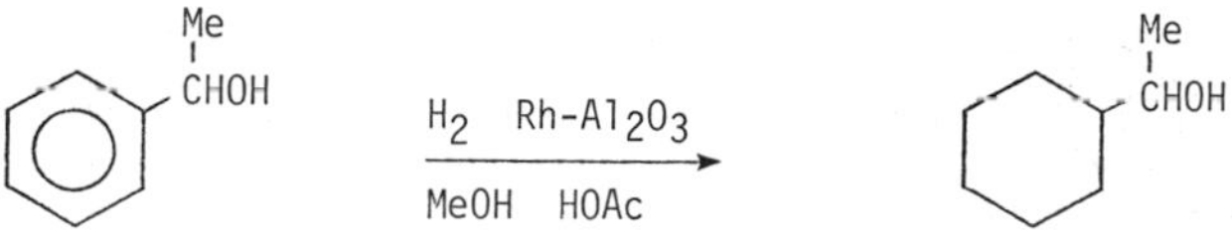
Me
CHOH
H_2 $Rh-Al_2O_3$
MeOH HOAc
Me
CHOH

80%

JOC (1962) 27 2288

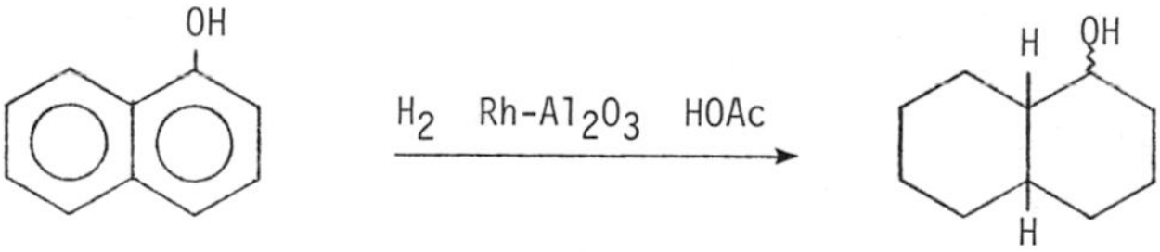
OH
H_2 $Rh-Al_2O_3$ HOAc
H OH
H

90%

JOC (1964) 29 3427

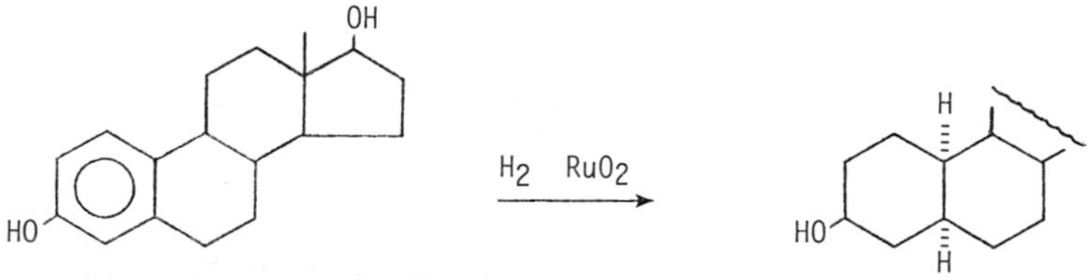
OH
HO
H_2 RuO_2
H
HO
H

JOC (1958) 23 1404

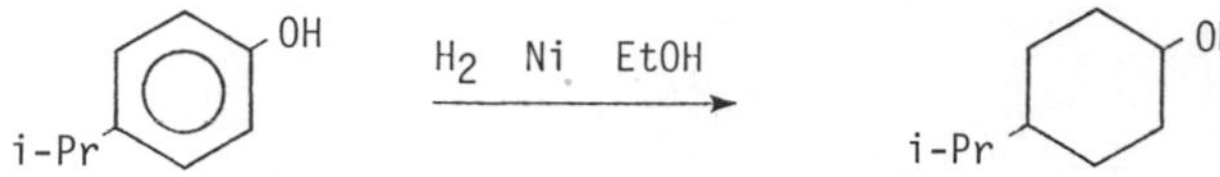

JACS (1949) 71 3889

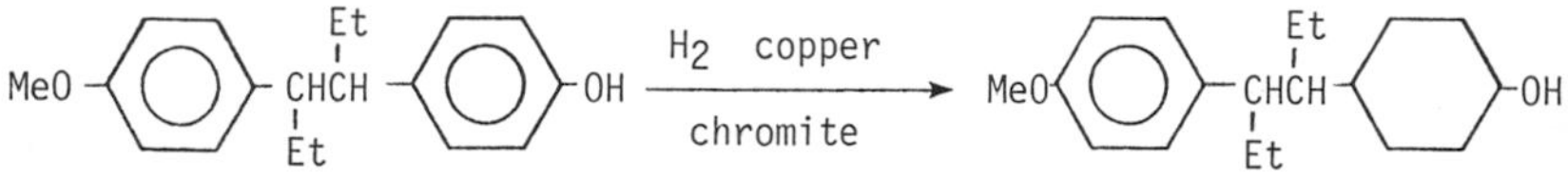

~52%

JACS (1948) 70 4127

Section 34   Alcohols and Phenols from Aldehydes

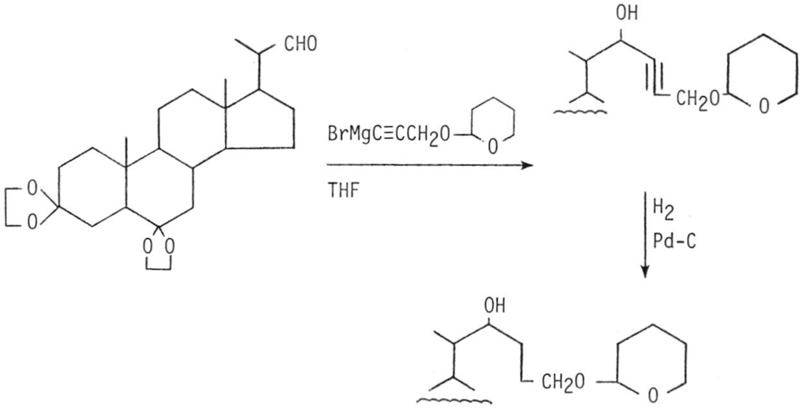

Chem Pharm Bull (1969) 17 690
Tetr Lett (1966) 3457

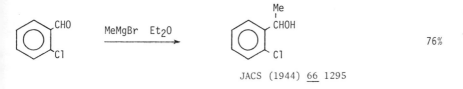

CH₂=CHCH₂Br   Mg   Et₂O

(Procedure for unstable Grignard reagents)

JOC (1963) 28 3269

PhCHO →[PhBr Li THF] Ph₂CHOH   96%

(One-step procedure)

Chem Comm (1970) 1160

MeMgBr   Et₂O   76%

JACS (1944) 66 1295

C₆H₁₃CHO →[H₂ Pt FeCl₃ / EtOH] C₆H₁₃CH₂OH

JACS (1923) 45 1071

H₂   PtO₂   FeSO₄ / EtOH

JACS (1940) 62 1478

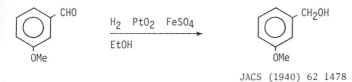

Me₂C=CHCH₂CH₂C(Me)=CHCHO →[H₂ Pt FeSO₄] Me₂C=CHCH₂CH₂C(Me)=CHCH₂OH   100%

JACS (1926) 48 477
     (1925) 47 3061

$C_6H_{13}CHO$ $\xrightarrow{\text{H}_2 \text{ (50 atmos)} \quad \text{RhCl}_3(\text{Ph}_3\text{P})_3}$ $C_6H_{13}CH_2OH$ 100%

Chem Comm (1965) 17

PrCHO $\xrightarrow[\text{RhCl}_3\cdot3\text{H}_2\text{O}]{\text{H}_2 \quad \text{CO (240 atmos)}}$ PrCH$_2$OH   70%

Ber (1966) 99 1086

$C_6H_{13}CHO$ $\xrightarrow{\text{Fe} \quad \text{HOAc} \quad \text{H}_2\text{O}}$ $C_6H_{13}CH_2OH$ 75-81%

Org Synth (1932) Coll Vol 1 304
JACS (1939) 61 2134

$C_6H_{13}CHO$ $\xrightarrow{\text{LiAlH}_4 \quad \text{Et}_2\text{O}}$ $C_6H_{13}CH_2OH$ 86%

Org React (1951) 6 469
JACS (1947) 69 1197

PhCHO $\xrightarrow{\text{LiAl(OBu-t)}_3\text{H} \quad \text{Et}_2\text{O}}$ PhCH$_2$OH

JACS (1958) 80 5372

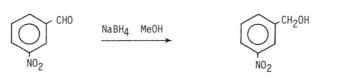

NaBH$_4$ MeOH   82%

JACS (1949) 71 122

$$PhCHO \xrightarrow{\text{LiBH}_3\text{CN} \quad \text{MeOH}} PhCH_2OH \qquad\qquad 78\%$$

JACS (1969) <u>91</u> 3996

$$PrCHO \xrightarrow{\text{B}_2\text{H}_6 \quad \text{THF}} PrCH_2OH$$

JACS (1960) <u>82</u> 681

$$PhCH=CHCHO \xrightarrow{\text{C}_5\text{H}_5\text{NBH}_3 \quad \text{C}_6\text{H}_6} PhCH=CHCH_2OH \qquad\qquad 84\%$$

JOC (1958) <u>23</u> 1561

$$PhCHO \xrightarrow{\text{Ph}_3\text{SnH}} PhCH_2OH \qquad\qquad 85\%$$

JACS (1961) <u>83</u> 1246

$$MeCH=CHCHO \xrightarrow{\text{Bu}_2\text{SnH}_2 \quad \text{Et}_2\text{O}} MeCH=CHCH_2OH \qquad\qquad 59\%$$

JACS (1961) <u>83</u> 1246

$$PrCHO \xrightarrow{\text{(i-PrO)}_3\text{Al} \quad \text{C}_6\text{H}_6} PrCH_2OH \qquad\qquad 36\%$$

Org React (1944) <u>2</u> 178

85-90%

JACS (1935) <u>57</u> 905
Org React (1944) <u>2</u> 94

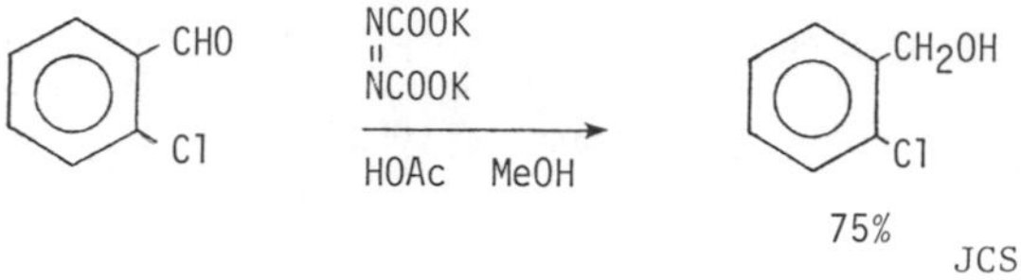

$$
\begin{array}{c} NCOOK \\ \| \\ NCOOK \end{array}
$$

HOAc   MeOH

(Rate of reduction
ArC=O > AliphC=O   )

75%

JCS $\underline{C}$ (1967) 1120

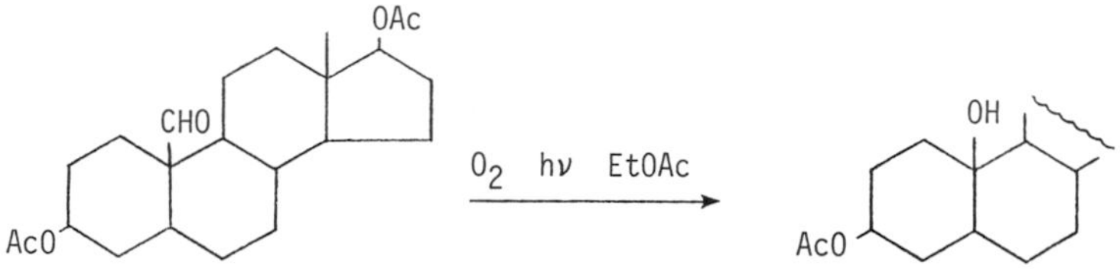

O_2 hν EtOAc

Tetr Lett (1966) 2507

20%

1 Pyridine oxide

CH_2Cl_2

2 Zn

21%

NaOH

MeOH

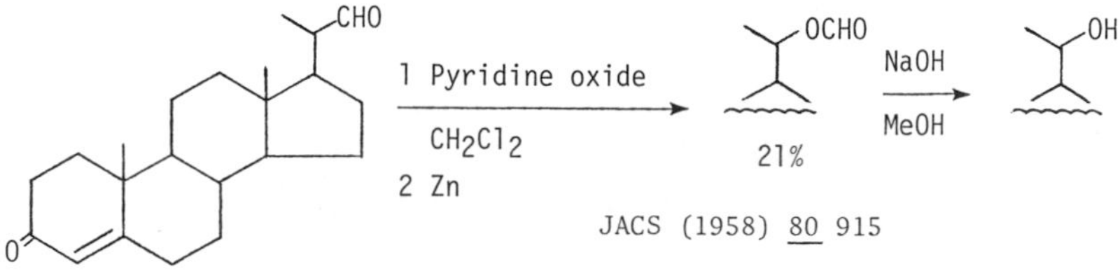

JACS (1958) $\underline{80}$ 915

$C_5H_{11}CH_2CHO$ $\xrightarrow{\text{H}_2\text{O}_2}$ $C_5H_{11}CH_2OCHO$ ----→ $C_5H_{11}CH_2OH$

Org React (1957) $\underline{9}$ 73
Ber (1941) $\underline{74B}$ 1552

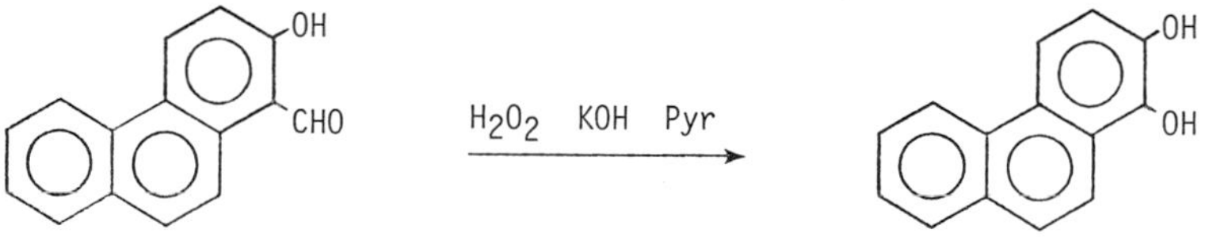

H_2O_2 KOH   Pyr

80%

Org React (1957) $\underline{9}$ 73
Ber (1940) $\underline{73B}$ 935

Alcohols may also be prepared from aldehydes by some of the methods
listed in section 42 (Alcohols and Phenols from Ketones)

Section 35   Alcohols and Phenols from Alkyls, Methylenes and Aryls
oooooooooooooooooooooooooooooooooooooooooooooooooooooo

No examples of the reaction RR' → ROH (R'=alkyl, aryl etc.) occur in the
literature. For reactions of the type RH → ROH (R=alkyl or aryl) see
section 41 (Alcohols and Phenols from Hydrides)

Section 36   Alcohols from Amides
ooooooooooooooooooooo

Ber (1955) 88 301

$PhCH_2OH$ 80%

JOC (1953) 18 1190

$C_{11}H_{23}CONH_2$

H₂ (200-300 atmos)   250°
────────────────────────────→ $C_{11}H_{23}CH_2OH$
Copper chromite   EtOH

JACS (1934) 56 2419

$C_{13}H_{27}CONH_2$

Electrolysis   LiCl   MeNH₂
────────────────────────────→ $C_{13}H_{27}CH_2OH$ 92%

JOC (1970) 35 1210

$PhCONHC_6H_{13}$

Electrolysis   Me₄NCl   MeOH
────────────────────────────→ $PhCH_2OH$ 69%

Ber (1965) 98 3462

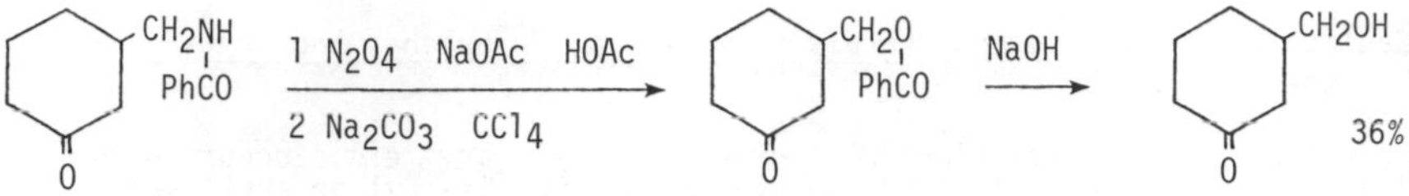

JOC (1969) 34 3834

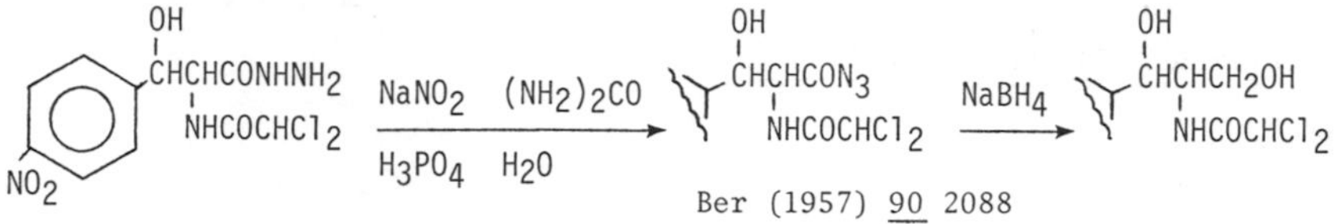

Ber (1957) 90 2088

Alcohols may also be prepared by conversion of amides into esters, followed by hydrolysis. See section 111 (Esters from Amides)

Section 37   Alcohols and Phenols from Amines
              °°°°°°°°°°°°°°°°°°°°°°°°°°°°°°°°°°°°°°

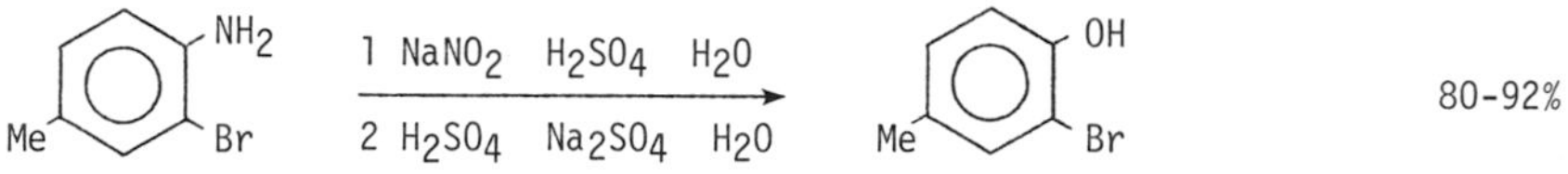

80-92%

Org Synth (1955) Coll Vol 3 130

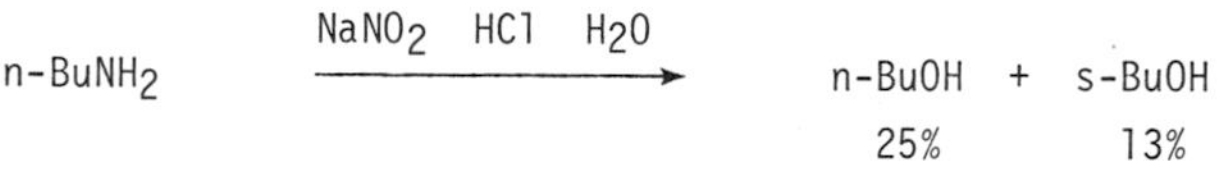

n-BuNH$_2$ $\xrightarrow{\text{NaNO}_2 \quad \text{HCl} \quad \text{H}_2\text{O}}$ n-BuOH + s-BuOH

                                              25%   13%

                                        JACS (1932) 54 3441

67%

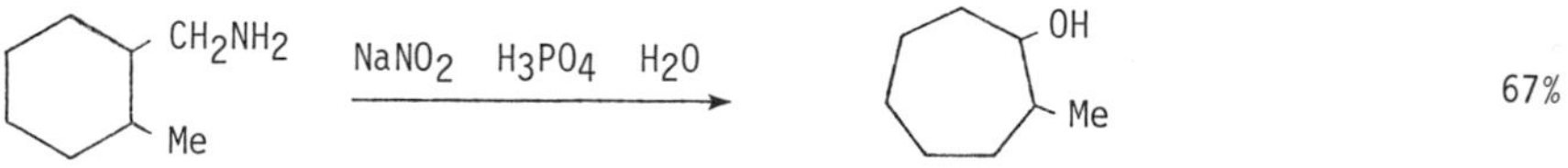

JOC (1965) 30 350

$$\xrightarrow[\text{dioxane} \quad H_2O]{\text{NaNO}_2 \quad \text{HOAc}}$$

96%

JCS (1959) 345

Chem Pharm Bull (1960) **8** 266

Section 111 contains further examples of the conversion of N-acyl amines into esters from which alcohols can be prepared by hydrolysis

Section 38   Alcohols and Phenols from Esters

$C_{15}H_{31}COOMe$ $\xrightarrow{\text{EtMgBr} \quad \text{Et}_2O}$ $C_{15}H_{31}\underset{\underset{OH}{|}}{C}Et_2$ 96%

JACS (1945) **67** 2239

PhCOOMe $\xrightarrow{\text{MeI} \quad \text{Li} \quad \text{THF}}$ $Ph\underset{\underset{Me}{|}}{\overset{\overset{Me}{|}}{C}}OH$ (One-step procedure)   74%

Chem Comm (1970) 1160

$C_{11}H_{23}COOEt$ $\xrightarrow{\text{Na} \quad \text{EtOH} \quad \text{toluene}}$ $C_{11}H_{23}CH_2OH$ 65-75%

Org Synth (1943) Coll Vol 2 372

$C_7H_{15}COOEt$ $\xrightarrow[\text{H}_2\text{O} \quad \text{Et}_2\text{O}]{\text{Na} \quad \text{NaOAc} \quad \text{HOAc}}$ $C_7H_{15}CH_2OH$ $> 90\%$

Rec Trav Chim (1923) <u>42</u> 1050

$C_5H_{11}COOMe$ $\xrightarrow[\text{250°}]{\text{H}_2 \text{ (200 atmos)} \quad \text{copper chromite}}$ $C_5H_{11}CH_2OH$ 92%

JACS (1932) <u>54</u> 1145
Org React (1954) <u>8</u> 1

PrCOOBu $\xrightarrow[\text{218°}]{\text{H}_2 \text{ (205 atmos)} \quad \text{ReO}_3}$ PrCH$_2$OH 85%

JOC (1963) <u>28</u> 2345

PhCOOEt $\xrightarrow{\text{Electrolysis} \quad \text{Me}_4\text{NCl} \quad \text{MeOH}}$ PhCH$_2$OH 89%

Ber (1965) <u>98</u> 3462

$C_{15}H_{31}COOEt$ $\xrightarrow{\text{LiAlH}_4 \quad \text{Et}_2\text{O}}$ $C_{15}H_{31}CH_2OH$ 98%

JACS (1947) <u>69</u> 1197
Org React (1951) <u>6</u> 469

$\xrightarrow{\text{LiBH}_4 \quad \text{THF}}$

91%

Carbohydrate Res (1967) <u>4</u> 504

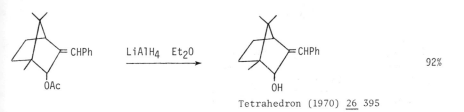

JACS (1956) 78 2582   84%

Ph(CH$_2$)$_2$COOMe $\xrightarrow[\text{reflux}]{\text{NaBH}_4 \quad \text{MeOH}}$ Ph(CH$_2$)$_2$CH$_2$OH   73%

JOC (1963) 28 3261

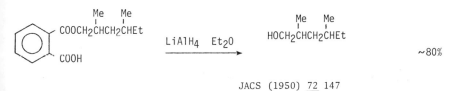

Tetrahedron (1970) 26 395   92%

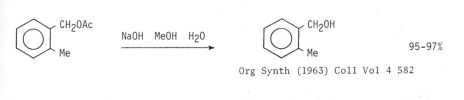

JACS (1950) 72 147   ~80%

Org Synth (1963) Coll Vol 4 582   95-97%

PhCOOEt $\xrightarrow{\text{NaOH} \quad \text{Me}_2\text{SO}}$ EtOH

JCS (1965) 1290

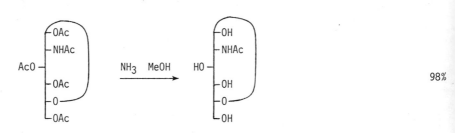

$$\underset{\substack{|\\ \text{COOEt}}}{\text{MeC}}=\text{CH(CH}_2)_2\underset{\substack{|\\ \text{Me}}}{\text{C}}=\text{CHCH}_2\text{OAc} \quad \xrightarrow{\text{K}_2\text{CO}_3 \quad \text{EtOH}} \quad \underset{\substack{|\\ \text{COOEt}}}{\text{MeC}}=\text{CH(CH}_2)_2\underset{\substack{|\\ \text{Me}}}{\text{C}}=\text{CHCH}_2\text{OH} \quad 98\%$$

JACS (1969) 91 4318

MeONa   MeOH

CHCl₃   60%

JCS (1967) 448

MeOK   MeOH   99%

JACS (1948) 70 314

NH₃   MeOH   98%

JCS (1956) 2042

HCOOBu $\xrightarrow{\text{NaH \quad diglyme}}$ BuOH   68%

Tetr Lett (1965) 1713

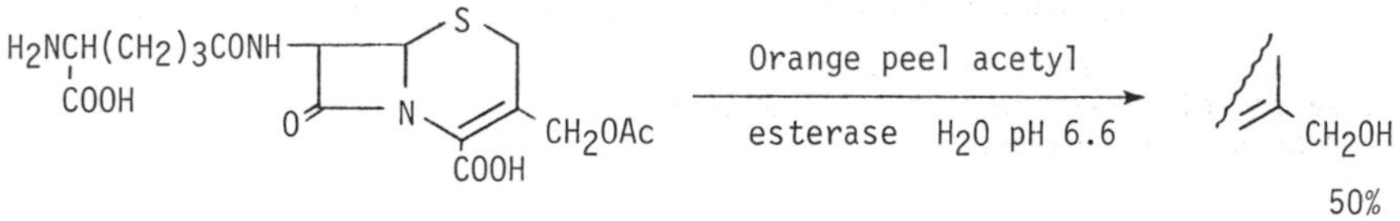

H₂NCH(CH₂)₃CONH ... CH₂OAc

$H_2NCH(CH_2)_3CONH$
$COOH$

$COOH$
CH_2OAc

Orange peel acetyl

esterase   H₂O pH 6.6

CH_2OH

50%

Biochem J (1961) 81 591

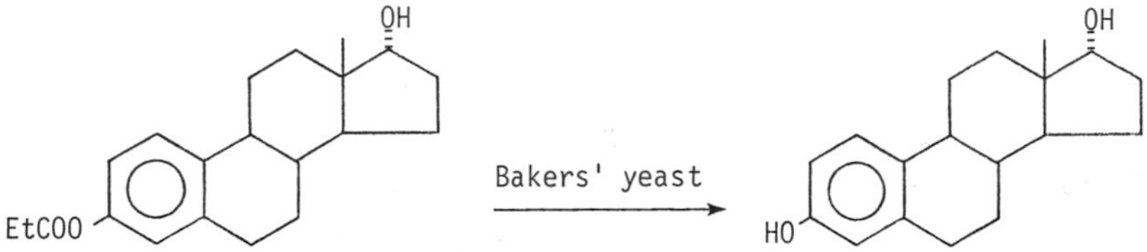

Bakers' yeast

EtCOO

HO

Ber (1938) 71B 2696

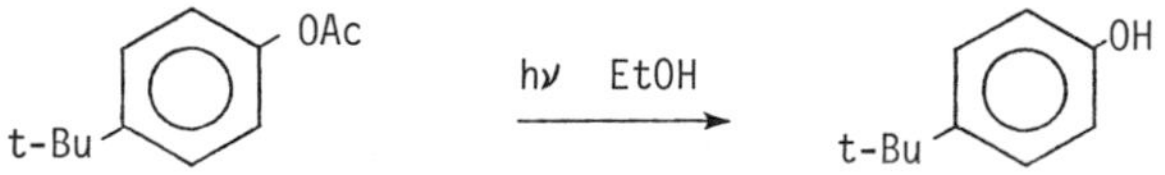

OAc

t-Bu

hν   EtOH

OH

t-Bu

34%

JCS (1965) 5162

AcO   COOMe

AcO   COOMe

HCl   MeOH

AcO''   H

HO'''   H

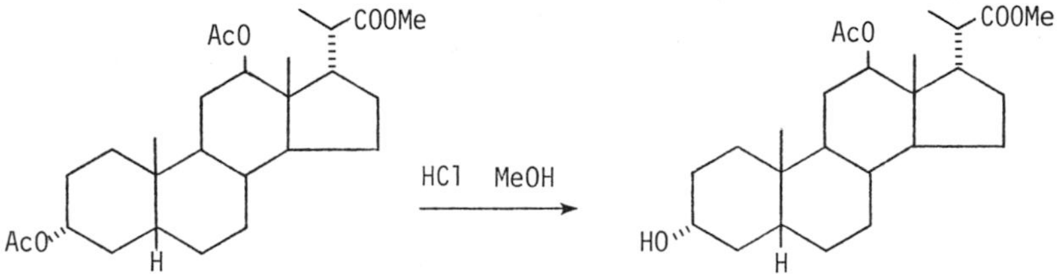

88%

Helv (1944) 27 713

Further examples of the hydrolysis and cleavage of esters are included in
section 23 (Carboxylic Acids from Esters) and section 45A (Protection of
Alcohols and Phenols)

Section 39   Alcohols and Phenols from Ethers and Epoxides

Alcohols and phenols from ethers by rearrangements . . . . page   92
Cleavage of ethers to alcohols and phenols 92-100
Alcohols from epoxides 100-102

Org React (1944) $\underline{2}$ 1   73%

$$PhCH_2OMe \xrightarrow{PhLi \quad Et_2O} PhCHOH \overset{Me}{|}$$

35%

Annalen (1942) $\underline{550}$ 260
JOC (1962) $\underline{27}$ 1933

$$Me_2C=CHCH_2OCH_2CH=CMe_2 \xrightarrow{t-BuLi \quad THF} Me_2C=CHCHCCH_2 \overset{Me}{\underset{HO \; Me}{|}}$$

31%

Tetr Lett (1970) 353

Review: The Cleavage of Ethers   Chem Rev (1954) $\underline{54}$ 615

JOC (1941) $\underline{6}$ 852   90%

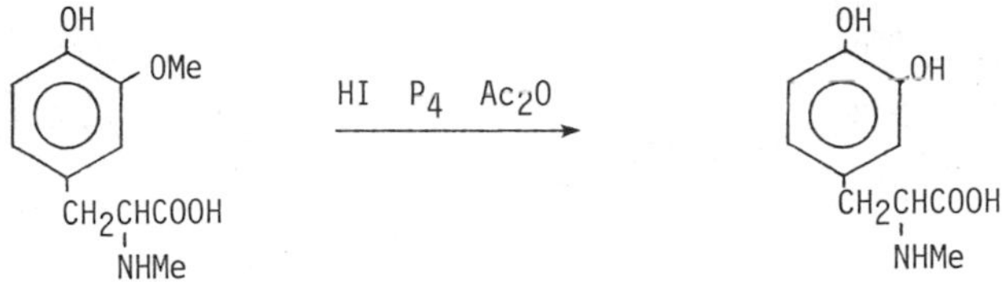

$$\xrightarrow{\text{HI \quad P}_4 \quad \text{Ac}_2\text{O}}$$

82%

Org Synth (1955) Coll Vol 3 586

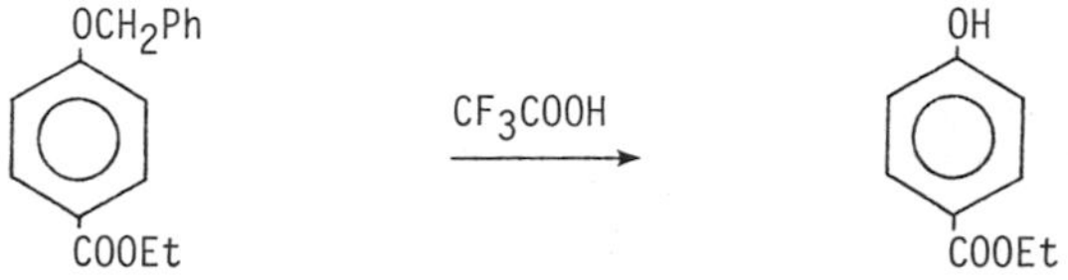

$$\xrightarrow{\text{CF}_3\text{COOH}}$$

JOC (1965) 30 2491

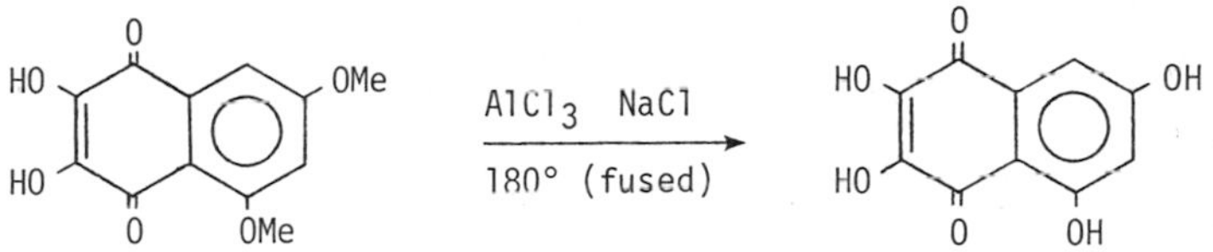

$$\xrightarrow[\text{180° (fused)}]{\text{AlCl}_3 \quad \text{NaCl}}$$

Low yield

JCS (1961) 1008

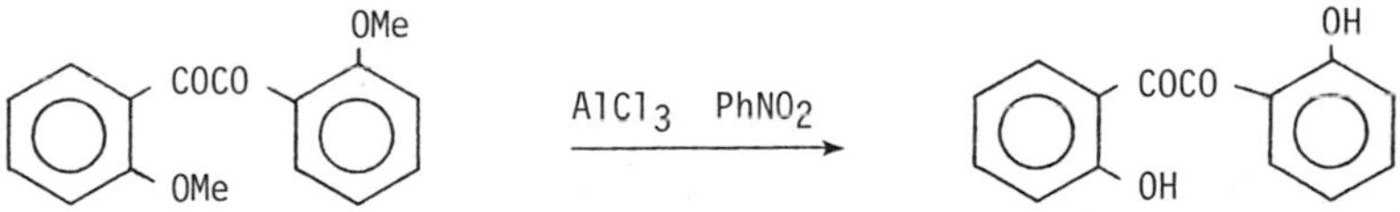

$$\xrightarrow{\text{AlCl}_3 \quad \text{PhNO}_2}$$

50-60%

Ber (1943) 76B 900

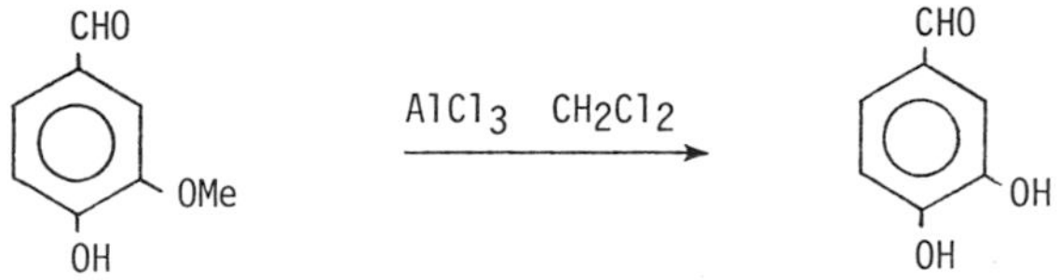

$$\xrightarrow{\text{AlCl}_3 \quad \text{CH}_2\text{Cl}_2}$$

87%

JOC (1962) 27 2037

$$\xrightarrow{\text{AlBr}_3 \quad \text{C}_6\text{H}_6}$$

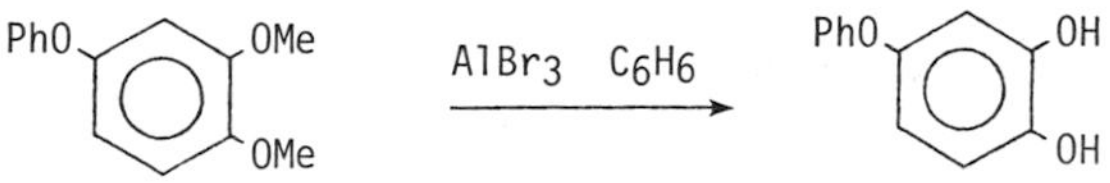

87%

Ber (1960) 93 2761

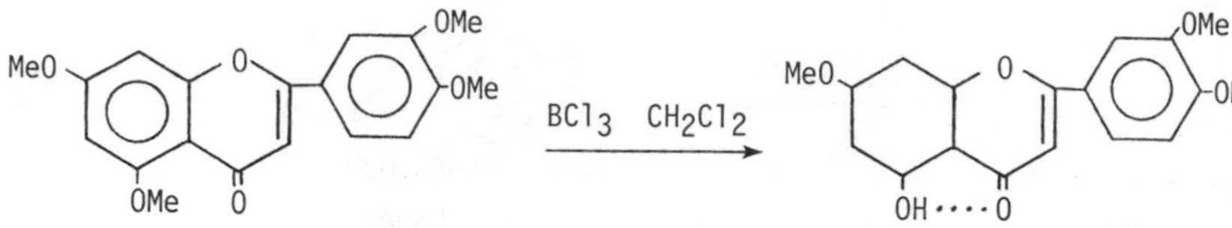

$$\xrightarrow{BCl_3 \quad CH_2Cl_2}$$

86%

Tetr Lett (1966) 4153
JCS (1962) 1260

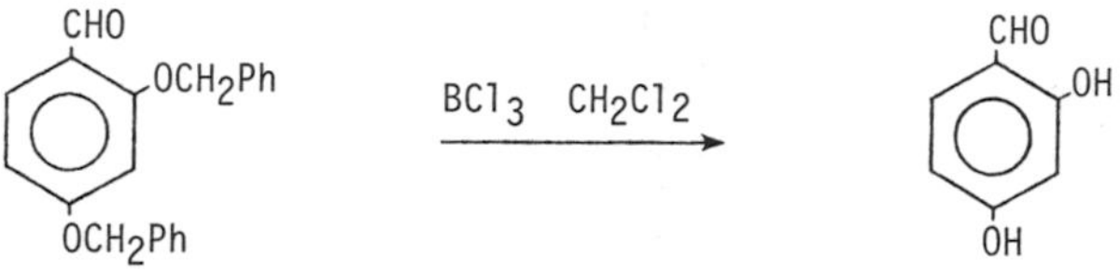

$$\xrightarrow{BCl_3 \quad CH_2Cl_2}$$

75%

Tetr Lett (1966) 4155

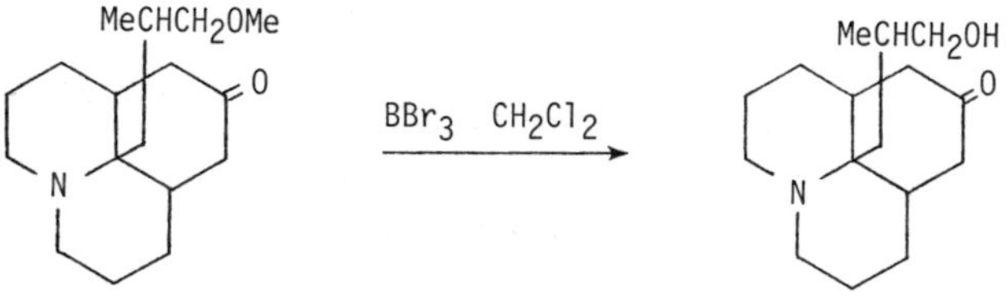

$$\xrightarrow{BBr_3 \quad CH_2Cl_2}$$

JACS (1968) <u>90</u> 1648

$$\xrightarrow{BBr_3 \quad CH_2Cl_2}$$

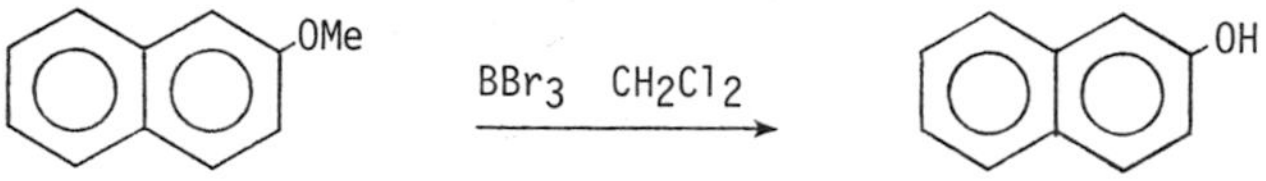

67%

Tetrahedron (1968) <u>24</u> 2289

PhOEt $\xrightarrow[\text{2 } H_2O]{\text{1 } BI_3}$ PhOH

Tetr Lett (1967) 4131

Pr_2O $\xrightarrow{BI_3}$ $(PrO)_2BI$ $\dashrightarrow$ PrOH

Tetr Lett (1967) 4131

Et_2O →[$C_5H_5NBH_2I$ C_6H_6 Et_2O]→ $C_5H_5NBH_2OEt$ ----→ EtOH

JACS (1968) 90 6260

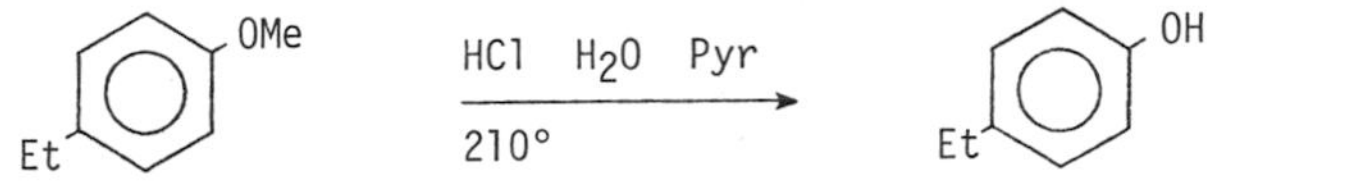

HCl H_2O Pyr
210°

Chem Ind (1967) 1138

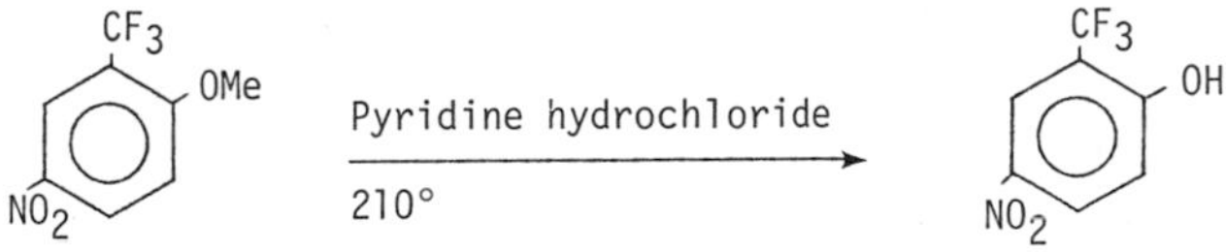

97%

Pyridine hydrochloride
210°

JOC (1962) 27 4660

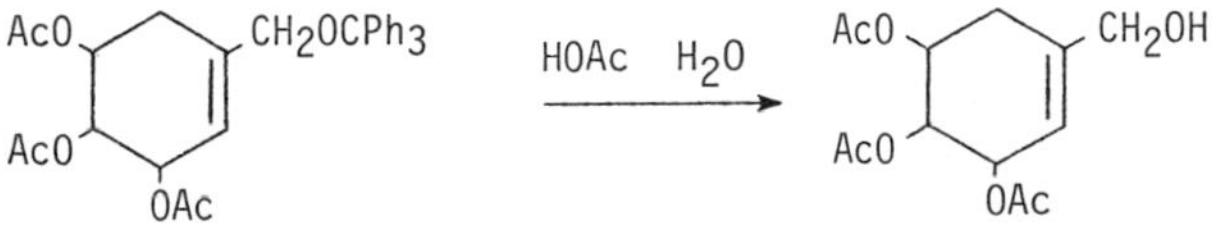

75%

HOAc H_2O

Ber (1956) 89 898

80%

Further examples of the cleavage of triphenylmethyl ethers are included in section 45A (Protection of Alcohols and Phenols)

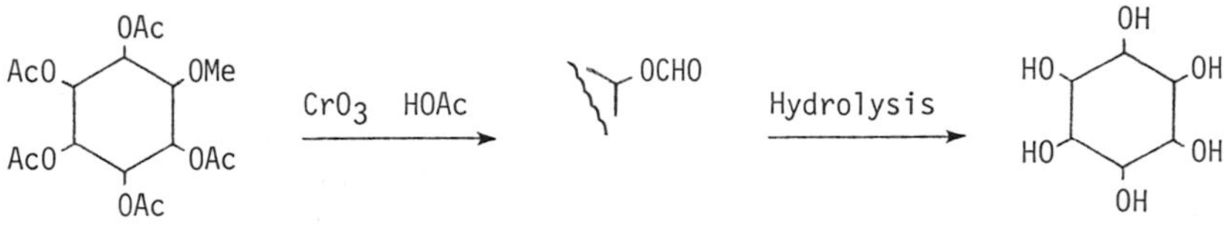

CrO_3 HOAc   Hydrolysis

Carbohydrate Res (1970) 12 147

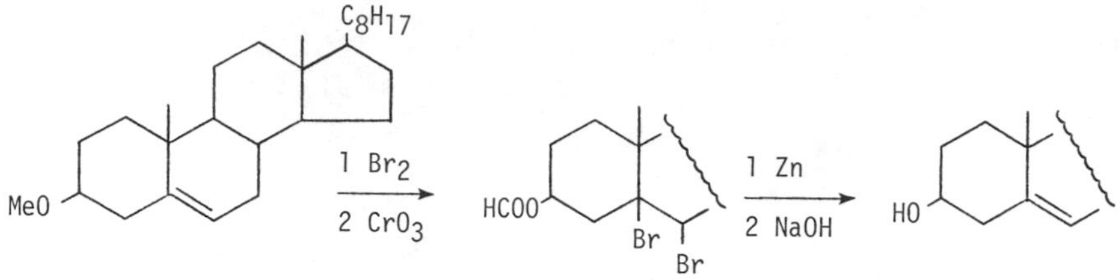

Chem Comm (1966) 752

$BuOCH_2Pr$ —($PhCOO_2Bu$-t; CuCl C_6H_6)→ $BuOCHPr$ (OOCPh) 50% - - - → BuOH

Acta Chem Scand (1961) 15 249
Tetr Lett (1960) (2) 4
Arkiv Kemi (1960) 16 287

$BuOCH_2Pr$ —(EtOOCN=NCOOEt, hν)→ $BuOCHPr$ (N(COOEt)–NHCOOEt) 60% —(H_2O)→ BuOH

Chem Comm (1965) 259

$EtOCH_2Me$ —(PhN=NOAc)→ EtOCHMe (OAc) - - -(H_2O)→ EtOH

JACS (1964) 86 3180

$BuOCH_2Ph$ —(H_2 Pd)→ BuOH 100%

Org React (1953) 7 263
JACS (1954) 76 3188

71%

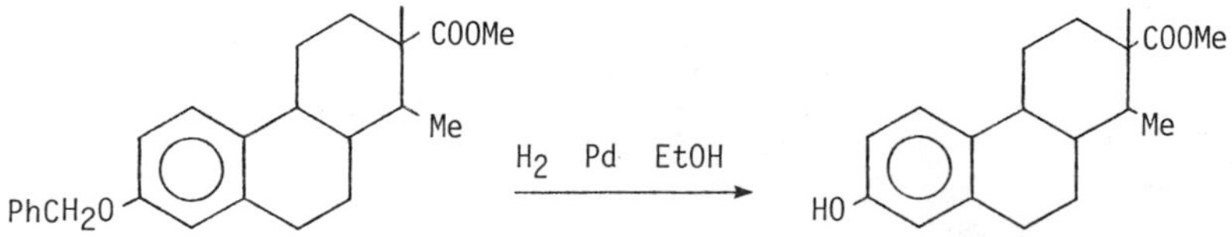

J Med Chem (1969) 12 192

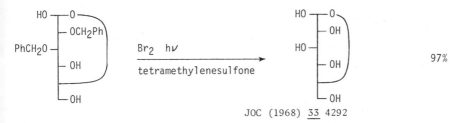

$$\xrightarrow[\text{tetramethylenesulfone}]{Br_2 \quad h\nu}$$

97%

JOC (1968) 33 4292

$$\xrightarrow{(Me_3Si)_2Hg \quad 160°}$$

JCS C (1967) 2188

$$\xrightarrow{LiI \quad collidine}$$

100%

Chem Comm (1969) 616
JACS (1950) 72 3396
For cleavage with KCN see Ber (1933) 66B 1623

$$\xrightarrow{EtS^- \quad DMF}$$

94%

Tetr Lett (1970) 1327

PhOMe $\xrightarrow[185°]{MeNH_2 \quad H_2O}$ PhOH

Chem Ind (1970) 1230
Aust J Chem (1970) 23 2539

PhOR $\xrightarrow{Ph_2PLi \quad THF}$ PhOH

83% (R=Me)
88% (R=PhCH_2)

No reaction (R=Et, i-Pr)

JCS (1965) 4120
Tetr Lett (1967) 3661
JACS (1970) 92 7232

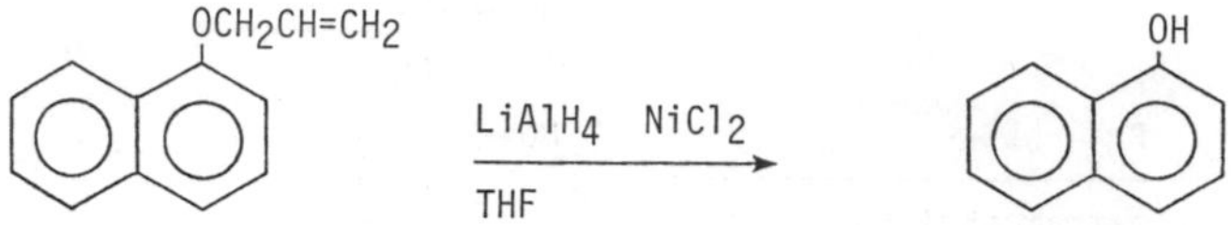

OCH2CH=CH2
LiAlH4 NiCl2
THF
OH

76%

JACS (1957) 79 5463

PhOMe $\xrightarrow{\text{Lithium diphenyl THF}}$ PhOH 80%

JOC (1963) 28 707

PhOR $\xrightarrow{\text{BuMgBr } CoCl_2 \text{ } Et_2O}$ PhOH

86% (R=$PhCH_2$)
43% (R=Ph)
No reaction (R=Me)

JOC (1952) 17 669

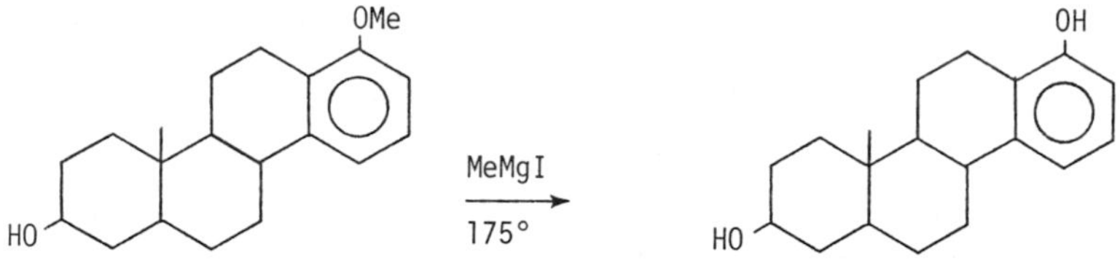

OMe
HO
MeMgI
175°
OH
HO

90%

JACS (1956) 78 6322

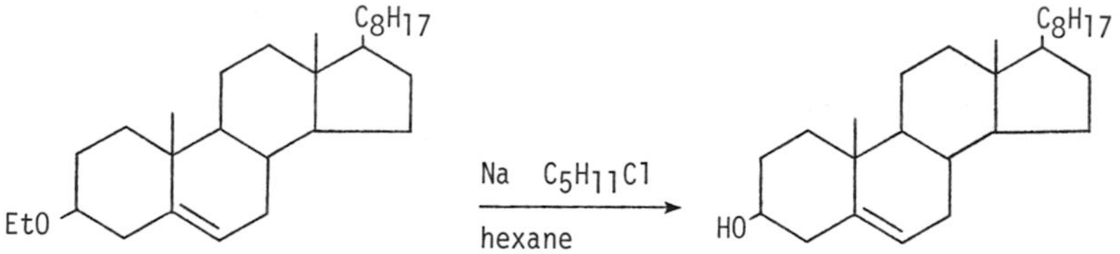

C8H17
EtO
Na C5H11Cl
hexane
C8H17
HO

74%

JACS (1951) 73 1263

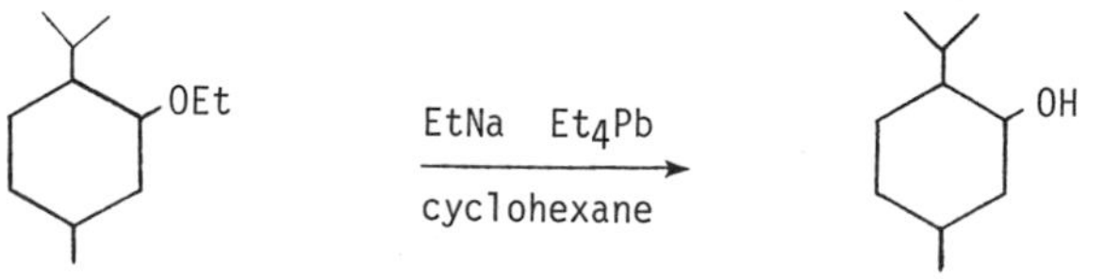

OEt
EtNa Et4Pb
cyclohexane
OH

72%

J Prakt Chem (1938) 151 61

$PhOCH_2CH=CH_2$ $\xrightarrow{\begin{array}{c}1\ Mg\ \ THF \\ \hline 2\ CO_2\end{array}}$ PhOH   86%

J Organometallic Chem (1969) 18 249

PhOR $\xrightarrow{NaNH_2\ \ piperidine}$ PhOH   82% (R=Me)
                                                                90% (R=PhCH_2)

Chem Ind (1957) 80
JACS (1951) 73 1437

$PhCH_2OEt$ $\xrightarrow{Li\ \ THF}$ EtOH

JOC (1961) 26 3723

PhOMe $\xrightarrow{K\ \ HMPA}$ PhOH   76-80%

Bull Soc Chim Fr (1966) 3344

PhOMe $\xrightarrow{Na\ \ Pyr}$ PhOH   94%

Ber (1943) 76B 156

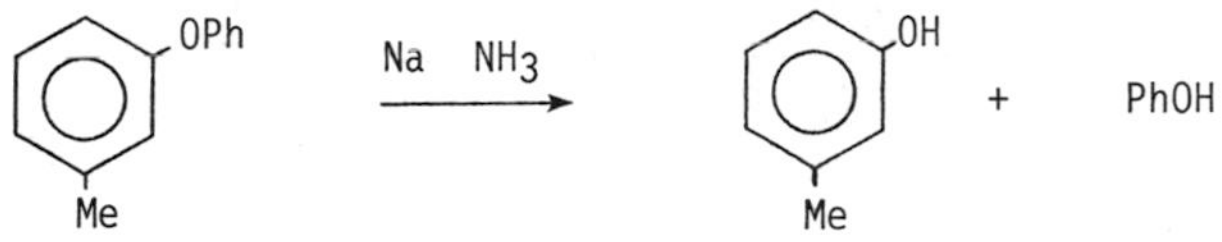

$\xrightarrow{Na\ \ NH_3}$

+   PhOH

JACS (1937) 59 1488

PhOMe $\xrightarrow{Li\ \ H_2NCH_2CH_2NH_2}$ PhOH   54%

JOC (1957) 22 891

JACS (1970) <u>92</u> 553

Alcohols and phenols may also be prepared by conversion of ethers into esters, followed by hydrolysis. See section 114 (Esters from Ethers)

Further methods for the cleavage of ethers to alcohols are included in section 45A (Protection of Alcohols and Phenols)

$$EtCH-CH_2 \quad \xrightarrow{MeMgCl} \quad \underset{45\%}{EtCHCH_2Me} \quad + \quad \underset{22\%}{EtCHCH_2OH}$$

	OH	Me
MeMgCl	EtCHCH$_2$Me 45%	EtCHCH$_2$OH 22%
Me$_2$Mg	82%	11%
MeLi	90%	-
Me$_2$CuLi	88%	-

JACS (1970) <u>92</u> 4979
Chem Rev (1951) <u>49</u> 413

1 Me$_2$Mg   dioxane

2 H$_2$SO$_4$ MeOH

41%

JACS (1960) <u>82</u> 3995
Bull Soc Chim Fr (1969) 4414

1 2-Lithio-1,3-dithiane

THF

2 Ni   EtOH

71%

Chem Comm (1970) 141 741

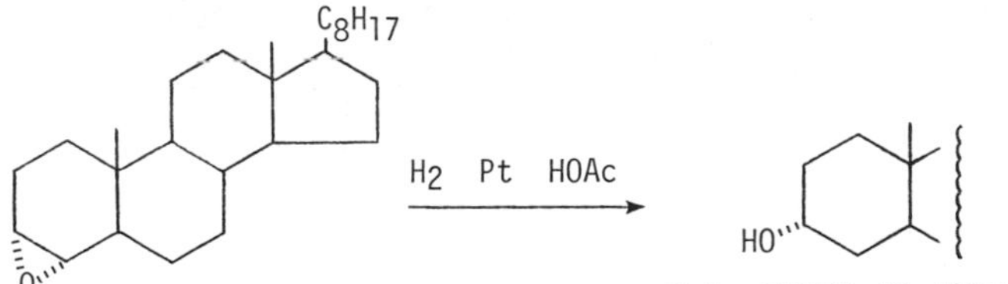

H2 Pt HOAc

Helv (1953) 36 1332
    (1943) 26 562

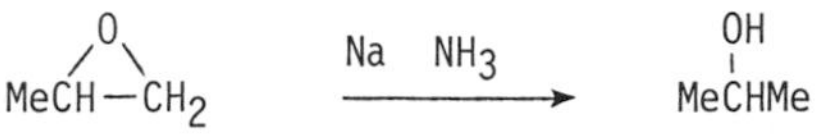

81%

Na NH3

OH
|
MeCHMe

Quart Rev (1958) 12 17

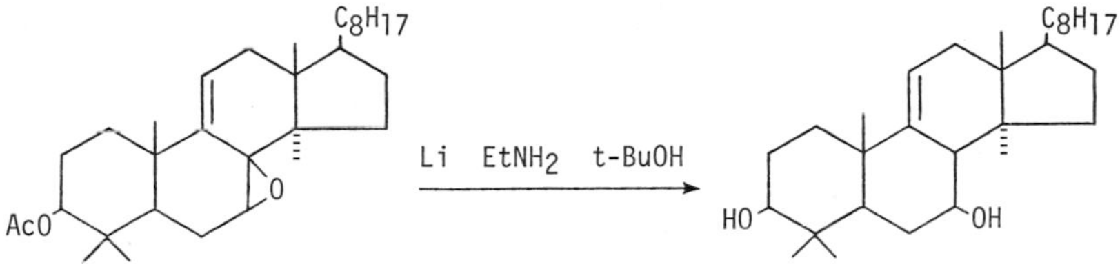

Li EtNH2 t-BuOH

JCS C (1968) 1581

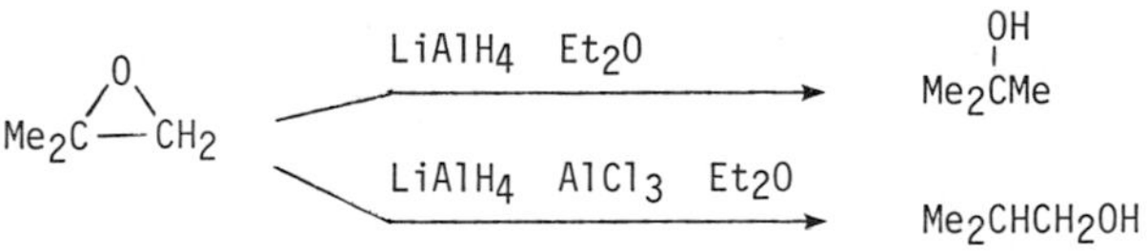

LiAlH4 Et2O

OH
|
Me2CMe   26%

LiAlH4 AlCl3 Et2O

Me2CHCH2OH   55%

JACS (1956) 78 3226
Org React (1951) 6 469

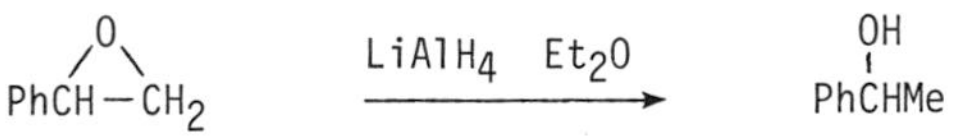

LiAlH4 Et2O

OH
|
PhCHMe   94%

JACS (1948) 70 3738

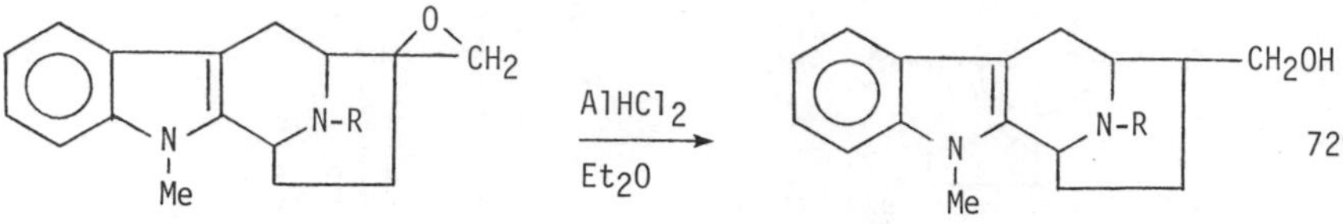

72%

Tetr Lett (1969) 901

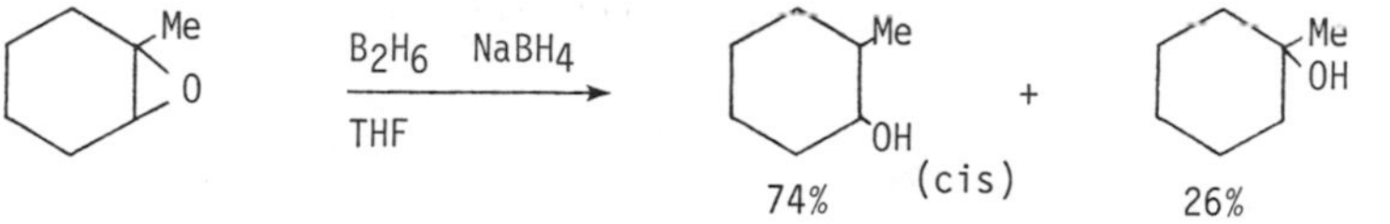

JACS (1968) 90 2686

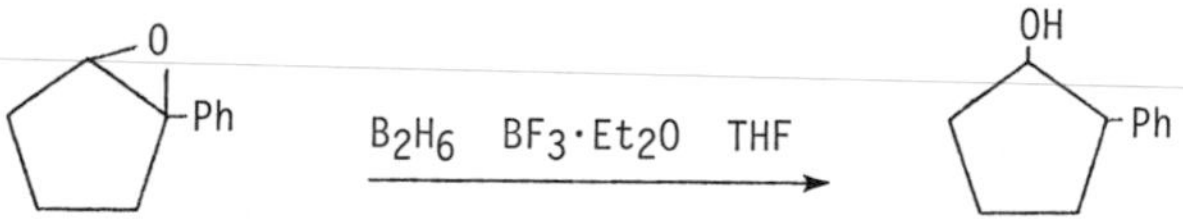

82%

Chem Comm (1968) 1549

Section 40 Alcohols and Phenols from Halides and Sulfonates

i-PrCl ----→ i-PrLi —(1 CH_2=CH_2; 2 Ph_2CO)→ i-$PrCH_2CH_2C(Ph)_2OH$

JACS (1969) 91 6362

BuBr ----→ BuLi —(CH_2=$CHCH_2OH$, $Me_2NCH_2CH_2NMe_2$)→ $BuCH(Me)CH_2OH$ 72%

Tetr Lett (1969) 325

$CH_2=CHCH_2Br$ $\xrightarrow{\begin{array}{c}1 \text{ Mg Et}_2O\\ \hline 2 \text{ (EtO)}_2CO\end{array}}$ $(CH_2=CHCH_2)_3COH$ 30%

Compt Rend (1968) C 267 773
Org Synth (1943) Coll Vol 2 602

PhBr $\xrightarrow{1 \text{ Mg Et}_2O}$ $\begin{array}{c}\text{2 PhCOOEt } C_6H_6\\ \text{2 Ph}_2CO \text{ } C_6H_6\end{array}$ Ph_3COH 89-93%

Org Synth (1955) Coll Vol 3 839

        $\xrightarrow{\begin{array}{c}1 \text{ Mg Et}_2O\\ \hline 2 \text{ MeCHO}\end{array}}$
82-88%

Org Synth (1955) Coll Vol 3 200

BuBr $\xrightarrow{\begin{array}{c}1 \text{ Mg Et}_2O\\ \hline 2 \text{ HCOOEt}\end{array}}$ Bu_2CHOH 83-85%

Org Synth (1943) Coll Vol 2 179

$C_6H_{13}F$ $\xrightarrow{\begin{array}{c}\text{Mg I}_2\\ \hline \text{THF}\end{array}}$ $C_6H_{13}MgF$ $\xrightarrow{Ph_2CO}$ $C_6H_{13}\overset{\displaystyle Ph}{\underset{\displaystyle Ph}{C}}OH$

95%

JACS (1970) 92 433

$\xrightarrow{\text{Mg Et}_2O}$ $\begin{array}{c}\overset{O}{CH_2-CH_2}\\ \\ CH_2-CH_2 \text{ THF}\\ CH_2-O\end{array}$

$CH_2(CH_2)_2OH$ 59%

$CH_2(CH_2)_3OH$ 37%

JOC (1968) 33 2991
Org Synth (1932) Coll Vol 1 306

95%

Rec Trav Chim (1965) 84 1200

For Grignard reaction with HCHO gas see JACS (1933) 55 1119
and JOC (1968) 33 3408

68%

Can J Chem (1962) 40 2175
JOC (1939) 4 318

PhBr $\xrightarrow[\text{2 O}_2]{\text{1 Mg Et}_2\text{O}}$ PhOH   46%

JACS (1955) 77 6032

$\underset{\text{Me}}{\text{C}_6\text{H}_{13}\text{CHCl}}$ $\xrightarrow[\text{2 O}_2]{\text{1 Mg Et}_2\text{O}}$ $\underset{\text{Me}}{\text{C}_6\text{H}_{13}\text{CHOOH}}$ $\dashrightarrow$ $\underset{\text{Me}}{\text{C}_6\text{H}_{13}\text{CHOH}}$

91%

JACS (1955) 77 6032
(1943) 65 501

t-BuCl $\xrightarrow[\text{2 HgCl}_2]{\text{1 Mg Et}_2\text{O}}$ t-BuHgCl $\xrightarrow{\text{O}_3\quad\text{CH}_2\text{Cl}_2}$ t-BuOH   50%

Tetr Lett (1970) 2679

PhCl $\xrightarrow[\substack{\text{2 (MeO)}_3\text{B} \\ \text{3 H}_2\text{O}_2\ \ \text{H}_2\text{O}}]{\text{1 Mg Et}_2\text{O}}$ PhOH   78%

JOC (1957) 22 1001

PhBr $\xrightarrow[\text{2 t-BuOOH}]{\text{1 Mg Et}_2\text{O}}$ PhOH   98%

JACS (1959) 81 4230

PhBr $\xrightarrow[\text{2 PhCOO}_2\text{Bu-t}]{\text{1 Mg Et}_2\text{O}}$ PhOBu-t $\xrightarrow{\text{Acid}}$ PhOH

Arkiv Kemi (1961) 17 393
Org Synth (1963) 43 55

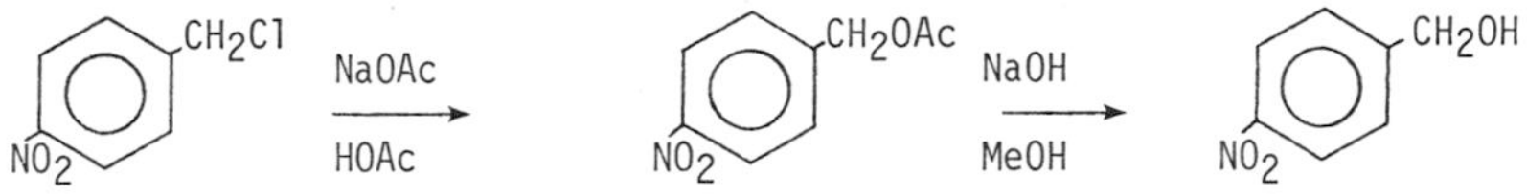

Org Synth (1955) Coll Vol 3 650 652

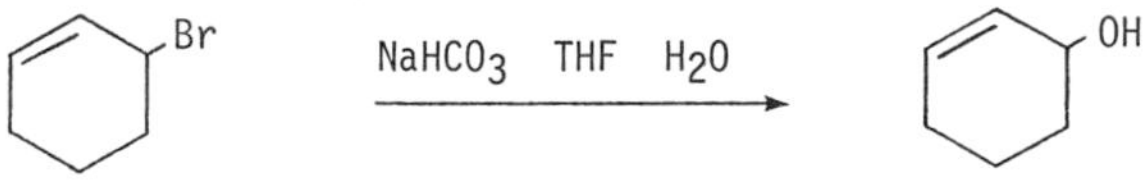

JACS (1961) 83 198
(1946) 68 751

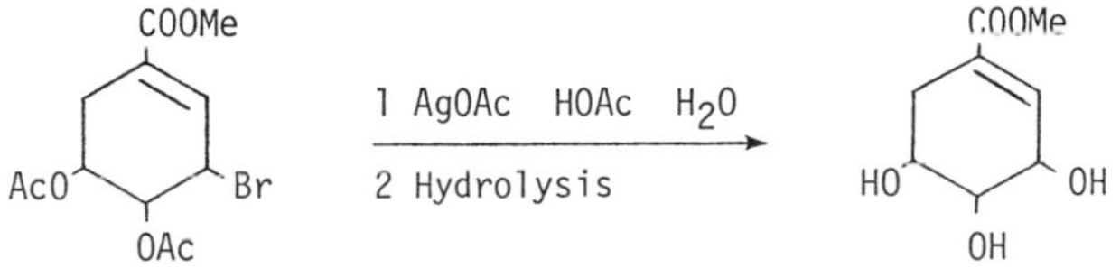

Ber (1964) 97 443

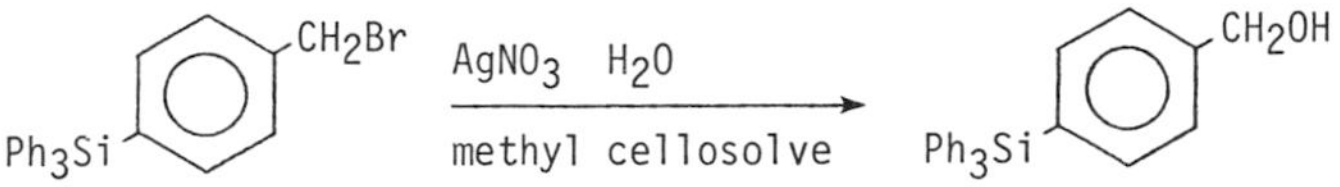

76%

JACS (1956) 78 1689
J Biol Chem (1946) 164 569

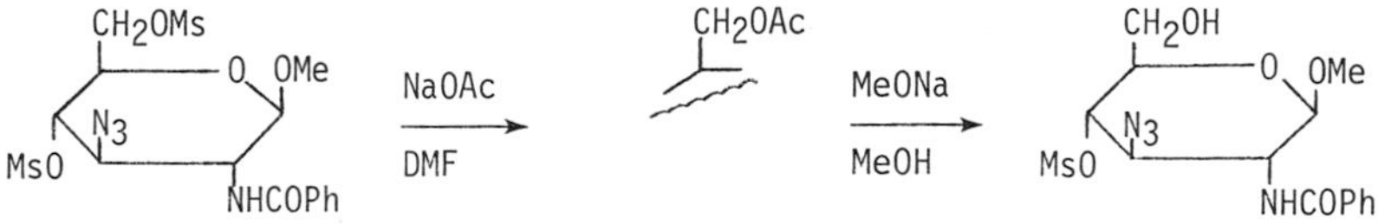

77%

Ber (1970) 103 37

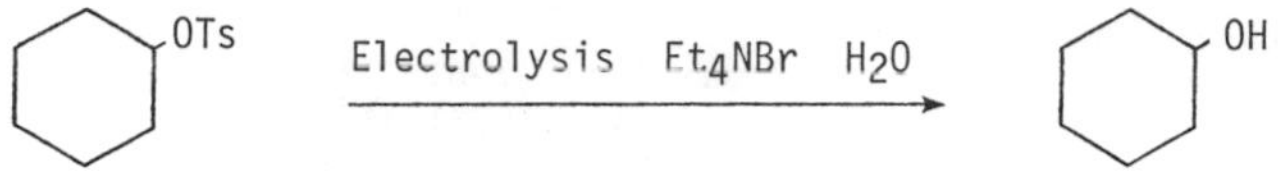

JOC (1968) 33 2716   96%

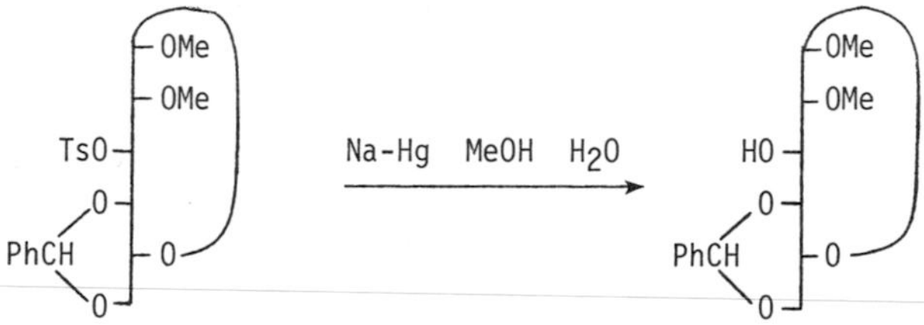

Helv (1945) 28 1164   88%

PhOTs $\xrightarrow{\text{Sodium-naphthalene\ \ THF}}$ PhOH   99%

JACS (1966) 88 1581

$C_8H_{17}OTs \xrightarrow{\text{Na\ \ NH}_3\ \ \text{Et}_2O} C_8H_{17}OH$ 56%

JOC (1956) 21 479

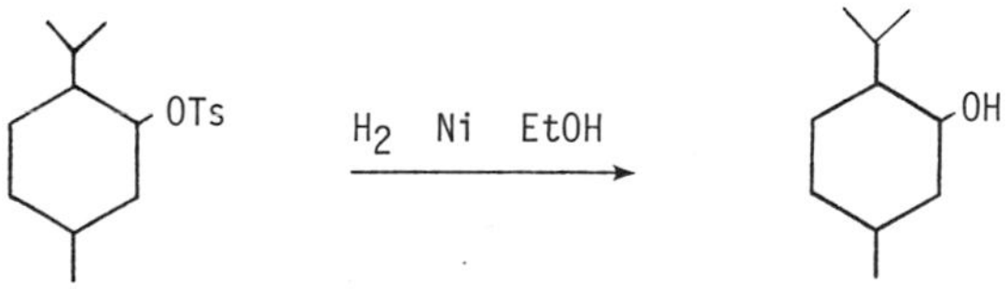

JCS (1949) S178

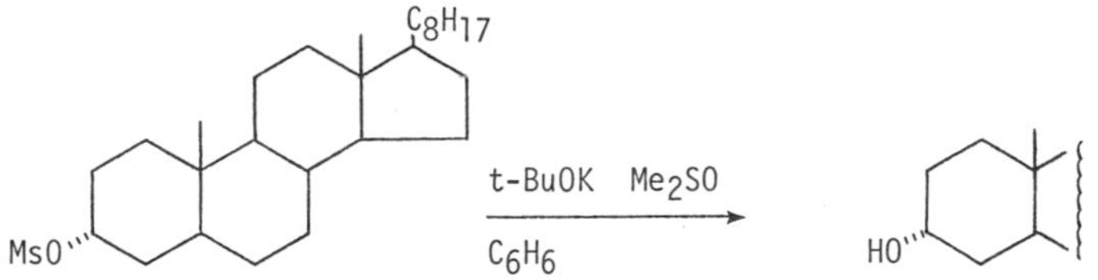

t-BuOK Me$_2$SO
──────────────→
C$_6$H$_6$

Tetr Lett (1964) 305

87%

Further examples of the cleavage of sulfonates to alcohols and phenols
are included in section 45A (Protection of Alcohols and Phenols)

Section 41   Alcohols and Phenols from Hydrides (RH)

PhH

MeCH─CH$_2$ AlCl$_3$
──────────────→
C$_6$H$_6$ CS$_2$

Me
|
PhCHCH$_2$OH   56%

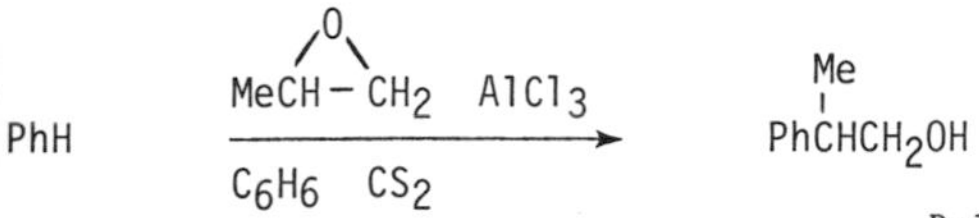

Bull Chem Soc Jap (1967) 40 2980

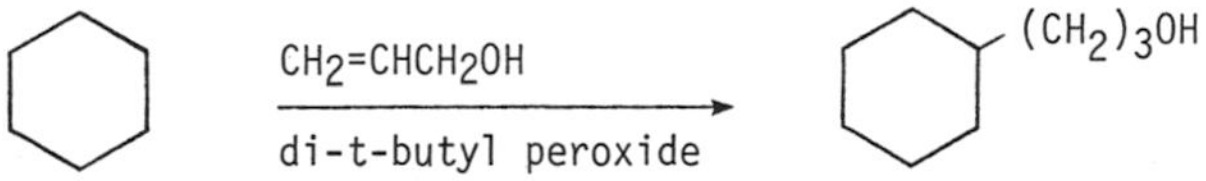

CH$_2$=CHCH$_2$OH
──────────────→
di-t-butyl peroxide

(CH$_2$)$_3$OH

38%

Neftekhimiya (1965) 5 554
(Chem Abs 63 16223)

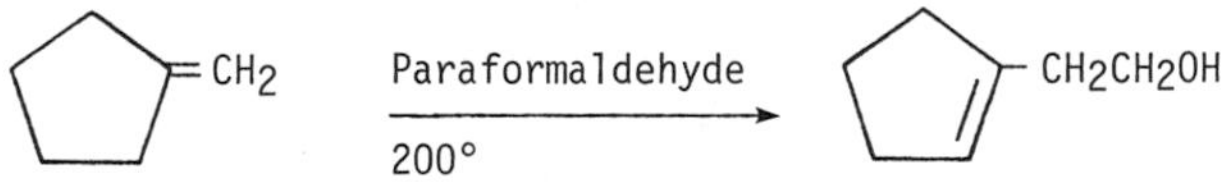

=CH$_2$

Paraformaldehyde
──────────────→
200°

CH$_2$CH$_2$OH

38%

JACS (1950) 72 2871
Chem Rev (1952) 51 505

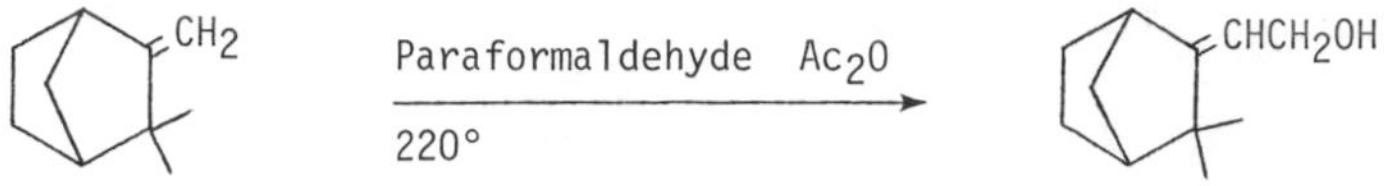
CH2
Paraformaldehyde Ac2O
220°
CHCH2OH

Aust J Chem (1967) 20 2033

C_7H_{16} —— 1 $(CF_3COO)_4Pb$; 2 NaOH H_2O ——> $C_7H_{15}OH$ (isomer mixture) ~45%

JACS (1967) 89 3662

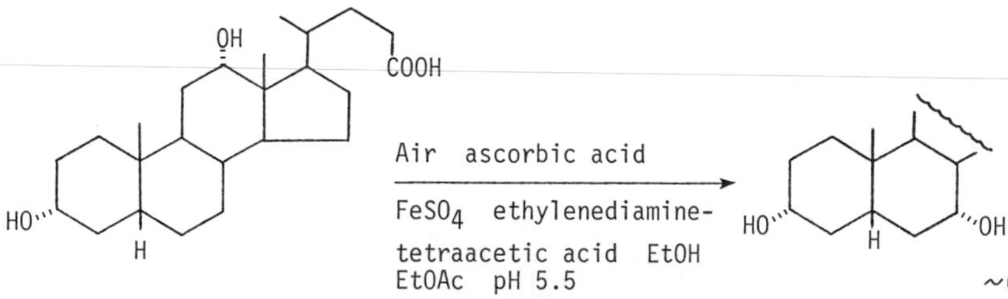
OH
COOH
HO
H
Air ascorbic acid
FeSO4 ethylenediamine-
tetraacetic acid EtOH
EtOAc pH 5.5
HO
H
OH

~61%

Steroids (1965) 5 451

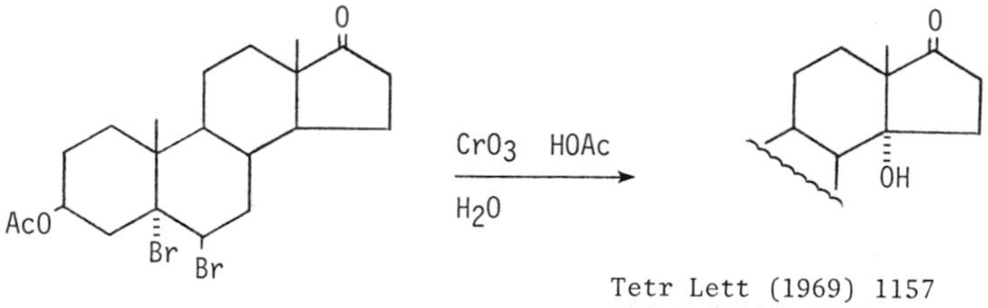
O
AcO
Br
Br
CrO3 HOAc
H2O
O
OH

30%

Tetr Lett (1969) 1157
JCS C (1968) 2346
JACS (1948) 70 3237

$BuCH(Me)(Et)$ —— $Na_2Cr_2O_7$ $HClO_4$; HOAc H_2O ——> $BuCOH(Me)(Et)$ 6%

JACS (1961) 83 423
Tetrahedron (1968) 24 4667

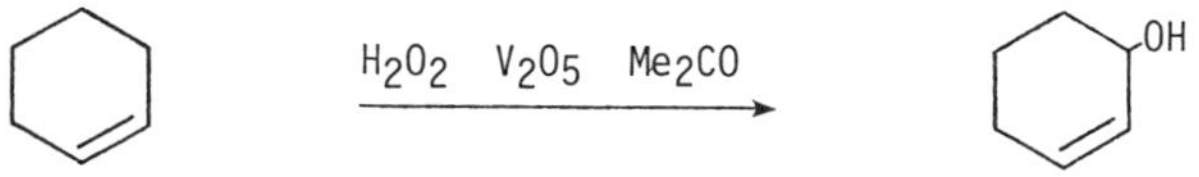
H2O2 V2O5 Me2CO
OH

JOC (1963) 28 2057

$Me_2C{=}CMe_2$ → (1 H_2O_2 NaOCl H_2O; 2 Reduction) → $CH_2{=}CMe{-}CMe_2OH$ 63%

$Me_2C{=}CMe_2$ → (O_2 hν rose bengal MeOH) → $CH_2{=}CMe{-}CMe_2OH$ 100%

JACS (1968) 90 975

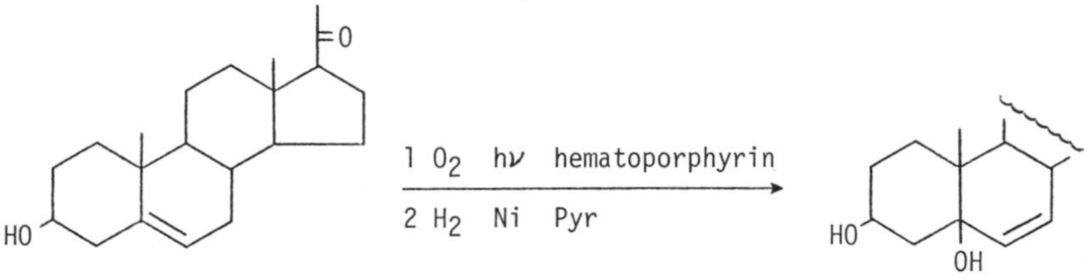

O
HO
1 O2 hν hematoporphyrin
2 H2 Ni Pyr
HO
OH

55%

Annalen (1958) 618 194 185

$PhCH_3$ —(Argentic picolinate, Me_2SO)→ $PhCH_2OH$

Tetr Lett (1967) 415

RH —(1 $(CF_3COO)_4Pb$; 2 NaOH H_2O)→ ROH (R=Ph or $PhCH_2$) ~45%

JACS (1967) 89 3662

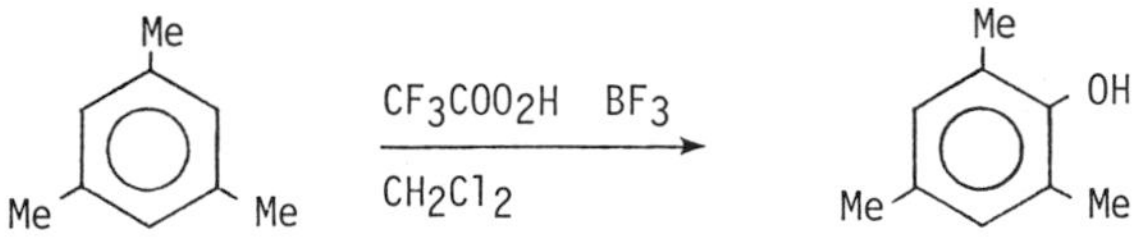

Me
Me
Me
CF3COO2H BF3
CH2Cl2
Me
OH
Me
Me

89%

JOC (1964) 29 2397
(1966) 31 153

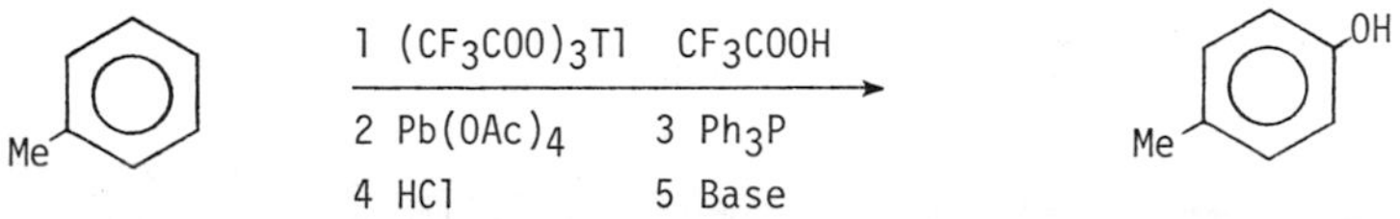

Me
1 (CF3COO)3Tl CF3COOH
2 Pb(OAc)4 3 Ph3P
4 HCl 5 Base
OH
Me

62%

JACS (1970) 92 3520

PhH $\xrightarrow{\text{H}_2\text{O}_2 \quad \text{V}_2\text{O}_5 \quad \text{t-BuOH}}$ PhOH   30%

JACS (1937) 59 2342
Tetrahedron (1968) 24 3475

35%

JACS (1949) 71 3889
Org React (1946) 3 141

Alcohols may also be prepared by conversion of hydrides into esters, followed by hydrolysis. See section 116 (Esters from Hydrides)

Section 42   Alcohols and Phenols from Ketones

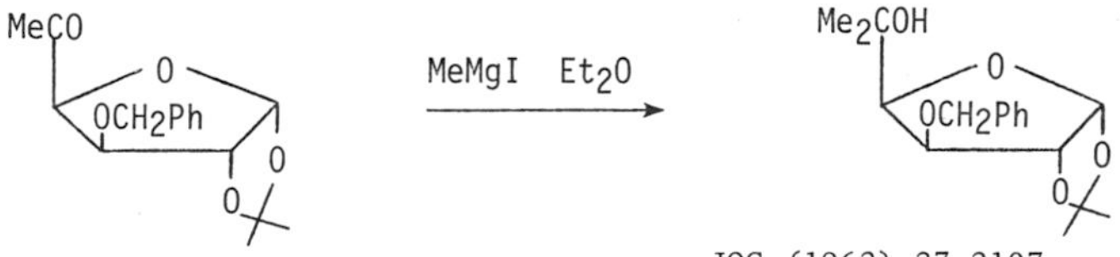

71%

JOC (1962) 27 2107

$(i\text{-Pr})_2\text{CO}$ $\xrightarrow[\text{Et}_2\text{O}]{\text{PrMgBr} \quad \text{LiClO}_4}$ $(i\text{-Pr})_2\underset{\underset{\text{Pr}}{|}}{\text{COH}}$ 70%

(Procedure for hindered ketones)

Chem Comm (1970) 470

Me_2CO $\xrightarrow{\text{PhBr Li THF}}$ Me_2COH | Ph (One-step procedure)   34%

Chem Comm (1970) 1160

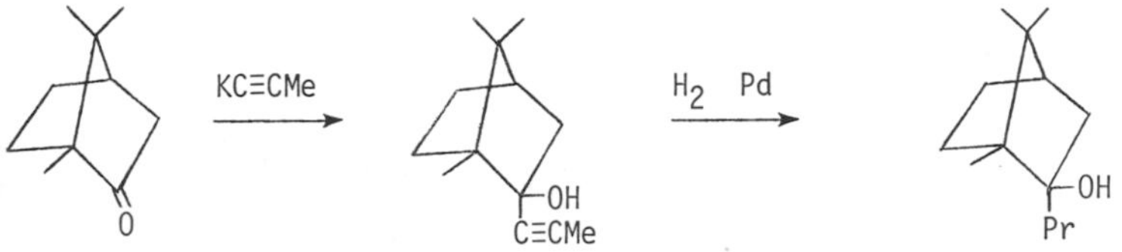

Tetr Lett (1965) 1619
JOC (1969) 34 3754

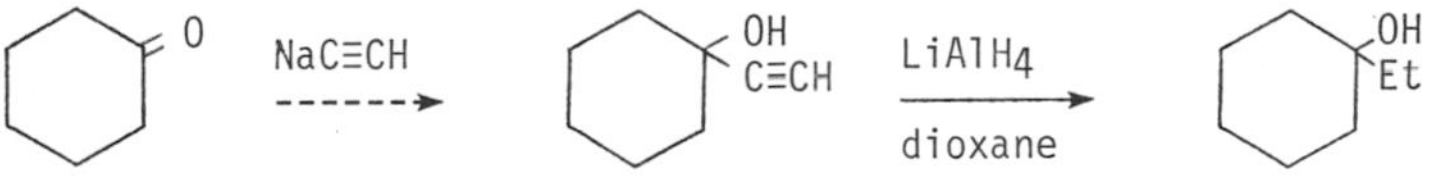

JCS (1954) 1854

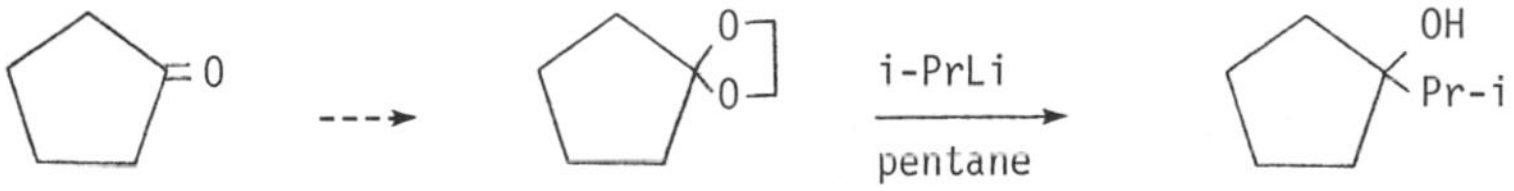

J Heterocyclic Chem (1969) 6 139

$i-PrCH_2COMe$ $\xrightarrow{\text{H}_2 \text{ Pd-C}}$ $i-PrCH_2CHMe$ with OH

(Pt Rh or Ru catalysts may also be used)

JOC (1959) 24 1855

For stereochemistry see Chem Rev (1957) 57 895

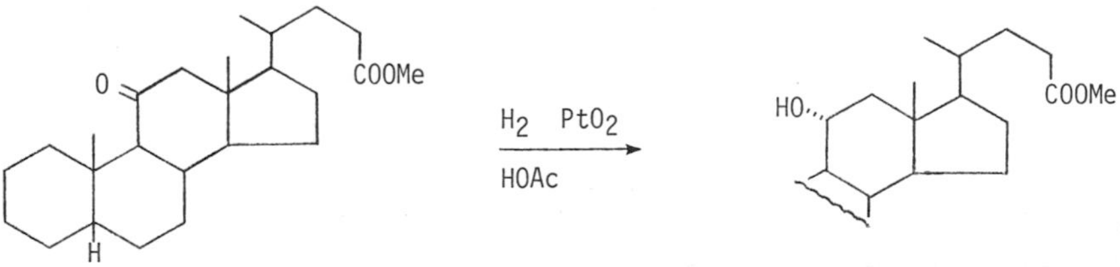

Helv (1943) 26 562
JCS (1954) 2487

PhCOMe $\xrightarrow{H_2\ \ RhH_2(PhPMe_2)_2^+\ ClO_4^-}$ PhCH(OH)Me

Chem Comm (1970) 567

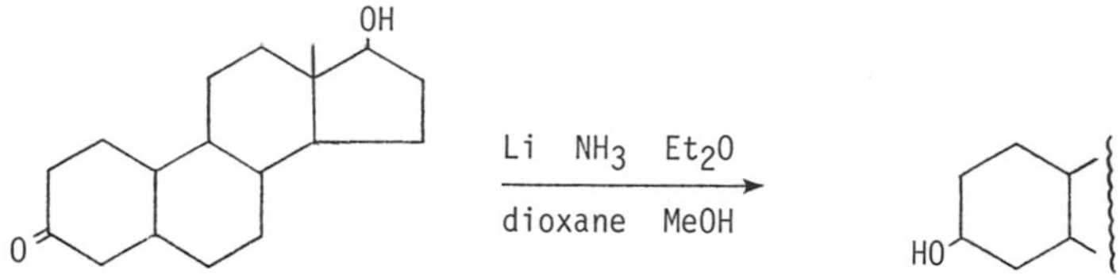
OH
O
Li NH3 Et2O
dioxane MeOH
HO

65%

JACS (1958) 80 6115
(1968) 90 6486

Pr_2CO $\xrightarrow{Li\ \ NH_2CH_2CH_2NH_2}$ Pr_2CHOH 30%

JOC (1957) 22 891

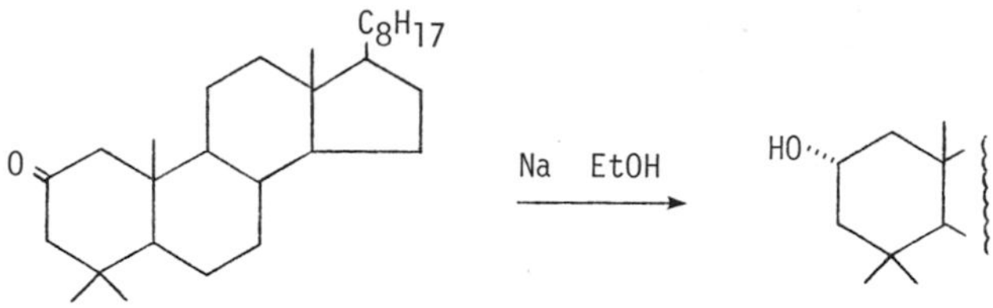
C8H17
O
Na EtOH
HO

93%

Bull Soc Chim Fr (1964) 2236
JCS C (1969) 968

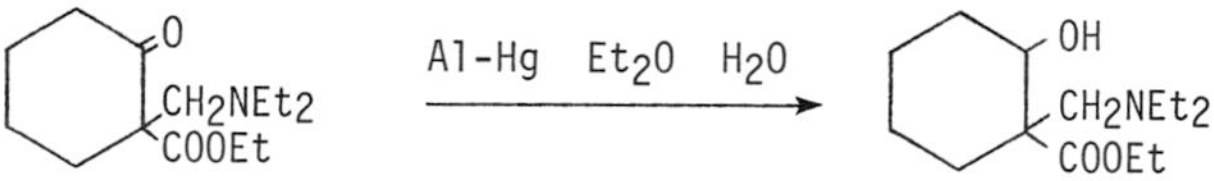
O
CH2NEt2
COOEt
Al-Hg Et2O H2O
OH
CH2NEt2
COOEt

74%

Arch Pharm (1942) 280 361

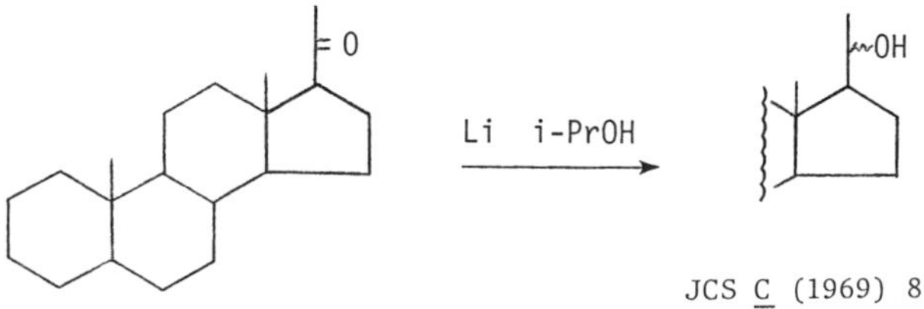
O
Li i-PrOH
OH

JCS C (1969) 804 968

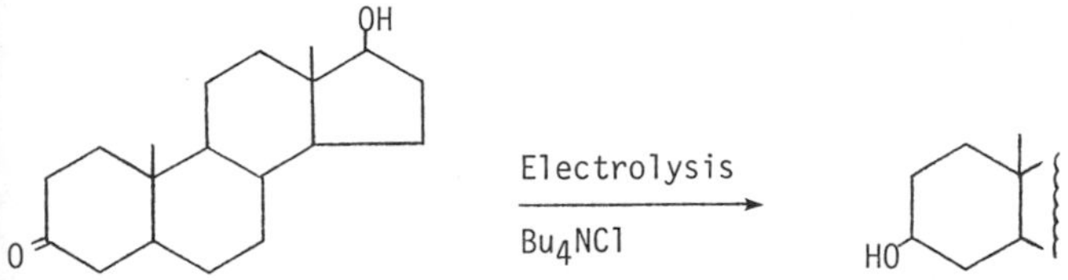
OH
O
Electrolysis
Bu4NCl
HO

100%

JOC (1961) 26 1738
(1970) 35 261

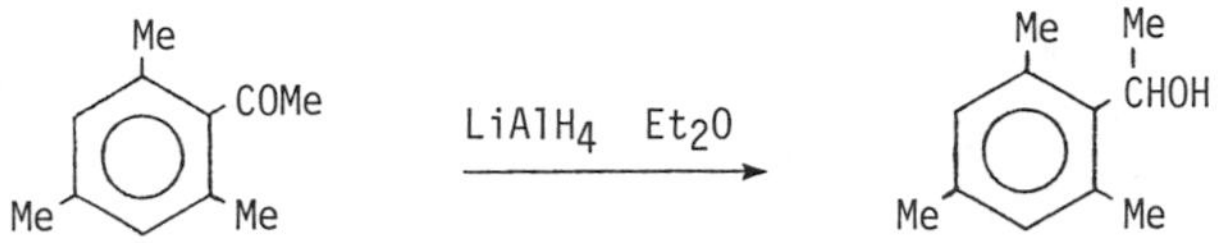
Me
COMe
Me
Me
LiAlH4 Et2O
Me
Me
CHOH
Me
Me

JACS (1947) 69 1197
Org React (1951) 6 469

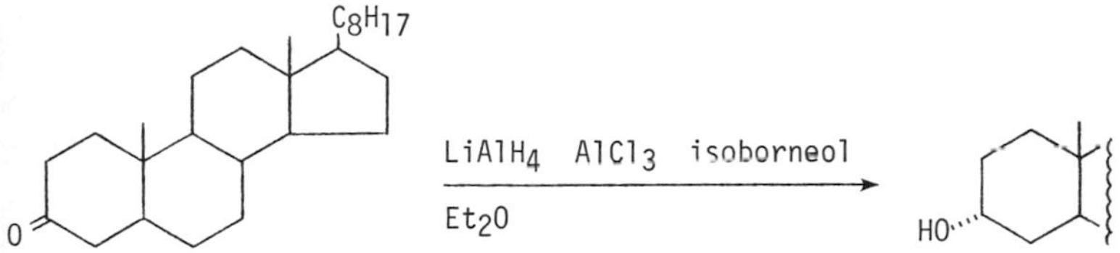
C8H17
O
LiAlH4 AlCl3 isoborneol
Et2O
HO

70%

JOC (1965) 30 3809

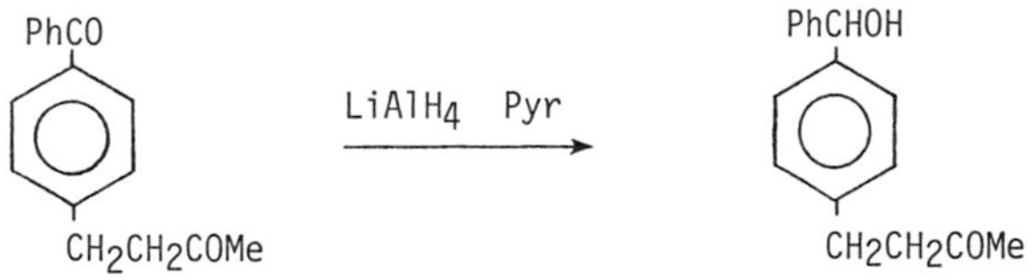
PhCO
CH2CH2COMe
LiAlH4 Pyr
PhCHOH
CH2CH2COMe

56%

JACS (1962) 84 1756

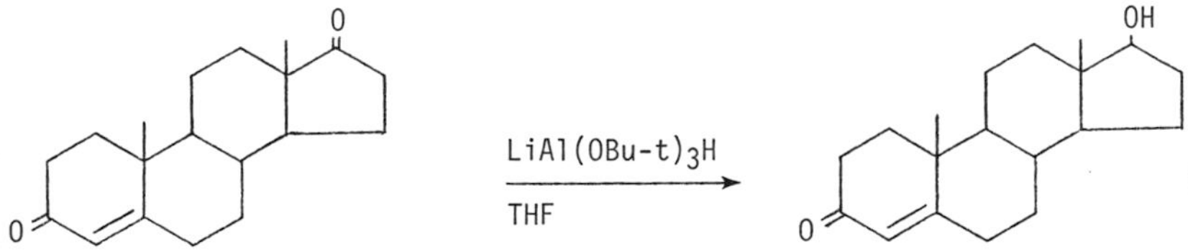
O
O
LiAl(OBu-t)3H
THF
OH
O

55%

Coll Czech (1959) 24 2284

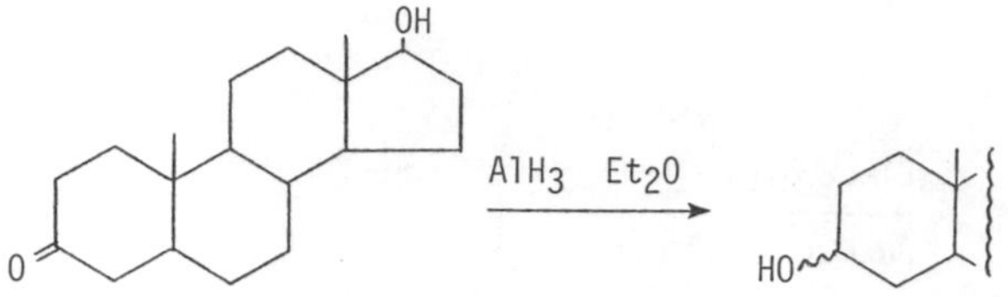

AlH_3 Et_2O

α/β = 34/66

JCS <u>B</u> (1967) 581

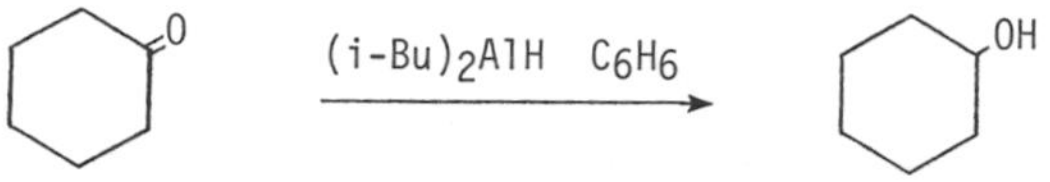

$(i-Bu)_2AlH$ C_6H_6

83%

JOC (1959) <u>24</u> 627

For reduction of unsaturated ketones see Chem Comm (1970) 213

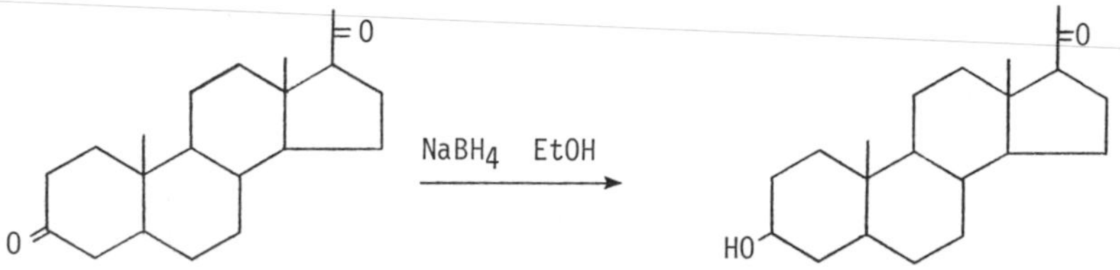

$NaBH_4$ EtOH

76%

JACS (1953) <u>75</u> 1286

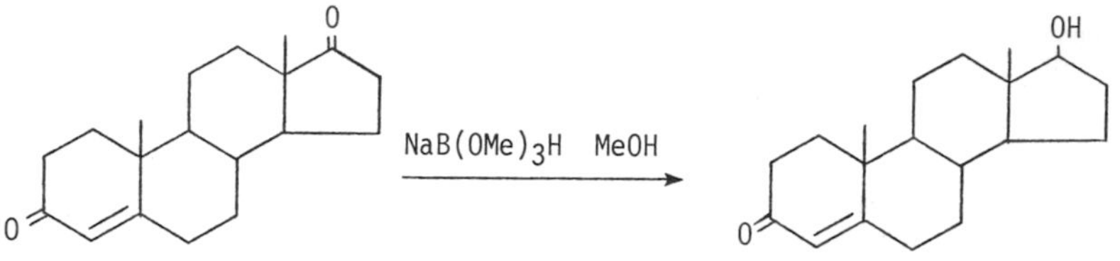

$NaB(OMe)_3H$ MeOH

22%

JCS (1955) 3426

THF

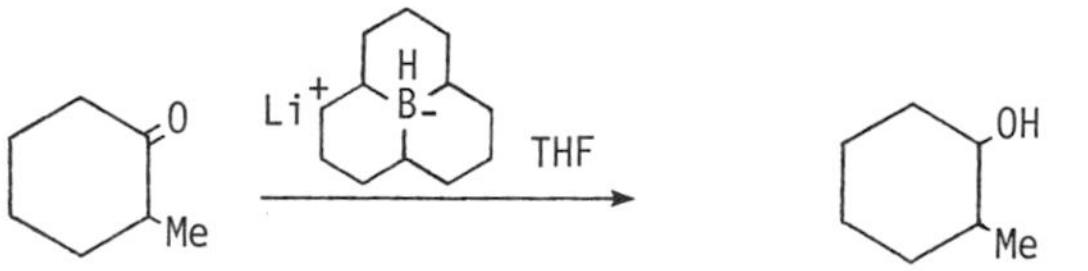

cis/trans=99/1

JACS (1970) <u>92</u> 709

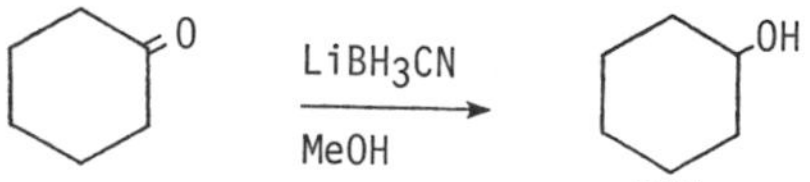

$LiBH_3CN$

MeOH

77%

($LiBH_3CN$ is stable to pH 3)

JACS (1969) <u>91</u> 3996

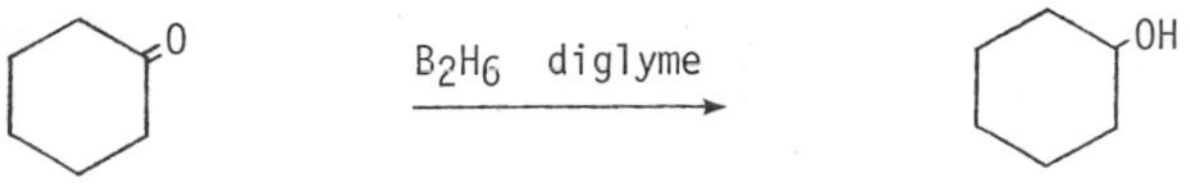

JACS (1960) 82 681

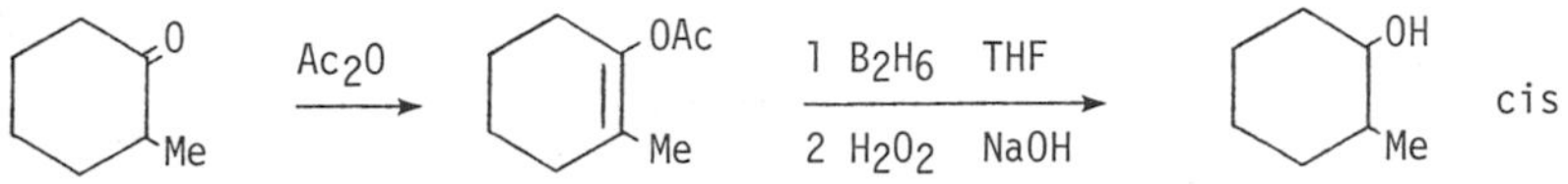

Tetr Lett (1968) 4937

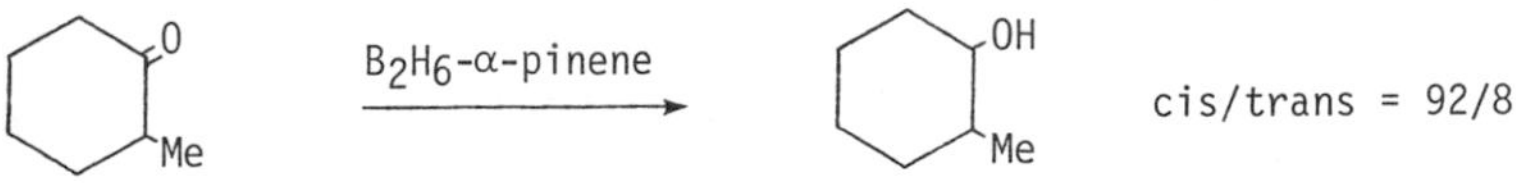

cis/trans = 92/8

JACS (1961) 83 3166

Ph$_2$CO $\xrightarrow{\text{Pyr-B}_2\text{H}_6 \quad \text{toluene}}$ Ph$_2$CHOH   ~83%

JOC (1958) 23 1561

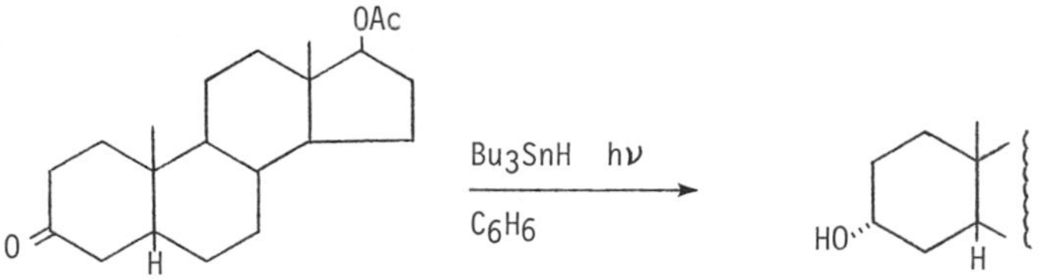

66%

Helv (1967) 50 2259
Tetr Lett (1968) 5385

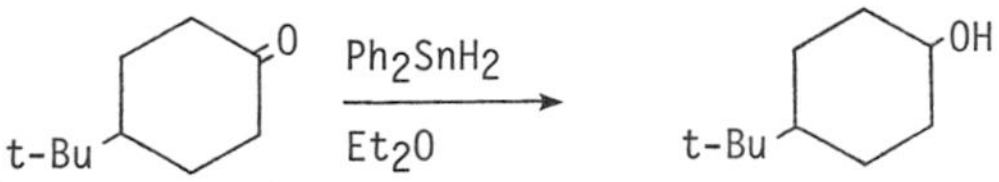

85-93%

JACS (1958) 80 3798
          (1961) 83 1246

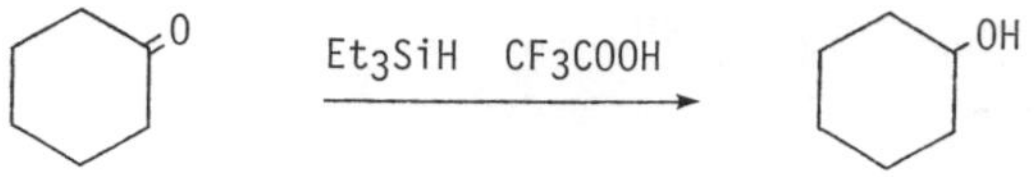

74%

Tetrahedron (1967) 23 2235

Ph_2CO $\xrightarrow{\text{NaH xylene}}$ Ph_2CHOH 83%

JACS (1946) 68 2647

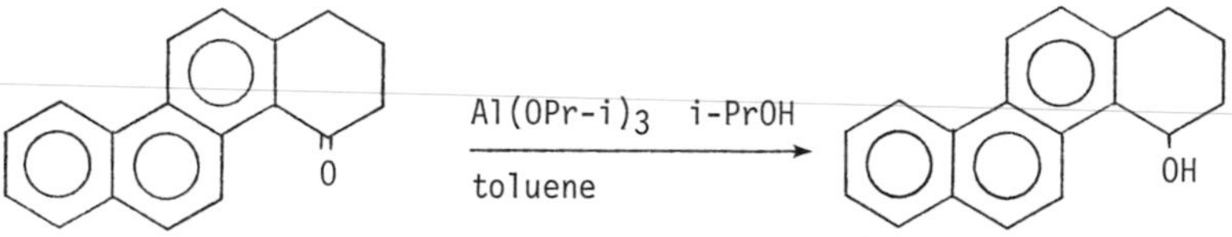

76%

JOC (1939) 4 456
Org React (1944) 2 178

PhCOMe $\xrightarrow{\text{i-PrOLi i-PrOH}}$ PhCH(Me)OH >90%

JCS C (1969) 804

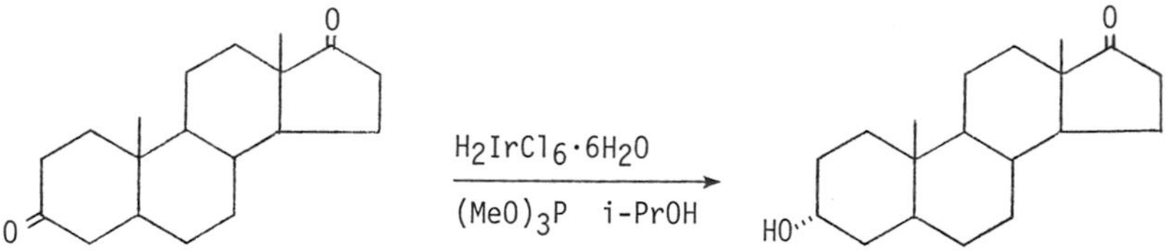

85%

JCS C (1969) 1653
(1970) 785
Chem Comm (1970) 162

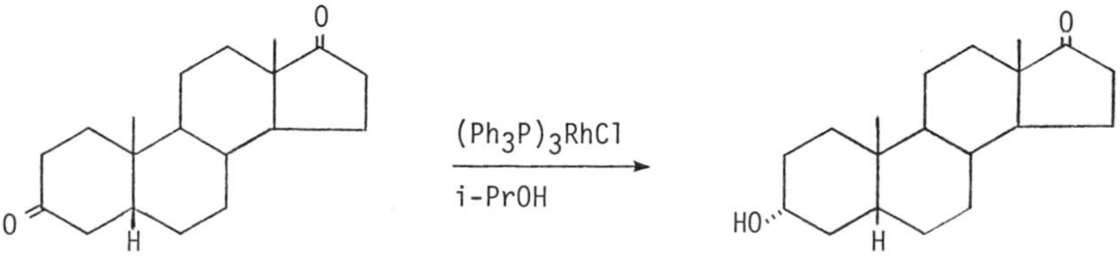

100%

Chem Comm (1970) 162

Ph_2CO —(KOH $HOCH_2CH_2OH$)→ Ph_2CHOH 93%

JOC (1967) 32 840
(1960) 25 1707

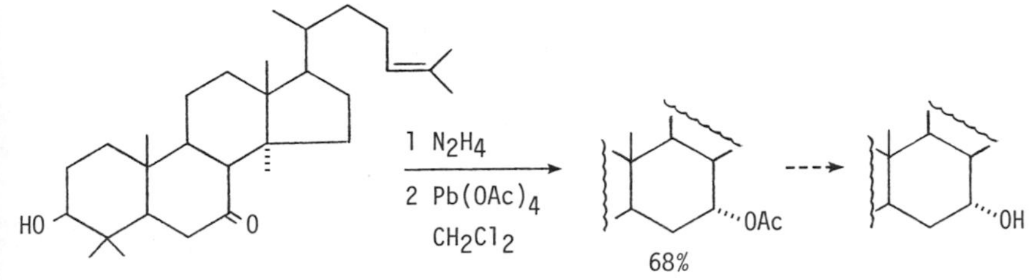
HO
O
1 N_2H_4
2 $Pb(OAc)_4$
CH_2Cl_2
OAc
68%
OH

Chem Comm (1969) 450

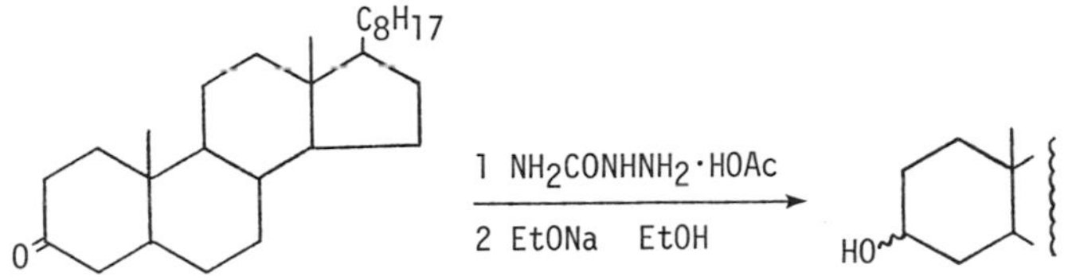
C_8H_{17}
O
1 $NH_2CONHNH_2 \cdot HOAc$
2 EtONa EtOH
HO

90%

JACS (1939) 61 1992

PhCOMe —(NCOONa=NCOONa HOAc MeOH)→ PhCH(Me)OH 31%

(Rate of reduction ArC=O > AliphC=O)

Chem Comm (1965) 71

Ph_2CO —(PhLi-Pyr Et_2O)→ Ph_2CHOH 62%

Can J Chem (1963) 41 1961

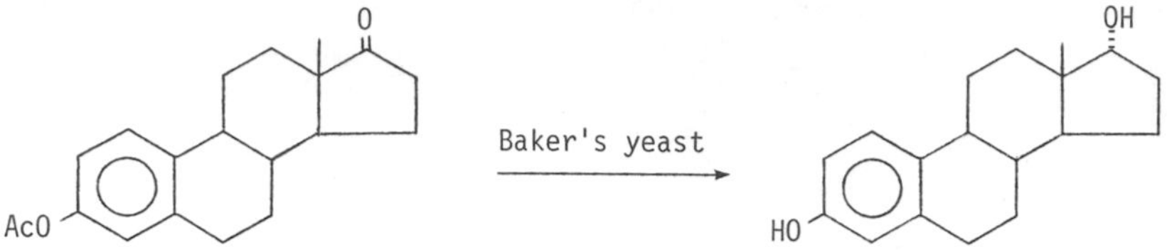

O
AcO
Baker's yeast
OH
HO

67%

Ber (1938) 71B 2696

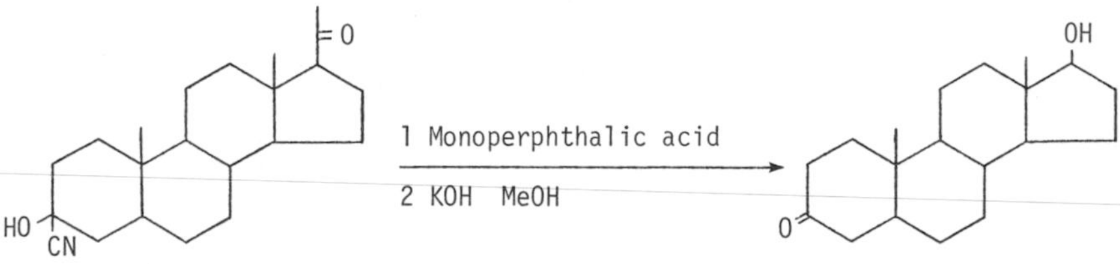

=O
HO
CN
1 Monoperphthalic acid
2 KOH MeOH
OH
O

Gazz (1961) 91 1250
(Chem Abs 56 10211)

EtCOPh $\xrightarrow{PhCOO_2H \quad CHCl_3}$ EtCOOPh ----→ PhOH

73%

JACS (1949) 71 14

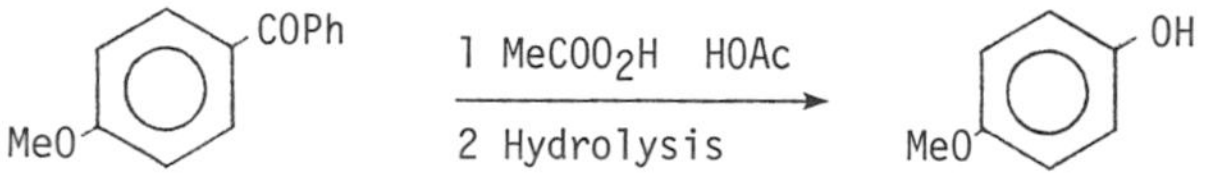

COPh
MeO
1 MeCOO2H HOAc
2 Hydrolysis
OH
MeO

86%

JACS (1950) 72 5515

Further examples of the Baeyer-Villiger degradation of ketones to alcohols and phenols via esters are included in section 117 (Esters from Ketones)

Some of the methods listed in section 34 (Alcohols and Phenols from Aldehydes) may also be applied to the preparation of alcohols from ketones

Section 43   Alcohols and Phenols from Nitriles
             ○○○○○○○○○○○○○○○○○○○○○○○○○○○○○○○○○○○○

No examples

Section 44   Alcohols from Olefins
             ○○○○○○○○○○○○○○○○○○○○○○○○

$C_6H_{13}CH=CH_2$ $\xrightarrow{\begin{array}{l}\text{1 NaBH}_4 \quad \text{BF}_3 \quad \text{diglyme} \\ \hline \text{2 CO} \\ \text{3 H}_2O_2 \quad \text{NaOH} \quad \text{H}_2O\end{array}}$ $(C_6H_{13}CH_2CH_2)_3COH$ 90%

JACS (1967) 89 2737
     (1970) 92 6648

$C_5H_{11}CH=CH_2$ $\xrightarrow{\begin{array}{l}\text{Di-t-butyl peroxide} \\ \hline \text{MeCH}_2OH\end{array}}$ $C_5H_{11}CH_2CH_2\underset{\underset{\text{Me}}{|}}{CHOH}$ 40-60%

Izv (1964) 894
(Chem Abs 61 5510)

$PrCH=CH_2$ $\xrightarrow{\begin{array}{l}\text{1 B}_2H_6 \quad \text{THF} \\ \hline \text{2 MeOCHCl}_2 \quad \text{MeLi} \quad \text{Et}_2O \\ \text{3 H}_2O_2 \quad \text{NaOH} \quad \text{H}_2O\end{array}}$ $(PrCH_2CH_2)_2CHOH$ + $PrCH_2\underset{\underset{\text{OH}}{|}}{CH}CH_2CH_2Pr$

                                                              21%   28%

Tetr Lett (1969) 2955

$C_6H_{13}CH=CH_2$ $\xrightarrow{\begin{array}{l}\text{1 B}_2H_6 \quad \text{THF} \\ \hline \text{2 CO} \quad \text{LiBH}_4 \\ \text{3 KOH} \quad \text{EtOH} \\ \text{4 H}_2O_2\end{array}}$ $C_6H_{13}(CH_2)_3OH$ 70%

JACS (1967) 89 2740

$PhCH=CH_2$ $\xrightarrow{\begin{array}{l}\text{HCHO} \\ \hline \text{H}_2SO_4\end{array}}$ Ph—[ring with O] $\xrightarrow{\begin{array}{l}\text{H}_2 \text{ (2,600 psi)} \\ \hline \text{copper chromite}\end{array}}$ $Ph(CH_2)_3OH$ ~74%

JACS (1950) 72 5314
Chem Rev (1952) 51 505

$C_5H_{11}CH=CH_2$ $\xrightarrow[\text{RhCl}_3 \cdot 3H_2O]{\text{CO} \quad H_2 \text{ (240 atmos)}}$ $C_5H_{11}(CH_2)_3OH$ 73%

Ber (1966) 99 1086

$\xrightarrow[\text{Co}]{\text{CO} \quad H_2 \text{ (450 atmos)}}$ 56%

JACS (1952) 74 4496

$BuCH=CH_2$ $\xrightarrow[\substack{2 \; CH_2=SMe_2 \quad Me_2SO \\ 3 \; H_2O_2 \quad NaOH \quad H_2O}]{1 \; B_2H_6}$ $Bu(CH_2)_3OH$ + $Bu(CH_2)_2OH$

Chem Comm (1967) 505

$i\text{-}PrCH_2CH=CH_2$ $\xrightarrow[\substack{2 \; H_2O_2 \quad NaOH \quad H_2O}]{1 \; NaBH_4 \quad BF_3 \cdot Et_2O \quad THF}$ $i\text{-}Pr(CH_2)_3OH$ 80%

Org React (1963) 13 1

$\xrightarrow[\substack{2 \; H_2O_2 \quad NaOH \quad H_2O}]{1 \; LiBH_4 \quad H_2SO_4 \quad THF}$ 22%

JOC (1963) 28 3551

$CH_2=CH(CH_2)_8COOH$ $\xrightarrow[\substack{2 \; H_2O_2}]{1 \; (i\text{-}PrCH)_2BH \quad THF \\ \overset{Me}{|}}$ $HOCH_2CH_2(CH_2)_8COOH$ 82%

JACS (1961) 83 486

$Me_2C=CHMe$ $\xrightarrow[\substack{2 \; H_2O_2 \quad NaOH \quad H_2O}]{1 \quad \text{(BH)} \quad THF}$ $Me_2CHCHMe \atop \quad\;\; \overset{|}{OH}$ ~95%

JACS (1968) 90 5281

$C_{10}H_{21}CH=CHC_{11}H_{23}$ $\xrightarrow{\text{1 NaBH}_4 \quad BF_3 \cdot Et_2O}$ $C_{22}H_{45}CH_2OH$ 51%

1 NaBH$_4$ BF$_3\cdot$Et$_2$O

bis-2-ethoxyethyl ether
185-190°

2 H$_2$O$_2$ NaOH   H$_2$O

JOC (1961) <u>26</u> 3657
JACS (1967) <u>89</u> 561 567

$PhCH=CHMe$ $\xrightarrow{\text{1 Et}_2\text{AlCl}}$ $PhCH_2\underset{\underset{OH}{|}}{CH}Me$ 75%

2 H$_2$O$_2$ NaOH

Tetr Lett (1970) 3471

Cr(OAc)$_2$

BuSH
Me$_2$SO

JACS (1966) <u>88</u> 3016

$MeCO(CH_2)_2CH=CMe_2$ $\xrightarrow{\text{1 HCOOH}}$ $MeCO(CH_2)_2CH_2\underset{\underset{Me}{|}}{\overset{\overset{Me}{|}}{C}}OH$ 80-95%

2 NaHCO$_3$ MeOH

Bol Inst Quim Univ Nac Aut Mex (1965) <u>17</u> 181
(Chem Abs <u>65</u> 8963)
JACS (1953) <u>75</u> 6212

$MeCO(CH_2)_2CH=CMe_2$ $\xrightarrow{\text{1 H}_2\text{SO}_4 \quad \text{H}_2\text{O}}$ $MeCO(CH_2)_2CH_2\underset{\underset{Me}{|}}{\overset{\overset{Me}{|}}{C}}OH$ 85%

2 NaOH   H$_2$O

JACS (1955) <u>77</u> 1617

$BuCH=CH_2$ $\xrightarrow{\text{1 Hg(OAc)}_2 \quad \text{THF} \quad \text{H}_2\text{O}}$ $BuCHMe$ 96%

1 Hg(OAc)$_2$ THF   H$_2$O

2 NaOH   H$_2$O

3 NaBH$_4$ NaOH   H$_2$O

$\underset{OH}{|}$

JACS (1967) <u>89</u> 1522
JOC (1970) <u>35</u> 1844

$$\xrightarrow[\text{MeOCH}_2\text{CH}_2\text{OMe}]{\text{H}_2\text{O} \quad h\nu \quad \text{xylene}}$$

50%

JACS (1967) 89 6788
Acc Chem Res (1969) 2 33

$$\text{Me}_2\text{C=CH(CH}_2)_2\overset{\overset{\text{Me}}{|}}{\text{CH}}\text{CH}_2\text{COOMe} \xrightarrow[\text{2 NaBH}_4]{\text{1 O}_3 \quad \text{MeOH}} \text{HOCH}_2(\text{CH}_2)_2\overset{\overset{\text{Me}}{|}}{\text{CH}}\text{CH}_2\text{COOMe}$$

72%

JACS (1968) 90 3525
Helv (1967) 50 2445

Section 45   Alcohols from Miscellaneous Compounds

$$\text{PhCH=CHCH}_2\text{OH} \xrightarrow{\text{H}_2 \quad \text{Pd}} \text{Ph(CH}_2)_3\text{OH}$$

55%

Annalen (1924) 439 276

$$\text{CH}_2\text{=CHCH}_2\text{OH} \xrightarrow[\text{NaOH}]{\text{ClCH}_2\text{CONHNH}_2 \cdot \text{HCl}} \text{Me(CH}_2)_2\text{OH}$$

70%

Chem Ind (1964) 839

$$\text{PhCH=CHCH}_2\text{OH} \xrightarrow{\text{LiAlH}_4 \quad \text{Et}_2\text{O}} \text{Ph(CH}_2)_3\text{OH}$$

93%

JACS (1948) 70 3484

$$\xrightarrow{\text{Ni} \quad \text{EtOH}}$$

Gazz (1963) 93 1028
(Chem Abs 60 3029)
JACS (1953) 75 1700

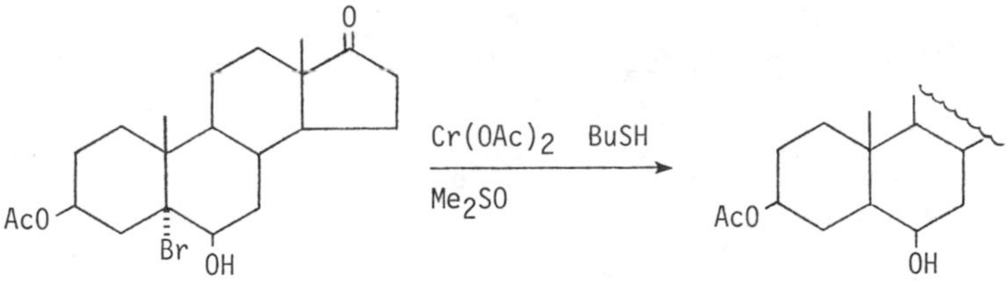
O
AcO
Br
OH
$Cr(OAc)_2$ BuSH
Me_2SO
AcO
OH

65%

JACS (1966) 88 3016

$PhCH{=}CHCHO \xrightarrow{H_2\ \ Pt} Ph(CH_2)_3OH$

JACS (1925) 47 3061

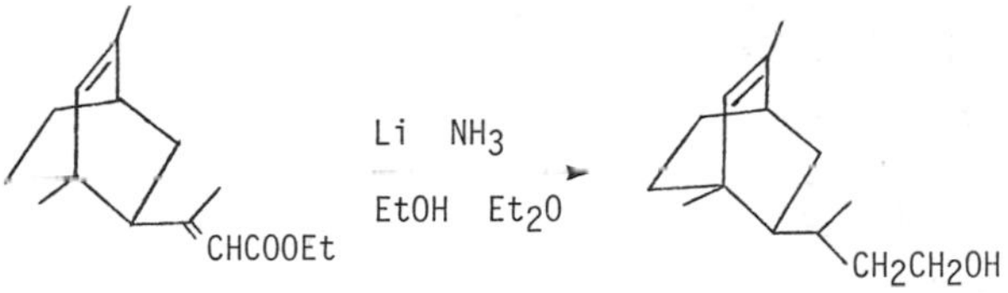
CHCOOEt
Li NH_3
EtOH Et_2O
CH_2CH_2OH

Tetr Lett (1970) 3219

$MeCH{=}CHCOOH \xrightarrow{B_2H_6} Me(CH_2)_3OH$

Hua Hsueh Hsueh Pao (1965) 31 376
(Chem Abs 64 8022)

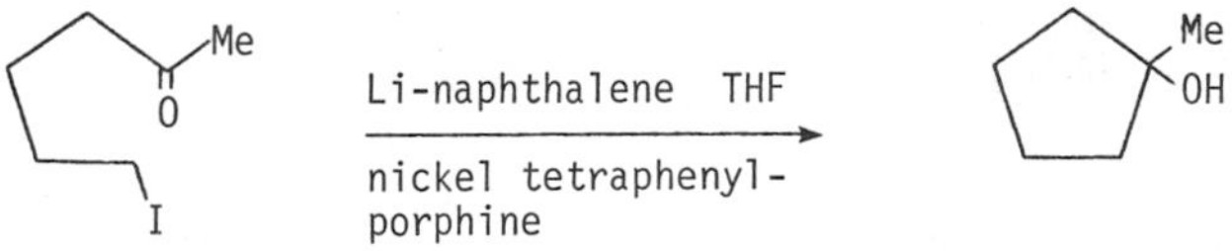
Me
O
I
Li-naphthalene THF
nickel tetraphenyl-
porphine
Me
OH

61%

JACS (1970) 92 395

Section 45A   Protection of Alcohols and Phenols

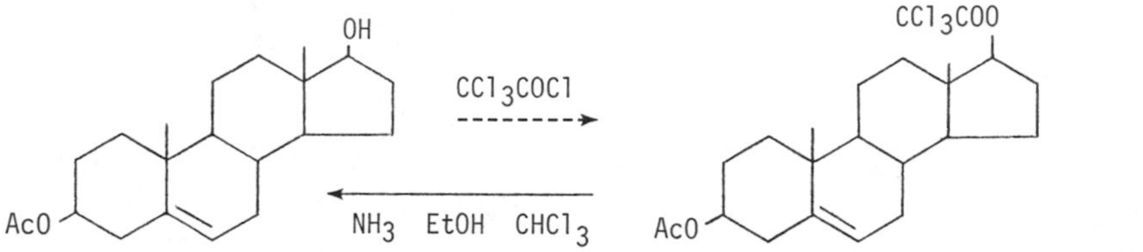

(Stable to CrO_3 and acid)

Coll Czech (1962) 27 2567
Tetr Lett (1968) 4681

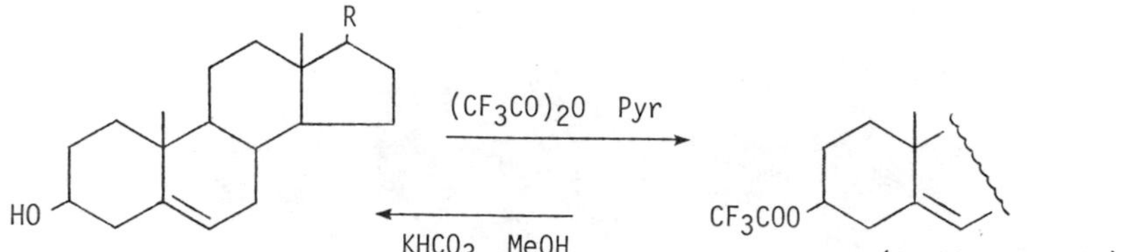

(Stable to CrO_3)

Helv (1954) 37 388

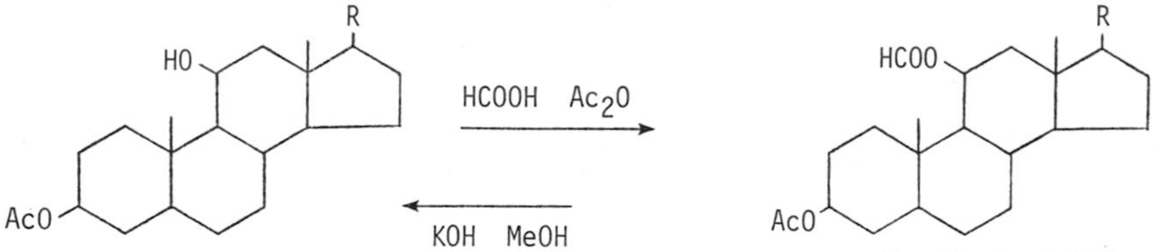

(Stable to CrO_3)

Helv (1954) 37 443

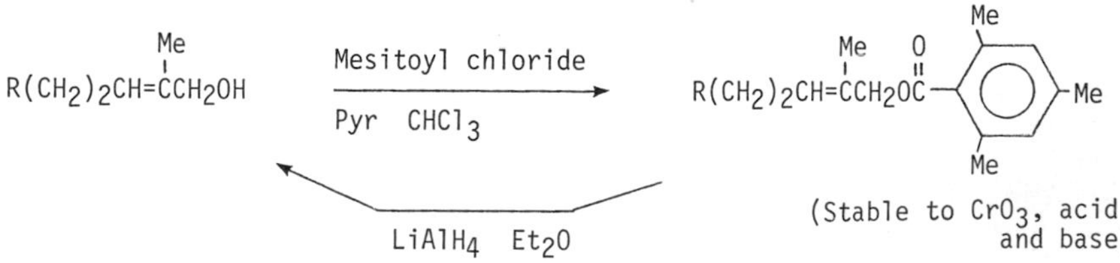

(Stable to CrO_3, acid and base)

JACS (1969) 91 4318
(1967) 89 2758

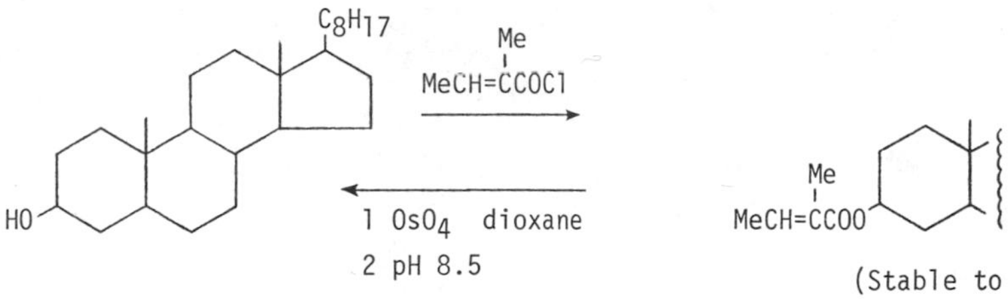
C8H17
HO
Me
MeCH=CCOCl
1 OsO4 dioxane
2 pH 8.5
Me
MeCH=CCOO

(Stable to acid)

JOC (1962) 27 3103

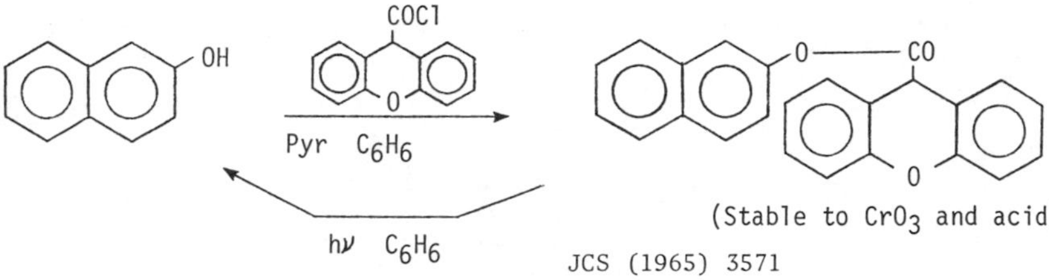
OH
COCl
O
Pyr C6H6
hν C6H6
O
CO
O

(Stable to CrO_3 and acid)

JCS (1965) 3571

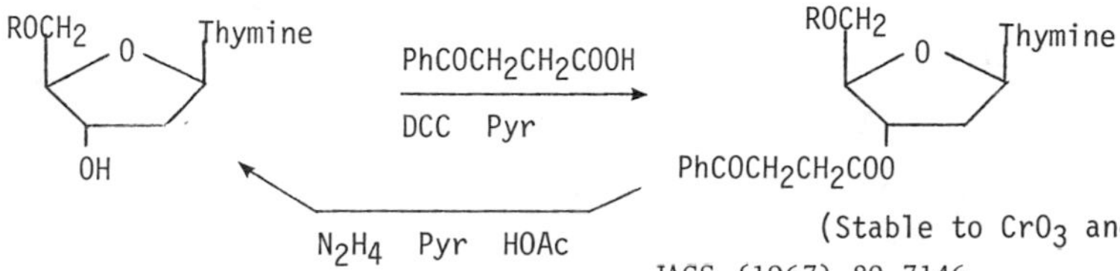
ROCH2
O
Thymine
OH
PhCOCH2CH2COOH
DCC Pyr
N2H4 Pyr HOAc
ROCH2
O
Thymine
PhCOCH2CH2COO

(Stable to CrO_3 and acid)

JACS (1967) 89 7146

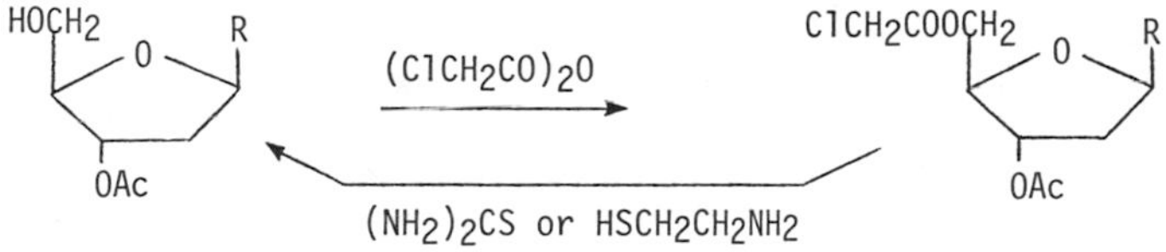
HOCH2
O
R
OAc
(ClCH2CO)2O
(NH2)2CS or HSCH2CH2NH2
ClCH2COOCH2
O
R
OAc

JOC (1970) 35 1940

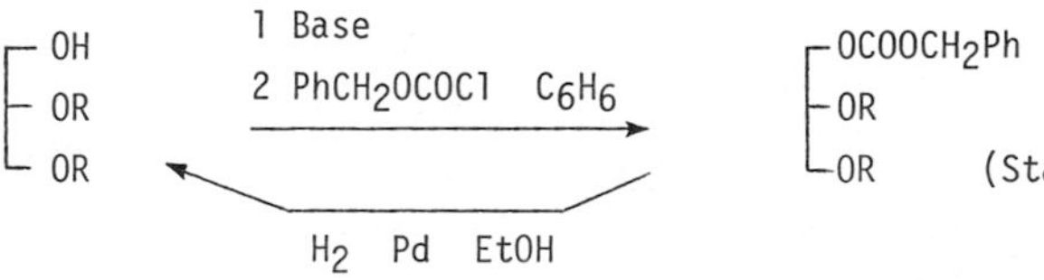
OH
OR
OR
1 Base
2 PhCH2OCOCl C6H6
H2 Pd EtOH
OCOOCH2Ph
OR
OR

(Stable to CrO_3 and acid)

JACS (1939) 61 3328

CCl_3CH_2OCOCl Pyr

$\xrightarrow{\hspace{2cm}}$

$\xleftarrow[\text{Zn   HOAc}]{}$

CCl_3CH_2OCOO

(Stable to CrO_3 and acid)

Tetr Lett (1967) 2555
JOC (1968) 33 3589

Further examples of the preparation and cleavage of esters are included
in section 108 (Esters from Alcohols and Phenols) and section 38 (Alcohols
and Phenols from Esters)

R

$HO\cdots$ H $\overset{\cdots}{O}H$

HNO_3 Ac_2O

$\xrightarrow{\hspace{2cm}}$

$\xleftarrow[\text{Zn   HOAc}]{}$

$NO_2O\cdots$ H $\overset{\cdots}{O}NO_2$

(Stable to CrO_3, acid and base)

Ber (1962) 95 1094
Chem Rev (1955) 55 485

$R[CH_2]_{10}CH_2OH$

$\xrightarrow[\hspace{1cm}]{H_3BO_3 \quad \text{toluene}}$

$\xleftarrow[H_2O]{}$

$(R[CH_2]_{10}CH_2O)_3B$

(Stable to Wittig reagents, acid and
base)

Rec Trav Chim (1953) 72 411
Tetr Lett (1969) 4155

OH

R

MsCl   Pyr

$\xrightarrow{\hspace{2cm}}$

$\xleftarrow[\text{NaOH   } H_2O]{}$

OMs

R

(Stable to CrO_3 and acid)

JACS (1957) 79 717

For cleavage of mesylates with PhLi see JCS C (1968) 2283

Further examples of the preparation and cleavage of sulfonates are included
in section 138 (Halides and Sulfonates from Alcohols and Phenols) and
section 40 (Alcohols and Phenols from Halides and Sulfonates)

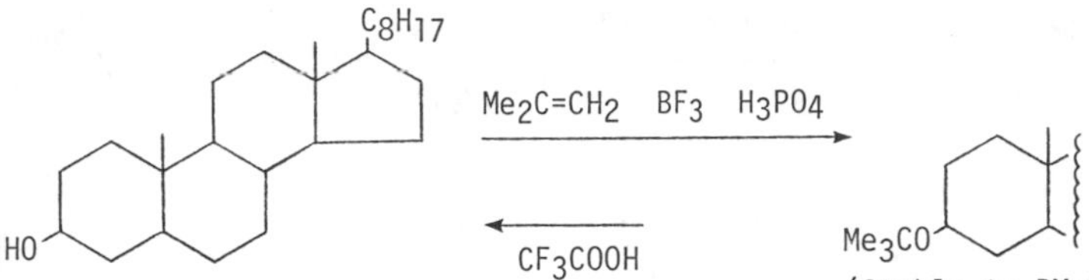
C_8H_{17}
$Me_2C=CH_2$ BF_3 H_3PO_4
HO
CF_3COOH
Me_3CO

(Stable to RMgX, $LiAlH_4$, CrO_3 and base)

JCS (1963) 755

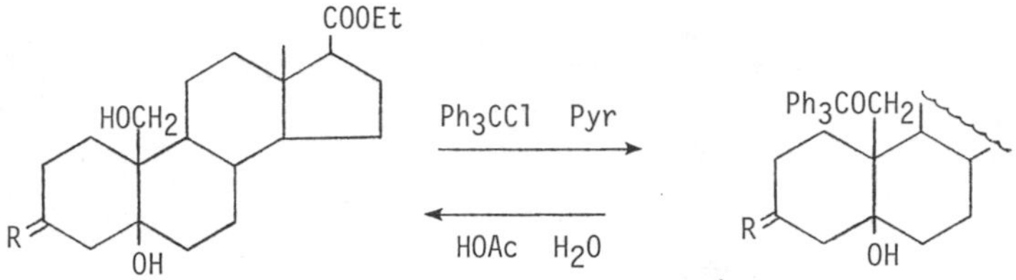
COOEt
$HOCH_2$
R
OH
Ph_3CCl Pyr
HOAc H_2O
Ph_3COCH_2
R
OH

(Stable to RMgX, $LiAlH_4$, CrO_3 and base)

JOC (1951) 16 349
(1950) 15 264

For hydrogenolysis of trityl ethers see Rec Trav Chim (1942) 61 373

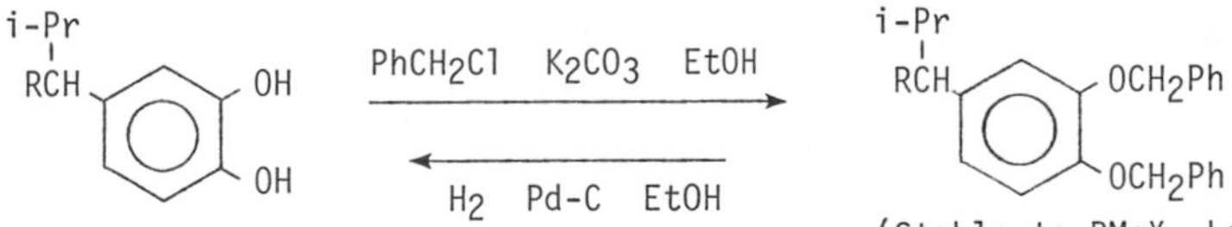
i-Pr
RCH
OH
OH
$PhCH_2Cl$ K_2CO_3 EtOH
H_2 Pd-C EtOH
i-Pr
RCH
OCH_2Ph
OCH_2Ph

(Stable to RMgX, $LiAlH_4$, CrO_3 acid and base)

J Med Chem (1967) 10 262
JACS (1954) 76 3188
Org React (1953) 7 263

Further examples of the cleavage of benzyl ethers are included in section 39 (Alcohols and Phenols from Ethers and Epoxides)

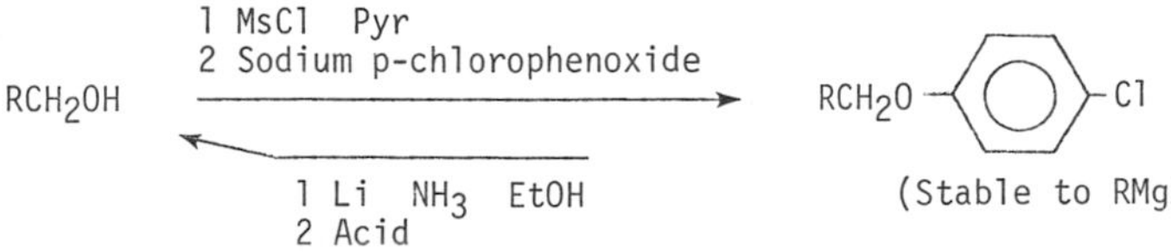
1 MsCl Pyr
2 Sodium p-chlorophenoxide
RCH_2OH
RCH_2O
Cl
1 Li NH_3 EtOH
2 Acid

(Stable to RMgX, $LiAlH_4$, CrO_3 acid and base)

JACS (1968) 90 1090

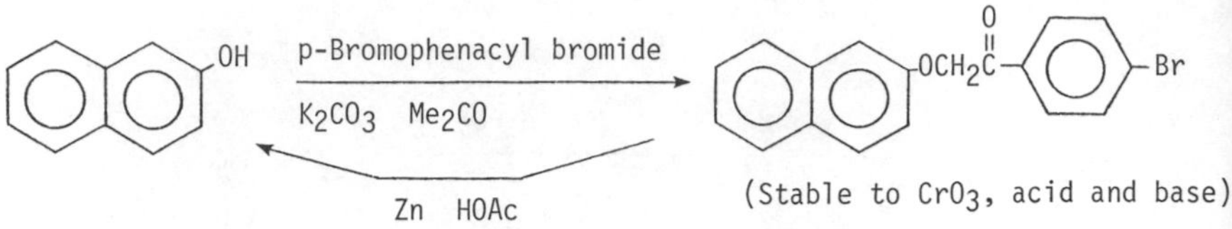
OH
p-Bromophenacyl bromide
K2CO3 Me2CO
Zn HOAc
O
OCH2C
Br
(Stable to CrO3, acid and base)

Tetr Lett (1970) 343

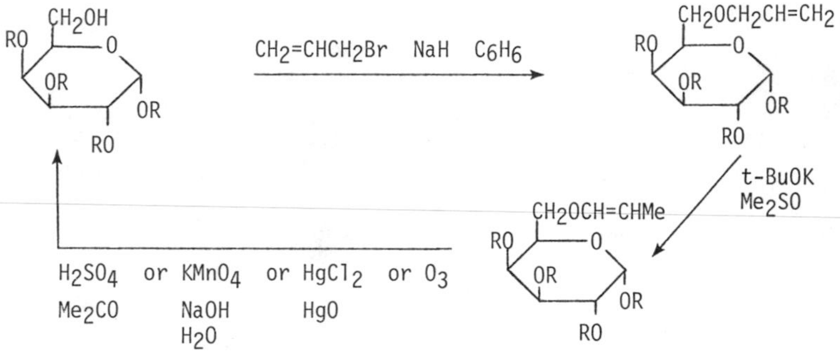
CH2OH
RO
O
OR
OR
RO
CH2=CHCH2Br NaH C6H6
CH2OCH2CH=CH2
t-BuOK
Me2SO
CH2OCH=CHMe
H2SO4 or KMnO4 or HgCl2 or O3
Me2CO NaOH HgO
H2O

JCS (1965) 2205
(1968) 1903

Further examples of the cleavage of ethers are included in section 39 (Alcohols and Phenols from Ethers and Epoxides)

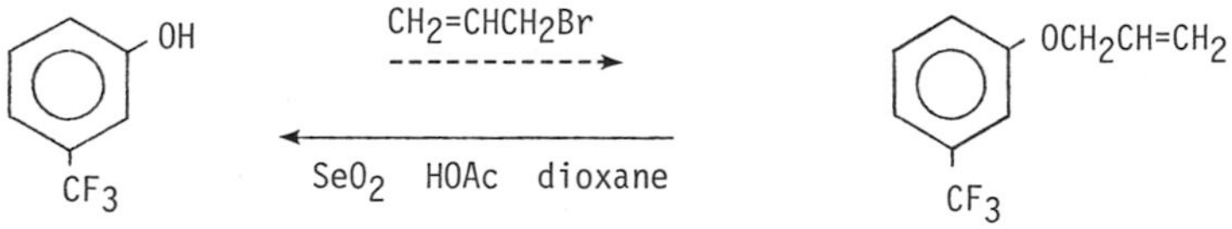
OH
CF3
CH2=CHCH2Br
SeO2 HOAc dioxane
OCH2CH=CH2
CF3

Tetr Lett (1970) 2885

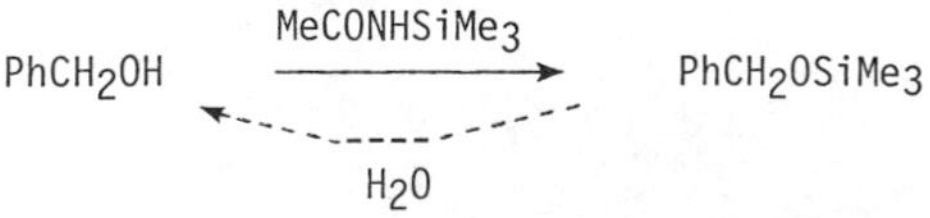
PhCH2OH
MeCONHSiMe3
H2O
PhCH2OSiMe3

Ber (1964) 97 2196

For trimethylsilyl ethers of 3ry alcohols see Chem Comm (1968) 466

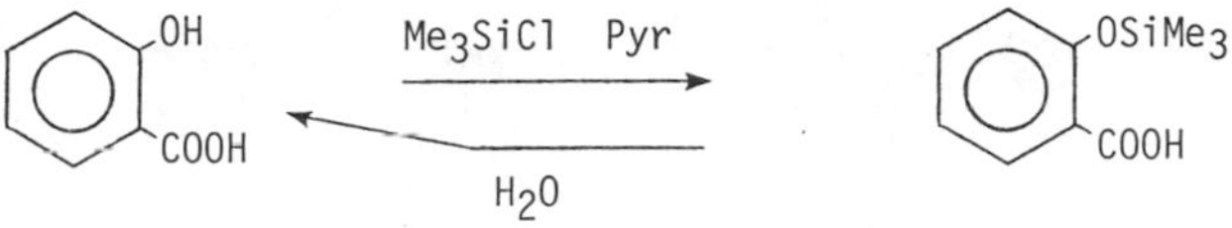

JOC (1957) <u>22</u> 592

For trimethylsilyl ethers of hindered phenols see JACS (1966) <u>88</u> 3390

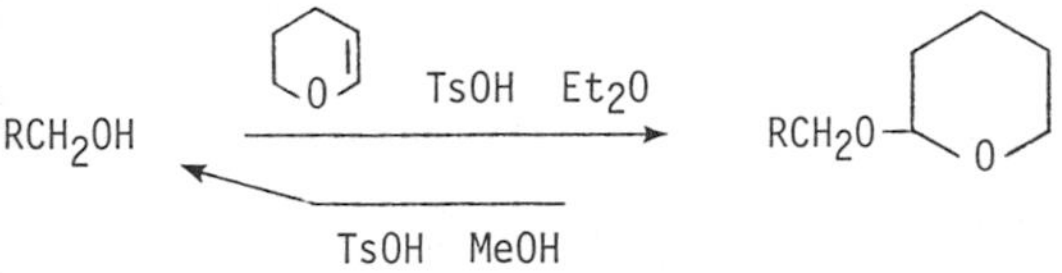

RCH_2OH TsOH Et_2O RCH_2O (Stable to RMgX, LiAlH$_4$, CrO$_3$ and base)

TsOH   MeOH

JACS (1969) <u>91</u> 4318

For tetrahydropyranyl ethers of 3ry alcohols see Tetrahedron (1961) <u>13</u> 241
and Steroids (1964) <u>4</u> 229

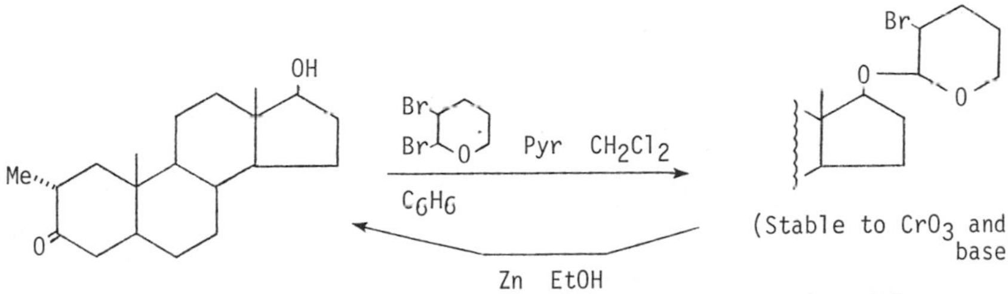

(Stable to CrO$_3$ and base)

Zn   EtOH

Steroids (1965) <u>6</u> 397

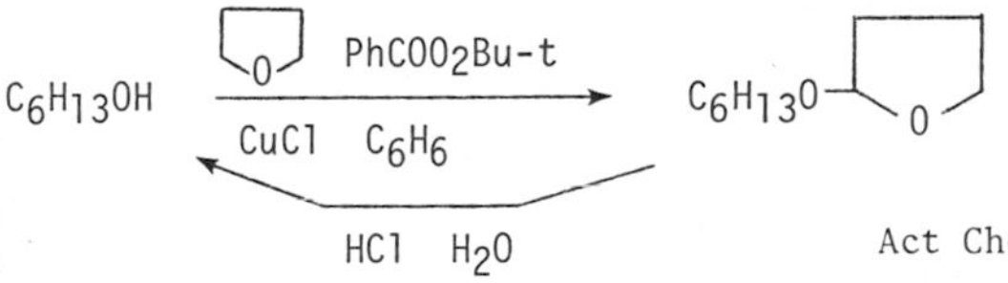

$C_6H_{13}OH$ PhCOO$_2$Bu-t $C_6H_{13}O$ (Stable to RMgX, LiAlH$_4$, CrO$_3$ and base)

CuCl C_6H_6

HCl H_2O

Act Chem Scand (1960) <u>14</u> 1854
(1961) <u>15</u> 249

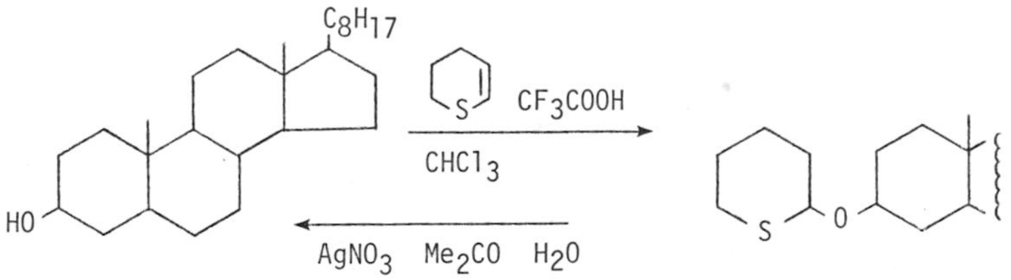

CF$_3$COOH

CHCl$_3$

AgNO$_3$ Me$_2$CO H_2O

JOC (1966) <u>31</u> 2333

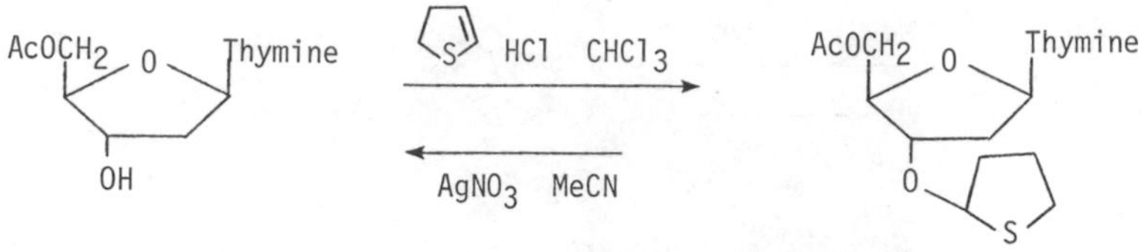

JOC (1966) 31 2333

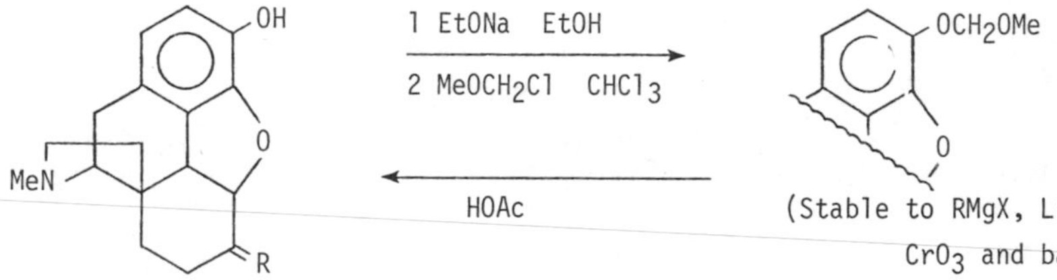

(Stable to RMgX, LiAlH$_4$, CrO$_3$ and base)

J Med Chem (1966) 9 1
JACS (1957) 79 5792

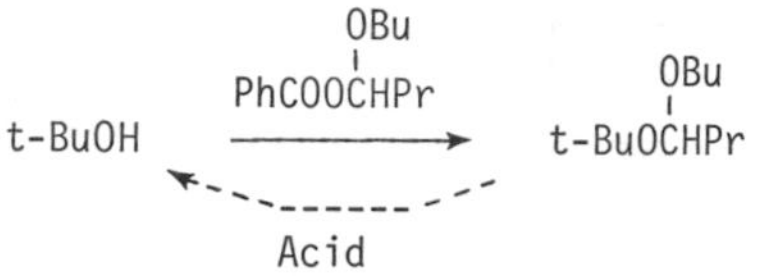

t-BuOH $\xrightarrow[\text{Acid}]{\text{PhCOOCHPr}}$ t-BuOCHPr   (Stable to RMgX, LiAlH$_4$, CrO$_3$ and base)

Tetrahedron (1961) 13 241

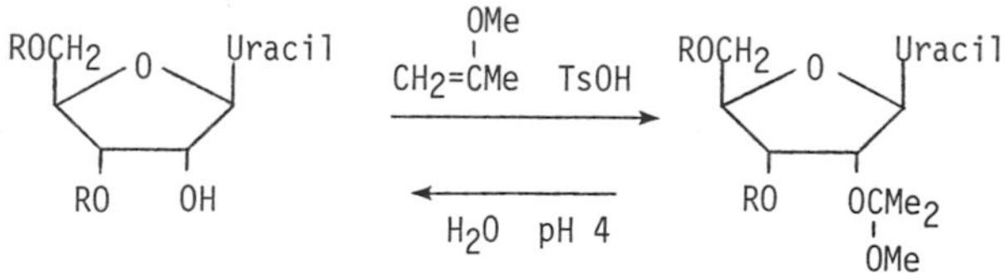

JACS (1967) 89 3366
Tetrahedron (1970) 26 1023

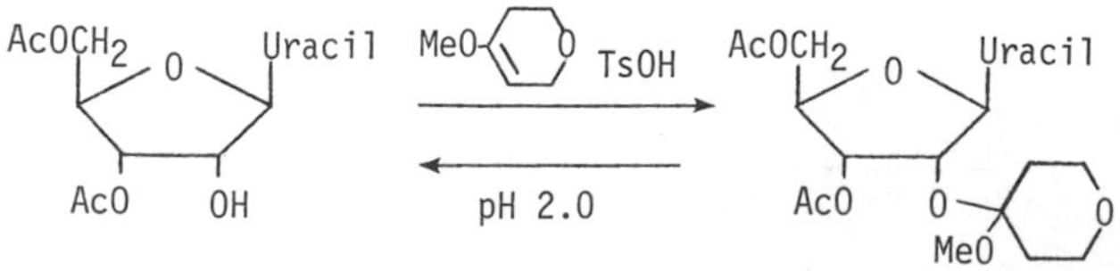

(Stable to RMgX,
LiAlH$_4$, CrO$_3$,
and base)

JACS (1967) 89 3366
Tetrahedron (1970) 26 1023

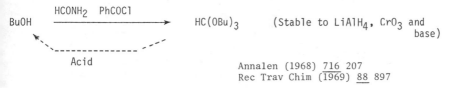

BuOH →[HCONH₂ PhCOCl] HC(OBu)₃   (Stable to LiAlH₄, CrO₃ and base)

Acid

Annalen (1968) 716 207
Rec Trav Chim (1969) 88 897

BuOH →[HC(SEt)₃ ZnCl₂] HC(OBu)₃

Acid

JACS (1948) 70 2268

HOCH₂—O—Uracil →[HC(OCH₂CH₂Cl)₃] [HOAc H₂O] CH₂OCH(OCH₂CH₂Cl)₂—O—Uracil

PhCOO   OCOPh   PhCOO   OCOPh

Tetr Lett (1969) 4443

Digitonin EtOH H₂O ⇄ Pyr

(Stable to CrO₃)

Digitonin

JACS (1960) 82 1257

Chapter 4   PREPARATION OF ALDEHYDES

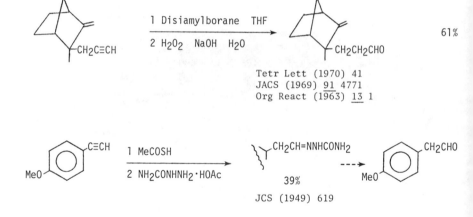

61%

Tetr Lett (1970) 41
JACS (1969) <u>91</u> 4771
Org React (1963) <u>13</u> 1

39%

JCS (1949) 619

Section 47   <u>Aldehydes from Carboxylic Acids, Acid Halides and Anhydrides</u>

Review: The Synthesis of Aldehydes from Carboxylic Acids

Org React (1954) <u>8</u> 218

PrCOOH

$$\xrightarrow[\begin{array}{c}\text{electrolysis} \\ \text{2 HCl} \quad \text{H}_2\text{O}\end{array}]{1 \quad \left[\begin{array}{c}\text{O} \\ \text{O}\end{array}\right]\text{CH(CH}_2)_2\text{COOH} \quad \text{KOH} \quad \text{MeOH}} \quad \text{Pr(CH}_2)_2\text{CHO}$$

Bull Chem Soc Jap (1965) <u>38</u> 922

Annalen (1961) <u>642</u> 121

55%

$$\text{PhCH}_2\text{COCl} \xrightarrow[\begin{array}{c}\text{2 EtSH} \quad \text{h}\nu \quad \text{C}_6\text{H}_6\end{array}]{1 \quad \text{CH}_2\text{N}_2 \quad \text{Et}_2\text{O}} \text{PhCH}_2\text{CH}_2\text{COSEt} \xrightarrow[\begin{array}{c}\text{2 HCl} \quad \text{H}_2\text{O} \quad \text{Et}_2\text{O}\end{array}]{1 \quad \text{Ni} \quad (\text{PhNHCH}_2)_2} \text{PhCH}_2\text{CH}_2\text{CHO}$$

Ber (1959) <u>92</u> 528
Org React (1954) <u>8</u> 218

$$\text{C}_6\text{H}_{13}\text{COOH} \xrightarrow[\begin{array}{c}\text{2 HCOOEt}\end{array}]{1 \quad (\text{i-Pr})_2\text{NLi} \quad \text{HMPA} \quad \text{THF}} \text{C}_6\text{H}_{13}\text{CHO} \qquad\qquad 65\%$$

Tetr Lett (1970) 699

$$\text{BuCOOH} \xrightarrow{\text{Li} \quad \text{MeNH}_2} \text{BuCHO} \qquad\qquad 66\%$$

JACS (1970) <u>92</u> 5774

36%

JCS (1943) 84
JOC (1963) <u>28</u> 3029

$C_8H_{17}COOH$ $\xrightarrow[\text{Na}_2\text{CO}_3 \quad \text{H}_2\text{O}]{\text{Na-Hg} \quad \text{H}_3\text{BO}_3 \quad \text{NaHSO}_3}$ $C_8H_{17}CHO$ 36%

J Soc Chem Ind (1943) 62 128

$Ph(CH_2)_2COOH$

1 $Me_2\overset{\overset{\displaystyle NH_2}{|}}{C}CH_2OH$

2 MeI   MeNO_2

3 NaBH_4   MeOH

4 NaOH   H_2O

$\longrightarrow$

$\xrightarrow[\text{H}_2\text{O}]{\text{HCl}}$ $Ph(CH_2)_2CHO$

J Heterocyclic Chem (1966) 3 531

1 N,N-Carbonyl diimidazole

THF

2 LiAl(OBu-t)_3H   THF

JOC (1970) 35 458
Annalen (1962) 654 119

1 ClCOOEt   NEt_3

C_6H_6

2 EtOMgCH(COOEt)_2
EtOH

$\xrightarrow[\text{MeOH}]{\text{NaBH}_4}$

79%

Ber (1965) 98 3040

1 ClCOOEt   NEt_3

2 NaSH

3 Ni   (PhNHCH_2)_2

HOAc   THF

$\xrightarrow[\text{CH}_2\text{Cl}_2]{\text{TsOH}}$

<47%

JOC (1966) 31 1922

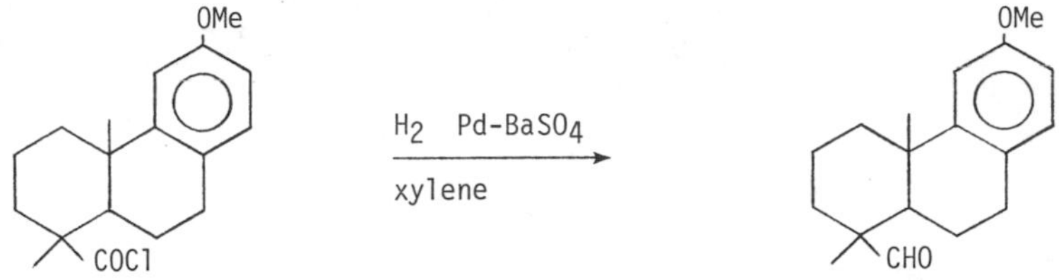

$$\xrightarrow[\text{xylene}]{\text{H}_2 \quad \text{Pd-BaSO}_4}$$

80%

JACS (1942) <u>64</u> 928
(1940) <u>62</u> 49
Org React (1948) <u>4</u> 362

PhCOCl $\xrightarrow{\text{Bu}_3\text{SnH}}$ PhCHO

JOC (1960) <u>25</u> 284

PhCOBr $\xrightarrow{\text{h}\nu \quad \text{Et}_2\text{O}}$ PhCHO   80%

Angew (1965) <u>77</u> 169
(Internat Ed <u>4</u> 146)

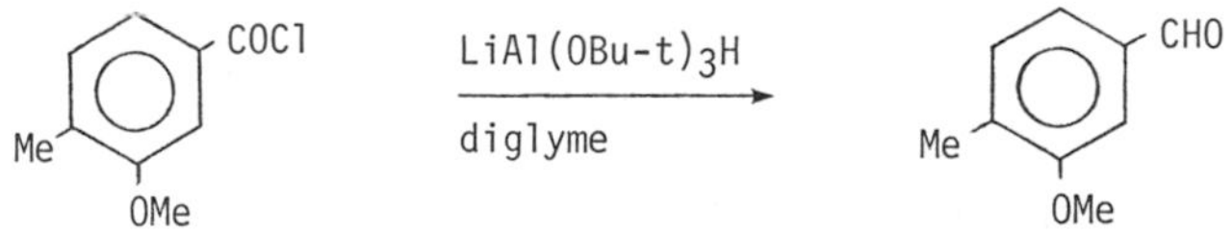

$$\xrightarrow[\text{diglyme}]{\text{LiAl(OBu-t)}_3\text{H}}$$

62%

J Med Chem (1970) <u>13</u> 26
JACS (1956) <u>78</u> 252

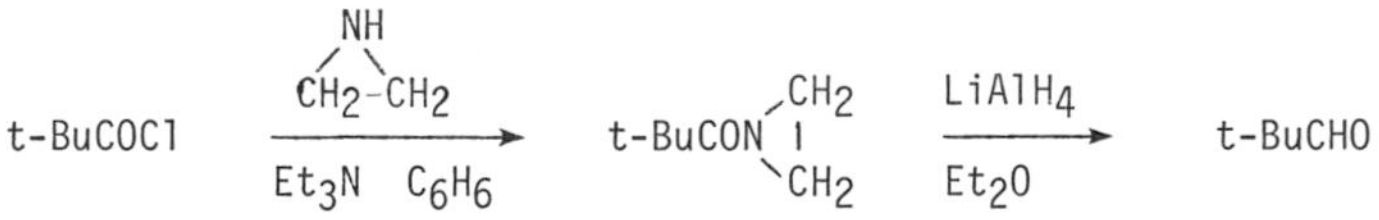

$$\text{t-BuCOCl} \xrightarrow[\text{Et}_3\text{N} \quad \text{C}_6\text{H}_6]{\overset{\text{NH}}{\underset{}{\text{CH}_2-\text{CH}_2}}} \text{t-BuCON}\overset{\text{CH}_2}{\underset{\text{CH}_2}{|}} \xrightarrow[\text{Et}_2\text{O}]{\text{LiAlH}_4} \text{t-BuCHO}$$

JACS (1961) <u>83</u> 4549

$$C_{15}H_{31}COCl \xrightarrow[\text{2 HOAc}]{\text{1 CH}_2\text{N}_2 \quad \text{Et}_2\text{O}} C_{15}H_{31}COCH_2OAc \xrightarrow[\substack{\text{i-PrOH} \\ \text{2 Pb(OAc)}_4 \\ \text{HOAc}}]{\text{1 Al(OPr-i)}_3} C_{15}H_{31}CHO$$

56%

Org React (1954) $\underline{8}$ 218

$$PhCOCl \xrightarrow{\text{EtSH} \quad \text{Pyr}} PhCOSEt \xrightarrow[\text{H}_2\text{O}]{\text{Ni} \quad \text{EtOH}} PhCHO$$

< 62%

JACS (1946) $\underline{68}$ 1455
Tetr Lett (1964) 1763
Org React (1954) $\underline{8}$ 218

$$PhCOCl \xrightarrow{\text{N}_2\text{H}_4} PhCONHNH_2 \xrightarrow[\substack{\text{2 Na}_2\text{CO}_3 \\ \text{ethylene glycol}}]{\text{1 PhSO}_2\text{Cl} \quad \text{Pyr}} PhCHO$$

~60%

Org React (1954) $\underline{8}$ 218
JOC (1961) $\underline{26}$ 3664

$$BuCOCl \xrightarrow[\text{2 H}_2\text{SO}_4 \quad \text{H}_2\text{O}]{\text{1 Quinoline} \quad \text{HCN} \quad \text{C}_6\text{H}_6} BuCHO$$

62%

JACS (1941) $\underline{63}$ 2021
Org React (1954) $\underline{8}$ 218

$$\xrightarrow[\text{chlorobenzene}]{\text{Pyridine N-oxide}}$$

69%

Tetr Lett (1965) 233

$$C_6H_{13}CH_2COCl \xrightarrow[\substack{\text{C}_6\text{H}_6 \\ \text{2 HCl}}]{\text{1 BuSH} \quad \text{ZnCl}_2} C_6H_{13}CH=C(SBu)_2 \xrightarrow[\text{2 N}_2\text{H}_4 \quad \text{MeOH}]{\text{1 Perphthalic acid} \quad \text{Et}_2\text{O}}$$

$$C_6H_{13}CH=NNH_2 \dashrightarrow C_6H_{13}CHO$$

Rec Trav Chim (1959) $\underline{78}$ 354

MeCHCH$_2$CH$_2$COOH

OH
|
MeCHCH$_2$CHCOOH   MeCHCH$_2$CHO

$\xrightarrow{\text{1 Bromination}}$
2 Hydrolysis

$\xrightarrow{\text{NaIO}_4}$

Tetr Lett (1968) 1725

Aldehydes may also be prepared from carboxylic acids via amide or ester
intermediates. See section 51 (Aldehydes from Amides) and section 53
(Aldehydes from Esters)

Section 48   Aldehydes from Alcohols

BuOH   ---->   BuOCH$_2$CH=CH$_2$ $\xrightarrow{\text{PrLi}}$ Bu(CH$_2$)$_2$CHO   29%

Tetr Lett (1969) 821

OH
|
i-PrCH$_2$CCH=CH$_2$ $\xrightarrow[\text{H}_3\text{PO}_4]{\text{C}_5\text{H}_{11}\text{CH}_2\text{CH(OEt)}_2}$ i-PrCH$_2$C=CHCH$_2$CHCHO   60%
| | |
Me   Me   C$_5$H$_{11}$

Helv (1967) 50 2095
Tetr Lett (1961) 493

OH
|
Me$_2$CCH=CH$_2$ $\xrightarrow[\text{H}_3\text{PO}_4]{\text{CH}_2\text{=CHOEt}}$ Me$_2$C=CH(CH$_2$)$_2$CHO   81%

Helv (1967) 50 2095
JACS (1961) 83 198

C$_7$H$_{15}$CH$_2$OH $\xrightarrow{\text{O}_2 \quad \text{Pt} \quad \text{EtOAc}}$ C$_7$H$_{15}$CHO   21%

JACS (1955) 77 190
Angew (1957) 69 600
For application to allylic alcohols see Helv (1957) 40 265

$$BuCH_2OH \xrightarrow{\text{Li} \quad NH_2CH_2CH_2NH_2} BuCHO \qquad\qquad 66\%$$

Applicable to benzylic alcohols

JOC (1962) 27 2662

Benzoquinone   Al(OBu-t)$_3$

Chem Comm (1969) 799

$$C_6H_{13}CH_2OH \xrightarrow{\text{CrO}_3\text{-Pyr} \quad CH_2Cl_2} C_6H_{13}CHO \qquad\qquad 93\%$$

Applicable to allylic and benzylic alcohols

Tetr Lett (1968) 3363
JOC (1961) 26 4814
(1970) 35 4000

CrO$_3$ H$_2$SO$_4$

DMF   H$_2$O

80%

Helv (1968) 51 772

$$PhCH=CHCH_2OH \xrightarrow[\text{Pyr·HCl} \quad Pyr]{\text{Na}_2Cr_2O_7} PhCH=CHCHO \qquad\qquad 80\%$$

Chem Ind (1969) 1594

$$EtCH_2OH \xrightarrow{\text{K}_2Cr_2O_7 \quad H_2SO_4 \quad H_2O} EtCHO \qquad\qquad 45\text{-}49\%$$

Org Synth (1943) Coll Vol 2 541
For application to allylic alcohols see Bull Soc Chim Fr (1933) 53 301
and Ber (1947) 80 137

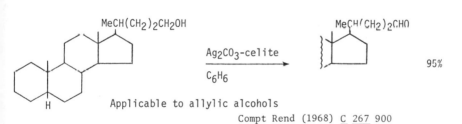

$$\text{cyclohexyl-}CH_2OH \xrightarrow{\text{t-Butyl chromate } C_6H_6} \text{cyclohexyl-}CHO \quad 40\%$$

Applicable to allylic and benzylic alcohols

Bull Chem Soc Jap (1965) <u>38</u> 1503 893 1141

$$PhCH{=}CHCH_2OH \xrightarrow[C_6H_6]{\text{Nickel peroxide}} PhCH{=}CHCHO \qquad 86\%$$

JOC (1962) <u>27</u> 1597
JCS (19?9) <u>C</u> 2173

$$Br{-}C_6H_4{-}CH_2OH \xrightarrow[\text{HOAc } H_2O]{(NH_4)_2Ce(NO_3)_6} Br{-}C_6H_4{-}CHO \qquad 93\%$$

JCS (1965) 5777
JOC (1967) <u>32</u> 2349   3865

MeCH(CH$_2$)$_2$CH$_2$OH

$$\xrightarrow[C_6H_6]{Ag_2CO_3\text{-celite}}$$

MeCH(CH$_2$)$_2$CHO   95%

Applicable to allylic alcohols

Compt Rend (1968) <u>C</u> <u>267</u> 900

$$i\text{-}PrCH_2OH \xrightarrow{AgO \quad H_3PO_4 \quad HOAc} i\text{-}PrCHO \qquad 30\%$$

Applicable to allylic alcohols

Tetr Lett (1967) 4193

$$BuCH_2OH \xrightarrow[Me_2SO \quad H_2O]{\text{Argentic picolinate}} BuCHO \qquad 74\%$$

Applicable to benzylic alcohols

Can J Chem (1969) <u>47</u> 1649

$$\xrightarrow{\text{MnO}_2 \quad \text{MeCN}}$$

76%

Proc Chem Soc (1964) 110
Chem Comm (1966) 121

$$\text{PhCH=CHCH}_2\text{OH} \xrightarrow{\text{MnO}_2 \quad \text{CCl}_4} \text{PhCH=CHCHO}$$

JCS (1952) 1094
JACS (1955) 77 4399

$$\text{PrCH}_2\text{OH} \xrightarrow{\text{Pb(OAc)}_4 \quad \text{Pyr}} \text{PrCHO}$$

70%

Applicable to allylic and benzylic alcohols

Tetr Lett (1964) 3071

$$\text{C}_5\text{H}_{11}\text{CH}_2\text{OH} \xrightarrow{\text{INO}_3\text{-Pyr} \quad \text{CHCl}_3} \text{C}_5\text{H}_{11}\text{CHO}$$

20%

Applicable to benzylic alcohols

JCS C (1970) 676

$$\xrightarrow{\text{N}_2\text{O}_4 \quad \text{CCl}_4}$$

95%

JCS (1955) 1110

$$\xrightarrow[\text{Me}_2\text{SO} \quad \text{Pyr} \quad \text{C}_6\text{H}_6]{\text{DCC} \quad \text{CF}_3\text{COOH}}$$

85%

Applicable to benzylic alcohols

JACS (1965) 87 5670
Chem Rev (1967) 67 247

$$\xrightarrow[\text{Et}_3\text{N}]{\text{SO}_3 \cdot \text{Pyr} \quad \text{Me}_2\text{SO}}$$

70%

JACS (1967) 89 5505

$$\text{C}_7\text{H}_{15}\text{CH}_2\text{OH} \quad ----\rightarrow \quad \text{C}_7\text{H}_{15}\text{CH}_2\text{OTs} \xrightarrow{\text{NaHCO}_3 \quad \text{Me}_2\text{SO}} \text{C}_7\text{H}_{15}\text{CHO} \qquad \sim 78\%$$

Applicable to benzylic alcohols

JACS (1959) 81 4113

$$\text{C}_5\text{H}_{11}\text{CH}_2\text{OH} \xrightarrow{\text{COCl}_2 \quad \text{Et}_2\text{O}} \text{C}_5\text{H}_{11}\text{CH}_2\text{OCOCl} \xrightarrow[\text{2 Et}_3\text{N}]{\text{1 Me}_2\text{SO}} \text{C}_5\text{H}_{11}\text{CHO} \qquad 68\%$$

JCS (1964) 1855

$$\text{PhCH}_2\text{OH} \xrightarrow{\text{KOCl} \quad \text{MeOH} \quad \text{H}_2\text{O}} \text{PhCHO} \qquad 77\%$$

JOC (1961) 26 1046

71%

Bull Soc Chim Belg (1951) 60 54
Compt Rend (1954) 238 2538
Chem Rev (1963) 63 21

$$\text{i-PrCH}_2\text{OH} \xrightarrow{\text{K}_2\text{S}_2\text{O}_8 \quad \text{AgNO}_3 \quad \text{H}_2\text{O}} \text{i-PrCHO} \qquad 31\%$$

JACS (1954) 76 6345
For application to benzylic alcohols see JCS (1960) 1332

$$C_{11}H_{23}CH_2OH \xrightarrow[\text{}]{\substack{NCOOEt \\ \| \\ NCOOEt \quad toluene}} C_{11}H_{23}CHO \qquad 20\%$$

Applicable to benzylic alcohols

JOC (1967) <u>32</u> 727

$$PhCH_2OH \xrightarrow[C_6H_6]{\text{4-Phenyl-1,2,4-triazoline-3,5-dione}} PhCHO \qquad 78\%$$

Chem Comm (1966) 744
JCS <u>C</u> (1969) 1474

$$\xrightarrow[\text{dicyanoquinone (DDQ)}]{\text{2,3-Dichloro-5,6-}}$$

Applicable only to allylic and benzylic alcohols

J Med Chem (1962) <u>5</u> 409
Chem Rev (1967) <u>67</u> 153

$$PhCH=CHCH_2OH \xrightarrow[CCl_4]{\text{Tetrachloro-1,2-benzoquinone}} PhCH=CHCHO \qquad 100\%$$

Applicable only to allylic and benzylic alcohols

JCS (1956) 3070

$$\xrightarrow[\substack{2 \; MeONa \\ 3 \; NaN_3 \quad DMF}]{1 \; MsCl \quad DMF}$$

$$\xrightarrow[\substack{2 \; h\nu \quad MeOH}]{1 \; Acetylation}$$

Carbohydrate Res (1968) <u>8</u> 366

$$\substack{(CH_2)_7COOH \\ | \\ CHOH \\ | \\ C_8H_{17}CHOH} \xrightarrow[\substack{EtOH \quad H_2O}]{KIO_4 \quad H_2SO_4} \substack{(CH_2)_7COOH \\ | \\ CHO} \qquad 76\%$$

Org React (1944) <u>2</u> 341

$BuOCH_2CHCH_2OH$ $\xrightarrow{\quad Pb(OAc)_4 \quad C_6H_6 \quad}$ $BuOCH_2CHO$ 57%
      |
      HO

JACS (1945) 67 39
JOC (1970) 35 249

$Ph(CH_2)_3CHCH(CH_2)_3Ph$ $\xrightarrow[CH_2Cl_2]{I_2 \quad HgO}$ $Ph(CH_2)_3CHO$ 99%
       | |
       HO OH

JCS C (1969) 383

$PhCHCH_2OH$ $\xrightarrow{\quad K_2S_2O_8 \quad AgNO_3 \quad H_2O \quad}$ $PhCHO$ 61%
  |
  OH

JACS (1954) 76 6345

$\xrightarrow{\quad O_2 \quad Co(OAc)_2 \quad PhCN \quad}$

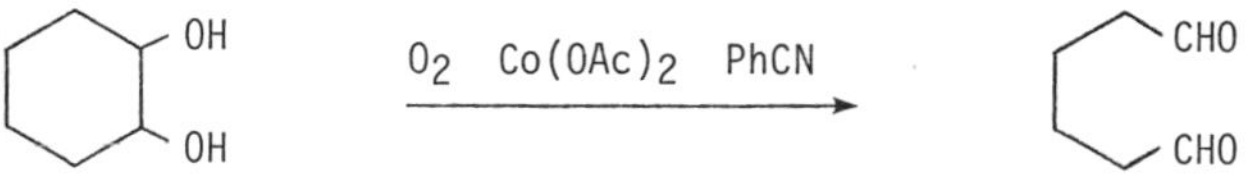

Tetr Lett (1968) 5689

$MeCHCHMe$ $\xrightarrow{\quad XeO_3 \quad H_2O \quad}$ $MeCHO$
  | |
  HO OH

JACS (1964) 86 2078

Section 49   Aldehydes from Aldehydes
              °°°°°°°°°°°°°°°°°°°°°°°°°°°°

PhCHO
$$\xrightarrow[\text{2 } HClO_4 \quad H_2O \quad Et_2O]{\text{1 } Ph_3P=CHO-\bigcirc-Me}$$
PhCH$_2$CHO   55%

Ber (1962) 95 2514

Tetr Lett (1969) 3665   36%

Carbohydrate Res (1969) 10 184   28%

$C_6H_{13}CHO$
$$\xrightarrow[\substack{Mg-Hg \quad Et_2O \\ \text{2 } EtONa}]{\text{1 } Cl_2CHCOOEt}$$
$C_6H_{13}CHCHCOOEt$
$$\xrightarrow[\text{2 } \triangle]{\text{1 } NaOH}$$
$C_6H_{13}CH_2CHO$

Compt Rend (1937) 204 272
Org React (1949) 5 413

$C_6H_{13}CHO$
$$\xrightarrow[HgCl_2 \quad THF]{ClCH_2OEt \quad Mg}$$
$C_6H_{13}CHCH_2OEt$
$$\xrightarrow{HCOOH}$$
$C_6H_{13}CH_2CHO$ 10%

Bull Soc Chim Fr (1959) 459

$C_5H_{11}CH_2CHO$ $\xrightarrow[\begin{array}{c}\text{2 EtMgBr}\\\text{3 BuI}\end{array}]{\text{1 t-BuNH}_2}$ $C_5H_{11}\underset{\underset{Bu}{|}}{C}HCH=NBu\text{-}t$ $\xrightarrow[\text{H}_2\text{O}]{\text{Acid}}$ $C_5H_{11}\underset{\underset{Bu}{|}}{C}HCHO$ 50%

JACS (1963) __85__ 2178

$PrCH_2CHO$ $\xrightarrow{\text{BuNHCH}_2\text{Pr-i}}$ $PrCH=CH\underset{\underset{Bu}{|}}{N}CH_2Pr\text{-}i$ $\xrightarrow[\text{2 Acid H}_2\text{O}]{\text{1 EtI}}$ $Pr\underset{\underset{Et}{|}}{C}HCHO$ 78%

Chem Comm (1967) 510

$\underset{\underset{(CH_2)_2COOMe}{|}}{Et}CHCHO$ $\xrightarrow{\text{Pyrrolidine}}$ $Et\underset{\underset{(CH_2)_2COOMe}{|}}{C}=CHN\begin{array}{c}\diagup\\\diagdown\end{array}$ $\xrightarrow[\text{2 Hydrolysis}]{\text{1 CH}_2=\text{CHCH}_2\text{Br}}$ $\overset{CH_2CH=CH_2}{\underset{\underset{(CH_2)_2COOMe}{|}}{Et\overset{|}{C}CHO}}$

Chem Comm (1965) 197
JACS (1963) __85__ 207
Angew (1960) __72__ 169

$PhCH_2CHO$ $\xrightarrow[\begin{array}{c}\text{2 KNH}_2\\\text{PhCH}_2\text{Cl}\quad\text{NH}_3\end{array}]{\text{1 PhNHNH}_2}$ $Ph\underset{\underset{CH_2Ph}{|}}{C}HCH=NNHPh$ $----\rightarrow$ $Ph\underset{\underset{CH_2Ph}{|}}{C}HCHO$

JACS (1967) __89__ 463

$Ph\underset{\underset{Me}{|}}{C}HCHO$ $\xrightarrow[\text{2 CN}^-]{\text{1 Br}_2}$ $PhC\overset{\overset{O}{\diagup\diagdown}}{-}\underset{\underset{Me}{|}}{C}HCN$ $\xrightarrow[\text{2 }\triangle]{\text{1 AlMe}_3}$ $Ph\underset{\underset{Me}{|}}{\overset{\overset{Me}{|}}{C}}CHO$

Tetr Lett (1970) 2947

$MeCH(CH_2)_2CHO$ $\xrightarrow[\begin{array}{c}\text{HOAc}\\\text{2 KOH}\quad\text{EtOH}\end{array}]{\text{1 IBr}\quad\text{HBr}}$ $Me\underset{\underset{\sim\sim\sim}{}}{\overset{\overset{HO\ OH}{|\ \ |}}{CHCH_2CHCHOEt}}$ $\xrightarrow[\begin{array}{c}\text{HOAc}\\\text{Me}_2\text{CO}\end{array}]{\text{NaIO}_4}$ $Me\underset{\underset{\sim\sim\sim}{}}{C}HCH_2CHO$

Chem Comm (1968) 851
Tetr Lett (1970) 5229

MeCH₂CHO ⟍
                    O₂ Cu(OAc)₂ Pyr Et₃N
                    ⟶   MeCHO
Me(CH₂)₅CHO ⟋

Rec Trav Chim (1966) **85** 437

Some of the methods included in section 57 (Aldehydes from Ketones) may
also be applied to the preparation of aldehydes from aldehydes

Section 50   Aldehydes from Alkyls

Reactions in which methyl groups are converted into aldehyde (RMe → RCHO)
are included in this section. For reactions in which a hydrogen is
replaced by aldehyde, RH → RCHO (R=alkyl, vinyl or aryl), see section 56
(Aldehydes from Hydrides)

$$\text{MeO-C}_6\text{H}_3(\text{CH}_3)(\text{NO}_2) \xrightarrow[\text{2 H}_2\text{O}]{1 \; CrO_2Cl_2 \;\; CS_2} \text{MeO-C}_6\text{H}_3(\text{CHO})(\text{NO}_2)$$

50%

JCS (1949) S230

$$\text{cyclohexyl-CH}_3 \xrightarrow[\text{2 H}_2\text{O}]{\substack{1 \; CrO_2Cl_2 \;\; Me_2C=CHMe \\ CCl_4}} \text{cyclohexyl-CHO}$$

25%

JACS (1951) **73** 221

$$\xrightarrow[\text{H}_2\text{SO}_4 \;\; HOAc]{CrO_3 \;\; Ac_2O} \quad \xrightarrow[\substack{EtOH \\ H_2O}]{H_2SO_4}$$

AcO...CH₃, COOEt, OAc → AcO...CH(OAc)₂, COOEt, OAc → AcO...CHO, COOEt, OAc

JOC (1968) **33** 4176
Org Synth (1943) Coll Vol 2 441

$$\text{PhCH}_3 \xrightarrow{Na_2S_2O_8 \;\; AgNO_3 \;\; H_2O} \text{PhCHO}$$

50%

JCS (1960) 1332

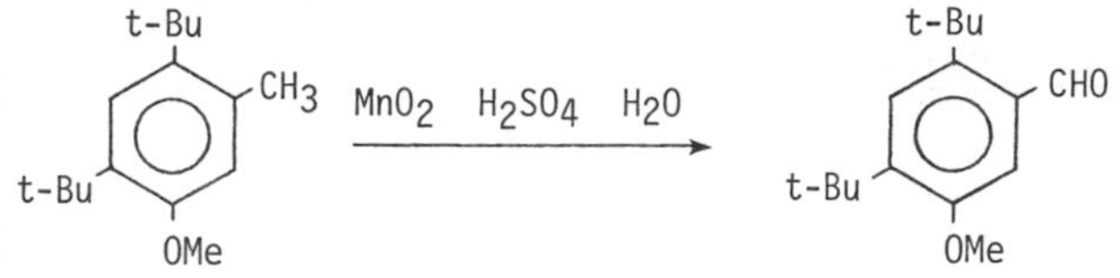

JOC (1951) 16 586

23%

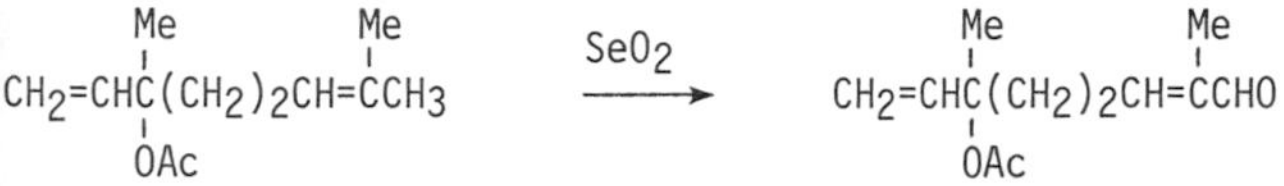

Chem Comm (1968) 1277
Org React (1949) 5 331

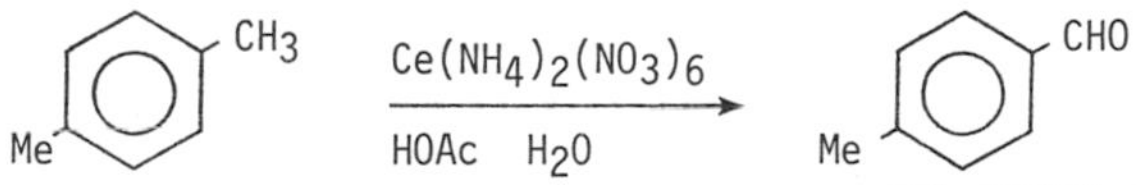

JOC (1966) 31 2033
JCS C (1970) 1630

73%

PhCH₃ →[Argentic picolinate Me₂SO] PhCHO

Tetr Lett (1967) 415

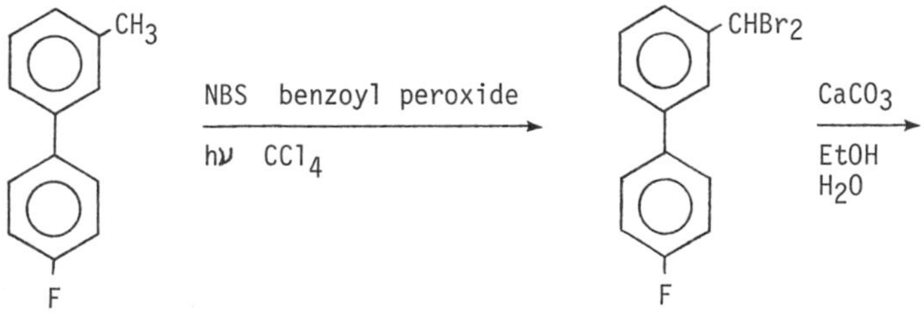

JACS (1966) 88 3318
      (1956) 78 1689
Org React (1954) 8 197

41%

Section 51   Aldehydes from Amides

$C_{15}H_{31}CONMe_2$ $\xrightarrow{\text{NaAlH}_4 \quad \text{THF}}$ $C_{15}H_{31}CHO$ 78%

Tetrahedron (1969) 25 5555

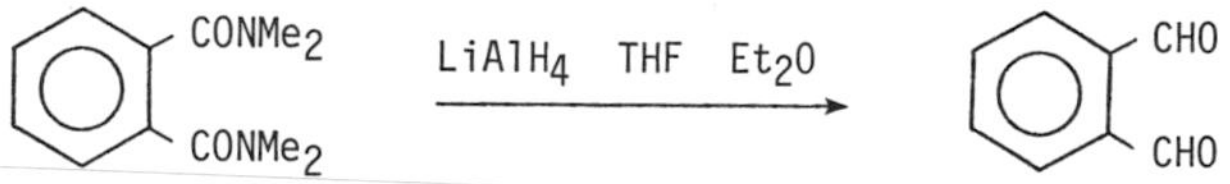

$\xrightarrow{\text{LiAlH}_4 \quad \text{THF} \quad \text{Et}_2\text{O}}$ ~70%

Ber (1951) 84 625

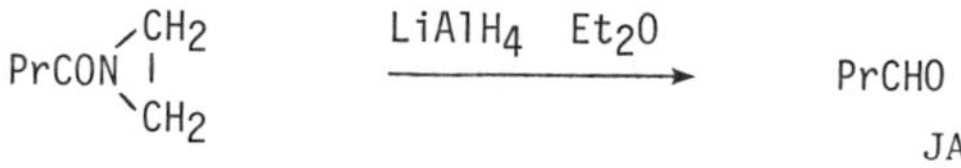

$\xrightarrow{\text{LiAlH}_4 \quad \text{Et}_2\text{O}}$ PrCHO   88%

JACS (1961) 83 4549

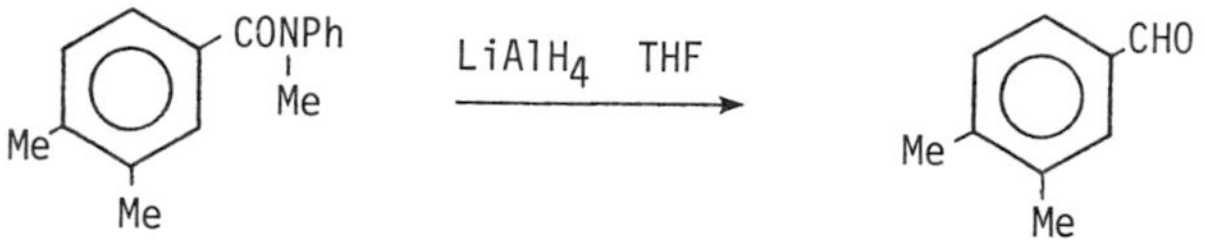

$\xrightarrow{\text{LiAlH}_4 \quad \text{THF}}$ 57%

Ber (1955) 88 301
Org React (1954) 8 218

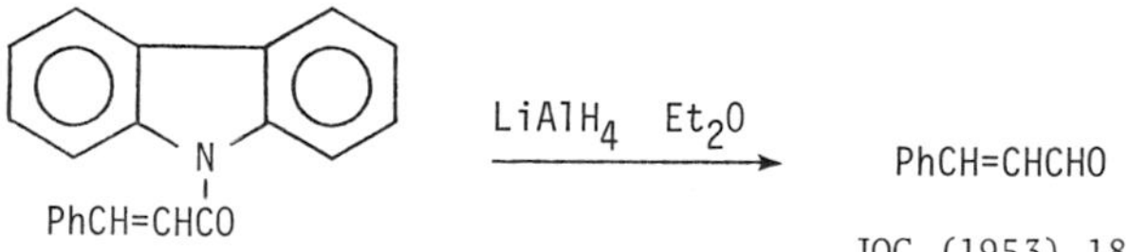

$\xrightarrow{\text{LiAlH}_4 \quad \text{Et}_2\text{O}}$ PhCH=CHCHO   45%

JOC (1953) 18 1190

$BuCONMe_2$ $\xrightarrow{\text{LiAl(OEt)}_3\text{H} \quad \text{Et}_2\text{O}}$ $BuCHO$ 92%

JACS (1964) 86 1089

C$_7$H$_{15}$CONPh $\xrightarrow{\text{(i-Bu)}_2\text{AlH Et}_2\text{O}}$ C$_7$H$_{15}$CHO   56%
        |
        Me
                                        Izv (1959) 2146
                                        (Chem Abs 54 10932)

C$_5$H$_{11}$CONHPh $\xrightarrow[\text{2 HOAc}]{\text{1 Na NH}_3}$ C$_5$H$_{11}$CHO   53%

                                        Aust J Chem (1955) 8 512

C$_5$H$_{11}$CONHMe $\xrightarrow[\text{MeNH}_2 \text{ EtOH}]{\text{Electrolysis LiCl}}$ C$_5$H$_{11}$CHO   50%

                                        JOC (1970) 35 1210

PhCONH$_2$ $\xrightarrow{\text{P}_2\text{S}_5 \text{ Pyr}}$ PhCSNH$_2$ $\xrightarrow{\text{Ni EtOH}}$ PhCHO

                                        JACS (1952) 74 3936
                                        Org React (1962) 12 356

PhCH=CHCONHPh $\xrightarrow[\text{toluene}]{\text{PCl}_5}$ PhCH=CHC=NPh $\xrightarrow[\text{Et}_2\text{O}]{\text{SnCl}_2 \text{ HCl}}$ PhCH=CHCHO   92%
                                       |
                                       Cl
                                        Org React (1954) 8 218

CONHNH$_2$ $\xrightarrow[\substack{\text{2 Na}_2\text{CO}_3 \\ \text{ethylene glycol}}]{\text{1 TsCl Pyr}}$ CHO   60%

                                        JOC (1961) 26 3664
                                        Org React (1954) 8 218

CONHNH$_2$ $\xrightarrow{\text{NaIO}_4 \text{ NH}_3 \text{ H}_2\text{O}}$ CHO   60-70%

                                        JACS (1952) 74 5796

PhCH$_2$NHAc $\xrightarrow{\text{Argentic picolinate Me}_2\text{SO}}$ PhCHO

Tetr Lett (1967) 415

FtNHAc $\xrightarrow{\text{K}_2\text{S}_2\text{O}_8 \quad \text{K}_2\text{HPO}_4 \quad \text{H}_2\text{O}}$ MeCHO   48%

JOC (1964) $\underline{29}$ 3632

Section 52   Aldehydes from Amines

1 NaNO$_2$ NaOAc HCl H$_2$O

2 HCH=NOH NaOAc CuSO$_4$
  Na$_2$SO$_3$

3 HCl H$_2$O   41%

JCS (1954) 1297

PhCH$_2$NH$_2$ $\xrightarrow{\text{1 2,4-Dinitrobenzaldehyde Pyr}}_{\text{2 HCl H}_2\text{O}}$ PhCHO   63%

JCS (1954) 209

C$_{11}$H$_{23}$CH$_2$NH$_2$

1 3,5-Dinitromesitylglyoxal C$_6$H$_6$

2 Diazabicyclononene Me$_2$SO THF
3 (COOH)$_2$ MeOH THF   C$_{11}$H$_{23}$CHO   34%

JACS (1969) $\underline{91}$ 1429

C$_5$H$_{11}$CH$_2$NH$_2$ $\xrightarrow{\text{Hexamine HCHO}}_{\text{HOAc H}_2\text{O}}$ C$_5$H$_{11}$CHO   42%

JCS (1953) 1737
Org React (1954) $\underline{8}$ 197

Angew (1965) 77 811
(Internat Ed 4 787)

60%

PhCH$_2$NEt$_2$
$\xrightarrow[\text{HOAc \quad H}_2\text{O}]{\text{NaNO}_2 \quad \text{NaOAc}}$
PhCHO

JACS (1967) 89 1147

MeCH$_2$NH$_2$
$\xrightarrow{\text{OF}_2 \quad \text{freon 11}}$
MeCH=NOH   --→   MeCHO

44%

JACS (1964) 86 1392

(PrCH$_2$)$_3$N
$\xrightarrow{\text{O}_3 \quad \text{H}_2\text{O}}$
PrCHO

47%

JCS (1964) 711

(PrCH$_2$)$_3$N
$\xrightarrow{\text{Br}_2 \quad \text{H}_2\text{O} \quad \text{pH 5}}$
PrCHO

84%

JACS (1968) 90 3502

(EtCH$_2$)$_3$N
$\xrightarrow{\text{NBS \quad dioxane \quad H}_2\text{O}}$
EtCHO

68%

JCS (1957) 4905
Chem Rev (1963) 63 21

PhCH$_2$CH$_2$NH$_2$
$\xrightarrow[\text{NaHCO}_3 \quad \text{Et}_2\text{O}]{\text{1 t-Butyl hypochlorite}}$
PhCH$_2$CHO

39%

2 EtONa EtOH
3 H$_2$SO$_4$ H$_2$O

JACS (1954) 76 5554

$PrCH_2NH_2$ $\xrightarrow[\text{H}_2\text{O} \quad \text{EtOH}]{\text{H}_2\text{O}_2 \quad \text{Na}_2\text{WO}_4}$ $PrCH=NOH$ ----> $PrCHO$

57%

Ber (1960) <u>93</u> 132

$PrCH_2NH_2$
$(PrCH_2)_2NH$ $\xrightarrow[\text{t-BuOH} \quad \text{H}_2\text{O}]{\text{KMnO}_4 \quad \text{MSO}_4}$ $PrCHO$
$(PrCH_2)_3N$

(M=Ca or Zn)

JOC (1967) <u>32</u> 3129

$PhCH_2NH_2$ $\xrightarrow{\text{MnO}_2}$ $PhCHO$ 34%

JACS (1955) <u>77</u> 4399

$PrCH_2NH_2$ $\xrightarrow{\text{Argentic picolinate} \quad \text{H}_2\text{O}}$ $PrCHO$ 14%

JCS (1965) 4962
Tetr Lett (1967) 415

Aldehydes may also be prepared by oxidation of N-acyl amines. See
section 51 (Aldehydes from Amides)

Section 53   Aldehydes from Esters
oooooooooooooooooooooooo

$PrCOOMe$ $\xrightarrow{\text{NaAlH}_4 \quad \text{THF}}$ $PrCHO$ 81%

Tetr Lett (1963) 2087

JOC (1966) <u>31</u> 283   70%

i-PrCOOEt $\xrightarrow[\text{hexane } Et_2O]{\text{(i-Bu)}_2AlH \quad \text{toluene}}$ i-PrCHO   80%

Tetr Lett (1962) 619
                        (1969) 1779

1 N_2H_4 H_2O
2 $PhSO_2Cl$ Pyr

Na_2CO_3
ethylene
glycol

87%

JCS (1936) 584
JOC (1961) 26 3664
Org React (1954) 8 218

1 $Me_2CCOOMe$ (Br)
Zn C_6H_6
2 KBH_4 MeOH

280°

87%

Bull Soc Chim Fr(1965) 1864

$Et(CH=CHCH_2)_3(CH_2)_6COOMe$ $\xrightarrow{\text{Na} \quad \text{xylene}}$ $Et(CH=CHCH_2)_3(CH_2)_6CHOH$
$Et(CH=CHCH_2)_3(CH_2)_6CO$

1 $LiAlH_4$
2 $Pb(OAc)_4$

$Et(CH=CHCH_2)_3(CH_2)_6CHO$ 68%

J Am Oil Chem Soc (1960) 37 425
(Chem Abs 54 23373)

$PrCH(COOEt)_2$ $\xrightarrow{N_2H_4}$ $PrCH(CONHNH_2)_2$ $\xrightarrow[\text{2 } H_2SO_4 \quad H_2O]{\text{1 } HNO_2 \quad H_2O}$ PrCHO   46%

Org React (1946) 3 337

Section 54   Aldehydes from Ethers and Epoxides

$(PrCH_2)_2O$ $\xrightarrow[\text{CuCl}]{\text{C}_6\text{H}_{13}\text{OH}\quad\text{t-butyl perbenzoate}}$ $PrCH(OC_6H_{13})_2$ ---→ PrCHO
                                                                              70%

Angew Chem (1961) 73 65
Act Chem Scand (1961) 15 249
Tetrahedron (1961) 13 241

$PrCH_2OBu$ $\xrightarrow[\text{}]{\overset{\overset{\text{NCOOEt}}{\|}}{\text{NCOOEt}}\quad h\nu}$ PrCHOBu $\xrightarrow{\text{H}_2\text{O}}$ PrCHO
                                              |
                                            NCOOEt
                                              |
                                            HNCOOEt

                                              60%   Chem Comm (1965) 259

$PrCH_2OBu$ $\xrightarrow{\text{Pb(OAc)}_4\quad h\nu}$ PrCHOBu $\xrightarrow{\text{Acid}}$ PrCHO
                                              |
                                             OAc   Annalen (1970) 735 47

$MeCH_2OEt$ $\xrightarrow[\text{2 Acid}]{\overset{\overset{\text{Ac}}{|}}{\text{1 PhNNO}}}$ MeCHO   50%

                                              JACS (1964) 86 3180

$MeOCH_2CH_2OMe$ $\xrightarrow[\text{NaOAc}\quad\text{H}_2\text{O}]{\text{p-Chlorobenzendiazonium chloride}}$ $MeOCH_2CHO$

                                              Angew (1958) 70 211

$MeCH_2OEt$ $\xrightarrow{\text{PhICl}_2\quad h\nu}$ MeCHOEt ----→ MeCHO
                                              |
                                             Cl

                                              79%
                                              Annalen (1969) 728 12

PhCH$_2$OMe $\xrightarrow{\text{Br}_2 \quad \text{HOAc} \quad \text{H}_2\text{O}}$ PhCHO   83%

JACS (1967) 89 3550

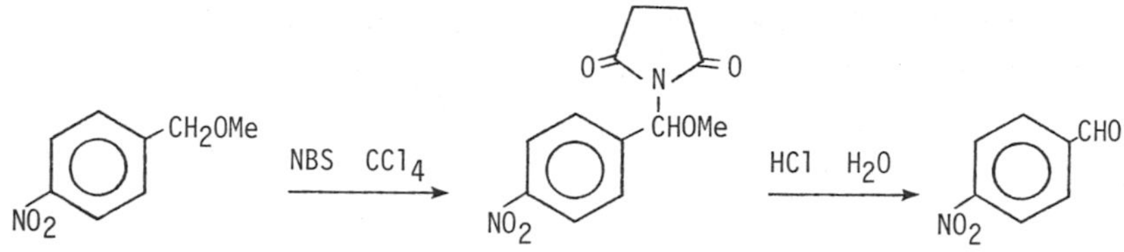

JOC (1958) 23 1490
Angew (1959) 71 349

PhCH=CHCH$_2$OMe $\xrightarrow{\text{SeO}_2 \quad \text{HOAc} \quad \text{dioxane}}$ PhCH=CHCHO   66%

Tetr Lett (1970) 2885

PrCH$_2$OMe $\xrightarrow[\text{2 Acid}]{\text{1 O}_2 \quad \text{Ph}_2\text{CO} \quad h\nu}$ PrCHO   55%

Ber (1963) 96 509

MeCH$_2$OCPh$_3$ $\xrightarrow{262°}$ MeCHO

JACS (1930) 52 753
      (1924) 46 2580

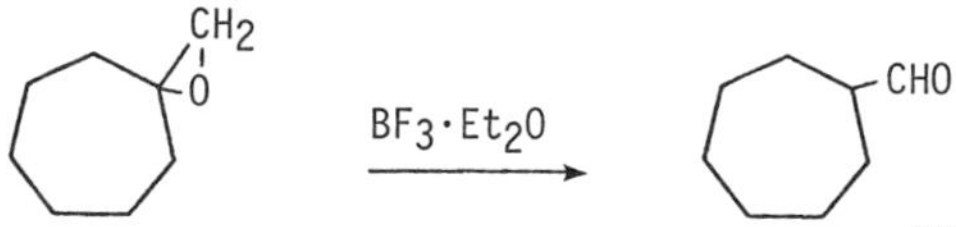

JACS (1962) 84 867
Chem Rev (1959) 59 737

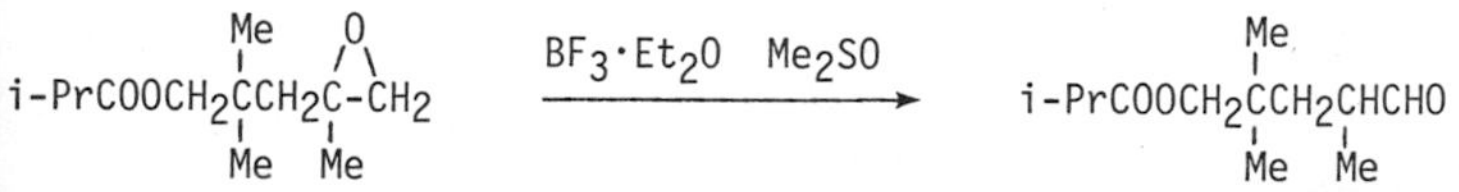

JOC (1965) 30 3480

$$C_6H_{13}\overset{O}{\overset{\wedge}{CHCH_2}} \xrightarrow{PrI \quad Me_2SO} C_6H_{13}CH_2CHO \qquad\qquad 80\%$$

Chem Comm (1968) 227

$$\begin{matrix} & \overset{Me}{|} & \overset{O}{\overset{\wedge}{}} \\ \Big((CH_2CH_2CH{=}C)_2CH_2CH_2\overset{}{CHCMe_2} \\ (CH{=}\underset{|}{C}CH_2CH_2)_2CH{=}CMe_2 \\ & \underset{Me}{} \end{matrix} \xrightarrow[Et_2O]{HIO_4} \begin{matrix} & \overset{Me}{|} \\ \Big((CH_2CH_2CH{=}C)_2CH_2CH_2CHO \\ (CH{=}\underset{|}{C}CH_2CH_2)_2CH{-}CMe_2 \\ & \underset{Me}{} \end{matrix}$$

Chem Comm (1968) 1067

Section 55   Aldehydes from Halides and Sulfonates
○○

JACS (1939) 61 2134

$$i\text{-}PrBr \xrightarrow[2\ MeOCH_2COMe]{1\ Mg\ \ Et_2O} i\text{-}Pr\underset{Me}{\overset{OH}{\underset{|}{\overset{|}{C}}}CH_2OMe} \xrightarrow{(COOH)_2} i\text{-}Pr\underset{Me}{\overset{}{\underset{|}{CHCHO}}}$$

JACS (1946) 68 2339

$$BuBr \dashrightarrow BuOCH_2CH{=}CH_2 \xrightarrow[pentane]{PrLi \quad THF} Bu(CH_2)_2CHO \qquad 29\%$$

Tetr Lett (1969) 821

$$CH_2{=}CH\underset{CH_2}{\overset{}{\underset{\|}{C}}}(CH_2)_2CH{=}\underset{Me}{\overset{}{\underset{|}{C}}}CH_2Br \xrightarrow[\substack{(i\text{-}Pr)_2NLi\ \ Et_2O \\ 2\ (COOH)_2\ \ H_2O}]{1\ CH_3CH{=}NBu\text{-}t} CH_2{=}CH\underset{CH_2}{\overset{}{\underset{\|}{C}}}(CH_2)_2CH{=}\underset{Me}{\overset{}{\underset{|}{C}}}(CH_2)_2CHO \quad 61\%$$

Helv (1967) 50 2440

PrI → PrCH₂CHO   65%

1 (structure) BuLi   THF
2 NaBH₄   THF   EtOH
3 (COOH)₂   H₂O

JACS (1969) 91 763 5886
     (1970) 92 6676

(cyclopentane dibromide) → (cyclopentane-CHO)   36%

1 (structure)   BuLi
2 NaBH₄

JACS (1969) 91 765

PhBr → PhCH₂CH=CH₂ → PhCH₂CHO

1 Mg   Et₂O
2 CH₂=CHCH₂Br

1 PhCOOAg   I₂
2 NaOH
3 Pb(OAc)₄

Helv (1934) 17 351

(phenanthrene bromide) → (phenanthrene-CHO)   40-42%

1 Mg   Et₂O
2 HC(OEt)₃
3 HCl   H₂O

Org Synth (1955) Coll Vol 3 701
          (1943) Coll Vol 2 323
Ber (1970) 103 643

(dibromobenzene) → (bromobenzaldehyde)   40%

1 Mg   Et₂O
2 HCOOEt

Annalen (1912) 393 215

(MeO-naphthalene-Br) → (MeO-naphthalene-CHO)   80%

1 Mg   THF
2 HCONMe₂

Chimia (1964) 18 141
JCS (1956) 4691

JOC (1970) $\underline{35}$ 711

JOC (1941) $\underline{6}$ 489

82%

$C_{12}H_{25}Br$ $\xrightarrow{\begin{array}{c} 1 \ Mg \ \ Et_2O \\ 2 \ \text{(imine)} \\ 3 \ HCl \ \ H_2O \end{array}}$ $C_{12}H_{25}CHO$ 73%

JACS (1955) $\underline{77}$ 5118

$\underset{Me}{EtCHBr}$ $----\rightarrow$ $\underset{Me}{EtCHLi}$ $\xrightarrow{\begin{array}{c} 1 \ t\text{-BuCH}_2\overset{Me}{\underset{Me}{\text{C}}}NC \\ Et_2O \\ 2 \ (COOH)_2 \ \ H_2O \end{array}}$ $\underset{Me}{EtCHCHO}$ 92%

JACS (1969) $\underline{91}$ 7778
      (1970) $\underline{92}$ 6675

$C_6H_{13}Br$ $\xrightarrow{\begin{array}{c} 1 \ Mg \ \ Et_2O \\ 2 \ p\text{-Dimethylaminobenzaldehyde} \\ 3 \ Cl^-N_2^+\text{—}\bigcirc\text{—}SO_3H \ \ H_2O \end{array}}$ $C_6H_{13}CHO$ 75%

JOC (1960) $\underline{25}$ 1691
      (1962) $\underline{27}$ 279

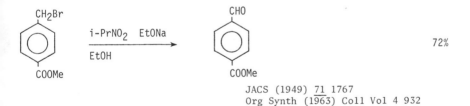

JOC (1941) 6 489

$(EtO)_2CHCH_2Br$ →[1 (dithiane) BuLi THF][2 $HgCl_2$] $(EtO)_2CHCH_2CHO$ <77%

Angew (1965) 77 1134
(Internat Ed 4 1075)

$C_9H_{19}Br$ →[1 $Na_2Fe(CO)_4$ Ph_3P THF][2 HOAc] $C_9H_{19}CHO$ 77%

JACS (1970) 92 6080

$PhCH_2Br$ →[$CH_2(COOEt)_2$][EtONa] $PhCH_2CH(COOEt)_2$ →[1 N_2H_4][2 HNO_2][3 HCl] $PhCH_2CHO$

Org React (1946) 3 337

→[i-PrNO$_2$ EtONa][EtOH]   72%

JACS (1949) 71 1767
Org Synth (1963) Coll Vol 4 932

$EtC=CH(CH_2)_2C=CHCH_2Br$ →[nitrocyclohexane][KOH i-PrOH H_2O] $EtC=CH(CH_2)_2C=CHCHO$
  Me   Me   Me   Me

Zh Obshch Khim (1959) 29 3965
(Chem Abs 54 21170)

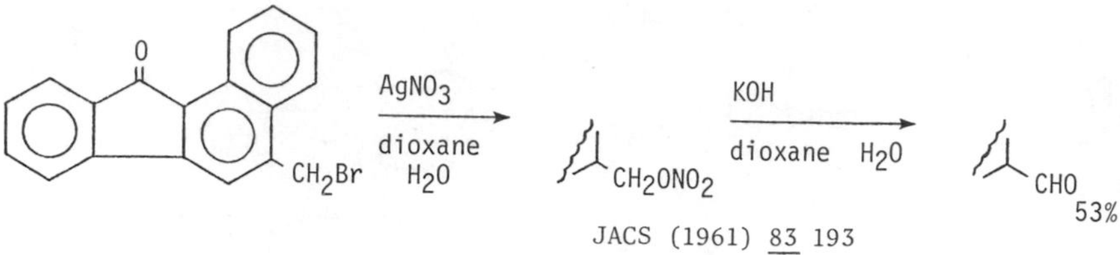

JACS (1961) 83 193

$C_7H_{15}CH_2X$ $\xrightarrow{\overset{+ -}{Me_3NO} \quad CHCl_3}$ $C_7H_{15}CHO$ 50% (X=I)
                                                                          55% (X=OTs)

Ber (1961) 94 1360

$C_6H_{13}CH_2Br$ $\xrightarrow[\text{MeCN}]{\text{AgOTs}}$ $C_6H_{13}CH_2OTs$ $\xrightarrow[\text{Me}_2SO]{\text{NaHCO}_3}$ $C_6H_{13}CHO$ 74%

JACS (1959) 81 4113

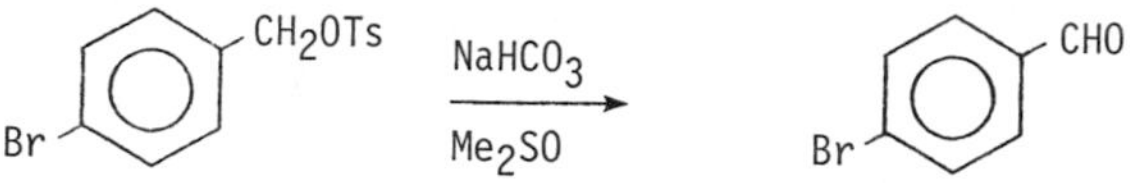

JACS (1959) 81 4113
Chem Rev (1967) 67 247

65%

$Me_2C=CH(CH_2CH_2\underset{Me}{C}=CH)_2CH_2Br$ $\xrightarrow[\text{2 p-Nitroso-} \atop \text{dimethylaniline}]{\text{1 Pyr}}$ $Me_2C=CH(CH_2CH_2\underset{Me}{C}=CH)_2CHO$

Helv (1941) 24 1039

$C_6H_{13}CH_2I$ $\xrightarrow[\text{2 HOAc}]{\text{1 Hexamine} \quad CHCl_3}$ $C_6H_{13}CHO$ 41%

Org React (1954) 8 197

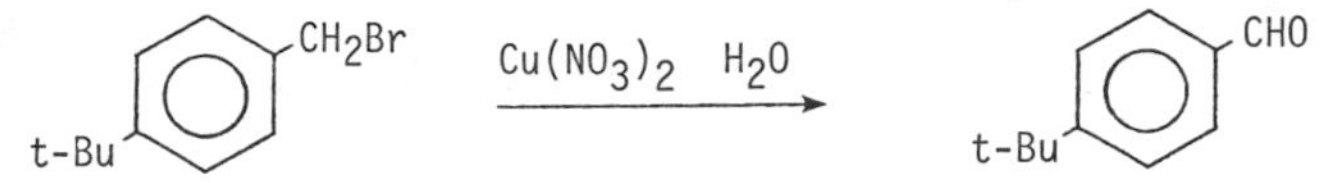

JCS (1935) 1847

$PhCH_2Cl$ → 1 Mg Et_2O / 2 O_2 / 3 NaOH →   PhCHO

JACS (1955) 77 6032

$C_7H_{15}CH_2I$ → NaN$_3$ / butyl carbitol → $C_7H_{15}CH_2N_3$ → 1 hν / 2 Acid → $C_7H_{15}CHO$ 50%

Tetrahedron (1965) 21 2877

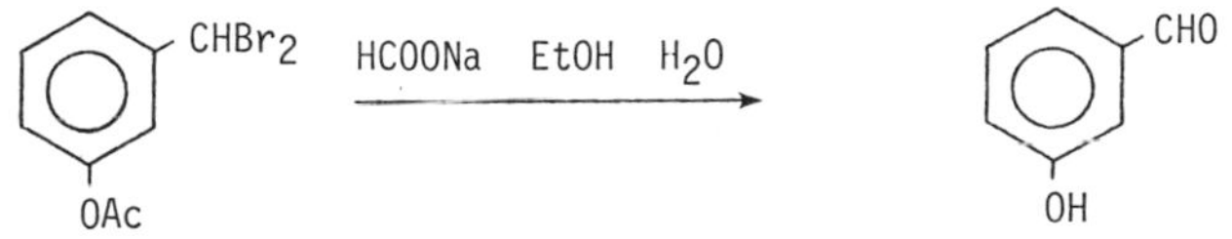

JCS (1955) 1628   70%

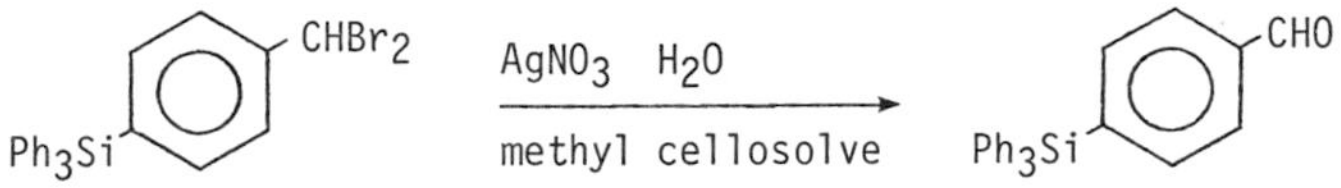

JACS (1956) 78 1689   67%

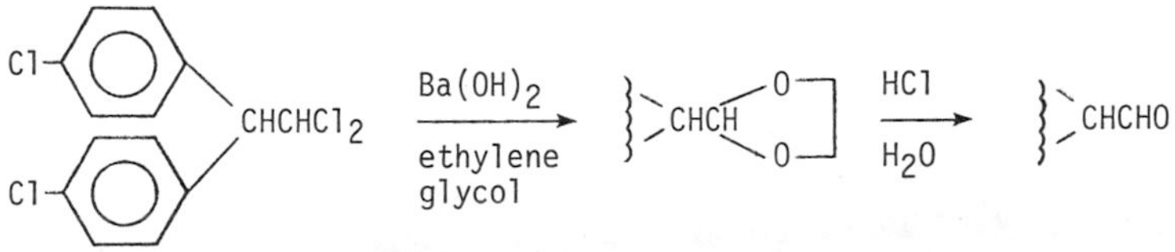

Ber (1959) 92 900   32%

Me$_2$CCH$_2$Br $\xrightarrow{\text{H}_2\text{O}}$ Me$_2$CHCHO   75%
   |
   Br
                                          JACS (1933) $\underline{55}$ 1136

Section 56   Aldehydes from Hydrides (RH)

Reactions in which a hydrogen is replaced by CHO, e.g. RH $\longrightarrow$ RCHO
(R=vinyl, ethinyl or aryl), are included in this section. For reactions
in which alkyl groups are oxidized to aldehyde, e.g. RCH$_3$ $\longrightarrow$ RCHO, see
section 50 (Aldehydes from Alkyls)

                    POCl$_3$ HCONMe$_2$ 41%

                    Z Chem (1967) $\underline{7}$ 346
                    Tetrahedron (1969) $\underline{25}$ 4535

                    POCl$_3$ HCONMe$_2$

                    Ber (1964) $\underline{97}$ 1252
                    JCS $\underline{C}$ (1969) 913

                    Ph$_3$P·Br$_2$ HCONMe$_2$ 45%

                    Annalen (1968) $\underline{718}$ 24

                         Me
                          |
                    POCl$_3$ HCONPh   71%

                    Org Synth (1963) Coll Vol 4 915
                              (1955) Coll Vol 3 98

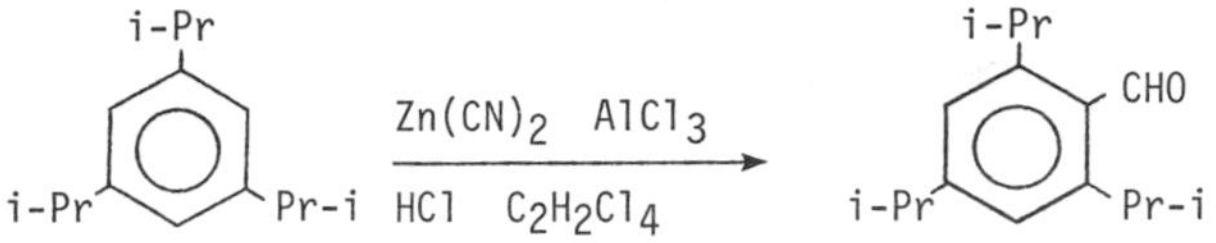

JACS (1942) 64 30

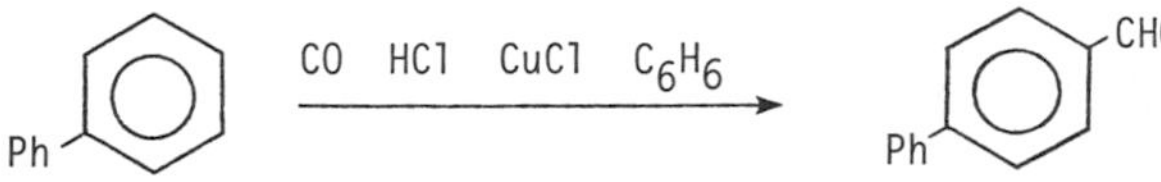

66%

Org React (1949) 5 290
Org Synth (1943) Coll Vol 2 583

73%

JACS (1960) 82 2380

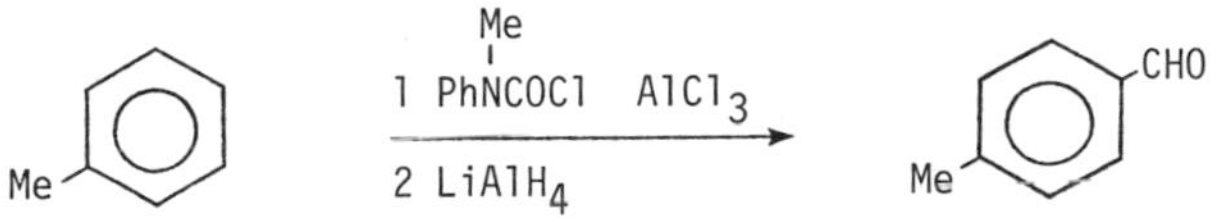

72%

Ber (1955) 88 301

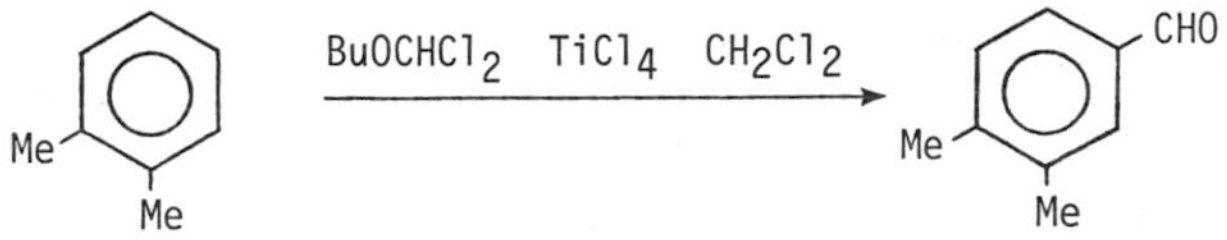

Ber (1960) 93 88

68%

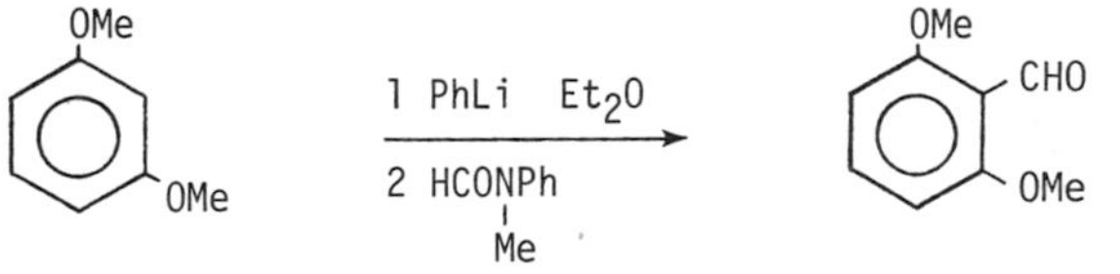

Org React (1954) 8 258

$$Me(CH_2)_3C{\equiv}CH \xrightarrow[2\ HCONMe_2]{1\ EtMgBr} Me(CH_2)_3C{\equiv}CCHO$$

JCS (1958) 1054

Section 57 Aldehydes from Ketones

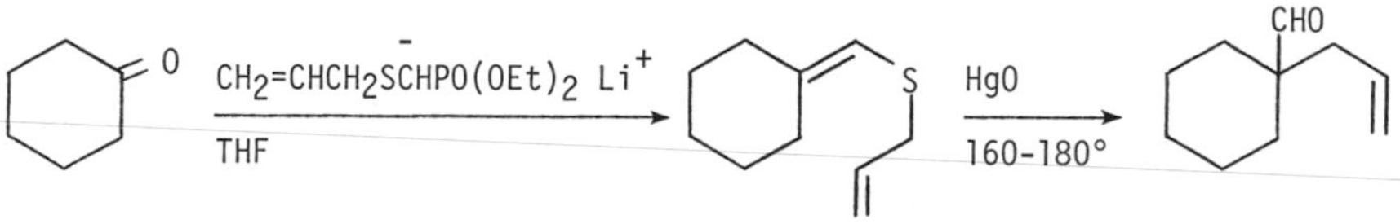
O
CH_2=$CHCH_2\bar{S}CHPO(OEt)_2$ Li^+
THF
S
HgO
160-180°
CHO

JACS (1970) 92 5522

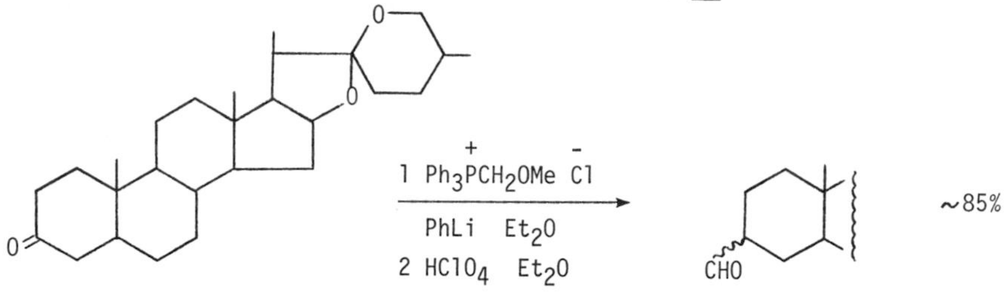
O
O
O
1 $Ph_3\overset{+}{P}CH_2OMe$ $\bar{C}l$
PhLi Et_2O
2 $HClO_4$ Et_2O
CHO
~85%

JACS (1958) 80 6150
Org React (1965) 14 270

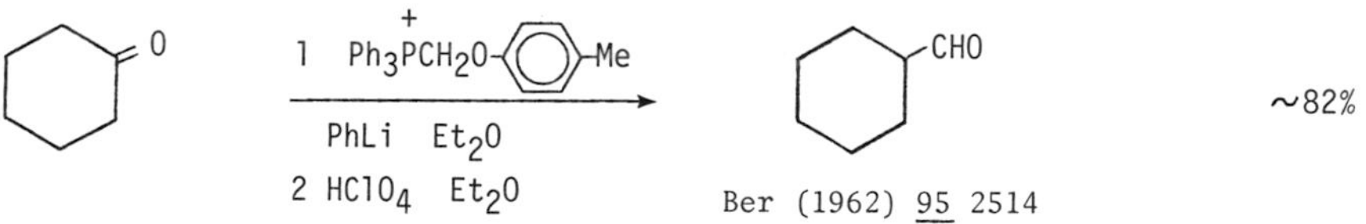
O
1 $Ph_3\overset{+}{P}CH_2O$-C6H4-Me
PhLi Et_2O
2 $HClO_4$ Et_2O
CHO
~82%

Ber (1962) 95 2514

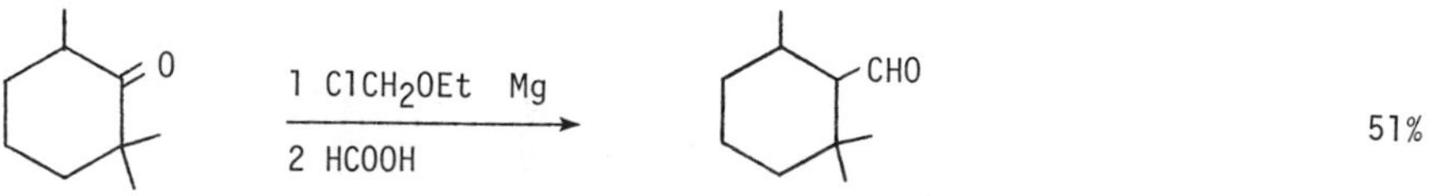
O
1 $ClCH_2OEt$ Mg
2 HCOOH
CHO
51%

JACS (1968) 90 3282
Compt Rend (1963) 256 1996

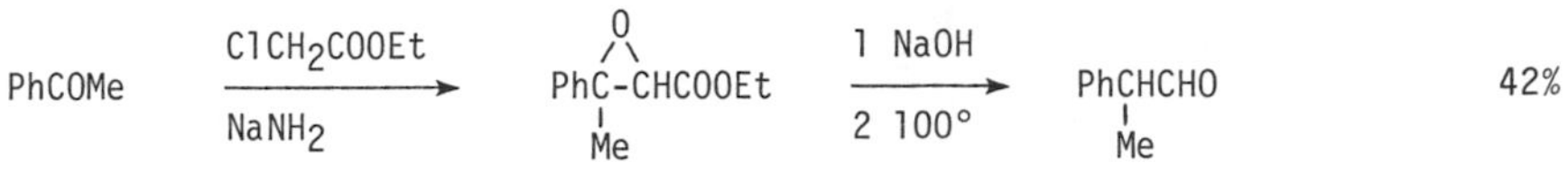

$$PhCOMe \xrightarrow[NaNH_2]{ClCH_2COOEt} PhC\text{-}CHCOOEt \xrightarrow[2\ 100°]{1\ NaOH} PhCHCHO$$

$$\begin{array}{c} O \\ PhC\text{-}CHCOOEt \\ | \\ Me \end{array}$$

$$PhCHCHO \quad 42\%$$
(with Me substituent)

Org React (1949) 5 413
JOC (1970) 35 1600
Chem Rev (1955) 55 283

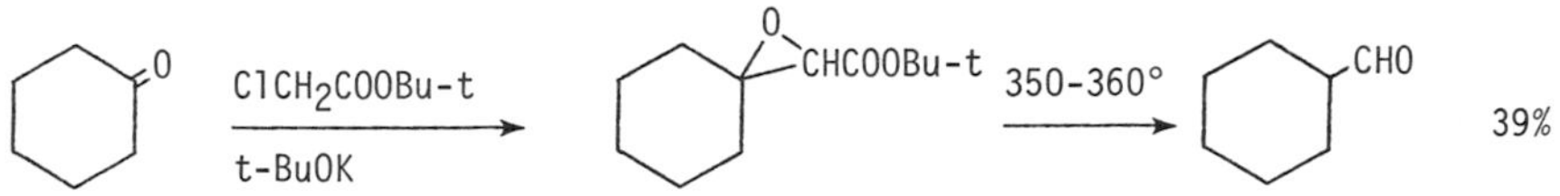

JACS (1963) 85 955   39%

$$Et_2CO \xrightarrow[C_6H_6]{BrCH_2COOEt\ \ Zn} Et_2CCH_2COOEt \xrightarrow[\substack{2\ NaNO_2\ \ HCl \\ 3\ KOH}]{1\ N_2H_4} Et_2CHCHO \quad 79\%$$

$$\begin{array}{c} Et_2CCH_2COOEt \\ | \\ OH \end{array}$$

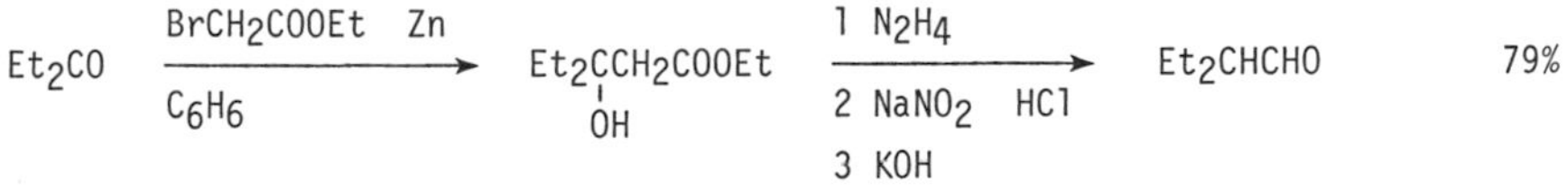

JACS (1951) 73 4199

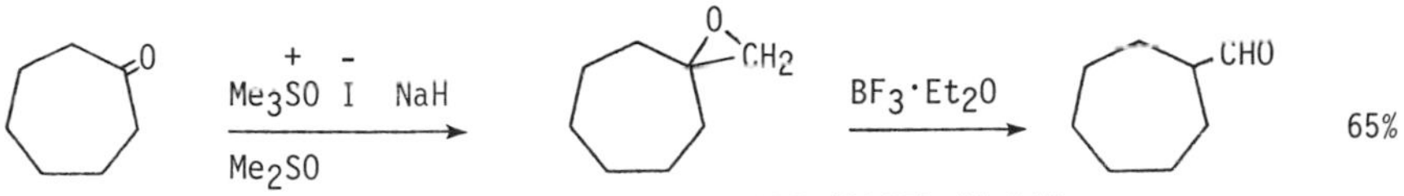

JACS (1962) 84 867   65%

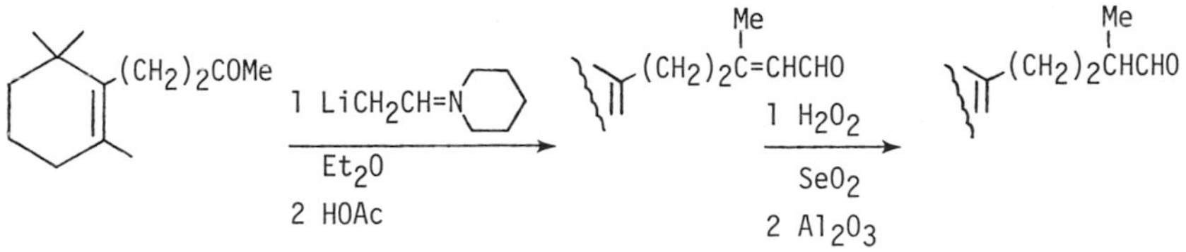

Tetr Lett (1970) 381

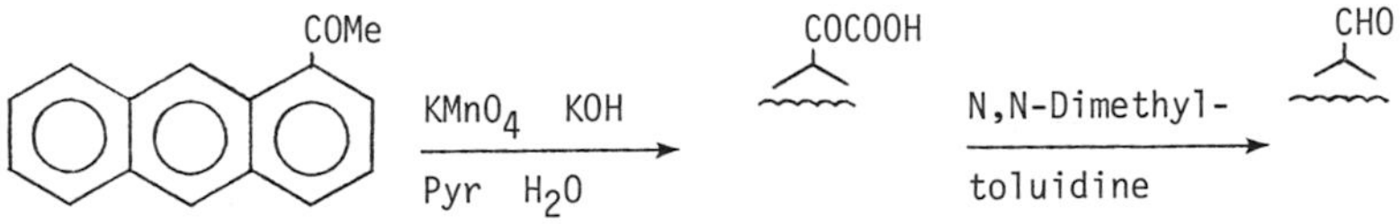

JACS (1965) 87 1247

$$PhCOCH_2Me \xrightarrow{\quad BuN_3 \quad H_2SO_4 \quad toluene \quad} PhCHO \quad + \quad MeCHO$$

62%   50%

JACS (1959) <u>81</u> 3369

Some of the methods included in section 49 (Aldehydes from Aldehydes) may also be applied to the preparation of aldehydes from ketones

Section 58   Aldehydes from Nitriles

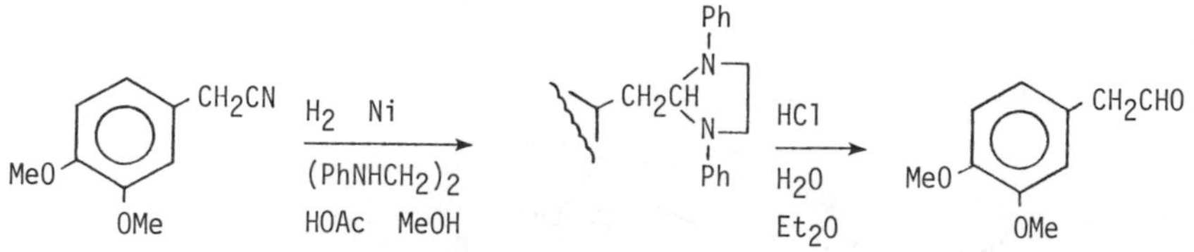

Ber (1957) <u>90</u> 617

$$PhCH_2CN \xrightarrow[\quad NaOAc \quad MeOH \quad H_2O \quad]{\quad 1 \quad H_2 \quad Ni \quad NH_2NHCONH_2 \cdot HCl \quad} PhCH_2CH=NNHCONH_2 \dashrightarrow PhCH_2CHO$$

Angew (1955) <u>67</u> 156
JOC (1962) <u>27</u> 850

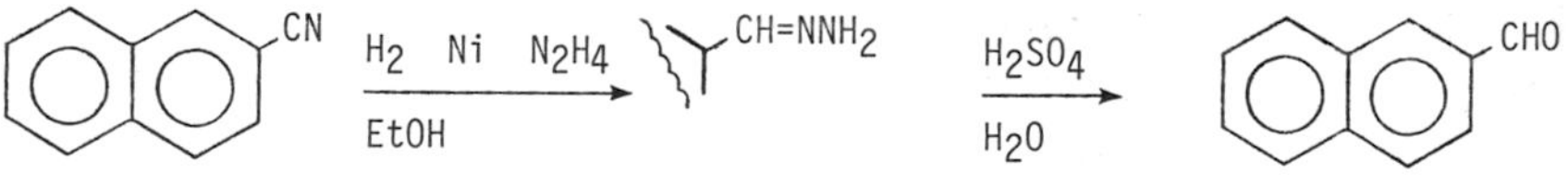

Gazz (1955) <u>85</u> 1705
(Chem Abs <u>50</u> 10037)

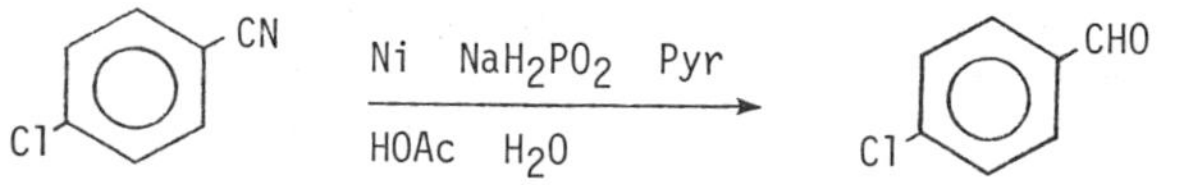

80%

JCS (1962) 3961

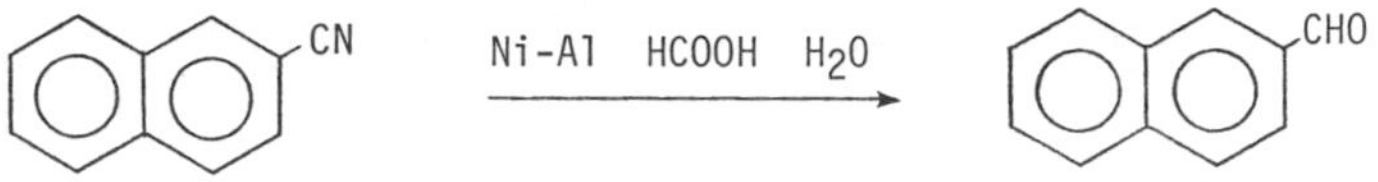

$$\xrightarrow{\text{Ni-Al} \quad \text{HCOOH} \quad H_2O}$$

95%

JCS (1965) 5775
    (1964) 5880

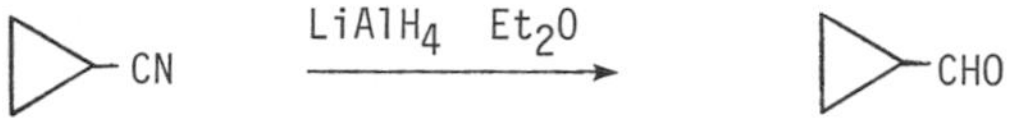

$$\xrightarrow{\text{LiAlH}_4 \quad \text{Et}_2O}$$

48%

JACS (1951) 73 4047
Org React (1954) 8 252

i-PrCN $\xrightarrow{\text{LiAl(OEt)}_3\text{H} \quad \text{Et}_2O}$ i-PrCHO

81%

JACS (1964) 86 1085

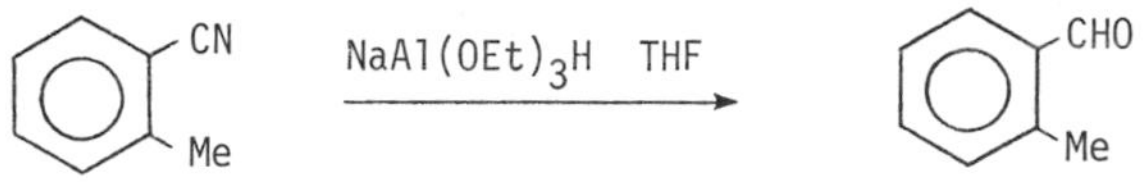

$$\xrightarrow{\text{NaAl(OEt)}_3\text{H} \quad \text{THF}}$$

88%

Annalen (1957) 607 24

PhCN $\xrightarrow{\text{(i-Bu)}_2\text{AlH} \quad C_6H_6}$ PhCHO

90%

JOC (1959) 24 627

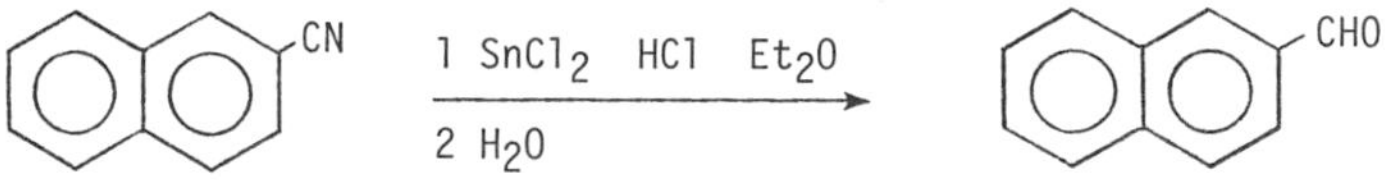

$$\xrightarrow[\text{2 } H_2O]{\text{1 SnCl}_2 \quad \text{HCl} \quad \text{Et}_2O}$$

73-80%

Org Synth (1955) Coll Vol 3 626

$C_{15}H_{31}CN$ $\xrightarrow[\text{2 } H_2O]{\text{1 SnCl}_2 \quad \text{HCl} \quad \text{Et}_2O}$ $C_{15}H_{31}CHO$

JCS (1925) 127 1874

$$MeSCH_2CH_2CN \xrightarrow{\text{MeOH   HCl}} MeSCH_2CH_2C(OMe)_3 \xrightarrow[\substack{C_6H_6 \\ \text{2 Acid}}]{\text{1 LiAlH}_4 \text{ Et}_2O} MeSCH_2CH_2CHO$$

JACS (1951) <u>73</u> 5005

$$PhCH_2CN \dashrightarrow PhCH_2 \overset{O}{\underset{N}{\diagup}} \xrightarrow[\text{2 Base}]{\text{1 NaBH}_4 \text{ THF   EtOH}} PhCH_2CHO$$

Chem Comm (1967) 1163

$$PhCN \xrightarrow[\text{Zn C}_6\text{H}_6]{\overset{\text{Me}}{\underset{\text{Me}}{BrCCOOEt}}} PhCOC\overset{Me}{\underset{Me}{C}}COOEt \xrightarrow[\text{2 260°}]{\text{1 H}_2 \text{ Ni EtOH}} PhCHO$$

Compt Rend (1953) <u>236</u> 826

$$PhCN \dashrightarrow[\text{H}_2\text{S   NH}_3]{} PhCSNH_2 \xrightarrow{\text{Ni EtOH}} PhCHO$$

JACS (1952) <u>74</u> 3936

Section 59   Aldehydes from Olefins

$$EtCH=CH_2 \xrightarrow[\substack{\text{2 MeCH=CHCHO} \\ \text{air   i-PrOH}}]{\text{1 B}_2\text{H}_6 \text{ THF}} Et(CH_2)_2\overset{}{\underset{Me}{C}}HCH_2CHO \qquad 90\%$$

JACS (1970) <u>92</u> 714

$$\begin{matrix} CH_2 \\ \| \\ CH \\ | \\ (CH_2)_8 \\ | \\ CN \end{matrix} \xrightarrow[\substack{\text{2 CO   LiAl(OBu-t)}_3\text{H} \\ \text{3 H}_2\text{O}_2 \text{ NaH}_2\text{PO}_4 \text{ K}_2\text{HPO}_4 \text{ H}_2\text{O}}]{\substack{1 \quad \text{BH} \\ \text{THF}}} \begin{matrix} CHO \\ | \\ (CH_2)_2 \\ | \\ (CH_2)_8 \\ | \\ CN \end{matrix}$$

JACS (1969) <u>91</u> 4606

$Me_2C=CH_2$ $\xrightarrow[\text{2 CO LiAl(OMe)}_3\text{H}]{\text{1 B}_2\text{H}_6 \quad \text{THF}}$ Me_2CHCH_2CHO 91%

JACS (1968) <u>90</u> 499 818

$C_6H_{13}CH=CH_2$ $\xrightarrow[\text{(PhO)}_3\text{P   toluene}]{\text{CO H}_2 \quad \text{(100 psi) Rh-C}}$ $C_6H_{13}(CH_2)_2CHO$ 89%

JOC (1969) <u>34</u> 327
Chem Comm (1967) 305
(1965)   17

PrCHCH=CH$_2$ | Me $\xrightarrow[\text{Co}_2\text{(CO)}_8 \quad \text{C}_6\text{H}_6]{\text{CO H}_2 \quad \text{(200 atmos)}}$ PrCH(CH$_2$)$_2$CHO | Me ~62%

JACS (1968) <u>90</u> 6817

45%

Tetr Lett (1969) 43
JOC (1968) <u>33</u> 805

$C_6H_{13}CH=CH_2$ $\xrightarrow[\text{2 HCl H}_2\text{O}]{\text{1   di-t-butyl peroxide}}$ $C_6H_{13}(CH_2)_2CHO$ ~20%

Tetr Lett (1968) 4339

$\xrightarrow{\text{Tl(NO}_3)_2 \quad \text{MeOH}}$

86%

Acc Chem Res (1970) <u>3</u> 338
Tetr Lett (1970) 5275

$(t\text{-}BuCH_2)_2C=CH_2$ $\xrightarrow[\text{2 Zn}]{\text{1 CrOCl}_2\quad CH_2Cl_2}$ $(t\text{-}BuCH_2)_2CHCHO$ 80%

JOC (1968) $\underline{33}$ 3970

JOC (1968) $\underline{33}$ 3953

$PhCH_2CH=CH_2$ $\xrightarrow{\text{DDQ}\quad C_6H_6}$ $PhCH=CHCHO$ 50%

Chem Comm (1969) 773

$C_8H_{17}CH=CH(CH_2)_7COOMe$ $\xrightarrow[\substack{\text{2 H}_2\quad Pd\text{-}C\quad MeOH\quad EtOAc \\ \text{(or Zn}\quad HOAc)}]{\text{1 O}_3\quad MeOH}$ $C_8H_{17}CHO$ 78%

For reduction of ozonides with tetracyanoethylene see

						JOC (1960) $\underline{25}$ 618
						JACS (1953) $\underline{75}$ 4952
For reduction of ozonides with				tetracyanoethylene see		Ber (1963) $\underline{96}$ 1564
"	"	"	"	"	$(MeO)_3P$	" JOC (1960) $\underline{25}$ 1031
"	"	"	"	"	$(Me_2N)_3P$	" Helv (1967) $\underline{50}$ 2387
"	"	"	"	"	Ph_3P	" JOC (1968) $\underline{33}$ 787
"	"	"	"	"	Me_2S	" Helv (1967) $\underline{50}$ 2445
"	"	"	"	"	dinitrophenylhydrazine	see JACS (1956) $\underline{78}$ 5601
"	"	"	"	"	Na_2SO_3	see JOC (1961) $\underline{26}$ 4912
"	selective ozonolysis					" JACS (1958) $\underline{80}$ 915
						(1969) $\underline{91}$ 4318
						(1958) $\underline{80}$ 915

$PhCH=CH_2$ $\xrightarrow[CCl_4]{\text{Bistriphenylsilyl chromate}}$ $PhCHO$

JOC (1970) $\underline{35}$ 774

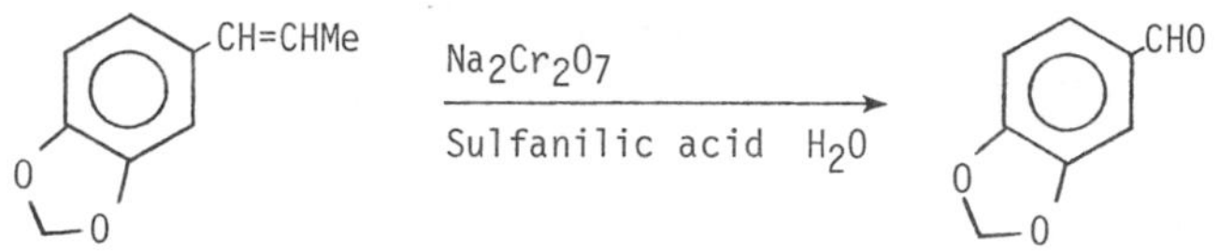

J Soc Chem Ind (1943) 62 90
JCS (1937) 369
Org React (1959) 10 1

$C_6H_{13}CH=CH_2$ $\xrightarrow{\text{RuO}_4 \quad \text{CCl}_4}$ $C_6H_{13}CHO$ 12%

JACS (1958) 80 6682

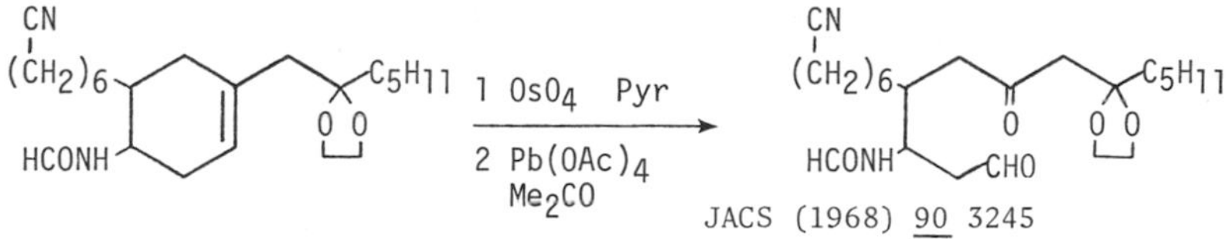

JACS (1968) 90 3245   43%

$C_{10}H_{21}CH=CH_2$ $\xrightarrow[\text{dioxane}]{\text{OsO}_4 \quad \text{NaIO}_4}$ $C_{10}H_{21}CHO$ 68%

JOC (1956) 21 478
Tetrahedron (1968) 24 3095

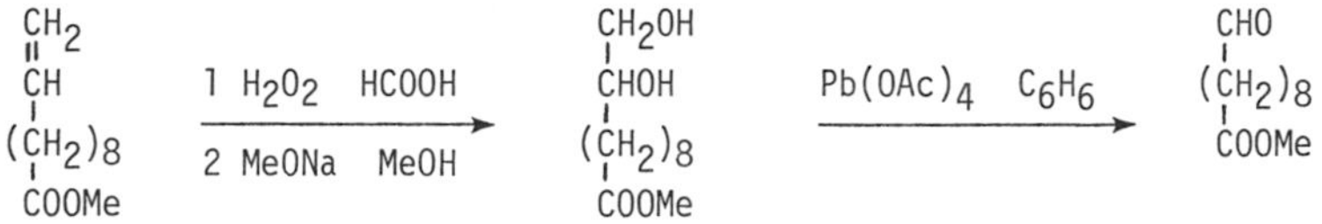

J Med Chem (1969) 12 911   51%

Helv (1925) 8 332   52%

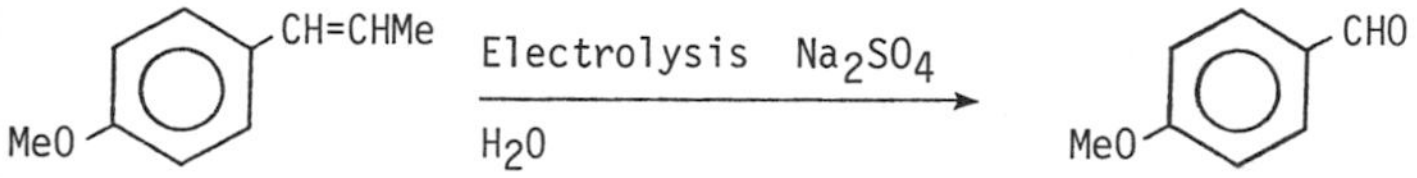

MeO —[benzodioxole ring]— CH=CHMe

$\xrightarrow[\begin{array}{c}\text{Pyr}\\ \text{2 KOH } H_2O\end{array}]{1\ C(NO_2)_4\ \ Me_2CO}$

$\sim$ CH=CMe / NO_2

$\xrightarrow[\text{2 HCl } H_2O]{1\ \text{Phenylethylamine}}$

$\sim$ CHO   79%

Can J Chem (1968) 46 75

Section 60   Aldehydes from Miscellaneous Compounds

$PrCH=CCHO$ / Et

$\xrightarrow[\text{2 HCl } H_2O]{1\ H_2\ \ Ni\ \ EtCHCH_2OH \overset{NH_2}{}}$

$PrCH_2CHCHO$ / Et

JACS (1955) 77 359

$MeCH=CHCHO$

$\xrightarrow{H_2\ \ (Ph_3P)_3RhCl\ \ C_6H_6}$

$Me(CH_2)_2CHO$

JCS C (1967) 270

[cyclohexene ring]—CH_2CH_2OH

$\xrightarrow{Fe(CO)_5}$

[cyclohexane ring]—CH_2CHO

Chem Comm (1968) 97

$Me_2C=CH(CH_2)_2C=CHCHO$ / Me

$\xrightarrow[\text{2 Hydrolysis}]{1\ PhCOO_2H}$

$Me_2\overset{O}{\overset{/\backslash}{C}-CH}(CH_2)_2CHCHO$ / Me

Org React (1957) 9 73
Tetr Lett (1970) 381

$PhCH=CHCN$

$\xrightarrow[\text{MeOH } H_2O]{Co_2(CO)_8\ \ HCl}$

$Ph(CH_2)_2CHO$ 50%

Chem Comm (1967) 1140

PhCH=CHCONH₂ $\xrightarrow[\text{2 Acid }\ H_2O]{\text{1 NaOCl }\ MeOH}$ PhCH₂CHO   ~70%

Org React (1946) 3 267

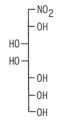 $\xrightarrow[\text{2 }H_2SO_4\ H_2O]{\text{1 NaOH }\ H_2O}$ 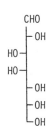   70%

JACS (1951) 73 4662
JOC (1943) 8 10
Chem Rev (1955) 55 137

PrCH₂NO₂ $\xrightarrow[\text{MgSO}_4\ H_2O]{\text{KMnO}_4\ KOH}$ PrCHO   97%

JOC (1962) 27 3699

C₁₀H₂₁CHCOOH $\xrightarrow{190\text{-}200°}$ C₁₀H₂₁CHO   96%
         |
        OH

J Soc Chem Ind (1943) 62 128

PhCHCOOH $\xrightarrow{\text{Electrolysis }\ MeOH}$ PhCHOMe $\xrightarrow{\text{HCl }\ H_2O}$ PhCHO
    | |
   OEt   OEt
                                                        71%

JOC (1962) 27 281

Section 60A   Protection of Aldehydes

i-PrCHO $\xrightarrow{\text{HC(OEt)}_3 \quad \text{NH}_4\text{Cl} \quad \text{EtOH}}$ i-PrCH(OEt)$_2$

$\xleftarrow{\hspace{1cm}\text{Acid}\hspace{1cm}}$

(Stable to RMgX, LiAlH$_4$, CrO$_3$
and base)

Bull Soc Chim Fr (1965) 1007
JCS (1953) 3864

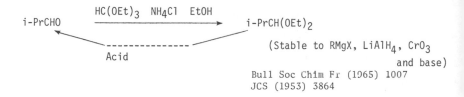

$\xrightarrow{\text{PhSO}_2\text{NHOH} \quad \text{MeOH}}$

$\xleftarrow{\text{PhSO}_2\text{NHOH} \quad \text{H}_2\text{O} \quad \text{dioxane}}$

JOC (1970) 35 1962

$\xrightarrow{\begin{array}{c}\text{Me}_2\text{SO}_4 \quad \text{NaOH}\\ \hline \text{MeOH} \quad \text{H}_2\text{O}\end{array}}$

$\xleftarrow{\text{Acid}}$

Ber (1958) 91 410

Me$_2$C=CH(CH$_2$)$_2$CHCH$_2$CHO
             |
             Me
$\xrightarrow{\text{HOCH}_2\text{CH}_2\text{OH} \quad \text{TsOH} \quad \text{C}_6\text{H}_6}$
$\xleftarrow{\hspace{1cm}\text{Acid}\hspace{1cm}}$
Me$_2$C=CH(CH$_2$)$_2$CHCH$_2$CH
             | O⌐
             Me |O⌐

(Stable to RMgX, LiAlH$_4$,
CrO$_3$ and base)

Tetrahedron (1959) 6 217

PhOCH$_2$CH$_2$CHO
$\xrightarrow{\text{2-Methyl-2,4-pentanediol} \quad \text{TsOH}}$
$\xleftarrow{\hspace{1cm}\text{Acid}\hspace{1cm}}$
PhOCH$_2$CH$_2$CH

JOC (1963) 28 594

$C_6H_{13}CHO$

$$C_6H_{13}CHO \xrightarrow[\text{HgCl}_2 \quad \text{dioxane} \quad \text{EtOH}]{\underset{\text{HSCH}_2}{\overset{\text{HSCH}_2}{\bigsqcup}}\text{Me} \quad \text{HCl} \quad \text{dioxane}} C_6H_{13}CH\underset{\text{SCH}_2}{\overset{\text{SCH}_2}{\bigsqcup}}\text{Me}$$

(Stable to LiAlH$_4$ and base)

JCS C (1966) 1005

$$\underset{\overset{|}{\text{OH}}}{\overset{\overset{\text{Me}}{|}}{PhCCHO}} \xrightarrow[\text{I}_2 \quad \text{NaHCO}_3 \quad \text{dioxane} \quad \text{H}_2\text{O}]{\text{MeSH}} \underset{\overset{|}{\text{OH}}}{\overset{\overset{\text{Me}}{|}}{PhCCH(SMe)_2}}$$

JOC (1969) 34 3618

For cleavage of thioketals with HgCl$_2$ see Ber (1958) 91 1043

$$PrCHO \xrightarrow[\text{HCl} \quad \text{H}_2\text{O}]{(PhNHCH_2)_2} PrCH\underset{\overset{|}{\text{Ph}}}{\overset{\overset{\text{Ph}}{|}}{\underset{\text{N}}{\overset{\text{N}}{\bigsqcup}}}}$$

Ber (1953) 86 1463

Piperidine   TsOH   C$_6$H$_6$

Acid

(Stable to LiAlH$_4$ and base)

JACS (1955) 77 1216

$$Me_2CHCHO \xrightarrow[\text{2 H}_3\text{PO}_4]{\text{1 HC(OEt)}_3 \quad \text{NH}_4\text{Cl} \quad \text{EtOH}} Me_2C=CHOEt$$

(Stable to RMgX, LiAlH$_4$, CrO$_3$ and base)

Acid

Bull Soc Chim Fr (1965) 1007

$$Me_2CHCHO \xrightarrow[\text{2 PhNMe}_2]{\text{1 MeOH HCl}} Me_2C=CHOMe$$

Acid

Ber (1956) 89 1468

$$NH_2CONHNH_2 \cdot HOAc \quad MeOH \quad H_2O$$

$$H_2SO_4 \quad H_2O \quad \text{pet ether}$$

JCS (1950) 3361

For cleavage of semicarbazones with HNO_2 see JACS (1934) 56 1794

$$\begin{array}{c} CHO \\ | \\ (CHOAc)_4 \\ | \\ CH_2OAc \end{array} \quad \xrightarrow[\text{NaNO}_2 \text{ HCl EtOH}]{\text{NH}_2\text{OH}} \quad \begin{array}{c} CH=NOH \\ | \\ (CHOAc)_4 \\ | \\ CH_2OAc \end{array}$$

JACS (1934) 56 1794

For cleavage of oximes with $(NH_4)_2Ce(NO_3)_6$ see Can J Chem (1969) 47 145

    "   "   "   "   " $NaHSO_3$ " JOC (1966) 31 3446

    "   "   "   "   " $Fe(CO)_5$ " JOC (1967) 32 2938

$$C_7H_{15}CHO \xrightarrow[\text{KHCO}_3 \text{ ethylene glycol } H_2O]{\text{2,4-Dinitrophenylhydrazine HCl}} C_7H_{15}CH=NNH$$

Aust J Chem (1968) 21 271

$$RCH=CHCHO \xrightarrow[\text{H}_2\text{SO}_4 \text{ EtOH } H_2O]{\text{Ac}_2\text{O} \text{ H}_3\text{PO}_4} RCH=CHCH(OAc)_2$$

JCS (1955) 1384

Some of the methods included in section 180A (Protection of Ketones) may also be applied to the protection of aldehydes

Chapter 5   PREPARATION OF ALKYLS METHYLENES AND ARYLS

This chapter lists the conversion of functional groups into Me, Et···, CH₂, Ph etc.

Section 61   <u>Alkyls and Methylenes from Acetylenes</u>

BuC≡CH $\xrightarrow[\begin{array}{l}\text{2 MeONa   THF}\\\text{3 MeI}\\\text{4 Hydrolysis}\end{array}]{\text{1 (i-Bu)}_2\text{AlH   THF}}$ Bu(CH₂)₂Me   74%

Tetr Lett (1966) 6021

$\left(\begin{array}{c}\text{C≡C}\\\text{(CH}_2)_8\end{array}\right)$ $\xrightarrow{\text{H}_2 \quad \text{PtO}_2 \quad \text{HOAc}}$ $\left(\begin{array}{c}\text{CH}_2\text{CH}_2\\\text{(CH}_2)_8\end{array}\right)$ 71%

JACS (1952) <u>74</u> 3636

For hydrogenation with Ni   catalyst see JCS (1952) 5032

"   "   " Ru   "   " JOC (1959) <u>24</u> 708

"   "   " Re₂S₇ "   " JACS (1954) <u>76</u> 1519

and JACS (1959) <u>81</u> 3587

177

BuC≡CH $\xrightarrow[\text{Ph}_3\text{P} \quad \text{MeOH}]{\text{H}_2 \quad \text{Rh}_2(\text{COOMe})_4}$ BuCH$_2$Me

Chem Comm (1969) 825
(1965) 131

C$_5$H$_{11}$C≡CH $\xrightarrow{\text{Li} \quad \text{NH}_2\text{CH}_2\text{CH}_2\text{NH}_2}$ C$_5$H$_{11}$CH$_2$Me

JOC (1957) 22 891

EtC≡CEt $\xrightarrow{\text{Na} \quad \text{HMPA} \quad \text{t-BuOH}}$ EtCH$_2$CH$_2$Et   79%

JOC (1970) 35 3565

PhC≡CPh $\xrightarrow{\text{N}_2\text{H}_4 \quad \text{air} \quad \text{Cu}^{2+} \quad \text{MeOH}}$ PhCH$_2$CH$_2$Ph   80%

Tetr Lett (1961) 347

Section 62   Alkyls from Carboxylic Acids

Reactions in which carboxyl groups are converted into alkyl e.g.
RCOOH ⟶ RCH$_3$, are included in this section. For the conversion
RCOOH ⟶ RH see section 152 (Hydrides from Carboxylic Acids)

Chem Comm (1967) 96
Advances in Org Chem (1960) 1 1

COOH
|
(CH₂)₄   Electrolysis MeCOOH MeOH →   Me
| |
COOMe   (CH₂)₄   40%
                                             |
                                            COOMe

JCS (1950) 3326

C₁₇H₃₅COOH ----→ C₁₇H₃₅CONHNHTs $\xrightarrow{\text{LiAlH}_4}$ C₁₇H₃₅Me   50-60%

Tetrahedron (1966) 22 487
Chim Ind (Milan) (1965) 47 62
(Chem Abs 62 10388)

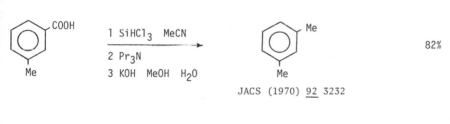

Hclv (1954) 37 1689
      (1946) 29 360

COOH ... Me structure

1 SiHCl₃ MeCN
——————————→
2 Pr₃N
3 KOH MeOH H₂O   82%

JACS (1970) 92 3232

C₁₇H₃₅COOH $\xrightarrow{\text{HI P}_4 \quad 200\text{-}250°}$ C₁₇H₃₅Me

Ber (1882) 15 1687

RCOOH $\xrightarrow{\text{H}_2 \text{ (120 atmos) MoS}_2 \quad 350°}$ RMe

Chem Listy (1956) 50 569
(Chem Abs 50 13854)

Section 63   Alkyls and Aryls from Alcohols
ooooooooooooooooooooooooooooooo

Reactions in which hydroxyl groups are replaced by alkyl or aryl e.g.
ROH ⟶ RPh, are included in this section. For the conversion ROH ⟶ RH
see section 153 (Hydrides from Alcohols and Phenols)

$$PhCH_2OH \quad \xrightarrow[\text{2 TiCl}_4 \text{ K } C_6H_6]{\text{1 NaH}} \quad PhCH_2CH_2Ph \qquad\qquad 51\%$$

JACS (1965) $\underline{87}$ 3277

JACS (1968) $\underline{90}$ 3284
Chem Comm (1969) 53

$$PrOH \quad \xrightarrow{\text{AlCl}_3 \ C_6H_6 \ H_2O} \quad PrPh$$

JOC (1940) $\underline{5}$ 253
JACS (1942) $\overline{64}$ 1576

63%

Org React (1946) $\underline{3}$ 1

Section 64   Alkyls from Aldehydes
              oooooooooooooooooooo

Reactions which convert the group CHO into CH_3 are included in this
section. For the conversion RCHO $\longrightarrow$ RH see section 154 (Hydrides from
Aldehydes)

$$C_6H_{13}CHO \xrightarrow{\quad H_2 \quad WS_2 \quad 320\text{-}330° \quad} C_6H_{13}Me$$

Chem Listy (1956) 50 565
(Chem Abs 50 13831)

38%

$$\xrightarrow{\begin{array}{c} H_2 \quad PdCl_2 \\ \hline H_2O \quad HOAc \end{array}}$$

JCS (1954) 4676
JACS (1956) 78 3769

86%

$$\xrightarrow{\quad NaBH_4 \quad Pd\text{-}C \quad i\text{-}PrOH \quad}$$

Can J Chem (1964) 42 514

89%

$$\xrightarrow{\quad LiAlH_4 \quad AlCl_3 \quad Et_2O \quad}$$

JCS (1957) 3755
Tetr Lett (1967) 1849

78%

$$\xrightarrow{\begin{array}{c} \text{Electrolysis} \\ \hline H_2SO_4 \quad \text{dioxane} \end{array}}$$

JACS (1967) 89 4789

90%

$$C_6H_{13}CHO \xrightarrow{\quad Zn \quad HCl \quad H_2O \quad} C_6H_{13}Me$$

Org React (1942) 1 155

72%

$$
\begin{array}{c}
\xrightarrow[\text{ethylene glycol}]{\text{1 N}_2\text{H}_4 \cdot \text{HCl} \quad \text{N}_2\text{H}_4} \\
\text{2 KOH}
\end{array}
$$

48%

Tetrahedron (1969) 25 1335
Org React (1948) 4 378

$$C_{11}H_{23}CHO \xrightarrow[\text{2 NaBH}_4 \quad \text{MeOH}]{\text{1 TsNHNH}_2} C_{11}H_{23}Me$$

Chem Ind (1964) 153 1689
Tetrahedron (1963) 19 1127

$$\xrightarrow[\text{2 I}_2 \quad \text{Et}_3\text{N} \quad \text{THF}]{\text{1 N}_2\text{H}_4 \quad \text{Et}_3\text{N}}$$

CH$_2$I$_2$ $\xrightarrow[\text{HOAc}]{\text{Zn}}$

JCS (1962) 470

$$C_5H_{11}CHO \xrightarrow[]{\text{EtSH} \quad \text{ZnCl}_2 \quad \text{Na}_2\text{SO}_4} C_5H_{11}CH(SEt)_2 \xrightarrow[]{\text{Ni} \quad \text{EtOH}} C_5H_{11}Me \quad 40\%$$

JACS (1944) 66 909
JOC (1968) 33 3551
Org React (1962) 12 356

Further examples of the conversion C=O ⟶ CH$_2$ are included in section 72
(Alkyls and Methylenes from Ketones)

Section 65   Alkyls and Aryls from Alkyls and Aryls

The conversion RR' ⟶ ArR (R,R'=alkyl), the epimerization, isomerization
and transposition of alkyl groups and the hydrogenation of aryl groups are
included in this section. For the conversions RR' ⟶ RH and ArR ⟶ ArH
(R,R'=alkyl) see section 155 (Hydrides from Alkyls)

$$t\text{-BuCH}_2\text{Pr-i} \xrightarrow[]{\text{PhH} \quad \text{AlCl}_3} t\text{-BuPh}$$

JACS (1935) 57 2415

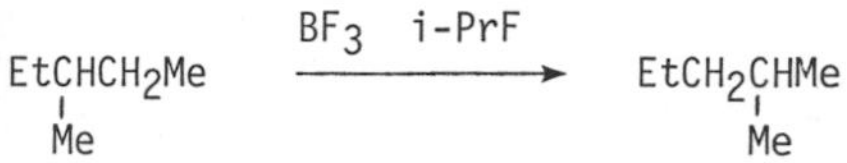

EtCHCH₂Me $\xrightarrow{\text{BF}_3 \quad \text{i-PrF}}$ EtCH₂CHMe   (+ higher mol wt compds)
  | |
  Me   Me

JACS (1951) 73 5013
JOC (1964) 29 94

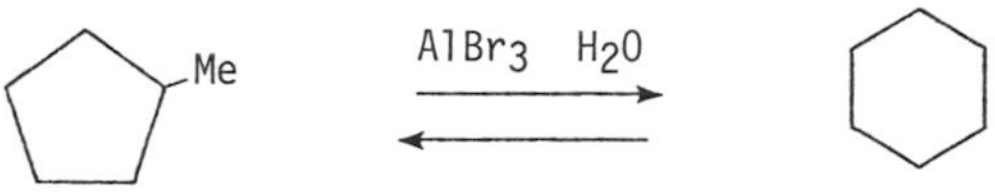

JACS (1948) 70 2773

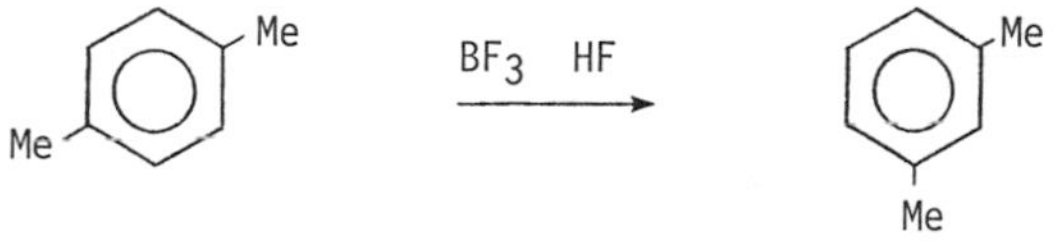

100%

JACS (1952) 74 6246
JCS (1952) 100

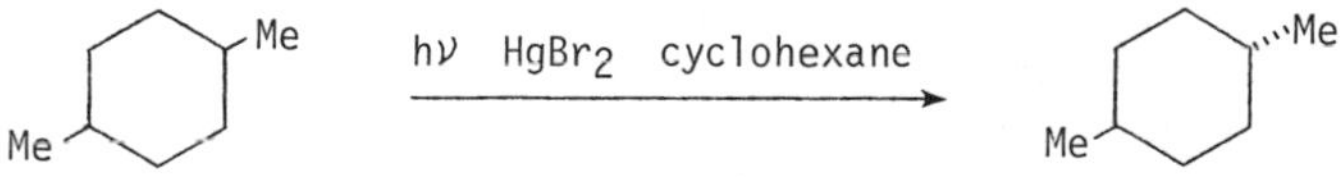

JACS (1970) 92 1094

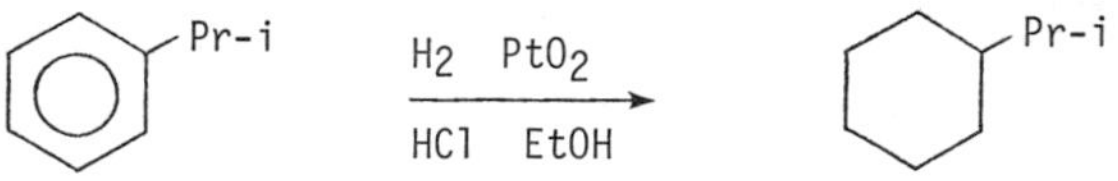

JACS (1936) 58 1594

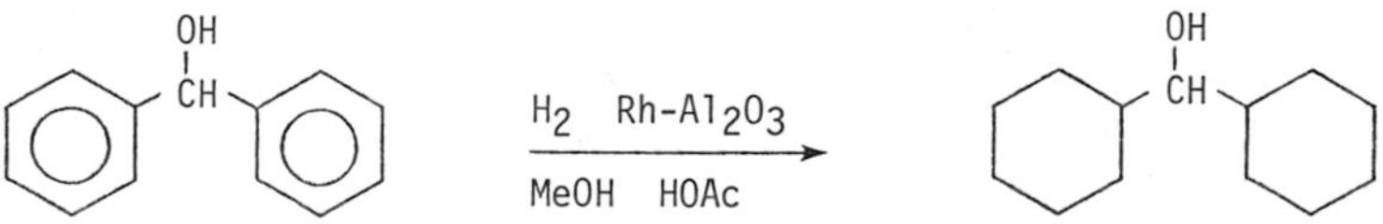

80%

JOC (1962) 27 2288

Section 66   Alkyls and Aryls from Amides

The conversions ArCONR'R" ⟶ ArMe and RCONHAr ⟶ ArAr' are included in this section. For the conversion RCONR'R" ⟶ RH see section 156 (Hydrides from Amides)

PhCON⟨ ⟩ $\xrightarrow[\text{copper chromite\quad dioxane}]{\text{H}_2\text{ (200-300 atmos)}}$ PhMe   79%

JACS (1934) 56 2419

$\xrightarrow[\text{2 C}_6\text{H}_6]{\text{1 HNO}_2\quad\text{HOAc}}$

JCS (1940) 361 369
Org React (1944) 2 224

Section 67   Alkyls and Aryls from Amines

Reactions in which an amine group is replaced by alkyl or aryl are included in this section. For the conversion RNH$_2$ ⟶ RH see section 157 (Hydrides from Amines)

$\xrightarrow[\text{2 NaOH}\quad\text{C}_6\text{H}_6\quad\text{H}_2\text{O}]{\text{1 NaNO}_2\quad\text{HCl}\quad\text{H}_2\text{O}}$ 34%

Org Synth (1932) Coll Vol 1 113
Org React (1944) 2 224
JCS (1962) 4257

$\xrightarrow{\text{NOCl}\quad\text{PhCl}}$ 15%

JACS (1969) 91 5073

PhCHNHCH$_2$Ph $\xrightarrow[\text{2 LiAlH}_4 \quad \text{Et}_2\text{O}]{\text{1 NaNO}_2 \quad \text{HCl} \quad \text{H}_2\text{O}}$ PhCHNCH$_2$Ph $\xrightarrow[\text{EtOH}]{\text{HgO}}$ PhCHCH$_2$Ph
| | |
Me   NH$_2$ Me
                                                                    |
                                                                    Me

Chem Pharm Bull (1970) __18__ 1137

Section 68   Alkyls and Aryls from Esters

This section lists the conversion of the group COOR to alkyl or aryl.
For the conversion RCOOR ⟶ RH see section 158 (Hydrides from Esters)

C$_{15}$H$_{31}$COOMe $\xrightarrow[\text{Et}_2\text{O}]{\text{EtMgBr}}$ C$_{15}$H$_{31}$COH $\xrightarrow[\text{2 H}_2 \quad \text{Pt} \quad \text{EtOH}]{\text{1 HCOOH}}$ C$_{15}$H$_{31}$CHEt   86%
                                            | |
                                            Et   Et
                                            |
                                            Et

JACS (1945) __67__ 2239

$\xrightarrow[]{\text{BF}_3 \quad \text{C}_6\text{H}_6}$ 37%

JACS (1937) __59__ 1204
Org React (1946) __3__ 1

Me
|
MeCH=CCH$_2$OCO $\xrightarrow[]{\text{EtMgBr} \quad \text{Et}_2\text{O}}$

Me
|
MeCH=CCH$_2$Et

JCS (1965) 702

$\xrightarrow[\text{260-300°}]{\text{H}_2 \quad \text{MoS}_2 \quad \text{cyclohexane}}$

+

Coll Czech (1958) __23__ 1322

Section 69   Alkyls and Aryls from Ethers

The conversion ROR ⟶ RR' (R'=alkyl or aryl) is included in this section.
For the hydrogenolysis of ethers (ROR ⟶ RH) see section 159 (Hydrides
from Ethers)

$$Et_2O \xrightarrow{\quad BF_3 \quad C_6H_6 \quad} EtPh \qquad (+ diethylbenzenes)$$ 25%

JACS (1938) <u>60</u> 125
Org React (1946) <u>3</u> 1

J Prakt Chem (1938) <u>151</u> 61

33%

Section 70   Alkyls and Aryls from Halides and Sulfonates

The replacement of halo and tosyloxy groups by alkyl or aryl groups is
included in this section. For the conversion RX ⟶ RH (X=halo or OTs)
see section 160 (Hydrides from Halides and Tosylates)

JCS (1950) 711
Chem Rev (1946) <u>38</u> 139

$$PhI \xrightarrow{\quad PhCu \quad Pyr \quad} PhPh$$ 60%

Tetr Lett (1968) 3307
JCS (1950) 7'1

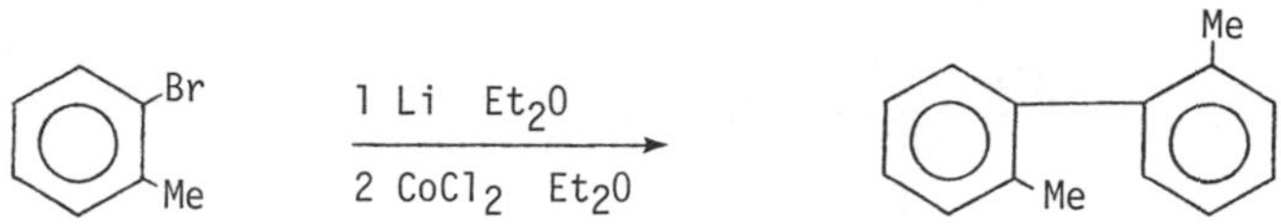

65%

Bull Soc Chim Fr (1964) 1331

PhCl → (PhLi piperidine Et$_2$O) → PhPh   61%

Angew (1957) 69 267

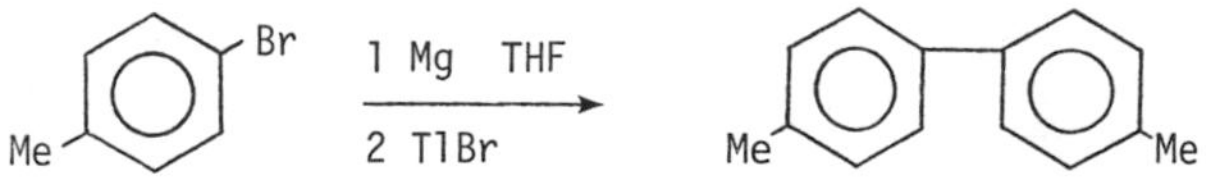

91%

JACS (1968) 90 2423
Tetrahedron (1970) 26 4041

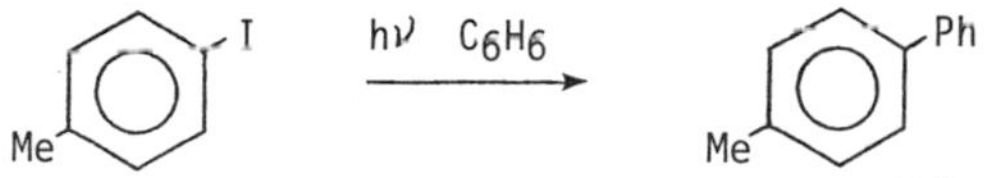

75%

JOC (1965) 30 2493

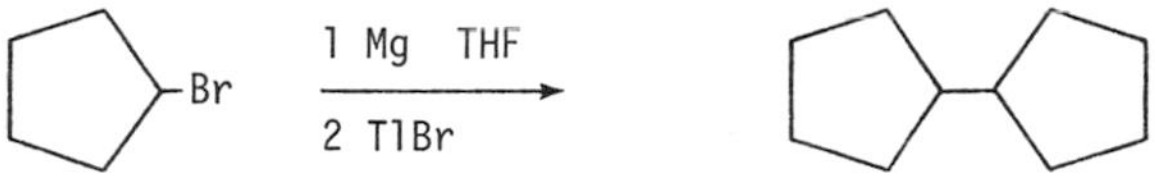

56%

JACS (1968) 90 2423

C$_6$H$_{13}$CHCl   → (LiCH$_2$CH=CH$_2$ / pentane) → C$_6$H$_{13}$CHCH$_2$CH=CH$_2$ 59%
       | |
       Me   Me

JOC (1970) 35 22
    (1954) 19 934

$$(MeO)_2CH(CH_2)_2\underset{Me}{C}=CHCH_2Cl \xrightarrow[\text{THF \quad HMPA}]{\overset{Me}{\underset{CH_2=CCH_2MgCl}{}}} (MeO)_2CH(CH_2)_2\underset{Me}{C}=CHCH_2CH_2\underset{Me}{C}=CH_2 \quad 95\%$$

Tetr Lett (1969) 1393

$$PhCH_2Cl \xrightarrow[\text{2 BuOTs}]{\text{1 Mg \quad Et_2O}} PhCH_2Bu \qquad\qquad 50\text{-}59\%$$

Org Synth (1943) Coll Vol 2 47
(1932) Coll Vol 1 471

$$EtCH(CH_2)_{10}I \xrightarrow{\text{C}_{18}\text{H}_{37}\text{I \quad Na \quad Et}_2\text{O}} EtCH(CH_2)_{10}C_{18}H_{37}$$
$$\underset{Me}{} \qquad\qquad\qquad\qquad\qquad\qquad \underset{Me}{}$$

JACS (1965) $\underline{87}$ 5452

$$BuBr \xrightarrow[\substack{\text{2 (ICuPBu}_3)_4 \\ \text{3 O}_2}]{\text{1 Li \quad THF}} BuBu \qquad\qquad 84\%$$

JACS (1967) $\underline{89}$ 5302

$$t\text{-BuCl} \xrightarrow{\text{FeCl}_3 \quad \text{C}_6\text{H}_6} t\text{-BuPh} \qquad\qquad 80\%$$

Org React (1946) $\underline{3}$ 1

$$PrBr \xrightarrow[\text{2 t-BuCl \quad HgCl}_2]{\text{1 Mg \quad Et}_2\text{O}} PrBu\text{-}t \qquad\qquad 21\%$$

JACS (1929) $\underline{51}$ 1483
(1938) $\underline{60}$ 2598

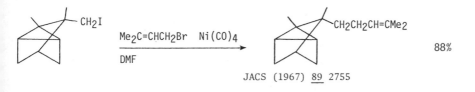

$Me_2C=CH(CH_2)_2\overset{Me}{\underset{|}{C}}=CH(CH_2)_2\overset{Me}{\underset{|}{C}}=CHCH_2Cl$

1 $(NH_2)_2CS$
2 Base O_2
3 Ph_3P
4 Benzyne
5 Li NH$_3$

$Me_2C=CH(CH_2)_2\overset{Me}{\underset{|}{C}}=CH(CH_2)_2\overset{Me}{\underset{|}{C}}=CHCH_2$
$Me_2C=CH(CH_2)_2\underset{\underset{Me}{|}}{C}=CH(CH_2)_2\underset{\underset{Me}{|}}{C}=CHCH_2$

Chem Comm (1969) 99

—CH$_2$I

$Me_2C=CHCH_2Br$ $Ni(CO)_4$
———————————————
DMF

—CH$_2$CH$_2$CH=CMe$_2$

88%

JACS (1967) 89 2755

MeCH=CHCH$_2$Cl

$Ni(CO)_4$ MeOH
————————————→

MeCH=CHCH$_2$CHCH=CH$_2$
$\underset{Me}{|}$

JACS (1951) 73 2654

$\underset{\underset{X}{O\ \ O}}{CH_2CH(CH_2)_2I}$

LiC≡CMe
————————→

$\underset{\underset{X}{O\ \ O}}{CH_2CH(CH_2)_2C≡CMe}$

H_2 Pd
————————→

$\underset{\underset{X}{O\ \ O}}{CH_2CH(CH_2)_2Pr}$

E. J. Corey, in press

I(CH$_2$)$_{10}$COOH

Bu$_2$CuLi Et$_2$O
————————————→

Bu(CH$_2$)$_{10}$COOH

76%

JACS (1968) 90 5615

RI

1 Me$_2$CuLi   Et$_2$O
————————————→
2 O$_2$

RMe   (R=C$_5$H$_{11}$ or Ph)

98-99%

JACS (1969) 91 4871

PhCH=CHBr $\xrightarrow{\text{Me}_2\text{CuLi} \quad \text{Et}_2\text{O}}$ PhCH=CHMe   81%

JACS (1967) $\underline{89}$ 3911 4245

$\underset{\underset{\text{Pr}}{|}}{\text{Me}_2\text{CCl}}$ $\xrightarrow{\text{Me}_3\text{Al}}$ $\underset{\underset{\text{Pr}}{|}}{\text{Me}_2\text{CMe}}$ 100%

JOC (1970) $\underline{35}$ 532

$C_{10}H_{21}I$ $\xrightarrow{\text{Me}_3\text{MnLi} \quad \text{Et}_2\text{O}}$ $C_{10}H_{21}Me$ 55%

Tetr Lett (1970) 315

PhCl $\xrightarrow{\text{NaH} \quad \text{Me}_2\text{SO}}$ $\underset{41\%}{PhCH_2SOMe}$ $\dashrightarrow[]{\text{Al-Hg}}$ PhMe

JACS (1965) $\underline{87}$ 1345

$C_{11}H_{23}X$ $\xrightarrow{\text{Me}_2\text{SO} \quad \text{NaH}}$ $C_{11}H_{23}CH_2SOMe$ $\xrightarrow[\text{EtOH}]{\text{Ni}}$ $C_{11}H_{23}Me$ 62%

(X=Br or OTs)

Aust J Chem (1966) $\underline{19}$ 521

$\xrightarrow{\text{Me}_2\text{CuLi} \quad \text{Et}_2\text{O}}$

65%

JACS (1967) $\underline{89}$ 3911 4245

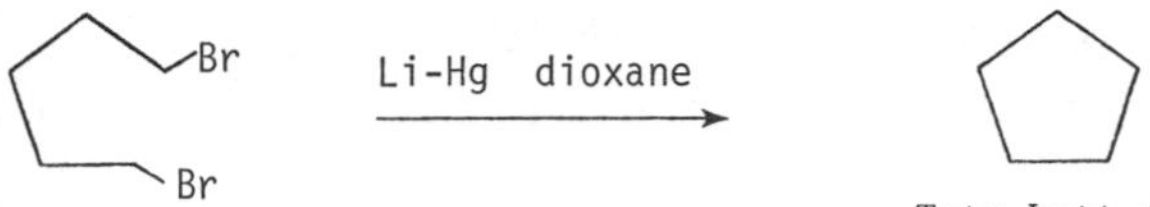
Br
Br
Li-Hg dioxane

76%

Tetr Lett (1967) 4925

Section 71 Alkyls and Aryls from Hydrides (RH)

This section lists examples of the reaction RH → RR' (R,R'=alkyl or aryl)

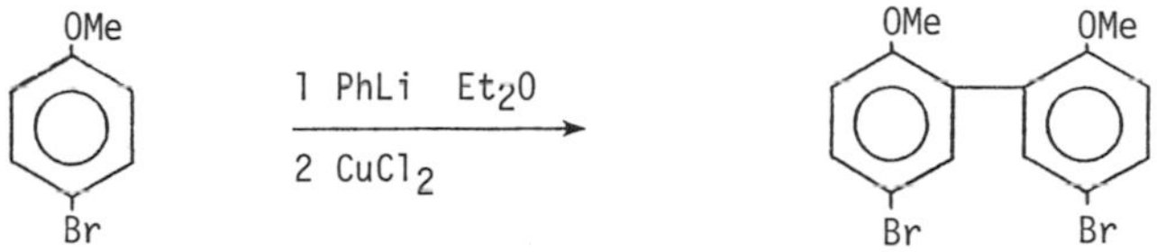
OMe
Br
1 PhLi Et_2O
2 $CuCl_2$
OMe
OMe
Br
Br

19%

JACS (1940) 62 1963

PhH $\xrightarrow[\text{2 } h\nu \text{ } C_6H_6]{\text{1 } Tl(OCOCF_3)_3 \text{ } CF_3COOH}$ PhPh 90%

JACS (1970) 92 6088

PhH $\xrightarrow{\text{n-PrCl } AlCl_3}$ PhPr-n + PhPr-i

JOC (1940) 5 253
Org React (1946) 3 1

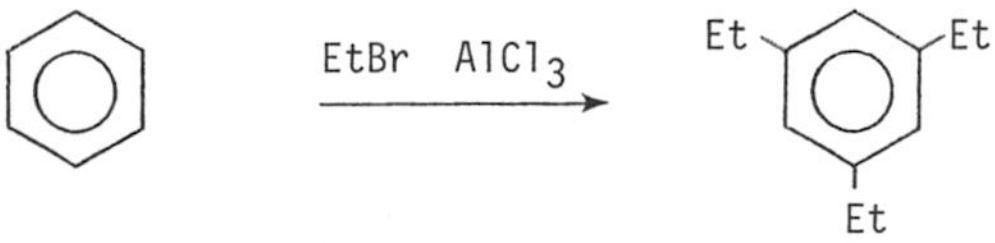
EtBr $AlCl_3$
Et
Et
Et

86%

Org React (1946) 3 1

$\xrightarrow{\text{i-PrLi \quad decalin \quad 165°}}$

JACS (1963) <u>85</u> 1356   20%

 $\xrightarrow{\text{NaH \quad Me}_2\text{SO}}$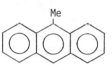

JOC (1966) <u>31</u> 248   33%

PhH $\xrightarrow{\text{MeHgI \quad h}\nu}$ PhMe

Proc Chem Soc (1961) 240

RH $\xrightarrow{\text{MeF-SbF}_3}$ RMe   (R=alkyl or aryl)

JACS (1969) <u>91</u> 2112

[thiophene]-CH$_3$ $\xrightarrow[\text{HMPA}]{\text{PhCH=CH}_2 \quad \text{t-BuOK}}$ [thiophene]-CH$_2$(CH$_2$)$_2$Ph

Tetr Lett (1968) 3723

Ph$_2$CH$_2$ $\xrightarrow[\text{2 C}_8\text{H}_{17}\text{Br}]{\text{1 NaNH}_2 \quad \text{NH}_3}$ Ph$_2$CHC$_8$H$_{17}$ 99%

JACS (1957) <u>79</u> 3142

PhCH$_3$ $\xrightarrow{\text{PrCl} \quad \text{Na}}$ PhCH$_2$Pr   43%

JACS (1941) <u>63</u> 327

Me$_2$CHMe $\xrightarrow{\text{CH}_2=\text{CH}_2 \quad \text{AlCl}_3 \quad \text{HCl}}$ Me$_2$CHCHMe$_2$

JACS (1945) <u>67</u> 1778

Section 72   Alkyls, Methylenes and Aryls from Ketones

The conversions R$_2$CO $\longrightarrow$ RR, R$_2$CH$_2$, R$_2$CHR, etc. (R=alkyl or aryl) are
included in this section. For the conversion R$_2$CO $\longrightarrow$ RH see section 162
(Hydrides from Ketones)

$\xrightarrow{\text{H}_2\text{O}_2}$

$\xrightarrow{150°}$ 44%

JACS (1968) <u>90</u> 817

EtCO(CH$_2$)$_{14}$COEt $\xrightarrow[\text{2 HI} \quad \text{P}_4 \quad 270°]{\text{1 C}_{16}\text{H}_{33}\text{C}\equiv\text{CMgBr}}$ EtCH(CH$_2$)$_{14}$CHEt   26%

$\qquad\qquad\qquad$ C$_{18}$H$_{37}$ $\quad$ C$_{18}$H$_{37}$

Helv (1937) <u>20</u> 1179

Me$_2$CO $\xrightarrow[\text{2 I}_2]{\text{1 BuMgBr} \quad \text{Et}_2\text{O}}$ Me$_2$C=CHPr $\xrightarrow[\text{catalyst}]{\text{H}_2}$ Me$_2$CHBu

JACS (1929) <u>51</u> 1483
      (1949) <u>71</u> 819
JOC (1948) <u>13</u> 239

Further examples of the conversion of ketones into olefins and of the
hydrogenation of olefins are included in section 207 (Olefins from Ketones)
and section 74 (Alkyls and Methylenes from Olefins)

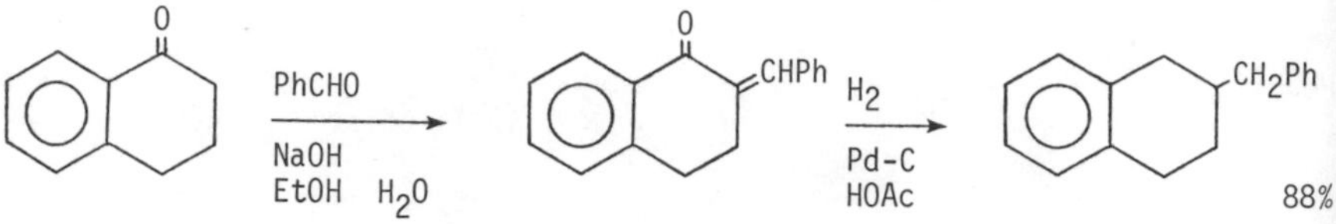
CHPh
CH2Ph
PhCHO
NaOH
EtOH H2O
H2
Pd-C
HOAc

88%

Org Prep and Procedures (1970) 2 37

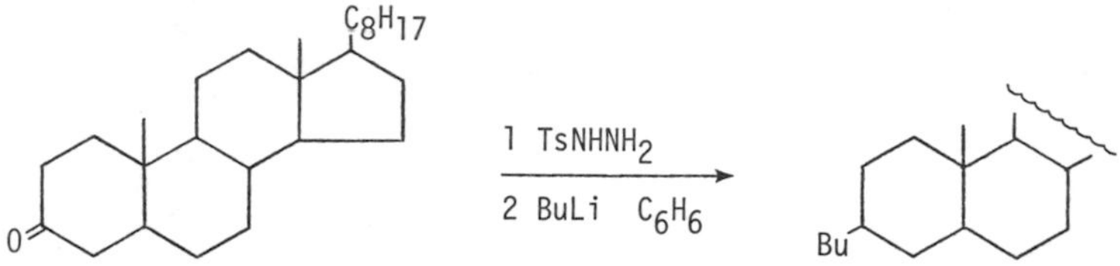
C8H17
1 TsNHNH2
2 BuLi C6H6
Bu

Chem Comm (1969) 1395

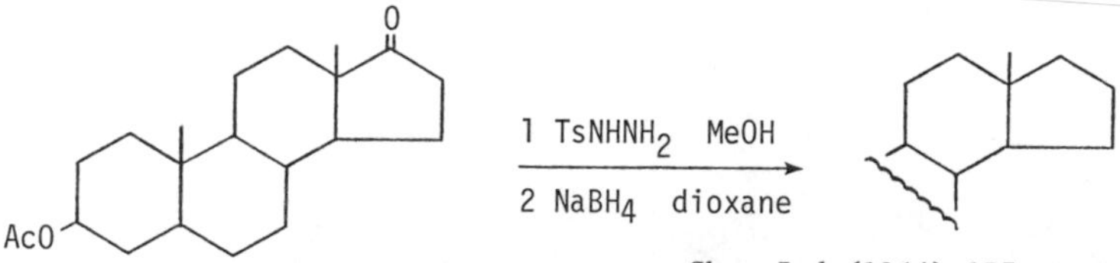
AcO
1 TsNHNH2 MeOH
2 NaBH4 dioxane

60-70%

Chem Ind (1964) 153
Tetrahedron (1963) 19 1127

PhCOEt —(N_2H_4 KOH; diethylene glycol)→ $PhCH_2Et$ 82%

JACS (1946) 68 2487
Org React (1948) 4 378

For reduction of hindered ketones with $N_2H_4 \cdot HCl$ etc. see
Chem Ind (1964) 1194

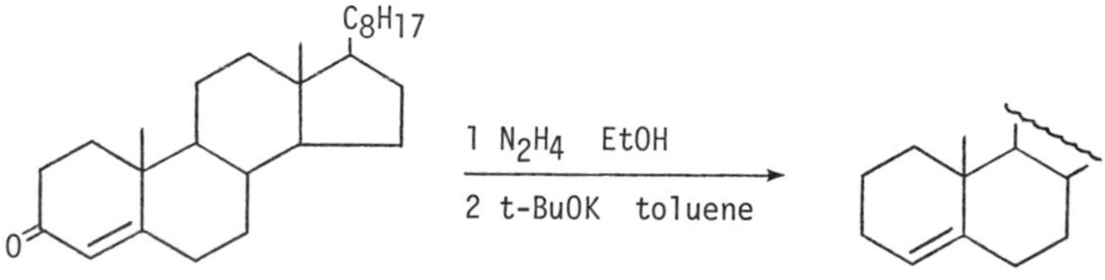
C8H17
1 N2H4 EtOH
2 t-BuOK toluene

~65%

JCS (1963) 1855
Tetrahedron (1970) 26 649

$$\text{1 } N_2H_4$$
$$\text{2 t-BuOK Me}_2\text{SO   25°}$$

JACS (1962) <u>84</u> 1734

For reduction of hydrazones with BuLi see JACS (1968) <u>90</u> 7287

$$N_2H_4 \cdot H_2O \quad KOH$$
$$\text{ethylene glycol}$$

JACS (1959) <u>81</u> 5834
        (1962) <u>84</u> 2938
JCS (1960) 4413

$$C_6H_{13}COMe \xrightarrow[Na_2SO_4]{EtSH \quad ZnCl_2} C_6H_{13}\underset{\underset{Me}{|}}{C}(SEt)_2 \xrightarrow[EtOH]{Ni \quad H_2O} C_6H_{13}CH_2Me \qquad 50\%$$

JACS (1944) <u>66</u> 909
Org React (1962) <u>12</u> 356

$$(C_{10}H_{21})_2CO \xrightarrow{Zn-Hg \quad HCl \quad H_2O} (C_{10}H_{21})_2CH_2 \qquad 32\%$$

Rec Trav Chim (1936) <u>55</u> 903
Org React (1942) <u>1</u> 155
JACS (1947) <u>69</u> 2350

$$\text{Zn   HCl   Et}_2\text{O}$$

75%

Chem Comm (1969) 919

$$\text{1 } B_2H_6 \quad \text{diglyme}$$
$$\text{2 Ac}_2\text{O}$$

85%

Tetrahedron (1964) <u>20</u> 957

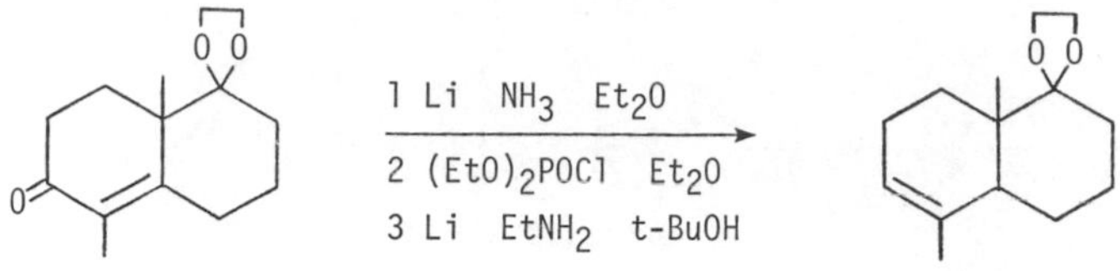

$$\begin{array}{l} \text{1 Li \quad NH}_3 \quad \text{Et}_2\text{O} \\ \text{2 (EtO)}_2\text{POCl} \quad \text{Et}_2\text{O} \\ \text{3 Li \quad EtNH}_2 \quad \text{t-BuOH} \end{array}$$

Tetr Lett (1969) 2145

50%

LiAlH$_4$ AlCl$_3$ Et$_2$O

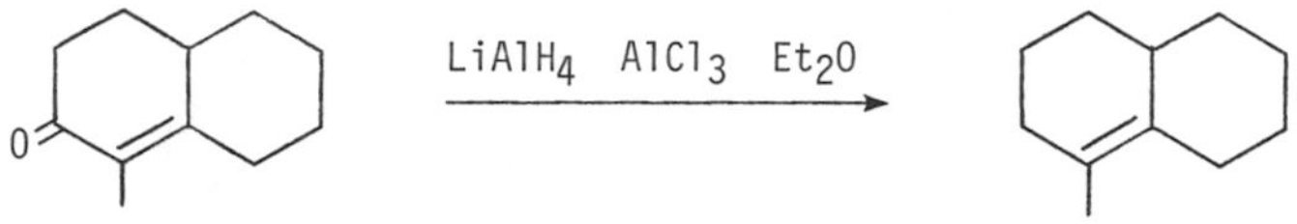

Tetrahedron (1965) 21 2641

For application to phenyl ketones see JCS (1957) 3755

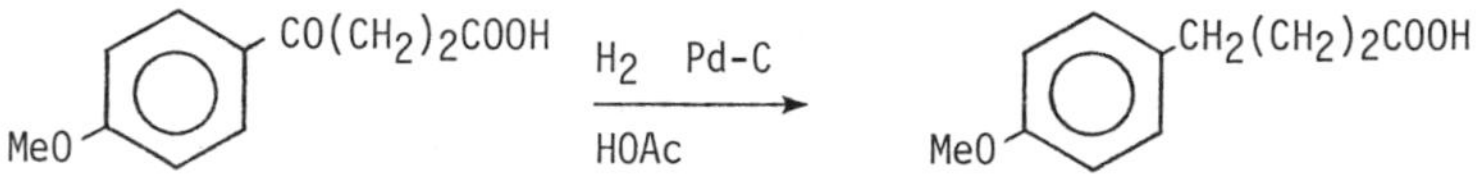

CO(CH$_2$)$_2$COOH   H$_2$ Pd-C

HOAc

MeO — CH$_2$(CH$_2$)$_2$COOH

JACS (1949) 71 1036
Org React (1953) 7 263

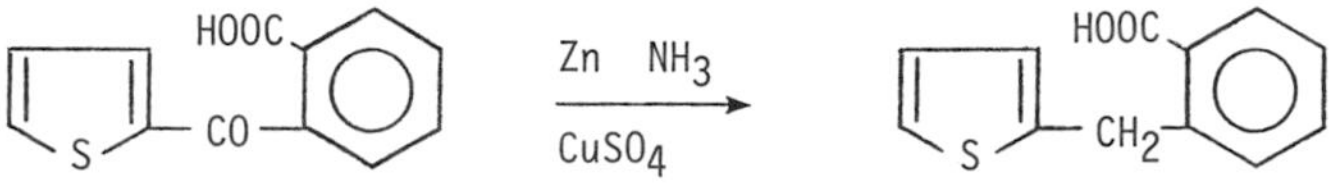

75%

HOOC   Zn   NH$_3$

CuSO$_4$

HOOC   CH$_2$

JOC (1970) 35 711

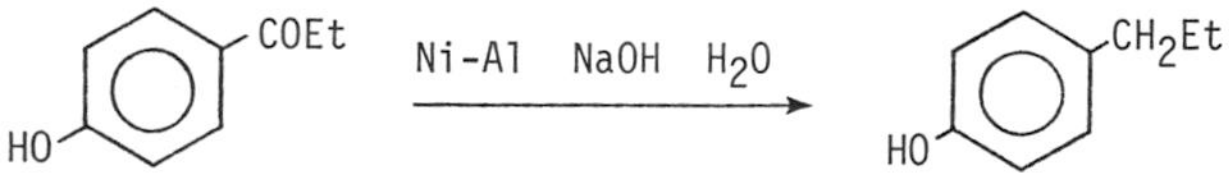

90%

COEt   Ni-Al   NaOH   H$_2$O

HO — CH$_2$Et

Org React (1953) 7 263

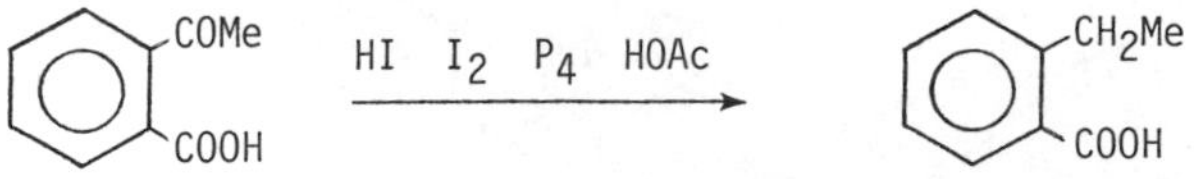

78%

COMe   HI   I$_2$ P$_4$ HOAc

COOH

CH$_2$Me
COOH

Ber (1959) 92 1705

94%

$$C_{10}H_{21}\overset{\overset{\displaystyle Pr}{|}}{\underset{\underset{\displaystyle Bu}{|}}{C}}CH_2COC_6H_{13} \quad \xrightarrow[\text{Et}_2\text{O}]{\text{LiAlH}_4} \quad C_{10}H_{21}\overset{\overset{\displaystyle Pr}{|}}{\underset{\underset{\displaystyle Bu}{|}}{C}}CH_2\underset{\underset{\displaystyle OH}{|}}{C}HC_6H_{13} \quad \xrightarrow[\text{2 H}_2 \quad \text{Ni}]{\text{1 KHSO}_4} \quad C_{10}H_{21}\overset{\overset{\displaystyle Pr}{|}}{\underset{\underset{\displaystyle Bu}{|}}{C}}CH_2CH_2C_6H_{13}$$

JOC (1959) $\underline{24}$ 1964

Electrolysis H_2SO_4

dioxane H_2O

97%

JACS (1967) $\underline{89}$ 4789

PhCOMe $\xrightarrow[]{\text{Et}_3\text{SiH} \quad \text{CF}_3\text{COOH}}$ PhCH$_2$Me   70%

Tetrahedron (1967) $\underline{23}$ 2235

(PhCH$_2$)$_2$CO $\xrightarrow{h\nu}$ PhCH$_2$CH$_2$Ph

JACS (1970) $\underline{92}$ 6076 6077

$\xrightarrow{h\nu}$

JACS (1942) $\underline{64}$ 80
(1961) $\underline{83}$ 4923

Section 73   Alkyls from Nitriles

The conversion RCN → RMe is included in this section. For the
replacement of CN by hydrogen, RCN → RH, see section 163 (Hydrides
from Nitriles)

Tetr Lett (1966) 4255
Tetrahedron (1961) 13 287

Section 74   Alkyls, Methylenes and Aryls from Olefins

The hydrogenation, alkylation, arylation, dimerization etc. of olefins,
forming alkanes or aryl-substituted alkanes, are included in this section.
For the conversion $R_2C=CR_2$ → RH or R_2CH_2 see section 164 (Hydrides from
Olefins)

$$BuCH=CH_2 \quad \xrightarrow[\text{2 KOH } AgNO_3 \text{ } H_2O]{\text{1 } NaBH_4 \text{ } BF_3 \cdot Et_2O \text{ diglyme}} \quad Bu(CH_2)_4Bu \qquad 66\%$$

JACS (1961) 83 1002
Chem Comm (1968) 938

$$EtCH=CHEt \quad \xrightarrow{C_6H_6 \text{ } H_2SO_4} \quad \underset{\underset{Ph}{|}}{EtCHCH_2Et}$$

JACS (1939) 61 1002
Org React (1946) 3 1

$C_{10}H_{21}CH=CH_2$ $\xrightarrow[\text{2 H}_2\text{O}]{\text{1 Pr}_3\text{Al}}$ $C_{10}H_{21}\underset{\underset{Pr}{|}}{CHMe}$

Angew (1952) <u>64</u> 323

t-BuLi   Et$_3$N

—Bu-t

JOC (1965) <u>30</u> 917

$C_5H_{11}CH=CH_2$ $\xrightarrow[\text{benzoyl peroxide}]{\text{CHCl}_2\text{SO}_2\text{Cl}}$ $C_5H_{11}\underset{\underset{Cl}{|}}{CHCH_2CHCl_2}$ $\xrightarrow[\text{dioxane}]{\text{LiAlH}_4}$ $C_5H_{11}(CH_2)_2Me$

$+ \ C_5H_{11}\underset{\underset{CH_2}{\diagdown\diagup}}{CHCH_2}$

Tetrahedron (1964) <u>20</u> 1613

$PhCH=CH_2$ $\overset{Br_2}{\dashrightarrow}$ $Ph\underset{\underset{Br}{|}}{CHCH_2Br}$ $\xrightarrow[\text{Me}_2\text{SO}]{\text{NaBH}_4}$ $PhCH_2Me$ 64%

Tetr Lett (1969) 3495

$BuCH=CH_2$ $\xrightarrow[\text{2 EtCOOH}]{\text{1 NaBH}_4 \ \ \text{BF}_3\cdot\text{Et}_2\text{O} \ \ \text{diglyme}}$ $BuCH_2Me$ 91%

JACS (1959) <u>81</u> 4108

$C_7H_{15}CH=CH_2$ $\xrightarrow[\substack{\text{2 Hg(OAc)}_2 \\ \text{3 NaBH}_4}]{\text{1 B}_2\text{H}_6 \ \ \text{THF}}$ $C_7H_{15}CH_2Me$

JACS (1970) <u>92</u> 3221

$C_6H_{13}CH=CH_2$ $\xrightarrow[\text{225°}]{\text{H}_2 \text{ (2000 psi) (i-Bu)}_3\text{B}}$ $C_6H_{13}CH_2Me$

JOC (1962) $\underline{27}$ 4368
JACS (1961) $\underline{83}$ 4672

$Ph_2C=CHMe$ $\xrightarrow[\text{2 H}_2\text{O}]{\text{1 Et}_2\text{AlCl}}$ Ph_2CHCH_2Me

Tetr Lett (1970) 3471

$C_6H_{13}CH=CH_2$ $\xrightarrow{\text{H}_2 \text{ Pt}}$ $C_6H_{13}CH_2Me$

JACS (1962) $\underline{84}$ 1495
Org Synth (1943) Coll Vol 2 191

For hydrogenation with Pd   catalyst see JACS (1960) $\underline{82}$ 6087
   "   "   " Ru   "   " JOC (1959) $\underline{24}$ 708
   "   "   " Re_2S_7 "   " JACS (1954) $\underline{76}$ 1519
                                                (1959) $\underline{81}$ 3587
   "   "   " Ni   "   " JOC (1970) $\underline{35}$ 1900

$PhSCH_2CH=CH_2$ $\xrightarrow{\text{H}_2 \text{ Ph}_3\text{PRhCl } \text{C}_6\text{H}_6}$ $PhS(CH_2)_2Me$ 93%

Tetr Lett (1967) 1935
JOC (1967) $\underline{32}$ 2013 3074

For hydrogenation with $Rh(COOMe)_4$ catalyst see Chem Comm (1969) 825
   "   "   " $RuClH(PPh_3)_3$ "   " Chem Comm (1967) 305

$C_8H_{17}CH=CH(CH_2)_7COOH$ $\xrightarrow{\text{N}_2\text{H}_4 \text{ air EtOH}}$ $C_8H_{17}(CH_2)_9COOH$ 90%

Chem Ind (1961) 433
JOC (1965) $\underline{30}$ 3980
Chem Rev (1965) $\underline{65}$ 51
For copper catalyzed diimide reduction see Tetr Lett (1961) 347
and JACS (1970) $\underline{92}$ 6635

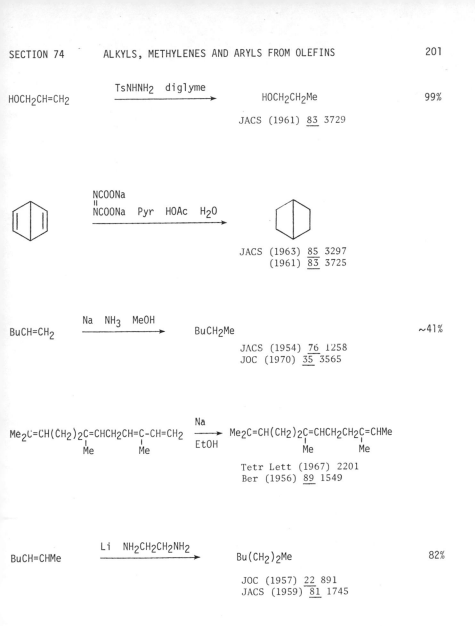

HOCH$_2$CH=CH$_2$ $\xrightarrow{\text{TsNHNH}_2 \quad \text{diglyme}}$ HOCH$_2$CH$_2$Me   99%

JACS (1961) 83 3729

NCOONa
||
NCOONa   Pyr   HOAc   H$_2$O

JACS (1963) 85 3297
      (1961) 83 3725

BuCH=CH$_2$ $\xrightarrow{\text{Na} \quad \text{NH}_3 \quad \text{MeOH}}$ BuCH$_2$Me   ~41%

JACS (1954) 76 1258
JOC (1970) 35 3565

Me$_2$C=CH(CH$_2$)$_2$C=CHCH$_2$CH=C-CH=CH$_2$ $\xrightarrow{\text{Na} \atop \text{EtOH}}$ Me$_2$C=CH(CH$_2$)$_2$C=CHCH$_2$CH$_2$C=CHMe
          | | | |
          Me   Me   Me   Me

Tetr Lett (1967) 2201
Ber (1956) 89 1549

BuCH=CHMe $\xrightarrow{\text{Li} \quad \text{NH}_2\text{CH}_2\text{CH}_2\text{NH}_2}$ Bu(CH$_2$)$_2$Me   82%

JOC (1957) 22 891
JACS (1959) 81 1745

Me$_2$C=CHMe $\xrightarrow{\text{Et}_3\text{SiH} \quad \text{CF}_3\text{COOH}}$ Me$_2$CHCH$_2$Me

(BuCH=CH$_2$ is not reduced)

Tetrahedron (1967) 23 2235

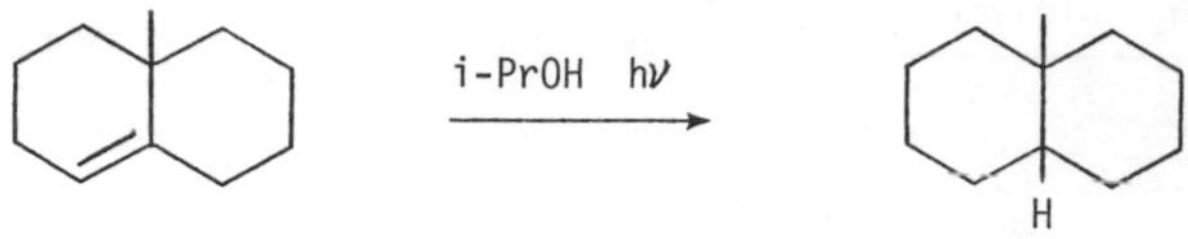

$$\xrightarrow{\text{i-PrOH } h\nu}$$

50%

Chem Comm (1968) 296

$$Me_2CHCH=CHCHMe_2 \xrightarrow[\text{2 } h\nu]{\text{1 } O_3} Me_2CHCHMe_2 \qquad 33\%$$

Tetr Lett (1968) 3291

Section 75   Alkyls and Methylenes from Miscellaneous Compounds
○○○

$$\begin{array}{c} PrCOCH(CH_2)_{12}COOMe \\ | \\ EtCOO \end{array} \xrightarrow[\text{2 NaBH}_4]{\text{1 TsNHNH}_2} PrCH_2CH_2(CH_2)_{12}COOMe \qquad 70\%$$

Chem Ind (1967) 2150

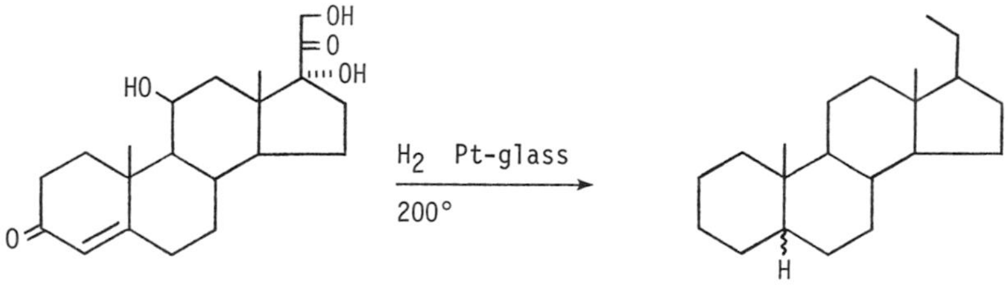

$$\xrightarrow[200°]{\text{H}_2 \quad \text{Pt-glass}}$$

Biochem J (1967) 105 40P

Chapter 6 PREPARATION OF AMIDES

Section 76 Amides from Acetylenes

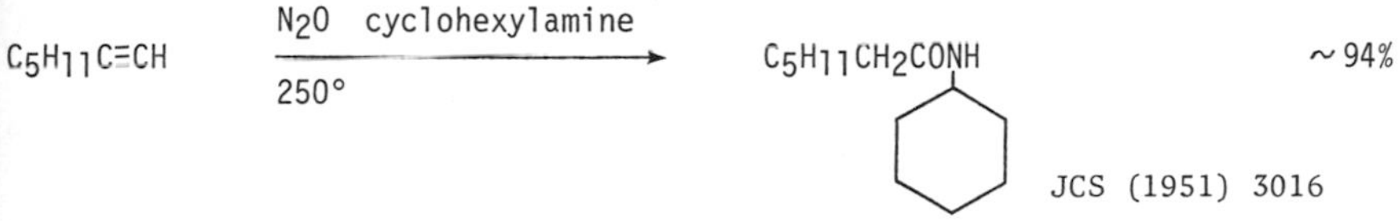

C5H11C≡CH
N2O cyclohexylamine
250°
C5H11CH2CONH
~94%

JCS (1951) 3016

$C_5H_{11}C{\equiv}CH$ —— S NH$_3$ H$_2$O Pyr ——→ $C_5H_{11}CH_2CONH_2$ 35%

JACS (1946) 68 2033

PhC≡CMe —— S NH$_3$ H$_2$O ——→ $PhCH_2CH_2CONH_2$ 72%

JACS (1946) 68 2029

$C_5H_{11}C{\equiv}CH$ —— 1 BuLi S; 2 Et$_2$NH t-BuBr ——→ $C_5H_{11}CH_2CSNEt_2$ ---→ $C_5H_{11}CH_2CONEt_2$

Rec Trav Chim (1968) 87 38

Section 77   Amides from Carboxylic Acids and Acid Halides
∘∘

JCS C (1970) 971

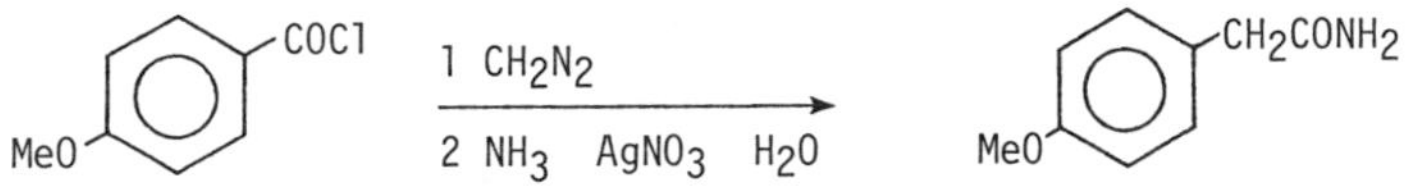

33%

Org React (1942) 1 38

57%

$PrCOCl \xrightarrow{CH_2N_2} PrCOCHN_2 \xrightarrow[C_6H_6]{PhNHMe \quad h\nu} PrCH_2CONMe$
                                                                    |
                                                                    Ph

Ber (1959) 92 528
JCS C (1970) 1208

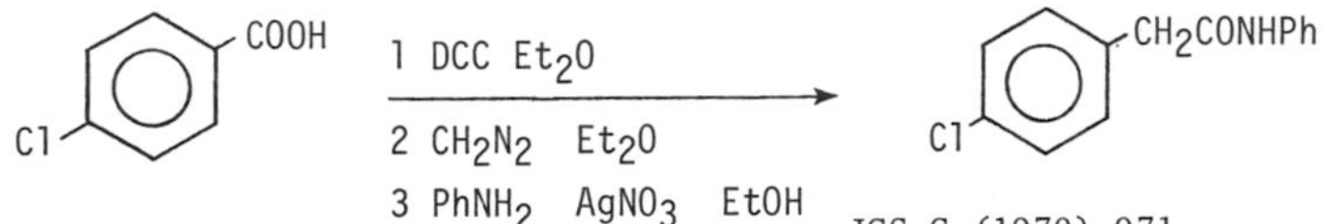

45%

$C_6H_{13}COOH \xrightarrow{NH_3 \quad 190°} C_6H_{13}CONH_2$

JACS (1931) 53 1879

75%

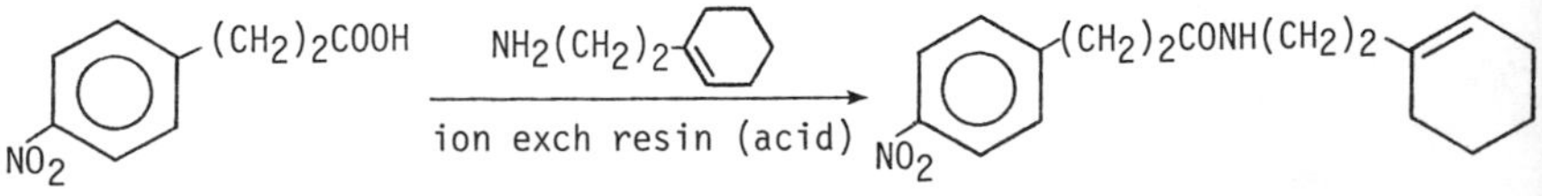

Helv (1961) 44 1546
JCS C (1969) 874

$C_6H_{13}COOH$ $\xrightarrow{\quad NH_2CONH_2 \quad 170\text{-}180° \quad}$ $C_6H_{13}CONH_2$ 68-74%

Org Synth (1963) Coll Vol 4 513

PhCOCl $\xrightarrow{\quad HCONMe_2 \quad 150° \quad}$ PhCONMe$_2$ 97%

JACS (1954) $\underline{76}$ 1372
    (1949) $\underline{71}$ 2215

i-PrCOCl $\xrightarrow{\quad NH_3 \quad H_2O \quad}$ i-PrCONH$_2$ 70%

Org Synth (1955) Coll Vol 3 490
JOC (1954) $\underline{19}$ 623

MeCH=CHCOCl $\xrightarrow{\quad NH_4OAc \quad Me_2CO \quad}$ MeCH=CHCONH$_2$ 63%

JCS (1962) 2824

$C_8H_{17}CH=CH(CH_2)_7COCl$ $\xrightarrow[\text{pet ether}]{\quad BuNH_2 \quad}$ $C_8H_{17}CH=CH(CH_2)_7CONHBu$ 83%

JACS (1949) $\underline{71}$ 2215
Helv (1959) $\underline{42}$ 2073

PhCH=CHCH$_2$CHCOOH
   |
   CH$_2$Ph
$\xrightarrow[\text{2 } NH_3]{\text{1 ClCOOEt   Et}_3N \quad CHCl_3}$
PhCH=CHCH$_2$CHCONH$_2$ 94%
       |
       CH$_2$Ph

J Med Chem (1968) $\underline{11}$ 534

JACS (1955) 77 6214   90%

JACS (1968) 90 4706
Compt Rend (1965) 260 2249
JACS (1968) 90 2448

Amides by reaction of acids with amines and HC≡COMe

							Rec Trav Chim (1955) 74 769
"	"	"	"	"	"	"	" HC≡CCH₂SMe₂Br
							JCS C (1969) 1904
"	"	"	"	"	"	"	" Me₂C(OMe)₂
							Chim Ther (1967) 2 195 (Chem Abs 67 108840)
"	"	"	"	"	"	"	" SiCl₄
							JOC (1969) 34 2766
"	"	"	"	"	"	"	" TiCl₄
							Can J Chem (1970) 48 983
"	"	"	"	"	"	"	" (PNCl₂)₃
							JOC (1968) 33 2979
"	"	"	"	"	"	"	" 2-iodo-1-methylpyridinium iodide
							JCS (1964) 4650
"	"	"	"	"	"	"	" SO₃·DMF
							JOC (1959) 24 368

Further examples of the reaction RCOOH + R₂NH ⟶ RCONR₂ are included in section 82 (Amides from Amines)

$C_{11}H_{23}COOH$ $\xrightarrow{\text{(Me}_2\text{N)}_3\text{PO 180-200°}}$ $C_{11}H_{23}CONMe_2$ 95%

Chem Ind (1966) 1529

$C_{11}H_{23}COONa$ $\xrightarrow{\text{Bu}_2\text{NCOCl}}$ $C_{11}H_{23}CONBu_2$ 78%

JOC (1963) 28 232

$PhCH_2COOH$ $\xrightarrow{\text{(BuNH)}_3\text{B}}$ $PhCH_2CONHBu$ 90%

Tetrahedron (1970) 26 1539

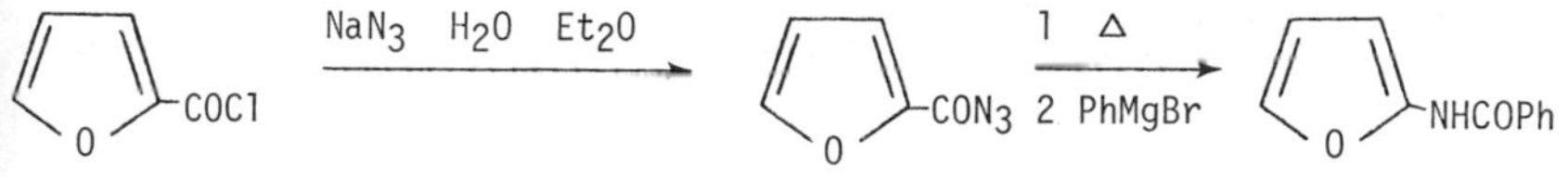

JACS (1938) 60 540   ~50%

Amides may also be prepared from carboxylic acids via ester intermediates.
See section 83 (Amides from Esters)

Section 78   Amides from Alcohols and Phenols

$$\xrightarrow[\text{xylene}]{\text{MeC(OMe)}_2\text{NMe}_2}$$

30%

CH₂CONMe₂

Tetr Lett (1968) 1899
Angew (1968) 80 626
(Internat Ed 7 629)

PhCH=CHCH₂OH $\xrightarrow[\text{NH}_3 \quad \text{Et}_2\text{O}]{\text{Nickel peroxide}}$ PhCH=CHCONH₂   85%

Chem Comm (1966) 17

PhCHCHMe₂ $\xrightarrow{\text{(NH}_4)_2\text{S}_x \quad 200°}$ PhCH₂CHCONH₂
| |
OH   Me

JACS (1946) 68 632 2033

$$\xrightarrow{\text{MeCN} \quad \text{H}_2\text{SO}_4 \quad \text{HOAc}}$$

JOC (1960) 25 331
Org React (1969) 17 213

$$\begin{array}{c} \text{Me} \\ | \\ \text{PhCH}_2\text{COH} \\ | \\ \text{Me} \end{array} \xrightarrow{\text{NaCN} \quad \text{H}_2\text{SO}_4 \quad \text{HOAc}} \begin{array}{c} \text{Me} \\ | \\ \text{PhCH}_2\text{CNHCHO} \\ | \\ \text{Me} \end{array}$$

92%

JOC (1961) 26 3002

$$\text{PhOH} \xrightarrow[\text{2 } 312\text{-}315°]{\substack{\text{Cl} \\ | \\ \text{1 PhN=CPh   MeONa}}} \text{Ph}_2\text{NCOPh}$$

Org React (1965) <u>14</u> 1

Section 79   <u>Amides from Aldehydes</u>

Arch Pharm (1957) <u>290</u> 218

$$\text{PhCHO} \xrightarrow{\text{PhCH}_2\text{CH}_2\text{N}_3 \quad \text{H}_2\text{SO}_4} \text{PhCONHCH}_2\text{CH}_2\text{Ph}$$ 10%

JACS (1955) <u>77</u> 951

$$\text{PhCHO} \xrightarrow[\text{CCl}_4]{\text{1 SO}_2\text{Cl}_2 \quad \text{benzoyl peroxide}} \text{PhCOCl} \xrightarrow{\text{PhNH}_2} \text{PhCONHPh}$$ 67%

Nippon Kagaku Z (1960) <u>81</u> 1450
(Chem Abs <u>56</u> 2370)

$$\text{C}_6\text{H}_{13}\text{CHO} \dashrightarrow \text{C}_6\text{H}_{13}\text{CH=NOH} \xrightarrow{\text{Ni   100°}} \text{C}_6\text{H}_{13}\text{CONH}_2$$ 100%

Org React (1960) <u>11</u> 1
JCS (1946) 599
JACS (1961) <u>83</u> 1983

PhCH=CHCHO $\xrightarrow[\text{Et}_2\text{O}]{\text{Nickel peroxide NH}_3}$ PhCH=CHCONH$_2$ 85%

Chem Comm (1966) 17

Ph(CH$_2$)$_2$CHO $\xrightarrow{\text{S   NH}_3\text{   Pyr   H}_2\text{O}}$ Ph(CH$_2$)$_2$CONH$_2$ 48%

JACS (1946) __68__ 2029

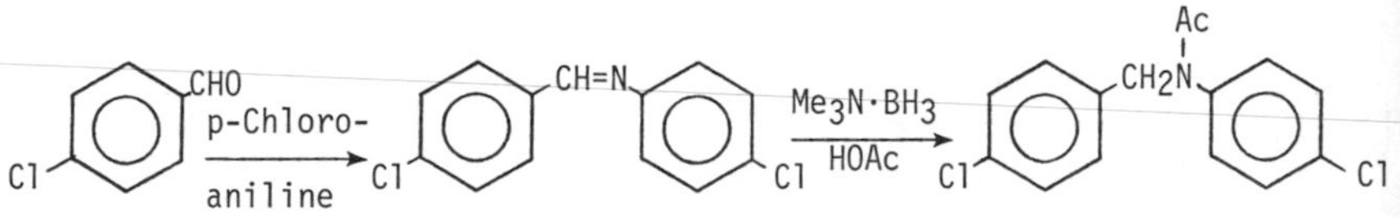

JOC (1962) __27__ 2640

Section 80   Amides from Alkyls

This section lists the conversion of alkyl groups into amide. For the conversion RH ⟶ RCONR$_2$ see section 86 (Amides from Hydrides)

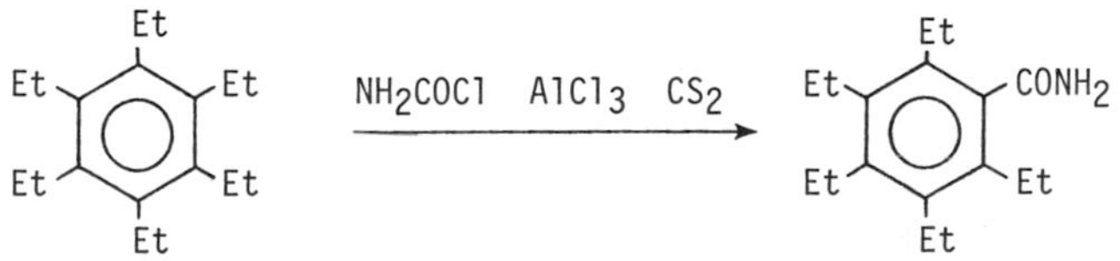

98%

Helv (1960) __43__ 1473

Section 81   Amides from Amides
             oooooooooooooooooo

$MeCH_2CONMe_2$ $\xrightarrow[\text{2 EtBr}]{\text{1 NaNH}_2 \quad \text{NH}_3}$ MeCHCONMe$_2$ 62%
          |
          Et

JOC (1966) <u>31</u> 982 989
Ber (1968) <u>101</u> 3113
JACS (1967) <u>89</u> 1647

31%   14%
Chem Ind (1965) 768

$MeCONH_2$ $\xrightarrow[\text{2 H}_2]{\text{1 MeCH=CH}_2 \quad \text{PdCl}_2 \quad \text{Na}_2\text{HPO}_4}$ MeCONHCHMe$_2$

Proc Chem Soc (1961) 370

$MeCONH_2$ $\xrightarrow[\text{H}_2\text{SO}_4 \quad \text{HOAc}]{\text{MeCH(OEt)}_2 \quad \text{H}_2 \quad \text{Pd-C}}$ MeCONHEt   45%

JOC (1962) <u>27</u> 2205

$MeCONH_2$ $\xrightarrow{\text{MeNH}_2 \cdot \text{HCl} \quad \Delta}$ MeCONHMe   75%

JACS (1943) <u>65</u> 1566

$$PhCH_2CONH_2 \xrightarrow{\quad PhNH_2 \quad BF_3 \quad \Delta \quad} PhCH_2CONHPh \qquad\qquad 99\%$$

JACS (1937) 59 1202

$$MeCONH_2 \xrightarrow[\quad H_3PO_4 \quad]{\quad Dihydropyran \quad} MeCONH-\!\!\!\left\langle \!\!\!\!\!\!\!\!\!\!\begin{array}{c} \\ O \end{array}\!\!\!\!\!\!\!\right\rangle \xrightarrow[\quad C_6H_{13}Br \quad]{\quad t\text{-}C_5H_{11}ONa \quad} MeCON-\!\!\!\left\langle \!\!\!\!\!\!\!\begin{array}{c} \\ O \end{array}\!\!\!\!\!\!\right\rangle$$

$$\big\downarrow Acid$$

$$MeCONHC_6H_{13}$$

Bull Soc Chim Fr (1964) 292

$$\xrightarrow[\quad 2 \; MeI \quad DMF \quad]{\quad 1 \; NaH \quad C_6H_6 \quad} \qquad\qquad > 66\%$$

JACS (1961) 83 1492
JOC (1949) 14 1099
      (1967) 32 3679

Examples of the N-alkylation of acetyl- and trifluoroacetyl-amines are included in section 97 (Amines from Amines)

$$PhCONHPh \xrightarrow{\quad PCl_5 \quad} \overset{Cl}{\underset{}{PhC\!=\!NPh}} \xrightarrow{\quad PhONa \quad} \overset{OPh}{\underset{}{PhC\!=\!NPh}} \xrightarrow{\quad 315° \quad} PhCONPh_2$$

Org React (1965) 14 1

$$PhCONBu_2 \xrightarrow[\quad 190\text{-}200° \quad]{\quad Pyridine \; hydrochloride \quad} PhCONHBu \qquad\qquad 30$$

Ber (1954) 87 1294

$PrCONMe_2$ $\xrightarrow{\quad K_2S_2O_8 \quad K_2HPO_4 \quad H_2O \quad}$ $PrCONHMe$ 44%

JOC (1964) <u>29</u> 3632

$CH_2=CH(CH_2)_7CONH_2$ $\xrightarrow{\quad Pb(OAc)_4 \quad C_6H_6 \quad HOAc \quad}$ $CH_2=CH(CH_2)_7NHAc$ 53%

Chem Comm (1965) 161
Aust J Chem (1968) <u>21</u> 185

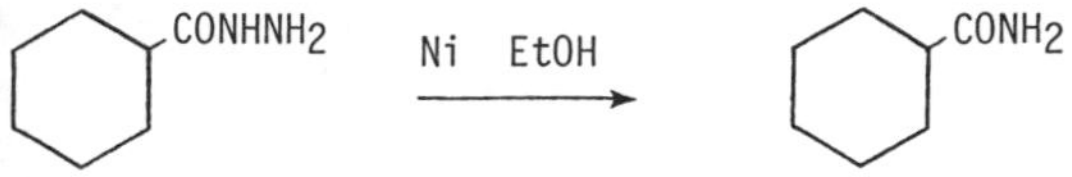

Ni   EtOH

JACS (1954) <u>76</u> 5774
JOC (1957) <u>22</u> 148

Section 82   Amides from Amines

MeCOCl   Pyr

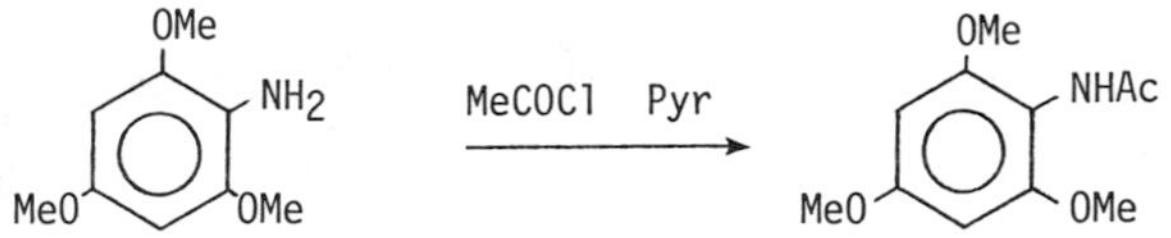

80%

JOC (1970) <u>35</u> 1219

$Ph_2C=CHCOCl$

NaOH H_2O Et_2O

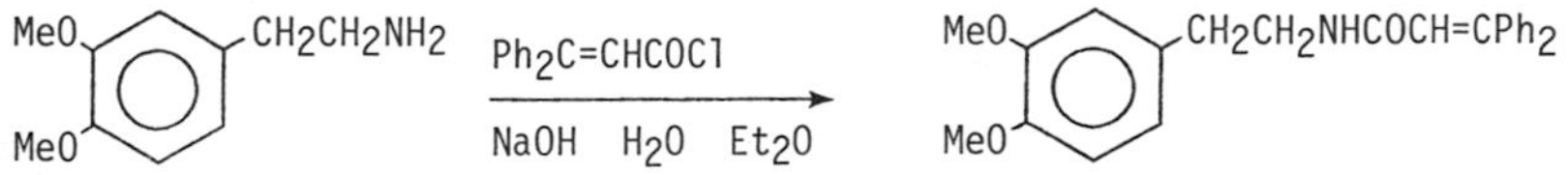

JOC (1970) <u>35</u> 825
Ber (1954) <u>87</u> 1760
For mild procedure see JCS (1962) 1445

CH$_2$CH$_2$Pr-i

i-PrCH$_2$CH$_2$ —⟨benzene ring⟩— NH$_2$ / Me

$\xrightarrow{\text{1 BuLi} \quad \text{2 Me}_2\text{CCOCl} \, / \, \text{Et}}$

CH$_2$CH$_2$Pr-i

i-PrCH$_2$CH$_2$ —⟨benzene ring⟩— NHCOCMe$_2$ / Et / Me

Tetr Lett (1964) 1597

i-PrCHNH$_2$ / COOH $\xrightarrow{\text{HCOOH} \quad \text{Ac}_2\text{O}}$ i-PrCHNHCHO / COOH   85-90%

JACS (1958) 80 1154
     (1968) 90 3245

PhNH$_2$ $\xrightarrow{\text{PhCOOH} \quad \text{PhSO}_2\text{Cl} \quad \text{Pyr}}$ PhCONHPh   94%

JACS (1955) 77 6214

PhCH$_2$NH$_2$ $\xrightarrow[\text{Me}_2\text{SO}]{\text{PhCOOH} \quad \text{Me}_2\overset{+}{\text{S}}\text{CH}_2\text{C}\equiv\text{CH} \; \overset{-}{\text{Br}}}$ PhCH$_2$NHCOPh   91%

JCS C (1969) 1904

⟨cyclohexyl⟩—NH$_2$ $\xrightarrow[\text{DCC} \quad \text{THF}]{\text{3,5-Dinitrobenzoic acid}}$ ⟨cyclohexyl⟩—NHCO—⟨benzene ring with NO$_2$ at 3 and 5 positions⟩   93%

Compt Rend (1965) 260 2249
JACS (1968) 90 4706

Further examples of the reaction R$_2$NH + RCOOH ⟶ RCONR$_2$ are included in section 77 (Amides from Carboxylic Acids and Acid Halides) and section 105A (Protection of Amines)

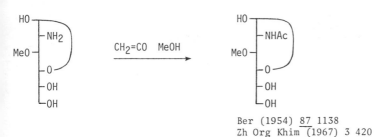

Ber (1954) <u>87</u> 1138
Zh Org Khim (1967) <u>3</u> 420
(Chem Abs <u>66</u> 115529)

PhNH₂ →[1 LiAlH₄ THF / 2 PhCOOEt] PhNHCOPh

    JOC (1962) <u>27</u> 1042
    JCS (1954) 1188

Further examples of the reaction R₂NH + RCOOR ⟶ RCONR₂ are included
in section 83 (Amides from Esters)

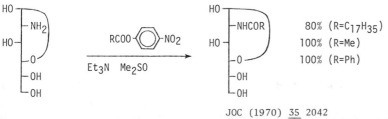

    JOC (1961) <u>26</u> 2563   97%

PhCH₂NH₂·HCl →[MeCONH₂ Δ] PhCH₂NHAc

    JACS (1943) <u>65</u> 1566

HO⌐
 ├─NH₂
HO┤
 ├─O
 ├─OH
 └─OH
     →[RCOO⟨⟩NO₂ / Et₃N Me₂SO]

HO⌐
 ├─NHCOR   80% (R=C₁₇H₃₅)
HO┤   100% (R=Me)
 ├─O   100% (R=Ph)
 ├─OH
 └─OH

    JOC (1970) <u>35</u> 2042
    Bull Chem Soc Jap (1963) <u>36</u> 754

$HOCH_2CH_2NH_2 \xrightarrow{CH_2=C(CN)OAc \quad CHCl_3} HOCH_2CH_2NHAc$

Chem Zvesti (1964) 18 218
(Chem Abs 61 14773)

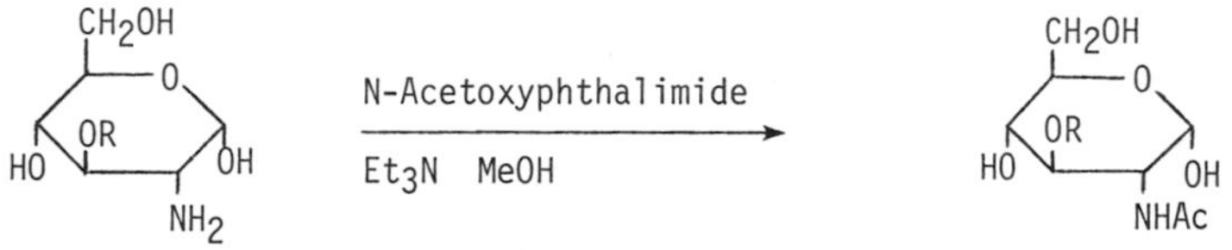
CH2OH
O
OR
HO
OH
NH2
N-Acetoxyphthalimide
Et3N MeOH
CH2OH
O
OR
HO
OH
NHAc

70%

JOC (1965) 30 448

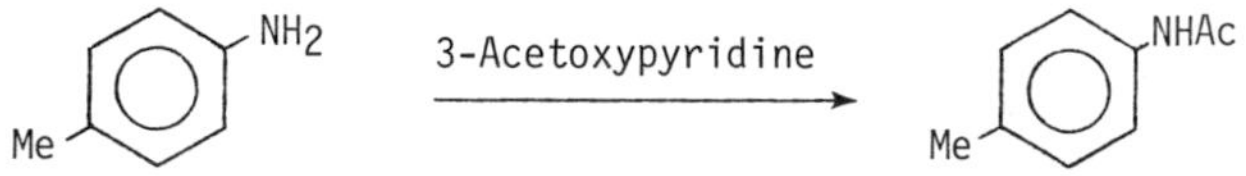
NH2
Me
3-Acetoxypyridine
NHAc
Me

88%

Bull Chem Soc Jap (1964) 37 864

$BuNH_2 \xrightarrow{CCl_3CHO \quad CHCl_3} BuNHCHO$ 83%

JACS (1952) 74 3933

$BuNH_2 \xrightarrow{CO \quad RhCl_2(CO)_4 \quad Me_3P \quad C_6H_6} BuNHCHO$ 96%

Tetr Lett (1969) 2329

$Me_2NH \xrightarrow{CO \quad CuCl} Me_2NCHO$ 73%

Bull Chem Soc Jap (1969) 42 2610

MeO—⟨benzene⟩—(CH₂)₂NH₂   N₂CHCO—⟨benzene⟩—OMe   →   MeO—⟨benzene⟩—(CH₂)₂NHCOCH₂—⟨benzene⟩—OMe

MeO—benzene—$(CH_2)_2NH_2$ N_2CHCO—benzene—OMe

$\xrightarrow{h\nu \quad THF}$

MeO—benzene—$(CH_2)_2NHCOCH_2$—benzene—OMe

Chem Ind (1969) 493   72%
JCS C (1970) 1208
Org React (1942) 1 38

$PhNMe_2$ $\xrightarrow[210-220°]{Ph(CH_2)_2COOH}$ $Ph(CH_2)_2CON\underset{Ph}{Me}$ 15%

Ber (1930) 63B 489

$PhNEt_2$ $\xrightarrow{Pb(OAc)_4 \quad Ac_2O \quad CHCl_3}$ $PhN\underset{Et}{Ac}$ 90%

Ber (1959) 92 288

Ac₂O   28%

Bull Soc Chim Fr (1964) 234

Ph_2NCH_2Me $\xrightarrow{KMnO_4 \quad NaHCO_3 \quad Me_2CO}$ Ph_2NAc 70%

JCS (1946) 454
JACS (1968) 90 1648

$PhN\underset{R}{Me}$ $\xrightarrow{MnO_2 \quad CHCl_3}$ $PhN\underset{R}{CHO}$ 83% (R=H)
80% (R=Me)

JCS (1957) 3032
(1966) 995

$$CrO_3 \quad Pyr$$

Tetrahedron (1967) $\underline{23}$ 4691

$$PhCH_2NMe_2 \quad \xrightarrow{O_2 \quad Pt \quad C_6H_6} \quad PhCH_2NCHO \atop Me$$

85%

Tetr Lett (1968) 4085
          (1970) 5049
Chem Comm (1969) 639

$$PrCH_2NBu_2 \quad \xrightarrow{O_3 \quad pentane} \quad PrCONBu_2$$

44%

JCS (1964) 711

$$PhCH_2CH_2NH_2 \quad \xrightarrow{S \quad (NH_4)_2S_x \quad dioxane} \quad PhCH_2CONH_2$$

32%

JACS (1953) $\underline{75}$ 740 5392

Section 83   Amides from Esters

$$C_8H_{17}CH=CH(CH_2)_7COOMe \quad \xrightarrow[230°]{C_{12}H_{25}NH_2} \quad C_8H_{17}CH=CH(CH_2)_7CONHC_{12}H_{25}$$

69%

JACS (1949) $\underline{71}$ 2215

For catalysis by 2-hydroxypyridine see JCS $\underline{C}$ (1969) 89
  "   "   " NaNH_2   " Chem Ind (1956) 277
  "   "   " MeONa   " JOC (1963) $\underline{28}$ 2915

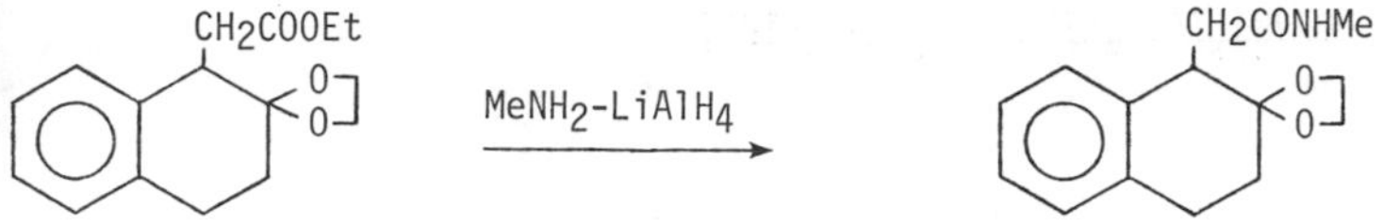

MeNH$_2$-LiAlH$_4$

61%

(Procedure for hindered esters)
Tetr Lett (1969) 1573
JCS (1954) 1188
Tetr Lett (1970) 1791

i-PrCOOMe → (HCONHR MeONa) i-PrCONHR

53% (R=H)
97% (R=Me)

JOC (1965) 30 2376

BuC≡CCOOMe → (NH$_3$ MeOH) BuC≡CCONH$_2$

JACS (1941) 63 1151

PhCOOEt → (NH$_3$ NH$_4$Cl) PhCONH$_2$

JACS (1938) 60 579

For catalysis by NaNH$_2$ see JACS (1955) 77 469
        "   "   " MeONa " JACS (1937) 59 1568
        "   "   " BuLi   " Tetr Lett (1970) 1791

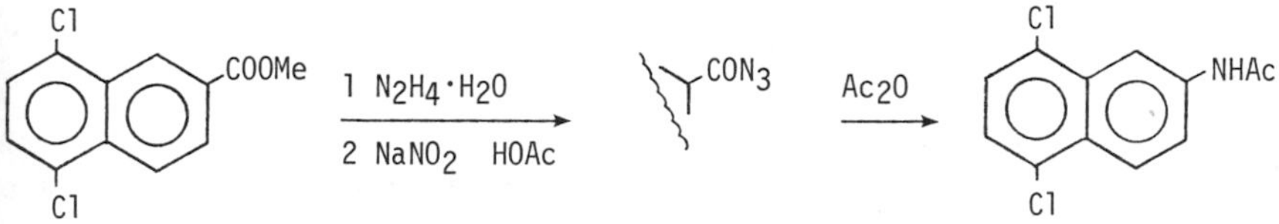

1 N$_2$H$_4$·H$_2$O
2 NaNO$_2$ HOAc

Ac$_2$O

Helv (1944) 27 883
Org React (1946) 3 337

HCOOCPh$_3$ → (MeCN H$_2$SO$_4$) AcNHCPh$_3$

JCS (1964) 5609
Org React (1969) 17 213

Section 84   Amides from Ethers and Epoxides
              ○○○○○○○○○○○○○○○○○○○○○○○○○○○○○○○○○○○○

t-BuOMe $\xrightarrow{\underset{\text{(CH}_2)_4\text{CN}}{\overset{\text{CN}}{|}}\quad \text{H}_2\text{SO}_4\quad \text{HOAc}}$ $\underset{\text{(CH}_2)_4\text{CONHBu-t}}{\overset{\text{CONHBu-t}}{|}}$ 75%

Org React (1969) <u>17</u> 213

$\xrightarrow[\text{MeCN}\quad \text{H}_2\text{O}]{\text{Electrolysis}\quad \text{Et}_4\text{NCN}}$ 9%

JACS (1969) <u>91</u> 4181

PhCHCH_2 (epoxide, O) $\xrightarrow{\text{S}\quad \text{(NH}_4)_2\text{S}_x\quad \text{dioxane}}$ $\text{PhCH}_2\text{CONH}_2$ 87%

JACS (1953) <u>75</u> 740

Section 85   Amides from Halides
              ○○○○○○○○○○○○○○○○○○○○○○○○

PhCH$_2$Cl $\xrightarrow[\text{2 LiI}\quad \text{lutidine}]{\text{1 MeCHCONMe}_2\quad \text{NaNH}_2\quad \text{HMPA}\atop \overset{\text{COOEt}}{|}}$ $\underset{\text{PhCH}_2\text{CHCONMe}_2}{\overset{\text{Me}}{|}}$ 46%

Ber (1968) <u>101</u> 4230

C$_5$H$_{11}$Br $\xrightarrow{\text{LiCH}_2\text{CONMe}_2\quad \text{THF}}$ C$_5$H$_{11}$CH$_2$CONMe$_2$ 64%

Ber (1968) <u>101</u> 3113
JOC (1966) <u>31</u> 982 989

$$\text{PhCH}_2\text{Cl} \xrightarrow[\text{Et}_2\text{O}]{\overset{\overset{\text{Li}}{|}}{\text{LiCH}_2\text{CONPh}}} \text{PhCH}_2\text{CH}_2\text{CONHPh} \qquad 69\%$$

JACS (1967) <u>89</u> 1647

$$\text{PhCH=CHBr} \xrightarrow[\text{MeOH}]{\text{Pyrrolidine} \quad \text{Ni(CO)}_4} \text{PhCH=CHCON} \diagup\!\!\!\bigcirc \qquad 82\%$$

JACS (1969) <u>91</u> 1233

$$\text{C}_7\text{H}_{15}\text{Br} \xrightarrow[\text{2 PhNCO}]{\text{1 Mg}} \text{C}_7\text{H}_{15}\text{CONHPh} \qquad 88\%$$

JACS (1947) <u>69</u> 2007
(1938) <u>60</u> 540

MeO—⟨benzene ring⟩—Br $\xrightarrow[\substack{\text{2 Ethyl cyclohexylidene-}\\ \text{carbamate}}]{\text{1 Mg}}$ MeO—⟨benzene ring⟩—CONH$_2$

Bull Res Council Israel (1952) <u>2</u> 72
(Chem Abs <u>48</u> 8727)

$$\text{PhCH}_2\text{CH}_2\text{Br} \xrightarrow{\text{S} \quad \text{NH}_3 \quad \text{H}_2\text{O} \quad 170°} \text{PhCH}_2\text{CONH}_2 \qquad 80\%$$

JACS (1953) <u>75</u> 740 5395

$$\text{C}_8\text{H}_{17}\text{Br} \xrightarrow{\text{HCONH}_2 \quad \text{NH}_3} \text{C}_8\text{H}_{17}\text{NHCHO} \qquad 91\%$$

JOC (1969) <u>34</u> 3204

For examples of the reaction RCONHR' + R"Hal ⟶ RCONR'R" (R'=H or alkyl) see section 81 (Amides from Amides)

Section 86   Amides from Hydrides (RH)

The conversions RH ⟶ RCH$_2$CONH$_2$, RCONR'R" or RNHAc are included in this section. For the reaction RR' ⟶ RCONH$_2$ (R=Ar, R'=alkyl) see section 80 (Amides from Alkyls)

$$\text{PhOMe} \xrightarrow{\text{ClCH}_2\text{CONH}_2 \quad h\nu \quad H_2O} \text{(2-CH}_2\text{CONH}_2\text{)-PhOMe}$$

Tetr Lett (1969) 2387   21%

$$\xrightarrow{\text{PhNCOCl} \quad \text{AlCl}_3}$$

Ber (1955) 88 301
JCS (1931) 2323

$$\xrightarrow{(\text{NH}_2)_2\text{CO} \quad \text{AlCl}_3}$$

JOC (1970) 35 2104   27%

$$\xrightarrow{\text{HCONH}_2 \quad h\nu \quad \text{Me}_2\text{CO}}$$

Tetr Lett (1963) 77   20%

$$\text{PhCH}_3 \xrightarrow{\text{HCONH}_2 \quad h\nu \quad \text{Me}_2\text{CO}} \text{PhCH}_2\text{CONH}_2$$ 23%

Tetr Lett (1963) 77

PhCH₃ →[Electrolysis MeCN / LiClO₄ H₂O] PhCH₂NHAc

Tetr Lett (1968) 2411

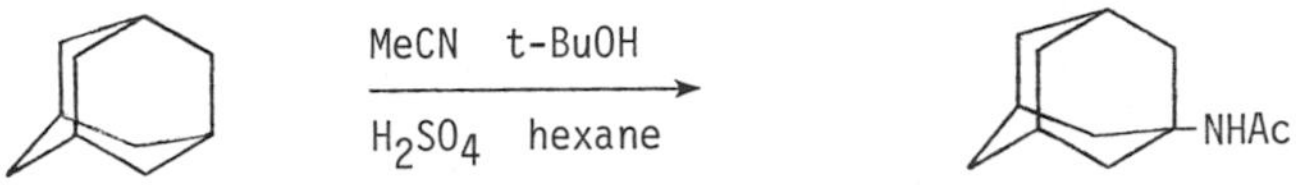

MeCN   t-BuOH
H₂SO₄   hexane

—NHAc   36%

Ber (1964) 97 3234
Org React (1969) 17 213

Section 87   Amides from Ketones

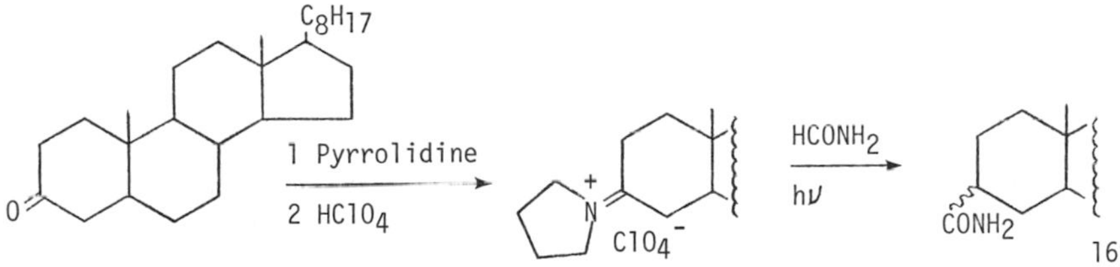

1 Pyrrolidine
2 HClO₄

ClO₄⁻

Helv (1967) 50 1759

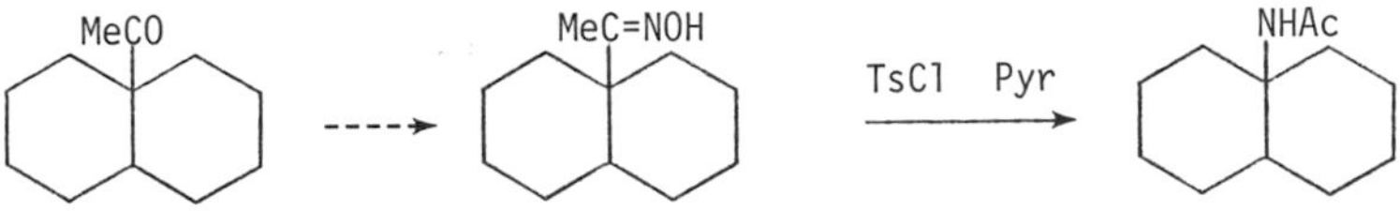

HCONH₂
hν

CONH₂   16%

MeCO ----→ MeC=NOH

TsCl   Pyr

NHAc   ~92%

JACS (1962) 84 1064

PhCOMe ---→ PhC=NOH
                |
                Me

CF₃COOH → PhNHAc   ~91%

Org React (1960) 11 1

$$\underset{\underset{Ph}{|}}{PhCOCMe_2} \dashrightarrow \underset{\underset{Ph}{|}}{\overset{\overset{NOH}{||}}{PhCCMe_2}} \xrightarrow{\text{Polyphosphoric acid}} PhCONH_2 \qquad \sim 63\%$$

JOC (1963) 28 278

$$\xrightarrow{h\nu \quad iPrOH} C_5H_{11}CONH_2 \qquad 44\%$$

$$\xrightarrow{h\nu \quad MeOH}$$

Can J Chem (1968) 46 3381

$$\xrightarrow[\substack{2\ Ph_3P \\ 3\ \Delta \\ 4\ H_2O}]{1\ Br_2} \qquad \sim 81\%$$

Tetr Lett (1965) 4541

$$\xrightarrow{PhNH_2} \xrightarrow[2\ Xylene\ \ reflux]{1\ MeCOO_2H\ \ Et_2O}$$

Ber (1958) 91 1057
Tetr Lett (1969) 2281

$$\xrightarrow{S \quad NH_3 \quad Pyr \quad H_2O}$$

40%

Org React (1946) 3 83

$$R_2CO \xrightarrow[EtOH]{N_2H_4 \cdot H_2O} R_2C=NNH_2 \xrightarrow[H_2SO_4 \quad H_2O]{NaNO_2} RCONHR \qquad \begin{array}{l} 40\% \ (R=i\text{-}Pr) \\ 72\% \ (R=Ph) \end{array}$$

JACS (1953) 75 5905

PhCH$_2$COMe $\xrightarrow{\text{NaN}_3 \quad \text{polyphosphoric acid}}$ PhCH$_2$NHAc   50%

JOC (1958) 23 1330
JCS (1942) 61
Org React (1946) 3 307

$\xrightarrow{\text{NaNH}_2 \quad \text{toluene}}$

88%

JACS (1953) 75 369
Org React (1957) 9 1

Amides may also be prepared by conversion of ketones into amines
followed by acylation. See section 102 (Amines from Ketones)

Section 88   Amides from Nitriles

PhCH$_2$CN $\xrightarrow{\text{Me}_2\text{C=CHEt} \quad \text{H}_2\text{SO}_4 \quad \text{H}_2\text{O}}$ PhCH$_2$CONHCMe$_2$ 69%
                                                            |
                                                            Pr

JACS (1948) 70 4045
Org React (1969) 17 213

MeCN $\xrightarrow[\text{EtCHOH} \quad \text{BF}_3]{\overset{\text{Me}}{|}}$ MeCONHCHEt$\overset{\text{Me}}{|}$ 65%

Acta Chem Scand (1968) 22 1787

PhCN $\xrightarrow{\text{(i-PrO)}_2\text{CH} \overset{+}{} \text{BF}_4^{-} \quad \text{CH}_2\text{Cl}_2}$ PhCONHPr-i

JOC (1969) 34 627

PhCH$_2$CN $\xrightarrow{\text{HCl \quad H}_2\text{O}}$ PhCH$_2$CONH$_2$

Org Synth (1963) Coll Vol 4 760
JACS (1948) 70 3091
Annalen (1968) 713 212

PhCH$_2$CN $\xrightarrow{\text{BF}_3 \cdot \text{AcOH}}$ PhCH$_2$CONH$_2$ 95%

JOC (1955) 20 1448

BuCN $\xrightarrow{\text{Et}_3\text{O}^+ \text{BF}_4^- \quad \text{CH}_2\text{Cl}_2}$ BuCONH$_2$ 93%

JOC (1969) 34 627

RCN $\xrightarrow{\text{MnO}_2 \quad \text{CH}_2\text{Cl}_2}$ RCONH$_2$ 40% (R=Me)
72% (R=Ph)

Chem Comm (1966) 121

$\xrightarrow{\text{Ni \quad Pyr \quad H}_2\text{O}}$

89%

Bull Chem Soc Jap (1966) 39 8
(1964) 37 1325

$\xrightarrow{\text{H}_2\text{O}_2 \quad \text{NaOH} \quad \text{EtOH} \quad \text{H}_2\text{O}}$

90-92%

Org Synth (1943) Coll Vol 2 586
JOC (1950) 15 800

$\xrightarrow{\text{Ion exch resin (basic) \quad H}_2\text{O}}$

89%

JOC (1960) 25 560

$$PhCN \xrightarrow{\text{NaOH \quad Me}_2\text{SO}} PhCONH_2 \qquad 96\%$$

JCS (1965) 1290

$$PhCH_2CN \xrightarrow{\text{H}_2 \quad \text{Ni} \quad \text{NaOAc} \quad \text{Ac}_2\text{O}} PhCH_2CH_2NHAc \qquad 97\%$$

JOC (1960) 25 1658

Amides may also be prepared by reduction of nitriles to amines followed by acylation. See section 103 (Amines from Nitriles)

Section 89   Amides from Olefins
              ००००००००००००००००००

$$BuCH=CH_2 \xrightarrow[\substack{+- \\ \text{2 Me}_2\text{SCHCONEt}_2}]{\text{1 B}_2\text{H}_6 \quad \text{THF}} Bu(CH_2)_3CONEt_2 \qquad \sim 50\%$$

JACS (1967) 89 6804

$$C_5H_{11}CH=CH_2 \xrightarrow[\text{di-t-butyl peroxide}]{\text{CH}_3\text{COONHEt}} C_5H_{11}(CH_2)_3CONHEt$$

Dokl (1964) 158 1127
(Chem Abs 62 2703)
JCS (1965) 1918

$$C_6H_{13}CH=CH_2 \xrightarrow[\text{di-t-butyl peroxide}]{\text{HCONMe}_2} C_6H_{13}(CH_2)_2CONMe_2 \qquad 56\%$$

Tetr Lett (1961) 238

BuCH=CH$_2$ $\xrightarrow[\text{Me}_2\text{CO} \quad \text{t-BuOH}]{\text{HCONH}_2 \quad h\nu}$ Bu(CH$_2$)$_2$CONH$_2$ 50%

JOC (1964) <u>29</u> 1855
     (1965) <u>30</u> 3361
Angew (1961) <u>73</u> 621

$\xrightarrow{\text{PhNH}_2 \quad \text{CO} \quad \text{Co}}$
CONHPh   70%

JACS (1952) <u>74</u> 4496

PhCH$_2$CH=CH$_2$ $\xrightarrow{\text{(NH}_4\text{)}_2\text{S}_x \quad \text{H}_2\text{O}}$ Ph(CH$_2$)$_2$CONH$_2$

JACS (1946) <u>68</u> 632 2025 2029

$\xrightarrow[\text{2 PhSH \quad Pyr \quad Me}_2\text{CO}]{\text{1 ClSO}_2\text{NCO}}$ NH   66%

JOC (1968) <u>33</u> 370
     (1970) <u>35</u> 2043
Annalen (1968) <u>718</u> 94

BuCH=CH$_2$ $\xrightarrow[\text{2 NaBH}_4 \quad \text{NaOH \quad H}_2\text{O}]{\text{1 Hg(NO}_3\text{)}_2 \quad \text{MeCN}}$ BuCHMe   70%
                                                          |
                                                        NHAc
JACS (1969) <u>91</u> 5647

C$_{12}$H$_{25}$CH=CH$_2$ $\xrightarrow{\text{NaCN \quad H}_2\text{SO}_4 \quad \text{MeCN}}$ C$_{12}$H$_{25}$CHMe   68%
                                                              |
                                                            NHAc

J Am Oil Chem Soc (1964) <u>41</u> 78
(Chem Abs <u>60</u> 6733)
JACS (1948) <u>70</u> 4045
Org React (1969) <u>17</u> 213

$MeCH=CH_2$ →[1 $MeCONH_2$ $PdCl_2$ / Na_2HPO_4 / 2 H_2] $Me_2CHNHAc$

Proc Chem Soc (1961) 370

Amides may also be prepared by conversion of olefins into amines
followed by acylation. See section 104 (Amines from Olefins)

Section 90   Amides from Miscellaneous Compounds
              ○○

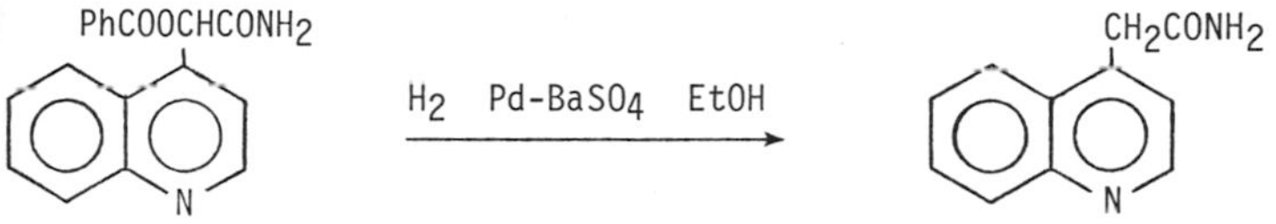

Arch Pharm (1957) 290 218   90%

$C_6H_{13}CH=CHCOOH$ →[S NH_3 Pyr H_2O] $C_6H_{13}CH_2CONH_2$ 41%

JOC (1947) 12 76

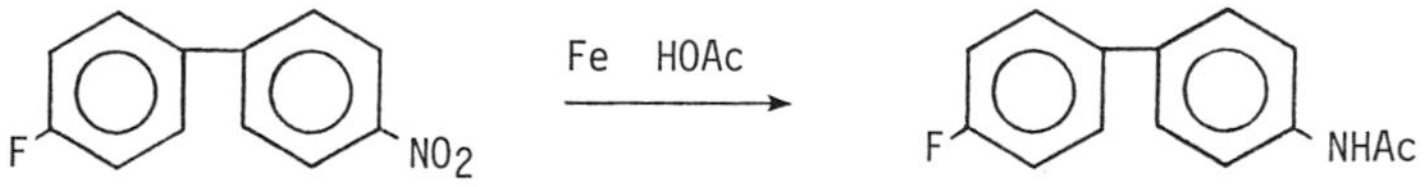

JACS (1966) 88 3318   71%

Chapter 7   PREPARATION
OF
AMINES

Section 91   Amines from Acetylenes

$$MeC{\equiv}CH \xrightarrow[\text{Zn(OAc)}_2]{Me_2NH \quad Cd(OAc)_2} \underset{NMe_2}{Me_2CC{\equiv}CMe} \xrightarrow[\text{HOAc}]{H_2 \quad Pt} \underset{NMe_2}{Me_2CCH_2CH_2Me}$$

JACS (1961) <u>83</u> 213 216

Section 92   Amines from Carboxylic Acids and Acid Halides

$$C_{11}H_{23}COOH{\cdot}NH_3 \xrightarrow[\text{copper chromite}]{H_2 \text{ (300 atmos)}} (C_{11}H_{23}CH_2)_2NH \qquad 79\%$$

JACS (1934) <u>56</u> 2419

$$C_9H_{19}COOH \xrightarrow{Li \quad MeNH_2} C_9H_{19}CH{=}NMe \xrightarrow[\text{MeNH}_2]{H_2 \quad Pd-C} C_9H_{19}CH_2NHMe \qquad 68\%$$

JACS (1970) <u>92</u> 5774

$C_{17}H_{35}COOH$ $\xrightarrow{\text{HN}_3 \quad \text{H}_2\text{SO}_4 \quad \text{C}_6\text{H}_6}$ $C_{17}H_{35}NH_2$ 96%

Org React (1946) <u>3</u> 307

$PhCH_2COCl$ $\xrightarrow[\text{2 HCl} \quad \text{H}_2\text{O}]{\text{1 NaN}_3 \quad \text{C}_6\text{H}_6}$ $PhCH_2NH_2$ 94%

Org React (1946) <u>3</u> 337
Nature (1963) <u>197</u> 787

Ph—△—COOH $\xrightarrow{\quad\quad}$ Ph—△—NH_2 77%

1 ClCOOEt   Et$_3$N   Me$_2$CO
2 NaN$_3$
3 Toluene 100°
4 HCl   H$_2$O   JOC (1961) <u>26</u> 3511

$\xrightarrow[\substack{\text{2 NaNO}_2 \quad \text{HOAc} \quad \text{H}_2\text{O} \\ \text{3 HCl} \quad \text{H}_2\text{O}}]{\text{1 N}_2\text{H}_4 \cdot \text{H}_2\text{O}}$

Rec Trav Chim (1921) <u>40</u> 285

$\xrightarrow[\text{polyphosphoric acid}]{\text{NH}_2\text{OH} \cdot \text{HCl}}$ 76%

JACS (1953) <u>75</u> 2014
Org React (1946) <u>3</u> 337
Chem Rev (1943) <u>33</u> 209

$C_5H_{11}COOH$ $\xrightarrow[\text{170-180°}]{\text{NH}_2\text{OSO}_3\text{H} \quad \text{mineral oil}}$ $C_5H_{11}NH_2$ 25%

JOC (1964) <u>29</u> 2576

PhCOOH $\xrightarrow{\text{MeNO}_2 \quad \text{polyphosphoric acid}}$ PhNH$_2$ 68%

JOC (1964) $\underline{29}$ 2576

43%   +   23%

Ber (1963) $\underline{96}$ 3359
Org React (1969) $\underline{17}$ 213

Section 93   Amines from Alcohols and Phenols

C$_{12}$H$_{25}$OH $\xrightarrow[\text{Cu-Ba-Cr oxide}]{\text{Et}_3\text{N} \quad \text{H}_2 \text{ (380 atmos)}}$ C$_{12}$H$_{25}$NEt$_2$ 70%

JACS (1952) $\underline{74}$ 4287

JCS (1946) 393
JACS (1933) $\underline{55}$ 345

Further examples of the preparation of amines from tosylates are
included in section 100 (Amines from Halides and Sulfonates)

PhCHOH $\xrightarrow{\text{KCN} \quad \text{H}_2\text{SO}_4 \quad \text{Bu}_2\text{O}}$ PhCHNCHO $\xrightarrow{\text{NaOH}}$ PhCHNH$_2$ 60%
| | |
Me   Me   Me

Org React (1969) $\underline{17}$ 213
JOC (1961) $\underline{26}$ 3002

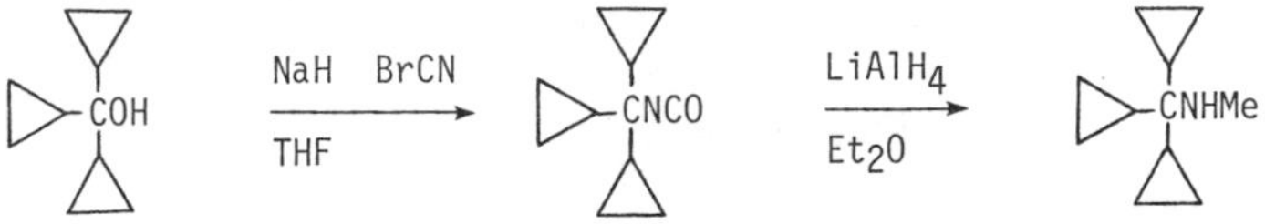

JOC (1968) 33 4054
JACS (1964) 86 4732

PhCHOH   1 NaH   MeOCH2CH2OMe   PhCHNMe2
  | ─────────────────────────> |
  Me   2 Me2NSO2Cl   Me
             3 60°   JACS (1965) 87 5261

NH3   SO2   H2O   150°   94-96%

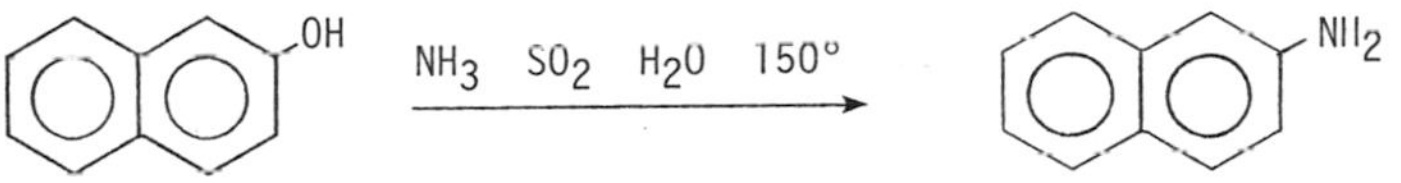

Org React (1942) 1 105

Section 94   Amines from Aldehydes
○○○○○○○○○○○○○○○○○○○○○○○○

PhCHO   MeNH2   PhCH=NMe   PhCH2MgBr   PhCHNHMe   84%
      ─────────>   ─────────> |
        C6H6   Et2O   CH2Ph

Org Synth (1963) Coll Vol 4 605

PhCHO   HCN   PhCHCN   H2   Pd-C   PhCH2CH2NH2   52%
      ─────────> | ─────────>
        KOH   OH   HCl   EtOH

JACS (1928) 50 3370

Tetr Lett (1967) 1201   50%

BuCHO $\xrightarrow{\text{HCOONH}_4 \quad \text{HCOOH}}$ $(\text{BuCH}_2)_3\text{N}$

Org React (1949) $\underline{5}$ 301

$C_6H_{13}CHO$ $\xrightarrow[\text{NaOAc \quad EtOH}]{\text{PhNH}_2 \quad H_2 \quad Ni}$ $C_6H_{13}CH_2NHPh$ 65%

Org React (1948) $\underline{4}$ 174

PhCHO $\xrightarrow[\text{EtOH}]{\text{NH}_3 \quad H_2 \quad Ni}$ $PhCH_2NH_2$ 89%

Org React (1948) $\underline{4}$ 174

JOC (1958) $\underline{23}$ 571

PhCHO $\xrightarrow[\text{2 } H_2 \quad PdO]{\text{1 } NH_2COOCH_2Ph}$ $PhCH_2NH_2$ 63%

JOC (1941) $\underline{6}$ 878

$$PhCHO \xrightarrow[\text{MeOH \quad pH 5-6}]{\text{EtNH}_2 \quad \text{LiBH}_3\text{CN}} PhCH_2NHEt \qquad 72\%$$

JACS (1969) <u>91</u> 3996

$$PhCHO \xrightarrow[\text{HOAc \quad EtOH}]{\text{PhNH}_2 \quad \text{NaBH}_4 \quad \text{NaOAc}} PhCH_2NHPh \qquad 83\%$$

JOC (1963) <u>28</u> 3259

JOC (1961) <u>26</u> 1437

J Biol Chem (1954) <u>211</u> 725

For reduction of oximes with H_2 and Pd-C see JACS (1928) <u>50</u> 3370

"   "   "   "   "   Ni   "   JCS <u>C</u> (1966) 531

"   "   "   "   "   Na and EtOH "   Org Synth (1943) Coll Vol 2 318

"   "   "   "   "   Na-Hg   "   JACS (1949) <u>71</u> 2257

$$PhCHO \xrightarrow{\text{NH}_2\text{OMe}} PhCH=NOMe \xrightarrow[\text{2 KOH \quad H}_2\text{O}]{\text{1 B}_2\text{H}_6 \quad \text{THF}} PhCH_2NH_2 \qquad 92\%$$

JOC (1969) <u>34</u> 1817

$$PhCHO \xrightarrow{\text{BuN}_3 \quad \text{H}_2\text{SO}_4 \quad \text{C}_6\text{H}_6} PhNHBu \qquad 22\%$$

JOC (1959) <u>24</u> 561

Section 95   Amines from Alkyls, Methylenes and Aryls
°°

No examples of the reaction RR' $\longrightarrow$ RNR$_2$ (R,R'=alkyl, aryl etc.) occur
in the literature. For the conversion RH $\longrightarrow$ RNR$_2$ see section 101
(Amines from Hydrides)

Section 96   Amines from Amides
°°°°°°°°°°°°°°°°°°°°°°

$$(i\text{-Pr})_2\text{NCHO} \xrightarrow{\quad \text{BuMgBr}\quad \text{Et}_2\text{O}\quad} (i\text{-Pr})_2\text{NCHBu}_2 \qquad 67\%$$

Monatsh (1951) $\underline{82}$ 330

$$C_6H_{13}CON\text{—}\bigcirc \xrightarrow[\text{dioxane}\ 250°]{H_2\ (300\ \text{atmos})\ \text{copper chromite}} C_6H_{13}CH_2N\text{—}\bigcirc \qquad 92\%$$

JACS (1934) $\underline{56}$ 2419

$$\xrightarrow{\quad \text{LiAlH}_4\quad \text{Et}_2\text{O}\quad} \qquad\qquad 84\%$$

Helv (1948) $\underline{31}$ 1397

$$\text{PhOCH}_2\text{CONH}_2 \xrightarrow{\quad \text{LiAlH}_4\quad \text{Et}_2\text{O}\quad} \text{PhOCH}_2\text{CH}_2\text{NH}_2 \qquad 80\%$$

Helv (1948) $\underline{31}$ 1397
Org React (1951) $\underline{6}$ 469
Org Synth (1963) Coll Vol 4 564

i-Pr(CH$_2$)$_2$CONEt$_2$ $\xrightarrow[\text{Et}_2\text{O}]{\text{LiAlH}_4 \quad \text{AlCl}_3}$ i-Pr(CH$_2$)$_2$CH$_2$NEt$_2$

Z. Chem (1966) 6 224

PhNHAc $\xrightarrow[\text{C}_6\text{H}_6]{\text{NaAlH}_2(\text{OCH}_2\text{CH}_2\text{OMe})_2}$ PhNHEt   84%

Tetr Lett (1968) 3303

PrCONH$_2$ $\xrightarrow{\text{NaBH}_4 \quad \text{CoCl}_2 \quad \text{MeOH}}$ PrCH$_2$NH$_2$ 70%

Tetr Lett (1969) 4555

C$_7$H$_{15}$CONMe$_2$ $\xrightarrow{\text{NaBH}_4 \quad \text{Pyr}}$ C$_7$H$_{15}$CH$_2$NMe$_2$ 41%

Chem Pharm Bull (1969) 17 98

PhCONHEt $\xrightarrow[\text{2 NaBH}_4 \quad \text{FtOH}]{\text{1 Et}_3\text{O}^+ \text{ BF}_4^- \quad \text{CH}_2\text{Cl}_2}$ PhCH$_2$NHEt   92%

Tetr Lett (1968) 61
Tetrahedron (1970) 26 803

C$_6$H$_{13}$CONMe$_2$ $\xrightarrow{\text{i-Bu}_2\text{AlH} \quad \text{Et}_2\text{O}}$ C$_6$H$_{13}$CH$_2$NMe$_2$ 75-95%

Izv (1959) 2146
(Chem Abs 54 10932)

PrCONEt$_2$ $\xrightarrow{\text{Et}_3\text{SiH} \quad \text{ZnCl}_2}$ PrCH$_2$NEt$_2$ 70%

Compt Rend (1962) 254 2357

$$C_5H_{11}CONHMe \xrightarrow{\text{B}_2\text{H}_6 \quad \text{THF}} C_5H_{11}CH_2NHMe \qquad 98\%$$

JACS (1964) <u>86</u> 3566
JOC (1968) <u>33</u> 3637

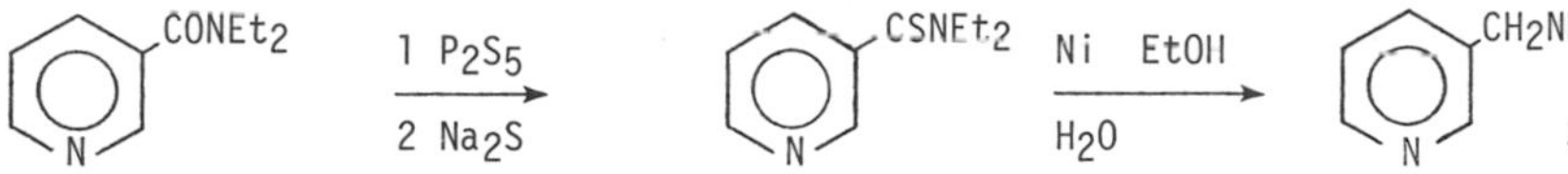

1 P$_2$S$_5$
2 Na$_2$S

Ni   EtOH
H$_2$O

38%

JOC (1951) <u>16</u> 131

Ca   NH$_3$

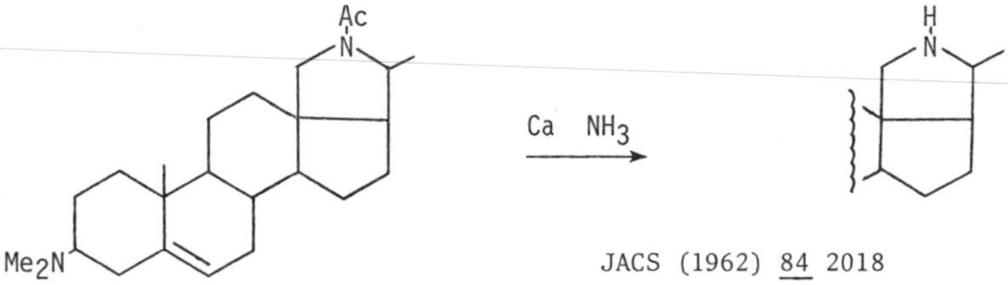

JACS (1962) <u>84</u> 2018

$$Et_2NCHO \xrightarrow{\text{NaH} \quad \text{diglyme}} Et_2NH \qquad 40\%$$

Tetr Lett (1965) 1713

MeCHCONHCH$_2$COOH
HCONH

H$_2$O$_2$ H$_2$O

MeCHCONHCH$_2$COOH
NH$_2$

Annalen (1960) <u>636</u> 140

KOH   MeOH   H$_2$O

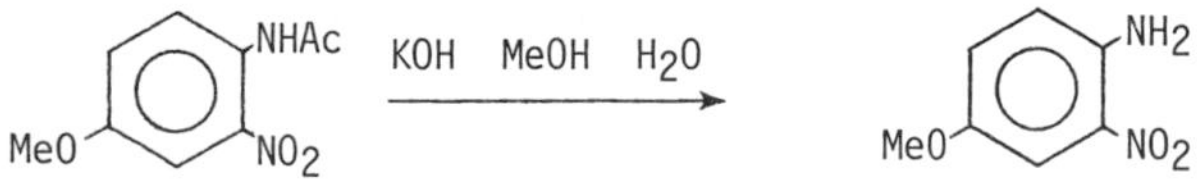

95-97%

Org Synth (1955) Coll Vol 3 661

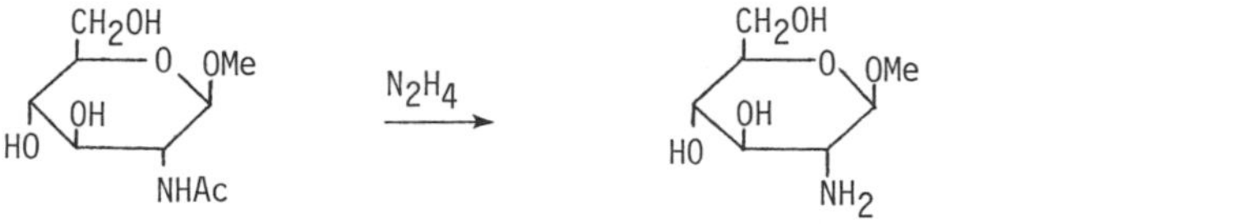

Bull Chem Soc Jap (1966) 39 185

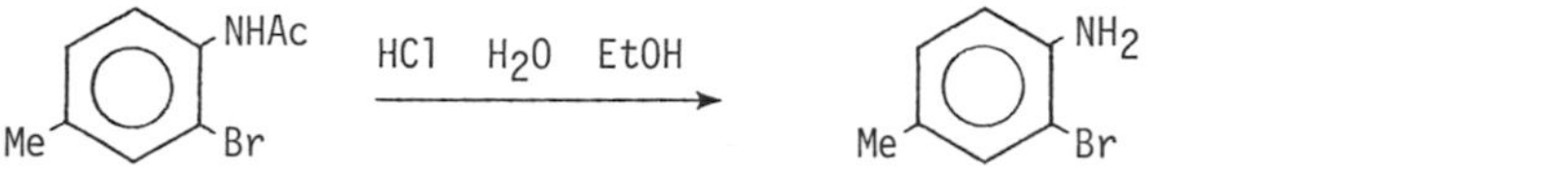

55%

Org Synth (1932) Coll Vol 1 111

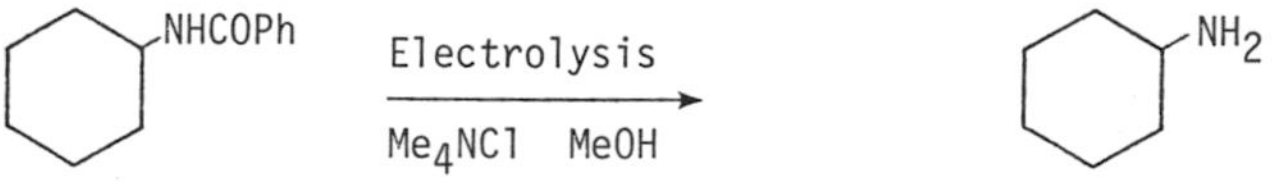

Ber (1965) 98 3462

71%

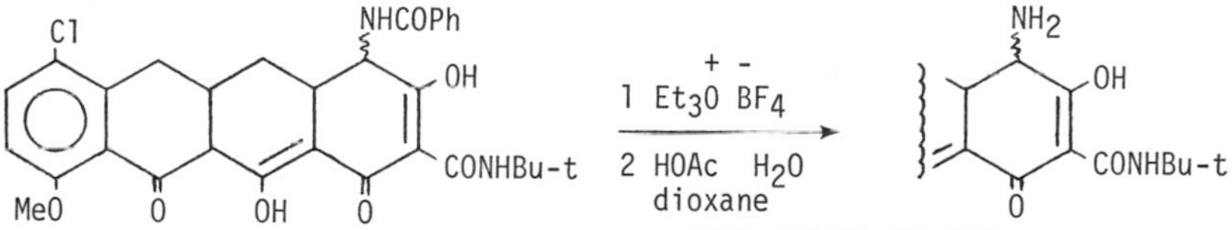

JACS (1965) 87 933
Helv (1968) 51 1108

$C_{15}H_{31}CONH_2$ $\xrightarrow[\text{2 CaO } H_2O]{\text{1 } Br_2 \text{ MeONa } MeOH}$ $C_{15}H_{31}NH_2$ 100%

Org React (1946) 3 267

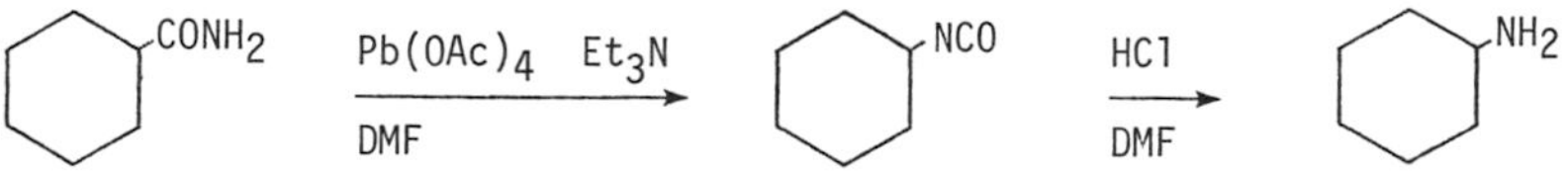

JACS (1965) 87 1141
Tetr Lett (1965) 4039

49%

PhCONHPh $\xrightarrow[\text{polyphosphoric acid}]{\text{NH}_2\text{OH·HCl}}$ PhNH$_2$ 76%

JACS (1953) $\underline{75}$ 2014

Section 97   Amines from Amines

Et$_2$NCH$_2$Me $\xrightarrow[\text{hexane}]{\text{BuLi   BuI}}$ Et$_2$NCHMe$\overset{|}{\underset{Bu}{}}$ 25

JOC (1966) $\underline{31}$ 2061

JACS (1957) $\underline{79}$ 5279

PrNH$_2$ $\xdashrightarrow{\text{PrCHO}}$ PrN=CHPr $\xrightarrow{\text{PhLi   Et}_2\text{O}}$ PrNHCHPr$\overset{|}{\underset{Ph}{}}$ 60%

Bull Soc Chim Fr (1964) 952

$(i-Pr)_2NH \cdot HCl \xrightarrow[\text{KCN} \quad H_2O]{\text{MeCHO}} (i-Pr)_2NCHCN \xrightarrow[\text{Et}_2O]{\text{MeMgCl}} (i-Pr)_2NCHMe_2$

Under the CHCN: $\underset{Me}{|}$

Monatsh (1962) __93__ 476

$(i-Pr)_2NH \xrightarrow{\text{HCOOH}} (i-Pr)_2NCHO \xrightarrow[\text{Et}_2O]{\text{EtMgCl}} (i-Pr)_2NCHEt_2$ 23%

Monatsh (1962) __93__ 476

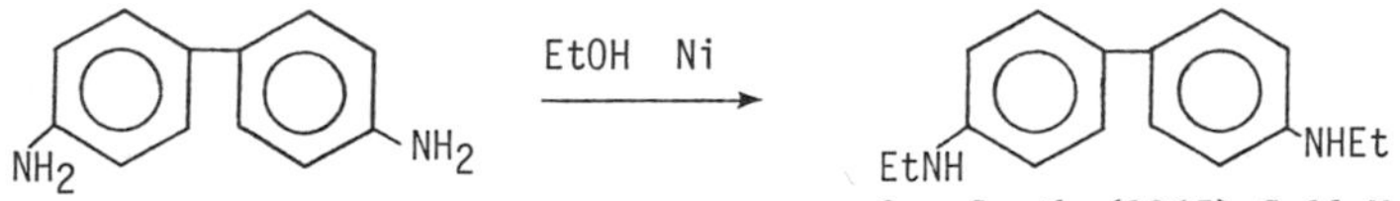

60-67%

EtNH

Org Synth (1963) Coll Vol 4 283
JACS (1954) __76__ 6174

$PhCH_2NH_2 \xrightarrow{\text{HCOOH} \quad \text{HCHO}} PhCH_2NMe_2$ 80%

Org React (1949) __5__ 301

$MeNHCO(CH_2)_4NH_2 \xrightarrow[H_2O]{\text{HCHO} \quad H_2 \quad Pd-C} MeNHCO(CH_2)_4NMe_2$ 60%

JCS __C__ (1969) 1358

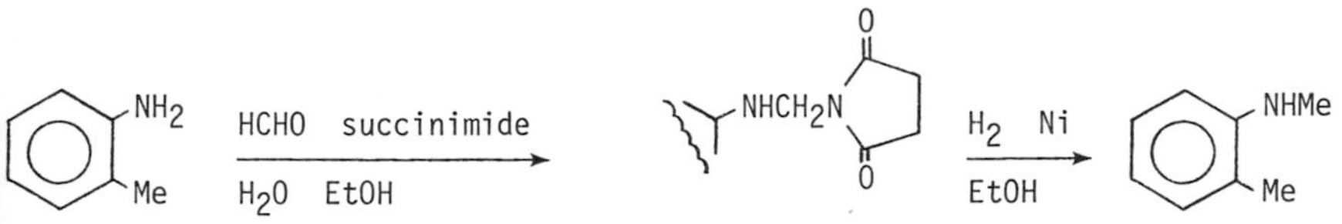

68%

Chem Pharm Bull (1967) __15__ 1339

PrNH$_2$ $\xrightarrow{\text{PrCHO   KOH}}$ PrN=CHPr $\xrightarrow{\text{H}_2 \text{ PtO}_2 \text{ EtOH}}$ PrNHCH$_2$Pr   65%

Org React (1948) <u>4</u> 174
Tetr Lett (1968) <u>2</u>639

C$_5$H$_{11}$NH$_2$ $\xrightarrow[\substack{2\ \text{H}_2\ \ \text{PtO}_2\ \ \text{HOAc} \\ 3\ \text{BuBr}\ \ \text{MeOH}}]{1\ \text{PhCHO}\ \ \text{C}_6\text{H}_6}$ C$_5$H$_{11}$NBu ($\overset{|}{\text{CH}_2\text{Ph}}$) $\xrightarrow[\text{HOAc}]{\text{H}_2\ \text{ PtO}_2}$ C$_5$H$_{11}$NHBu

JACS (1941) <u>63</u> 1964

Cyclohexanone / HCOOH   HOAc   86%

Rev Chim (Bucharest) (1968) <u>19</u> 360
(Chem Abs <u>69</u> 105967)
Org React (1949) <u>5</u> 301

BuNH$_2$ $\xrightarrow[\text{NaOAc   HOAc   H}_2\text{O}]{\text{Me}_2\text{CO   NaBH}_4}$ BuNHCHMe$_2$ 63%

JOC (1963) <u>28</u> 3259

For further examples of the reductive alkylation of amines with aldehydes and ketones see section 94 (Amines from Aldehydes) and section 102 (Amines from Ketones)

$\xrightarrow[\text{2 NaNH}_2\ \ \text{NH}_3]{1\ \text{Ph}_3\text{PBr}_2\ \ \text{Et}_3\text{N}}$ ...N=PPh$_3$ $\xrightarrow[\text{2 KOH EtOH}]{1\ \text{EtI}}$ ...NHEt   ~67%

JOC (1970) <u>35</u> 2826

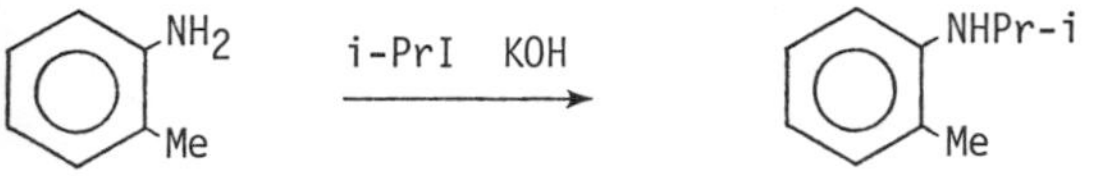

NHPr-i (from o-toluidine with i-PrI, KOH)   92%

JACS (1960) 82 6163
Ber (1952) 85 1056

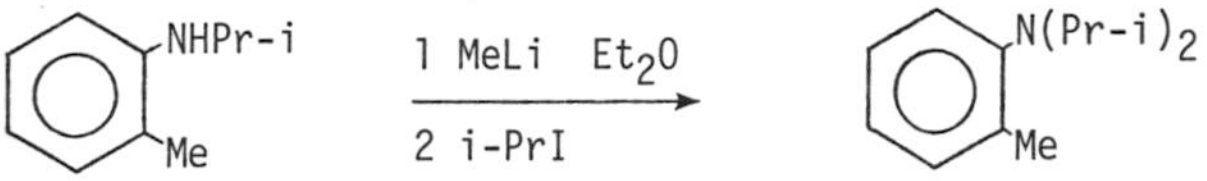

1 MeLi Et₂0
2 i-PrI

JACS (1960) 82 6163
       (1964) 86 2813

t-BuNH₂ → 1 PhLi Et₂0 / 2 Me₂CC≡CH (Cl) → t-BuN=CHCH=CMe₂ → H₂ Ni / EtOH → t-BuNHCH₂CH₂CHMe₂

JOC (1961) 26 3772

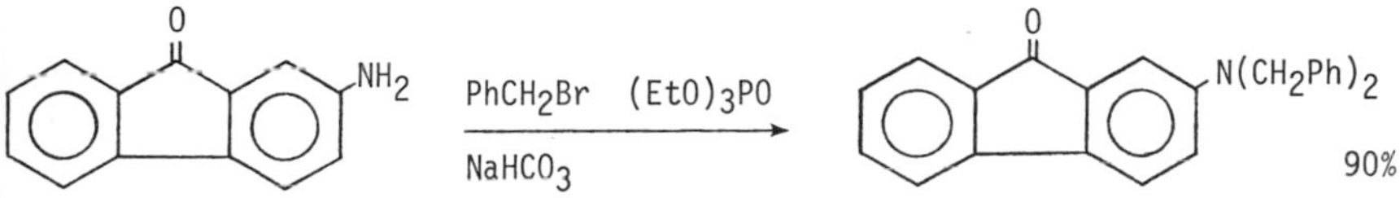

PhCH₂Br (EtO)₃PO
NaHCO₃

JOC (1962) 27 3639   90%

PhCH₂NH₂ → MeI NaHCO₃ / MeOH → PhCH₂NMe₃⁺I⁻ → LiAlH₄ / THF → PhCH₂NMe₂   72%

JACS (1960) 82 4651
Ber (1957) 90 395

Ph₂NH → PhI Cu K₂CO₃ / PhNO₂ → Ph₂NPh   82%

Org Synth (1941) Coll Vol 1 544
JCS (1946) 5

$$PhCH=CH-\underset{}{\bigcirc}-NH_2 \xrightarrow[\begin{array}{c}2 \text{ Na toluene}\\3 \text{ Me}_2SO_4\end{array}]{1 \text{ TsCl Pyr}} \overset{Ts}{\underset{}{\bigg|}}NMe \xrightarrow[H_2O]{HCl \quad HOAc} PhCH=CH-\underset{}{\bigcirc}-NHMe \quad 56\%$$

Annalen (1956) <u>598</u> 174
JOC (1968) <u>33</u> 1142
Ber (1953) <u>86</u> 1246

$$BuNH_2 \xrightarrow[2 \text{ MeI } K_2CO_3 \text{ Me}_2CO]{1 \text{ PhCOCH}_2SO_2Cl \text{ CH}_2Cl_2 \text{ Pyr}} \underset{Me}{BuNSO_2CH_2COPh} \xrightarrow[HOAc]{Zn \quad HCl} BuNHMe$$

Tetr Lett (1970) 345

$$PhNH_2 \xrightarrow[2 \text{ EtI KOH Me}_2CO]{1 \text{ (CF}_3CO)_2O} \underset{Et}{PhNCOCF_3} \xrightarrow[]{KOH \quad H_2O} PhNHEt \quad 83\%$$

JCS <u>C</u> (1969) 2223

$$MeO-\underset{}{\bigcirc}-NH_2 \xrightarrow[\begin{array}{c}2 \text{ Me}_2SO_4 \text{ NaNH}_2\\\text{toluene}\end{array}]{1 \text{ Ac}_2O \text{ NaOH } H_2O} \overset{Me}{\underset{}{\bigg|}}NAc \xrightarrow[EtOH]{KOH \quad H_2O} MeO-\underset{}{\bigcirc}-NHMe$$

Ber (1954) <u>87</u> 1760
JOC (1949) <u>14</u> 1099

$$\underset{}{\bigcirc}NH \xrightarrow[2 \text{ CH}_2=CH_2]{1 \text{ Na Pyr}} \underset{}{\bigcirc}NEt \quad 77\text{-}83\%$$

Org Synth (1963) <u>43</u> 45
Bull Chem Soc Jap (1967) <u>40</u> 2991

$$BuNH_2 \xrightarrow[2 \text{ H}_2]{\begin{array}{c}1 \text{ MeCH}=CH_2 \text{ PdCl}_2\\Na_2HPO_4 \text{ THF}\end{array}} BuNHCHMe_2$$

Proc Chem Soc (1961) 370

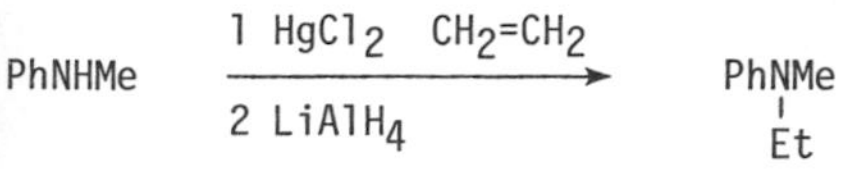

PhNHMe $\xrightarrow[\text{2 LiAlH}_4]{\text{1 HgCl}_2 \quad \text{CH}_2=\text{CH}_2}$ PhNMe
Et

Compt Rend (1966) C 262 1591
Tetr Lett (1967) 5165
        (1969) 2289

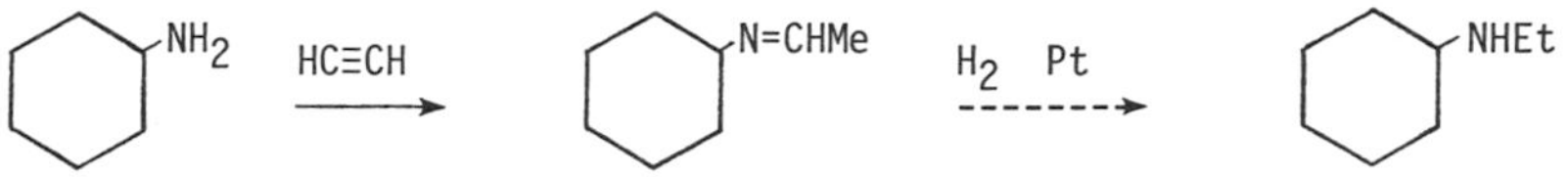

JACS (1961) 83 213

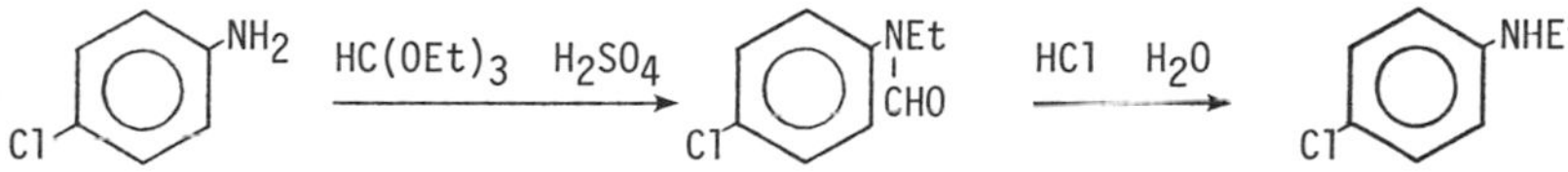

~74%

Org Synth (1963) Coll Vol 4 420
JACS (1956) 78 4778

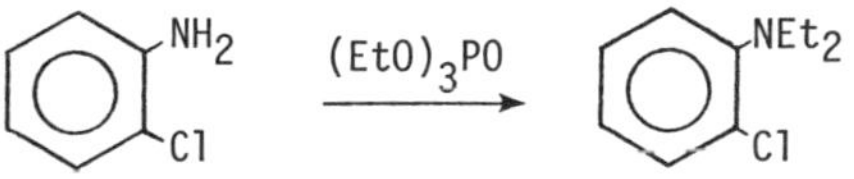

91%

JACS (1946) 68 895

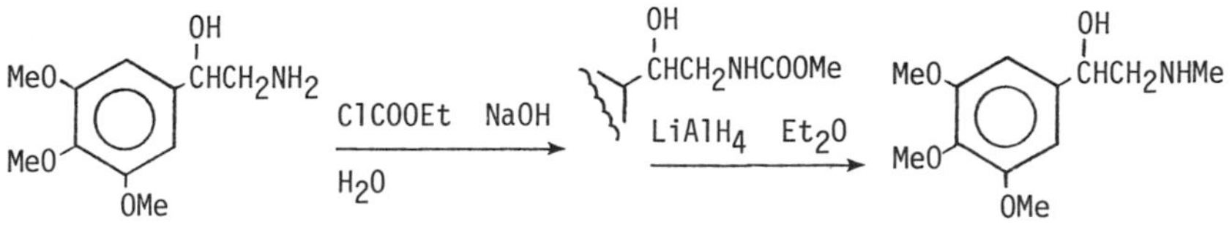

J Med Chem (1963) 6 227
JOC (1965) 30 2483

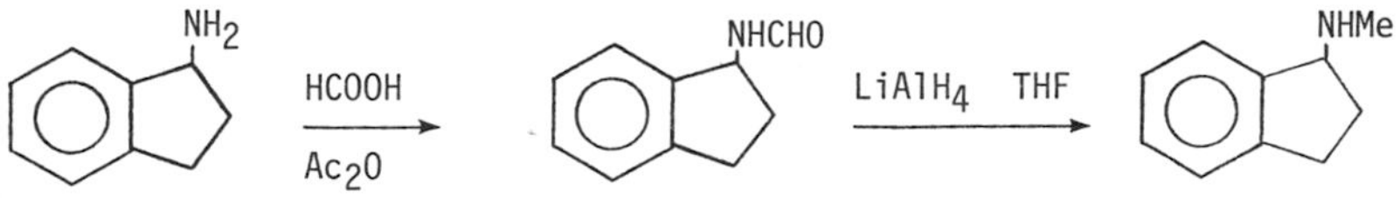

J Med Chem (1966) 9 830

Further examples of the preparation of N-alkyl amines by reduction of amides are included in section 96 (Amines from Amides)

$$\text{(piperidine) NH} \xrightarrow[\text{2 MeCOOEt}]{\text{1 LiAlH}_4 \quad \text{THF}} \text{(piperidine) NEt} \qquad 80\%$$

JOC (1962) 27 1042

$$\text{(cyclohexyl)} NH_2 \xrightarrow{\text{Ni} \quad \text{toluene}} \text{(cyclohexyl)} NH \text{(cyclohexyl)} \qquad 82\%$$

Annalen (1961) 644 23

$$\text{PhNMe}_2 \xrightarrow[\text{150-160°}]{\text{C}_5\text{H}_{11}\text{Br}} \begin{array}{c} \text{PhNMe} \\ | \\ \text{C}_5\text{H}_{11} \end{array}$$

Ber (1881) 14 622
Org React (1953) 7 198

$$\text{BuNH(CH}_2)_3\text{CH}_3 \xrightarrow[\text{2 H}_2\text{SO}_4 \quad \text{H}_2\text{O}]{\text{1 Cl}_2 \quad \text{NaOH} \quad \text{H}_2\text{O}} \text{BuN} \text{(pyrrolidine)} \qquad 70\text{-}80\%$$

Org Synth (1955) Coll Vol 3 159
Chem Rev (1963) 63 55

$$\text{(phenyl)NHC}_{16}\text{H}_{33} \xrightarrow{\text{CoCl}_2 \quad 212°} \text{(phenyl)NH}_2 \quad \text{C}_{16}\text{H}_{33}$$

JCS (1937) 1119
Tetrahedron (1961) 14 208

$$(\text{C}_6\text{H}_{13})_2\text{NCH}_2\text{Ph} \xrightarrow{\text{H}_2 \quad \text{PtO}_2 \quad \text{HOAc}} (\text{C}_6\text{H}_{13})_2\text{NH} \qquad 100\%$$

Org React (1953) 7 263
JACS (1950) 72 3410
      (1941) 63 1964

$$\xrightarrow[\text{CHCl}_3]{\text{BrCN} \quad \text{K}_2\text{CO}_3}$$ NCN $$\xrightarrow[]{\text{LiAlH}_4 \quad \text{THF}}$$ NH   49%

JACS (1967) 89 1942
(1955) 77 4079
Org Synth (1955) Coll Vol 3 608
Org React (1953) 7 198

Bu$_3$N $$\xrightarrow[]{\text{ClCOOPh} \quad \text{CH}_2\text{Cl}_2}$$ Bu$_2$NCOOPh $$\xrightarrow[]{\text{Acid}}$$ Bu$_2$NH

85%

JCS C (1967) 2015

(i-Pr)$_2$NEt $$\xrightarrow[\text{CuCl}]{\text{i-PrONO} \quad \text{O}_2}$$ (i-Pr)$_2$NNO --→ (i-Pr)$_2$NH

Angew (1970) 82 876
(Internat Ed 9 892)
JACS (1967) 89 1147

PhNMe$_2$ $$\xrightarrow[]{\text{HBr} \quad 150°}$$ PhNHMe

JOC (1963) 28 3144

Pr$_3$N $$\xrightarrow[]{\text{NBS} \quad \text{dioxane} \quad \text{H}_2\text{O}}$$ Pr$_2$NH   87%

JCS (1957) 4905
JACS (1968) 90 3502
Chem Rev (1963) 63 21

$$\xrightarrow[\text{MeOH}]{\text{O}_2 \quad h\nu \quad \text{methylene blue}}$$ 80%

Tetr Lett (1970) 3649

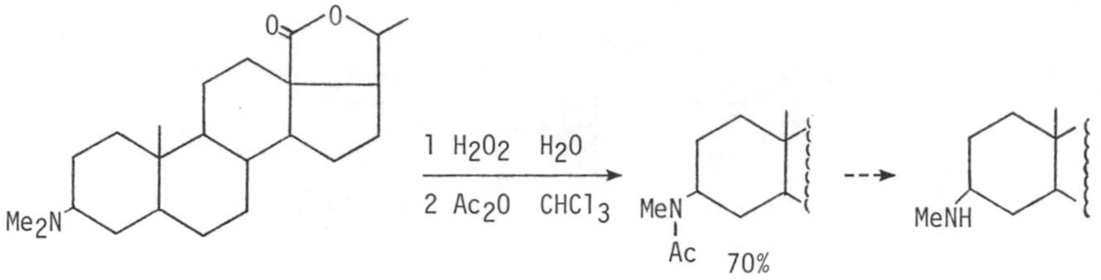
Me2N
1 H2O2 H2O
2 Ac2O CHCl3
MeN
Ac
70%
MeNH

Tetrahedron (1967) 23 4681

$(i\text{-}Pr)_2NH \xrightarrow{t\text{-}BuOOH} i\text{-}PrNH_2$ 88%

JOC (1960) 25 2114

Further reactions which may be used for the dealkylation of amines are included in section 82 (Amides from Amines) and section 81 (Amides from Amides)

$Et_3\overset{+}{N}CH_2Ph\ \ \overset{-}{Cl} \xrightarrow{PhSH\ \ NaOH} Et_3N$ 22%

J Med Chem (1969) 12 694
Tetr Lett (1966) 1375

$Bu\overset{+}{N}Me_3\ \ \overset{-}{I} \xrightarrow{HOCH_2CH_2NH_2} BuNMe_2$

Ber (1957) 90 395

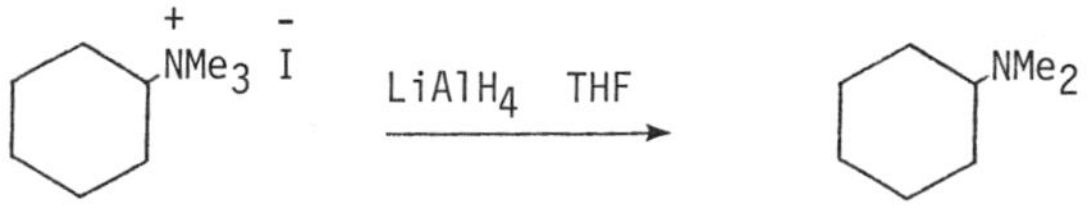
NMe3 I
LiAlH4 THF
NMe2

JACS (1960) 82 4651

Section 98   Amines from Esters

$$PhCH_2COOR \xrightarrow[\text{2 PhCOCl \quad MeONa}]{\text{1 NH}_2\text{OH \quad base}} PhCH_2CONHOCOPh \xrightarrow[\text{THF}]{\text{LiAlH}_4 \quad \text{AlCl}_3} PhCH_2CH_2NH_2$$

Bull Soc Chim Fr (1960) 509

Amines may also be prepared by conversion of esters into amides followed by reduction. See section 83 (Amides from Esters) and section 96 (Amines from Amides)

$$PhCH_2COOEt \xrightarrow[\text{2 NaNO}_2 \quad \text{HCl \quad H}_2\text{O}]{\text{1 N}_2\text{H}_4 \cdot \text{H}_2\text{O \quad EtOH}} PhCH_2CON_3 \xrightarrow[\text{H}_2\text{O}]{\text{HCl \quad HOAc}} PhCH_2NH_2 \qquad \sim 50\%$$

Org React (1946) 3 337

$$t\text{-BuOAc} \xrightarrow{\text{HCN \quad H}_2\text{SO}_4} t\text{-BuNH}_2$$

Org React (1969) 17 213

Section 99   Amines from Ethers

54%

Rec Trav Chim (1966) 85 56

Section 100   Amines from Halides and Sulfonates

PhBr $\xrightarrow[\text{2 ClCH}_2\text{CH}_2\text{NMe}_2]{\text{1 Mg Et}_2\text{O}}$ PhCH$_2$CH$_2$NMe$_2$ 13%

J Med Chem (1966) $\underline{9}$ 790

PhBr $\xrightarrow[\text{2 EtCN}]{\text{1 Mg Et}_2\text{O}}$ $\underset{\overset{|}{\text{Et}}}{\text{PhC=NH}}$ $\xrightarrow{\text{LiAlH}_4 \quad \text{THF}}$ $\underset{\overset{|}{\text{Et}}}{\text{PhCHNH}_2}$ 80%

JACS (1953) $\underline{75}$ 5898

BuBr $\xrightarrow[\text{2 (i-Pr)}_2\text{NCHO}]{\text{1 Mg Et}_2\text{O}}$ (i-Pr)$_2$NCHBu$_2$ 67%

Monatsh (1951) $\underline{82}$ 330

MeCl $\xrightarrow[\text{2 (i-Pr)}_2\text{NCHMe}]{\text{1 Mg Et}_2\text{O}}$ (i-Pr)$_2$NCHMe$_2$ 54%

$\quad\quad\quad\quad\quad\quad\quad\quad\overset{|}{\text{CN}}$

Monatsh (1962) $\underline{93}$ 476

CH$_2$=CHCH$_2$Cl $\xrightarrow[\text{2 BuOCH}_2\text{NEt}_2]{\text{1 Mg Et}_2\text{O}}$ CH$_2$=CHCH$_2$CH$_2$NEt$_2$

JCS (1923) $\underline{123}$ 532

C$_5$H$_{11}$Br $\xrightarrow[\text{2 NH}_2\text{OMe}]{\text{1 Mg Et}_2\text{O}}$ C$_5$H$_{11}$NH$_2$ 65%

JCS (1946) 781

BuBr $\xrightarrow[\text{2 ClNH}_2]{\text{1 Mg Et}_2\text{O dioxane}}$ BuNH$_2$ JACS (1941) 63 1692   97%
                                                                    (1936) 58 27

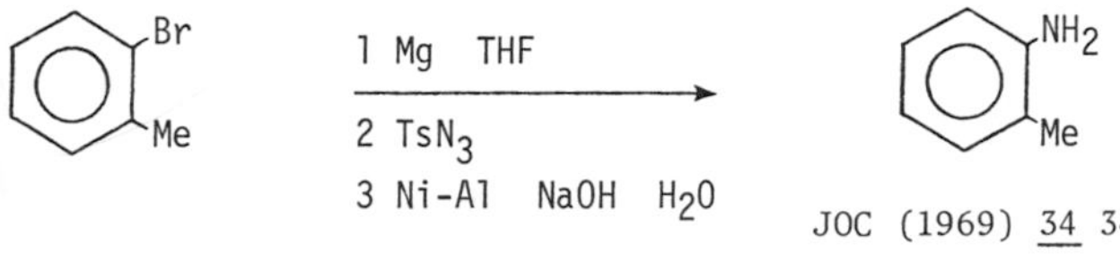

$\xrightarrow[\substack{\text{2 TsN}_3 \\ \text{3 Ni-Al   NaOH   H}_2\text{O}}]{\text{1 Mg   THF}}$ 82%

JOC (1969) 34 3430

C$_7$H$_{15}$Br $\xrightarrow{\text{NH}_3 \quad \text{MeOH}}$ C$_7$H$_{15}$NH$_2$ 47%

JACS (1932) 54 1499 3441

i-PrBr $\xrightarrow{\text{Et}_2\text{NH   HOCH}_2\text{CH}_2\text{OH}}$ Et$_2$NPr-i   60%

JACS (1932) 54 4457

PrBr $\xrightarrow{\text{PhNH}_2}$ PrNHPh   70%

JCS (1930) 992

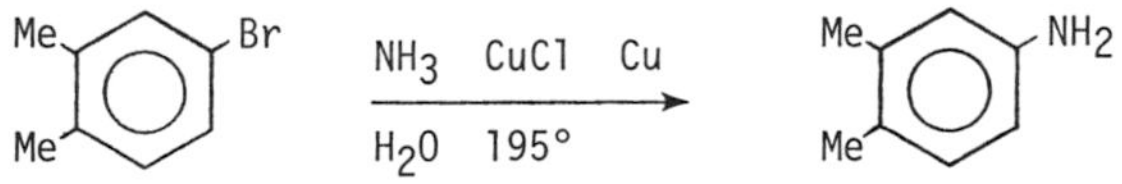

$\xrightarrow[\text{H}_2\text{O   195°}]{\text{NH}_3 \quad \text{CuCl   Cu}}$ 79%

Org Synth (1955) Coll Vol 3 307

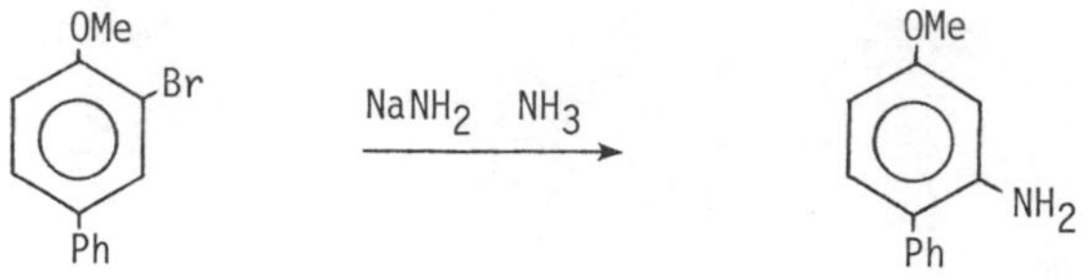

60%

JOC (1957) 22 500
Ber (1964) 97 1994
Chem Rev (1962) 62 81

PhI $\xrightarrow{\text{PhLi} \quad \text{Me}_3\text{N}}$ PhNMe$_2$ 60%

Ber (1943) 76B 109

Further examples of the reaction RHal + R$_2$NH $\rightarrow$ RNR$_2$ (R=alkyl, aryl etc.)
are included in section 97 (Amines from Amines)

$\text{C}_{16}\text{H}_{33}\text{OTs} \xrightarrow{\text{BuNH}_2 \quad \text{toluene}} \text{C}_{16}\text{H}_{33}\text{NHBu} + (\text{C}_{16}\text{H}_{33})_2\text{NBu}$
                                       51%   33%

JACS (1933) 55 345
JCS (1955) 694

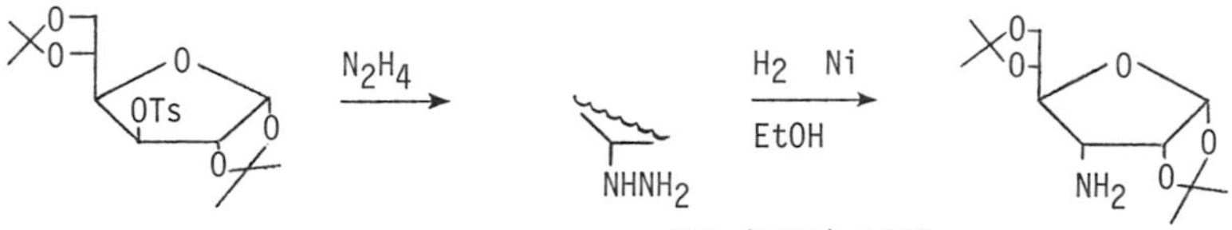

JCS (1961) 1643
Ber (1970) 103 475

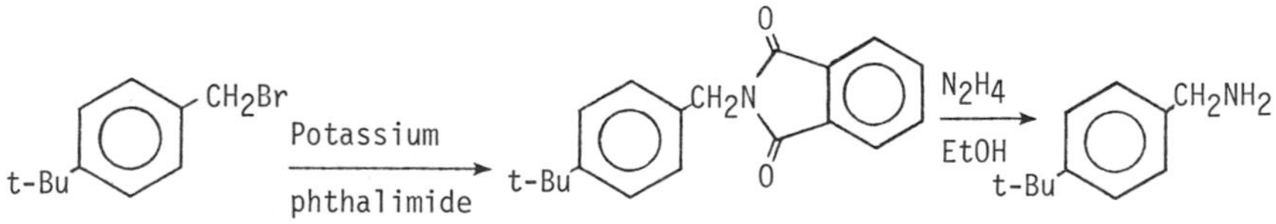

JCS (1935) 1847
JACS (1950) 72 2787
Angew (1968) 80 986
(Internat Ed 7 919)

R⟨benzene⟩Br → (1 Potassium phthalimide, CuBr DMA, 2 Hydrolysis) → R⟨benzene⟩NH₂

Chem Comm (1969) 578

PhCH₂Br —(Saccharin K₂CO₃)→ ⟨benzene fused CO–NCH₂Ph / SO₂⟩ —(1 KOH, 2 BuOTs, 3 HCl H₂O)→ PhCH₂NHBu   74%

J Pharm Soc Jap (1953) 73 1319
                 (1955) 75 153 159

⟨benzene ring with I and Cl⟩ —(N-Acetyl-p-anisidine K₂CO₃, copper bronze PhNO₂)→ ⟨ring⟩NH⟨ring⟩OMe, Cl   24%

JCS (1946) 5

⟨tricyclic⟩C=CHCH₂CH₂Br —(TsNMeNa)→ ⟩C=CHCH₂CH₂NMe(Ts) —(HBr, phenol HOAc)→ ⟩C=CHCH₂CH₂NHMe

Coll Czech (1967) 32 2826

BuX —(1 (PhS)₂NLi THF, 2 HCl H₂O)→ BuNH₂   60% (X=Br)
                                              78% (X=OTs)
                                            Tetr Lett (1970) 3411

PhCH₂Cl —(Hexamethylenetetramine, NaI EtOH)→ PhCH₂NH₂   82%

JACS (1939) 61 3585
Org React (1954) 8 197

$$C_5H_{11}Br \xrightarrow[\text{2 NaOH}]{\text{1 }(NH_2)_2C=NH \quad EtOH \quad H_2O} C_5H_{11}NH_2 \qquad 71\%$$

Tetr Lett (1969) 13

$$Me_2C=CH(CH_2)_3Br \xrightarrow[\substack{\text{2 KCN} \quad MeOH \\ \text{3 HOAc} \quad H_2O}]{\text{1 } NH_2CN \quad MeSOCH_2^- \quad Na^+ \quad Me_2SO} [Me_2C=CH(CH_2)_3]_2NH \qquad 55\%$$

Tetr Lett (1969) 3327
Tetrahedron (1970) 26 1275

$$\xrightarrow{NaN_3 \quad H_2O \quad CCl_4} \quad \xrightarrow{LiAlH_4 \quad Et_2O}$$

JACS (1970) 92 6302
(1951) 73 5865
Helv (1958) 41 181

$$\xrightarrow[DMF]{NaN_3} \qquad \xrightarrow[Et_2O]{LiAlH_4}$$

34%

JOC (1962) 27 2925
JACS (1969) 91 2961

$$\xrightarrow[\text{2 NaOH}]{\text{1 MeCN} \quad H_2SO_4}$$

65%

J Med Chem (1963) 6 760
Org React (1969) 17 213

$$t\text{-BuCl} \xrightarrow{NCl_3 \quad AlCl_3} t\text{-BuNH}_2 \qquad 90\%$$

JOC (1969) 34 911

Chem Eng News (1970) March 9 39
JACS (1964) 86 4732

Bu(CH$_2$)$_4$Cl $\xrightarrow{\text{NaN}_3}$ Bu(CH$_2$)$_4$N$_3$ $\xrightarrow{h\nu \quad \text{THF}}$ Bu—⟨ NH

JCS (1962) 622

Section 101   <u>Amines from Hydrides (RH)</u>
ooooooooooooooooooooooooo

PhH $\xrightarrow{\text{CH}_2\text{=CHCH}_2\text{NHBu} \quad \text{AlCl}_3}$ PhCHCH$_2$NHBu   66%
                                              |
                                              Me

JACS (1943) 65 674 762

PhH $\xrightarrow[]{\substack{\text{NH} \\ / \backslash \\ \text{MeCH-CH}_2 \quad \text{AlCl}_3}}$ PhCHCH$_2$NH$_2$ +   PhCH$_2$CHNH$_2$
                              | |
                              Me 12%   Me 4%

J Heterocyclic Chem (1968) 5 339

EtCHMe$_2$ $\xrightarrow{\text{HCN} \quad \text{t-BuOH} \quad \text{H}_2\text{SO}_4}$ EtCMe$_2$ 45%
                                              |
                                              NH$_2$

Ber (1964) 97 3234
Org React (1969) 17 213

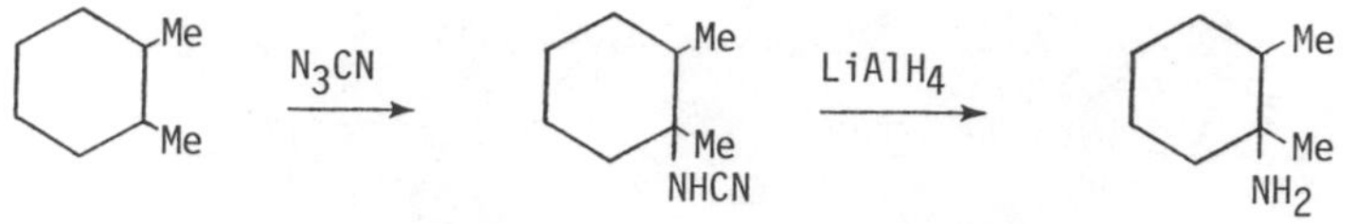

JACS (1967) 89 3177

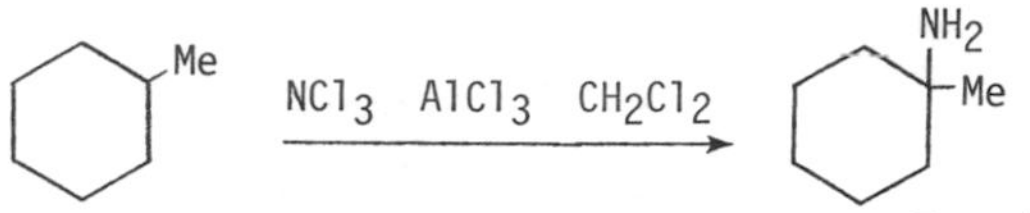

Tetrahedron (1967) 23 3563

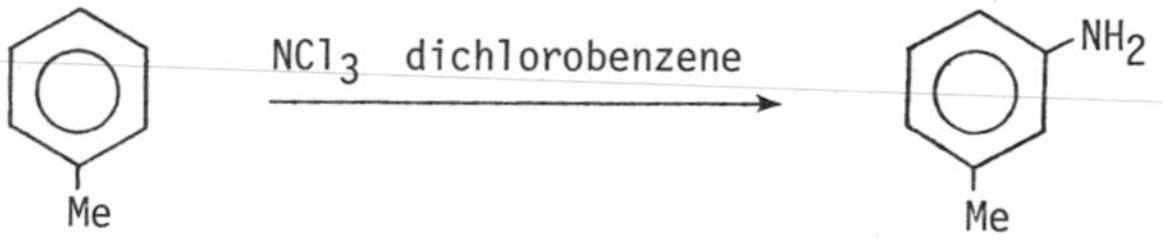

68%

JACS (1966) 88 100
     (1964) 86 1650

36%

PhH $\xrightarrow{\text{Me}_2\text{NCl} \quad \text{H}_2\text{SO}_4 \quad \text{Na}_2\text{SO}_4}$ PhNMe$_2$

Ber (1966) 99 1347 1361

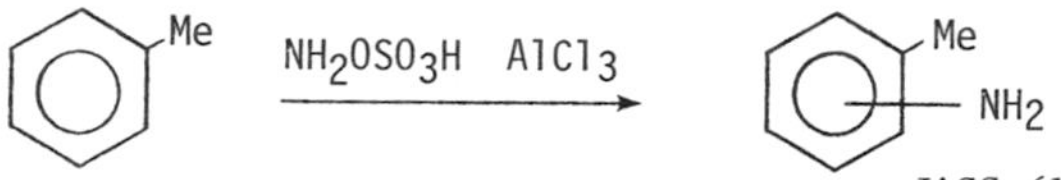

73%

JACS (1961) 83 221 743

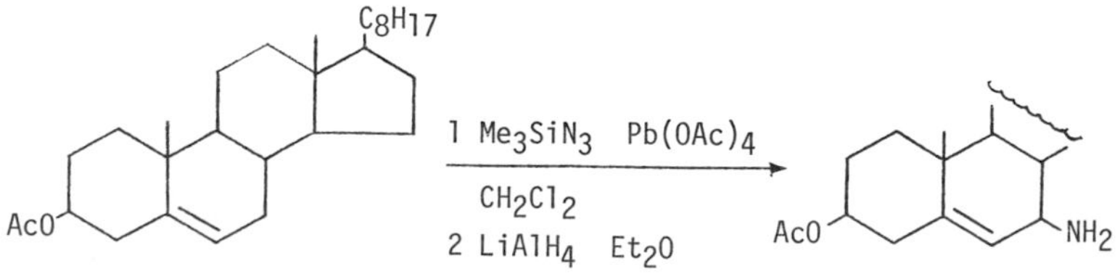

50%

Tetrahedron (1970) 26 1417

49%

$$Al(NO_3)_3 \cdot 9H_2O \xrightarrow{120-140°} \qquad \xrightarrow[\text{MeOH}]{H_2 \quad Ni}$$

JACS (1953) <u>75</u> 369

18%

$$\xrightarrow{HNO_3 \quad H_2SO_4} \qquad \xrightarrow[\text{EtOH}]{Fe \quad HCl \quad H_2O}$$

JCS (1939) 1299

52%

PhH $\xrightarrow[\text{2 NH}_3 \quad h\nu]{\text{1 (CF}_3\text{COO)}_3\text{Tl} \quad CF_3COOH}$ PhNH$_2$

Acc Chem Res (1970) <u>3</u> 338

$$\xrightarrow[\text{2 NH}_2\text{OMe}]{\text{1 BuLi} \quad Et_2O}$$

Org React (1954) <u>8</u> 258

54%

$$\xrightarrow{N_2H_4 \quad h\nu \quad t\text{-BuOH}}$$

Tetrahedron (1966) <u>22</u> 483

45%

Amines may also be prepared by conversion of hydrides (RH) into amides
(RNHCOR, RCONR$_2$ etc.) followed by hydrolysis or reduction. See section
86 (Amides from Hydrides) and section 96 (Amines from Amides)

Section 102 Amines from Ketones

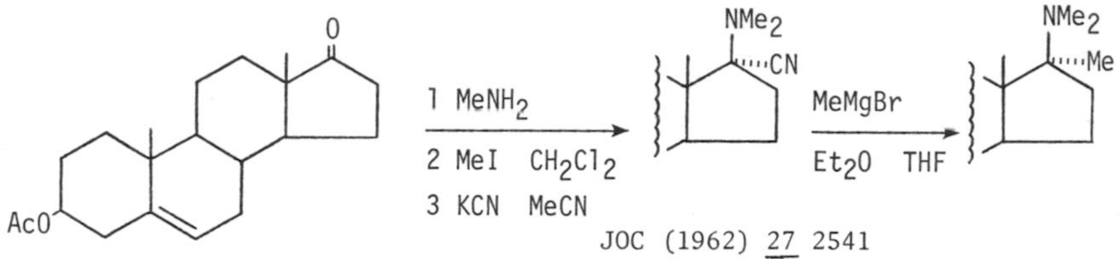
AcO
O
1 MeNH2
2 MeI CH2Cl2
3 KCN MeCN
NMe2
CN
MeMgBr
Et2O THF
NMe2
Me

JOC (1962) 27 2541
(1965) 30 3203

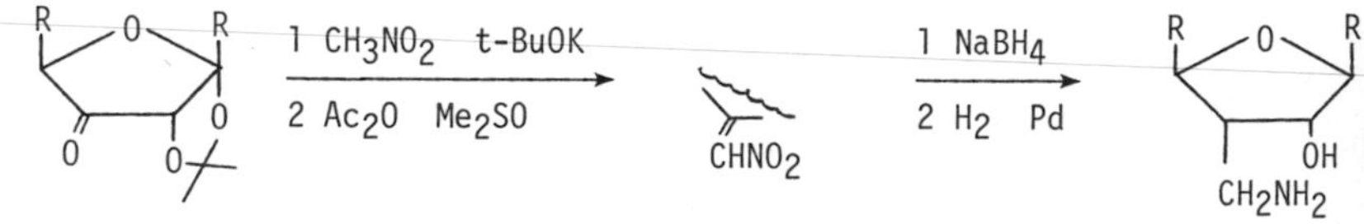
R
O
R
O
O
O
1 CH3NO2 t-BuOK
2 Ac2O Me2SO
CHNO2
1 NaBH4
2 H2 Pd
R
O
R
OH
CH2NH2

Tetr Lett (1970) 1063

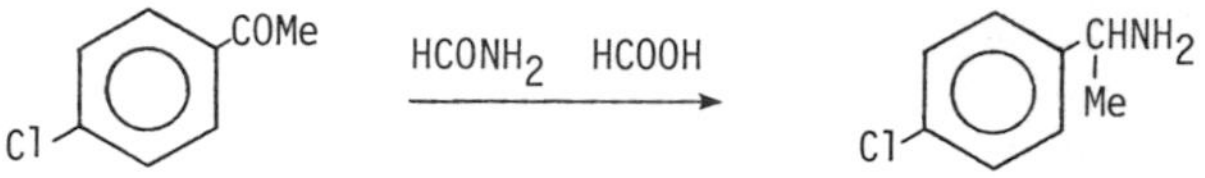
COMe
Cl
HCONH2 HCOOH
CHNH2
Me
Cl

82%

Org React (1949) 5 301

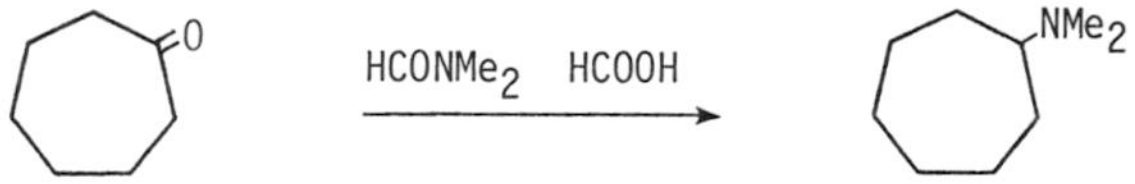
O
HCONMe2 HCOOH
NMe2

~71%

JOC (1968) 33 1647

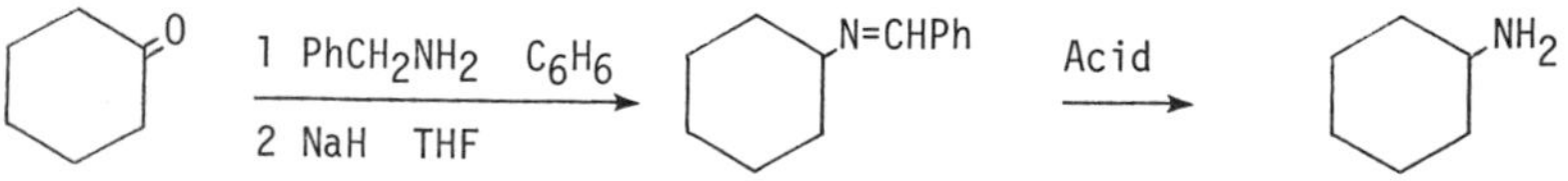
O
1 PhCH2NH2 C6H6
2 NaH THF
N=CHPh
Acid
NH2

57%

Can J Chem (1970) 48 570

Me_2CO $\xrightarrow[\text{HOAc \quad EtOH}]{\text{PhNH}_2 \quad \text{NaBH}_4 \quad \text{NaOAc}}$ $Me_2CHNHPh$ 91%

JOC (1963) $\underline{28}$ 3259

$(CH_2)_{11}$ CO $\xrightarrow[\text{MeOH \quad pH 5-6}]{\text{Me}_2\text{NH} \quad \text{LiBH}_3\text{CN}}$ $(CH_2)_{11}$ $CHNMe_2$ 96%

JACS (1969) $\underline{91}$ 3996

$\xrightarrow[\text{MeOH \quad pH 5-6}]{\text{NH}_3 \quad \text{LiBH}_3\text{CN}}$

48%

NH_2

JACS (1969) $\underline{91}$ 3996

$\xrightarrow{\text{Pyrrolidine}}$

$\xrightarrow[\text{2 MeOH}]{\text{1 B}_2\text{H}_6 \quad \text{THF}}$

Tetr Lett (1970) 2849
JOC (1965) $\underline{30}$ 3203

$\xrightarrow[\text{2 NaBH}_4]{\text{1 PhCH}_2\text{NH}_2}$ $PhCH_2NH$ $\xrightarrow[\text{EtOH}]{\text{H}_2 \quad \text{Pd-C}}$

NH_2

25%

Bull Soc Chim Fr (1963) 798

$$Ph_2CO \xrightarrow{\text{PhNH}_2 \quad \text{ZnCl}_2} Ph_2C=NPh \xrightarrow{\text{LiAlH}_4 \quad \text{Et}_2O} Ph_2CHNHPh \qquad 62\%$$

JOC (1958) <u>23</u> 535

For reduction of imines with H_2 and Pt see Bull Soc Chim Fr (1964) 753

    "   "   "   "   " H_2 " Pd " JCS <u>C</u> (1970) 1303

    "   "   "   "   " H_2 " Ni " JOC (1962) <u>27</u> 2209

    "   "   "   "   by electrolysis " JOC (1970) <u>35</u> 261

For review of reduction of Schiff's bases see Org React (1948) <u>4</u> 174

$$Ph_2CO \dashrightarrow Ph_2CN_2 \xrightarrow{\text{Et}_2NH \quad h\nu} Ph_2CHNEt_2 \qquad 23\%$$

Annalen (1958) <u>614</u> 19

$$\xrightarrow{\text{Me}_2\text{NNH}_2}$$

MeC=NNMe$_2$

1 MeI MeCN
2 NaH Me$_2$SO
3 NaSH
4 Ni

J Heterocyclic Chem (1964) <u>1</u> 53

$$PhCOEt \xrightarrow{\text{NH}_2\text{OH}} \underset{\text{Et}}{PhC=NOH} \xrightarrow[\text{EtOH} \quad \text{H}_2\text{O}]{\text{Ni} \quad \text{NaH}_2\text{PO}_2} \underset{\text{Et}}{PhCHNH_2} \qquad 78\%$$

JCS <u>C</u> (1966) 531

For reduction of oximes with Ni-Al and NaOH see JCS <u>C</u> (1966) 655

    "   "   "   "   "   Zn   " NH$_3$ " Monatsh (1963) <u>94</u> 677

    "   "   "   "   "   Na and NH$_3$ see Zh Obshch Khim (1965) <u>35</u> 125
                                              (Chem Abs <u>62</u> 13068)

    "   "   "   "   "   LiAlH$_4$ see JOC (1952) <u>17</u> 294

    "   "   "   "   "   B$_2$H$_6$ " JOC (1969) <u>34</u> 1817

    "   "   "   "   " H_2 and Rh " JOC (1962) <u>27</u> 2209

    "   "   "   "   by electrolysis   " JACS (1967) <u>89</u> 6374

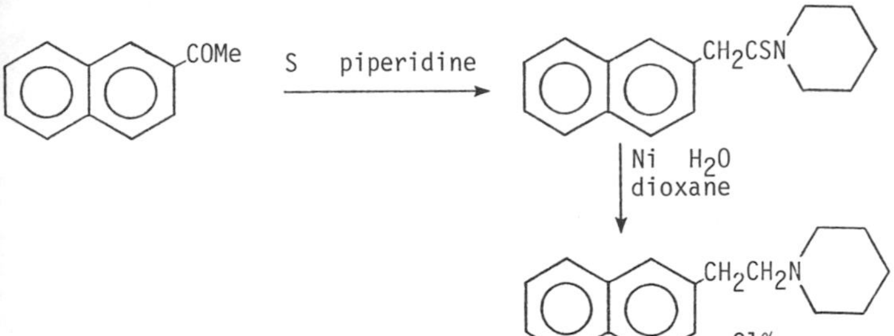

21%

JOC (1951) 16 131

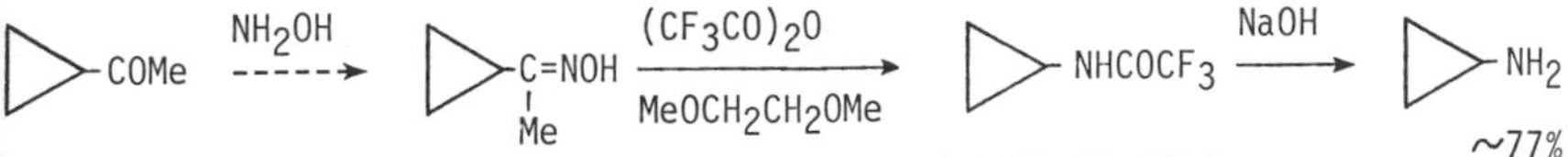

~77%

JACS (1957) 79 6522

$Ph_2CO \xrightarrow[\text{polyphosphoric acid}]{NH_2OH \cdot HCl} PhNH_2$ 66%

JACS (1953) 75 2014

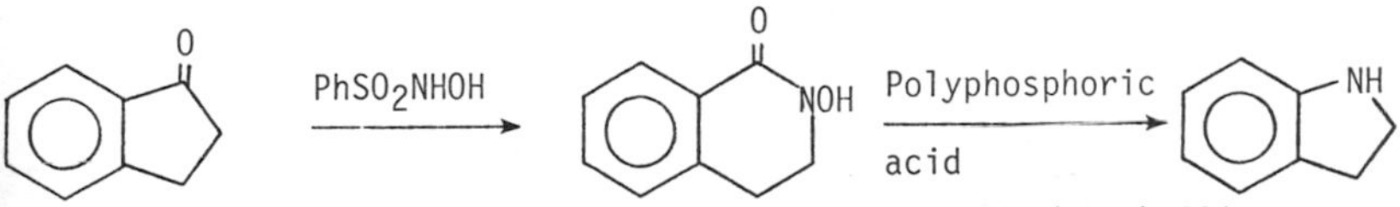

45%

Proc Chem Soc (1963) 224

Amines may also be prepared by conversion of ketones into amides followed by hydrolysis or reduction. See section 87 (Amides from Ketones) and section 96 (Amines from Amides)

Some of the reactions listed in section 94 (Amines from Aldehydes) may also be applied to the preparation of amines from ketones

Section 103   Amines from Nitriles

$C_5H_{11}CN$ $\xrightarrow[\text{2 LiAlH}_4 \quad \text{THF}]{\text{1 PhMgBr} \quad \text{Et}_2O}$ $C_5H_{11}\underset{\underset{Ph}{|}}{C}HNH_2$ JACS (1953) $\underline{75}$ 5898   54%

$PhCN$ $\xrightarrow[\substack{\text{2 EtOH} \\ \text{3 NaOH} \quad H_2O}]{\text{1 (i-PrO)}_2\overset{+}{C}H \quad \overset{-}{B}F_4 \quad CH_2Cl_2}$ $PhC=NPr\text{-}i$ $\underset{OEt}{|}$ $\xrightarrow{\text{NaBH}_4 \quad \text{EtOH}}$ $PhCH_2NHPr\text{-}i$ 94%

JOC (1969) $\underline{34}$ 627

$PhCN$ $\xrightarrow[\text{HCl} \quad \text{EtOH}]{H_2 \quad \text{Pd-C}}$ $PhCH_2NH_2$ JACS (1928) $\underline{50}$ 3370
JCS (1942) 426

$C_6H_{13}O(CH_2)_2CN$ $\xrightarrow[\text{NH}_3 \quad \text{EtOH}]{H_2 \quad \text{Rh-Al}_2O_3}$ $C_6H_{13}O(CH_2)_2CH_2NH_2$ 90%

JACS (1960) $\underline{82}$ 2386

$\xrightarrow[\text{EtOH} \quad H_2O]{\text{Ni} \quad \text{NaH}_2PO_2 \quad \text{NaOH}}$ 82%

JCS $\underline{C}$ (1966) 531
Bull Chem Soc Jap (1967) $\underline{40}$ 1548

$BuCN$ $\xrightarrow{\text{Na} \quad \text{EtOH} \quad \text{toluene}}$ $BuCH_2NH_2$ 76%

JACS (1934) $\underline{56}$ 1614
Ber (1942) $\underline{75B}$ 991

$$\text{o-Me-C}_6\text{H}_4\text{-CN} \xrightarrow{\text{LiAlH}_4 \ \ \text{Et}_2\text{O}} \text{o-Me-C}_6\text{H}_4\text{-CH}_2\text{NH}_2 \qquad 88\%$$

JACS (1948) 70 3738
Org React (1951) 6 469

$$\text{PhCH}_2\text{CN} \xrightarrow{\text{LiAlH}_4 \ \ \text{AlCl}_3 \ \ \text{Et}_2\text{O}} \text{PhCH}_2\text{CH}_2\text{NH}_2 \qquad 83\%$$

JACS (1955) 77 2544

$$\text{PhCN} \xrightarrow{(\text{i-Bu})_2\text{AlH} \ \ \text{C}_6\text{H}_6} \text{PhCH=NH} \xrightarrow{\text{H}_2 \ \ \text{Pd}} \text{PhCH}_2\text{NH}_2 \qquad 57\%$$

JOC (1959) 24 627

$$\text{p-Me-C}_6\text{H}_4\text{-CN} \xrightarrow[\text{diglyme}]{\text{NaBH}_4 \ \ \text{AlCl}_3} \text{p-Me-C}_6\text{H}_4\text{-CH}_2\text{NH}_2 \qquad 85\%$$

JACS (1956) 78 2582

$$\text{C}_7\text{H}_{15}\text{CN} \xrightarrow{\text{NaBH}_4 \ \ \text{CoCl}_2 \ \ \text{MeOH}} \text{C}_7\text{H}_{15}\text{CH}_2\text{NH}_2 \qquad 80\%$$

Tetr Lett (1969) 4555

$$\xrightarrow{\text{B}_2\text{H}_6 \ \ \text{THF}} \qquad 85\%$$

JACS (1960) 82 681

$$\xrightarrow[\text{H}_2\text{O}]{\text{H}_2\text{S} \ \ \text{NaOH}} \qquad \xrightarrow{\text{H}_2 \ \ \text{Pd-C}}$$

Chem Pharm Bull (1961) 9 119
JOC (1951) 16 131

PhCN $\xrightarrow{\text{MeOH HCl}}$ PhC=NH·HCl $\xrightarrow[\text{H}_2\text{SO}_4 \quad \text{H}_2\text{O}]{\text{Electrolysis}}$ PhCH$_2$NH$_2$ 76%
                              |
                             OMe

JACS (1935) <u>57</u> 772

PhCN $\xrightarrow[\text{polyphosphoric acid}]{\text{NH}_2\text{OH·HCl}}$ PhNH$_2$ JACS (1953) <u>75</u> 2014   20%

Amines may also be prepared by conversion of nitriles into amides followed by hydrolysis or reduction. See section 88 (Amides from Nitriles) and section 96 (Amines from Amides)

Section 104   Amines from Olefins
              ○○○○○○○○○○○○○○○○○○○○

$\xrightarrow{\text{MeCH}_2\text{NEt}_2 \quad h\nu}$

Chem Comm (1969) 753   11%

$\xrightarrow[\text{H}_2\text{SO}_4 \quad \text{HOAc}]{\text{Me}_2\text{NH} \quad \text{paraformaldehyde}}$

Tetr Lett (1966) 6483   19%

$\xrightarrow[\text{2 NH}_2\text{OSO}_3\text{H}]{\text{1 NaBH}_4 \quad \text{BF}_3\text{·Et}_2\text{O} \quad \text{diglyme}}$

JACS (1966) <u>88</u> 2870
JOC (1967) <u>32</u> 3199   45%

$MeCH=CH_2$ $\xrightarrow[\text{2 NaBH}_4]{\text{1 Hg(OAc)}_2 \quad \text{PhNH}_2}$ $Me_2CHNHPh$ 70%

Tetr Lett (1967) 5165
(1969) 2289
Compt Rend (1966) C 262 1591

$\xrightarrow[\text{2 NaBH}_4 \quad \text{NaOH} \quad \text{H}_2\text{O}]{\text{1 Hg(NO}_3)_2 \quad \text{MeCN}}$ NHAc ---→ NH2

JACS (1969) 91 5647

$\xrightarrow[\text{2 H}_2\text{O}]{\text{1 NaCN} \quad \text{H}_2\text{SO}_4 \quad \text{HOAc}}$ Me / NH2   58%

Tetrahedron (1967) 23 3563
Org React (1969) 17 213

$\xrightarrow[\text{2 MeOH}]{\text{1 AgCNO} \quad \text{I}_2 \quad \text{Et}_2\text{O}}$ NHCOOMe / I $\xrightarrow[\text{Et}_2\text{O}]{\text{Zn} \quad \text{HOAc}}$ NHCOOMe $\xrightarrow[]{\text{Acid}}$ NH2

JACS (1970) 92 1326

$MeCH=CH_2$ $\xrightarrow[\text{2 H}_2]{\text{1 BuNH}_2 \quad \text{PdCl}_2 \quad \text{Na}_2\text{HPO}_4 \quad \text{THF}}$ $Me_2CHNHBu$

Proc Chem Soc (1961) 370

$C_8H_{17}CH=CH_2$ $\xrightarrow[\text{2 H}_2 \quad \text{Ni} \quad \text{NH}_3]{\text{1 O}_3 \quad \text{MeOH}}$ $C_8H_{17}CH_2NH_2$ 60%

JOC (1962) 27 2392

Amines may also be prepared by conversion of olefins into amides followed
by hydrolysis or reduction. See section 89 (Amides from Olefins) and
section 96 (Amines from Amides)

Section 105   Amines from Miscellaneous Compounds
○○

$$EtCHNO_2 \quad \xrightarrow{\text{LiAlH}_4 \quad \text{Et}_2\text{O}} \quad EtCHNH_2$$

Me (below first structure)   Me (below second structure)   85%

JACS (1948) 70 3738
(1951) 73 1293

$$PrNO_2 \quad \xrightarrow[\text{Et}_2\text{O}]{\text{Pd} \quad \text{cyclohexene}} \quad PrNH_2 \qquad 55\%$$

JCS (1954) 3586

For reduction of nitro compounds with H_2 and Pd-C see
       Ber (1953) 86 939

   "   "   "   "   "   " H_2 and Ni see
       Org Synth (1955) Coll Vol 3 59

   "   "   "   "   "   "   Fe and HCl see
       Org Synth (1943) Coll Vol 2 160

   "   "   "   "   "   "   Al-Hg see
       JACS (1968) 90 3245

Section 105A   Protection of Amines
○○○○○○○○○○○○○○○○○○○○○○○○○

Amine protecting groups which are generally applicable are included in this
section. For a review of the protection of amino acids and peptides see
       Advances in Org Chem (1963) 3 159

HNMe / OR ... $\xrightarrow[\text{K}_2\text{CO}_3 \quad \text{MeOH} \quad \text{H}_2\text{O}]{(\text{CF}_3\text{CO})_2\text{O}}$ ... CF_3CONMe / OR

Me   O   OR   Me   O   OR

JOC (1965) 30 1287
JACS (1953) 75 3473

$$NH_2CH_2CONHR \quad \underset{\text{HCl} \quad \text{MeOH}}{\overset{\text{HCOOH} \quad \text{Ac}_2\text{O}}{\rightleftarrows}} \quad HCONHCH_2CONHR$$

Chem Ind (1953) 107
Helv (1960) 43 1751

O_2N-C$_6$H$_4$-NH$_2$

$$\xrightarrow[\substack{1\ (NH_2)_2CS\ \ EtOH \\ 2\ H_2O}]{ClCH_2COCl}}$$

O_2N-C$_6$H$_4$-NHCOCH$_2$Cl

JACS (1968) <u>90</u> 4508

HCl·NH$_2$CHCONHCH$_2$COOH
     |
    i-Bu

$$\xrightarrow[\text{PhNHNH}_2\ \ \text{HOAc}]{\text{Diketene\ \ Et}_3\text{N\ \ EtOH}}$$

MeCOCH$_2$CONHCHCONHCH$_2$COOH
                |
              i-Bu

Tetr Lett (1965) 605

Further examples of the preparation and cleavage of acyl derivatives of amines are included in section 82 (Amides from Amines) and section 96 (Amines from Amides)

PhCH$_2$CHNH$_2$
     |
    COOH

$$\xrightarrow[\text{N}_2\text{H}_4\cdot\text{H}_2\text{O\ \ EtOH}]{\text{Phthalic anhydride}}$$

PhCH$_2$CHN(phthalimide)   (Stable to acid)
     |
    COOH

JACS (1952) <u>74</u> 3822
Rec Trav Chim (1960) <u>79</u> 688

Cl-C$_6$H$_3$(R)-NH$_2$

$$\xrightarrow[\text{NaOH\ \ MeOH}]{\text{Ac}_2\text{O}}$$

Cl-C$_6$H$_3$(R)-NAc$_2$ (Stable to NBS)

Can J Chem (1955) <u>33</u> 1819

HN( )NH (piperazine)

$$\xrightarrow[\text{HCl\ \ H}_2\text{O}]{\text{ClCOOEt\ \ NaOH\ \ H}_2\text{O}}$$

HN( )NCOOEt

JCS (1929) 39
JACS (1954) <u>76</u> 1164

$$PhNH_2 \xrightleftharpoons[HBr]{ClCOOCH_2Ph} PhNHCOOCH_2Ph$$

JOC (1952) <u>17</u> 1564

For the hydrogenolysis of benzyloxycarbonyl groups see
JACS (1957) <u>79</u> 1636
and Org React (1953) 7 263
For the photolytic cleavage of benzyloxycarbonyl groups see
Tetr Lett (1962) 697

$$HCl \cdot NH_2CH_2COOMe \xrightleftharpoons[HBr \quad HOAc]{N_3COOBu\text{-}t \quad Et_3N \quad EtOAc} t\text{-}BuOCONHCH_2COOMe$$

Helv (1959) <u>42</u> 2622
JACS (1957) <u>79</u> 4686

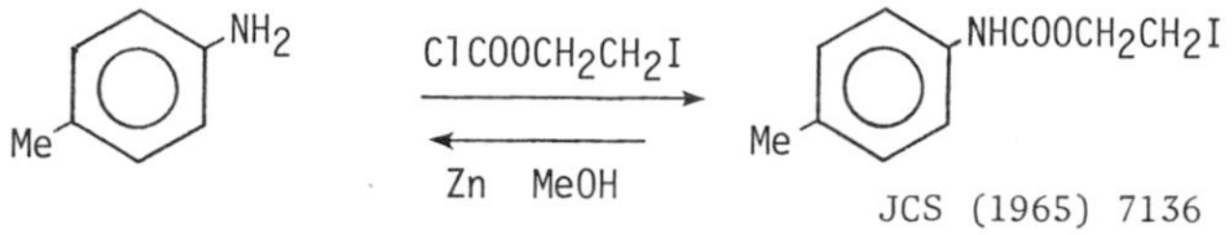

$$\xrightleftharpoons[Zn \quad MeOH]{ClCOOCH_2CH_2I}$$

JCS (1965) 7136

$$\xrightleftharpoons[Zn \quad MeOH]{ClCOOCH_2CCl_3}$$

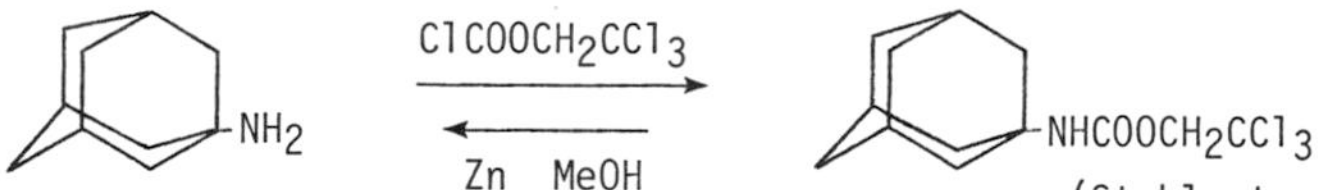

(Stable to acid, base and CrO_3)

Tetr Lett (1967) 2555

$$PhNH_2 \xrightleftharpoons[KOH \quad EtOH]{ClCOOCH_2CH_2Ts} PhNHCOOCH_2CH_2Ts$$

Proc Chem Soc (1962) 363

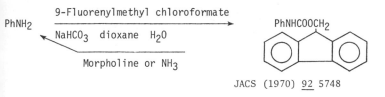

9-Fluorenylmethyl chloroformate

PhNH$_2$ $\xrightarrow{\text{NaHCO}_3 \quad \text{dioxane} \quad \text{H}_2\text{O}}$ PhNHCOOCH$_2$

Morpholine or NH$_3$

JACS (1970) 92 5748

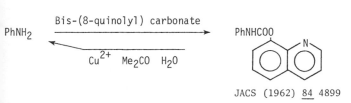

Bis-(8-quinolyl) carbonate

PhNH$_2$ $\xrightarrow{\text{Cu}^{2+} \quad \text{Me}_2\text{CO} \quad \text{H}_2\text{O}}$ PhNHCOO

JACS (1962) 84 4899

(C$_6$H$_{13}$)$_2$NH $\xrightarrow[\text{H}_2 \quad \text{Pt} \quad \text{HOAc}]{}$ (C$_6$H$_{13}$)$_2$NCH$_2$Ph

(Stable to acid, base and RMgX)

JCS (1940) 1307
Org React (1953) 7 263
JACS (1950) 72 3410

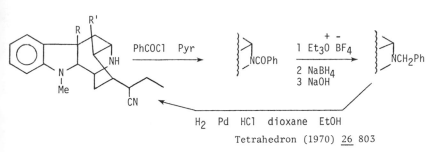

PhCOCl   Pyr

1 Et$_3$O$^+$ BF$_4$$^-$
2 NaBH$_4$
3 NaOH

H$_2$ Pd   HCl   dioxane   EtOH

Tetrahedron (1970) 26 803

HCl·NH$_2$CH$_2$COOR $\xrightarrow[\substack{\text{HOAc} \quad \text{H}_2\text{O} \\ \text{or H}_2 \quad \text{Pd EtOH}}]{\text{Ph}_3\text{CCl} \quad \text{Et}_3\text{N} \quad \text{CHCl}_3}$ Ph$_3$CNHCH$_2$COOR

(Stable to base)

JACS (1956) 78 1359

Br—C6H4—NH2
$\xrightarrow[\text{2 ClSiMe}_3]{\text{1 EtMgBr Et}_2\text{O}}$
$\xleftarrow{\text{MeOH}}$
Br—C6H4—N(SiMe3)2   (Stable to RLi)

JCS <u>C</u> (1966) 1706

C6H4(NHR)(R)
$\xrightarrow[\text{HCl HOAc}]{\text{TsCl Pyr}}$
C6H4(N(Ts)R)(R)

JACS (1949) <u>71</u> 1901
      (1952) <u>74</u> 2006

BuNH2
$\xrightarrow[\text{Zn HCl HOAc}]{\text{PhCOCH}_2\text{SO}_2\text{Cl Pyr CH}_2\text{Cl}_2}$
BuNHSO2CH2COPh

(Stable to acid, base and CrO3)
Tetr Lett (1970) 345

(piperidine with CH2CH2COOH and Pr, NH)
$\xrightarrow[\text{CuCl HCl H}_2\text{O}]{\text{NaNO}_2 \text{ HBr H}_2\text{O}}$
(piperidine with CH2CH2COOH and Pr, N–NO)

JACS (1946) <u>68</u> 146
      (1957) <u>79</u> 2215

(sugar: CH2OR, O, OR, RO, OR, NH2)
$\xrightarrow[\text{HCl Me}_2\text{CO}]{\text{MeO-C}_6\text{H}_4\text{-CHO}}$
N=CH—C6H4—OMe

JACS (1956) <u>78</u> 1393

Chapter 8 PREPARATION OF ESTERS

Section 106 Esters from Acetylenes

$$BuC{\equiv}CH \xrightarrow[2\ ClCOOEt]{1\ LiNH_2\ \ dioxane} BuC{\equiv}CCOOEt \xrightarrow{H_2\ \ Pd} BuCH_2CH_2COOEt$$

25% Annalen (1958) 614 37

$$C_5H_{11}C{\equiv}CH \xrightarrow[2\ TsCl]{1\ Na} C_5H_{11}C{\equiv}CCl \xrightarrow{EtONa} C_5H_{11}CH_2COOEt$$

Annalen (1931) 16 309

$$C_5H_{11}C{\equiv}CH \xrightarrow[250°]{N_2O\ \ MeOH} C_5H_{11}CH_2COOMe \quad 57\%$$

JCS (1951) 3016

$$PhC{\equiv}CH \xrightarrow[BF_3 \cdot Et_2O]{HOAc\ \ HgO} PhCH{=}CHOAc \xrightarrow{KOOCN{=}NCOOK} PhCH_2CH_2OAc$$

JOC (1965) 30 3985
JACS (1934) 56 1802

Section 107   Esters from Carboxylic Acids, Acid Halides and Anhydrides

$$C_5H_{11}COOH \xrightarrow[\text{electrolysis}]{\overset{\displaystyle COOH}{\underset{\displaystyle |}{(CH_2)_4COOMe}} \; MeOH \; H_2O} C_5H_{11}(CH_2)_4COOMe \qquad 48\%$$

JCS (1950) 3326
Advances in Org Chem (1960) $\underline{1}$ 1

$$\xrightarrow[\text{2 Ag}_2\text{O} \quad \text{EtOH}]{\text{1 CH}_2\text{N}_2 \quad \text{Et}_2\text{O}}$$

Org React (1942) $\underline{1}$ 38

73-82%

$$\xrightarrow[\text{2 h}\nu \quad \text{MeOH}]{\text{1 Me}_2\overset{+}{S}\overset{-}{OCH_2} \quad \text{THF}}$$

JACS (1964) $\underline{86}$ 1640

80%

$$CH_3COOH \xrightarrow{\overset{\displaystyle NH_2}{\underset{\displaystyle |}{Me_2CCH_2OH}}} CH_3C\overset{O}{\underset{N}{\diagdown}} \xrightarrow[\substack{\text{2 BuBr} \\ \text{3 H}_2\text{SO}_4 \quad \text{EtOH}}]{\text{1 BuLi}} BuCH_2COOEt$$

JACS (1970) $\underline{92}$ 6644

Further methods for the alkylation and homologation of carboxylic acids
and esters are included in section 17 (Carboxylic Acids, Acid Halides
and Anhydrides from Carboxylic Acids and Acid Halides) and section 113
(Esters from Esters)

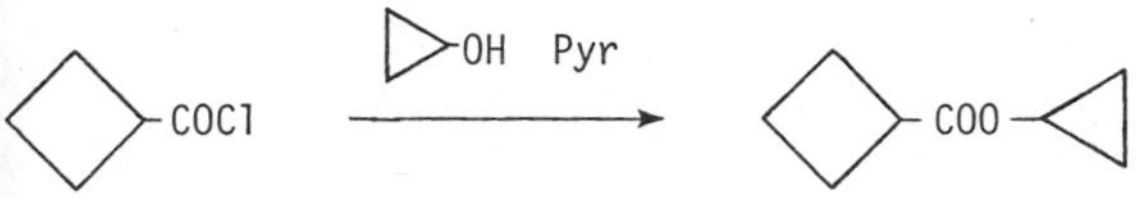

55%

JACS (1951) <u>73</u> 5487

t-BuCH$_2$COOH $\xrightarrow{\text{EtOH H}_2\text{SO}_4}$ t-BuCH$_2$COOEt   77%

JACS (1938) <u>60</u> 2790

C$_8$H$_{17}$COOH $\xrightarrow[\text{molecular sieves}]{\text{MeOH H}_2\text{SO}_4}$ C$_8$H$_{17}$COOMe

Chem Ind (1968) 1568
              (1967)   825

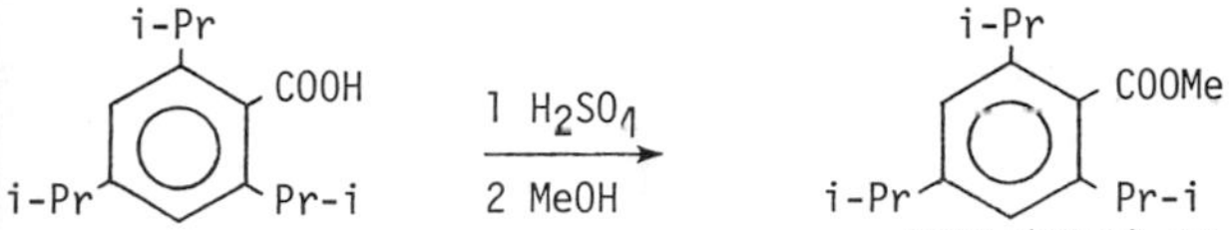

81%

JACS (1941) <u>63</u> 2431

C$_{12}$H$_{25}$COOH $\xrightarrow{\text{EtOH HCl}}$ C$_{12}$H$_{25}$COOEt

Org Synth (1943) Coll Vol 2 292
              (1932) Coll Vol 1 246

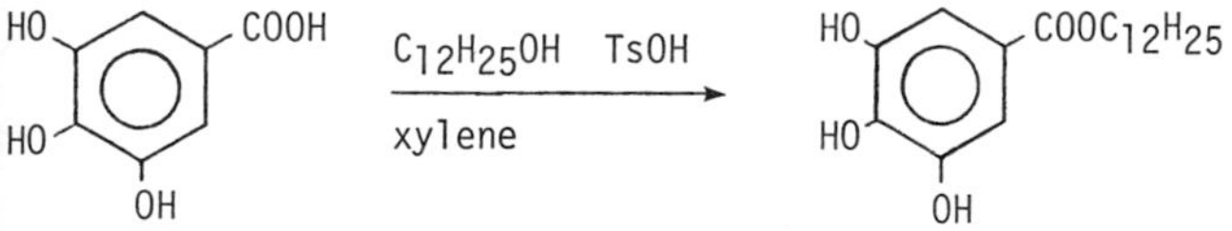

Rec Trav Chim (1951) <u>70</u> 277
JACS (1945) <u>67</u> 902
Acta Chem Scand (1959) <u>13</u> 1407

$$ \text{(R=Me, Et, i-Pr)} \qquad 75\text{-}81\% $$

Tetr Lett (1970) 4011
JCS (1965) 5770

$$ \underset{|}{CH_2COOH} \xrightarrow[]{PhOH \quad POCl_3} \underset{|}{CH_2COOPh} $$
$$ CH_2COOH \qquad\qquad CH_2COOPh \qquad 62\text{-}67\% $$

Org Synth (1963) Coll Vol 4 390 178

$$ PhCOOH \xrightarrow[CHCl_3]{PhOH \quad P_2O_5\text{-}Et_2O} PhCOOPh \qquad 90\% $$

Chem Ind (1964) 2102

$$ \underset{NH_2}{PhCHCOOH} \xrightarrow{PhCH_2OH \quad SOCl_2} \underset{NH_2}{PhCHCOOCH_2Ph} \qquad 90\% $$

JOC (1965) 30 3575
Annalen (1961) 640 136

MeCHCH2CH2COOH → MeCHCH2CH2COOMe

MeOH   MeCOCl   100%

Org Synth (1955) Coll Vol 3 237

$$ C_{15}H_{31}COOH \xrightarrow[\underset{X}{\underset{O \; O}{HOCH_2CHCH_2}}]{(CF_3CO)_2O} \underset{\underset{X}{O \; O}}{C_{15}H_{31}COOCH_2CHCH_2} $$

Applicable to hindered acids

JCS (1965) 4594
JOC (1965) 30 927

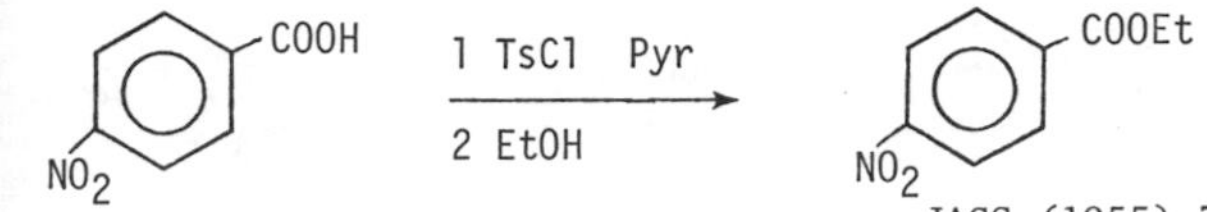

JACS (1955) 77 6214

90%

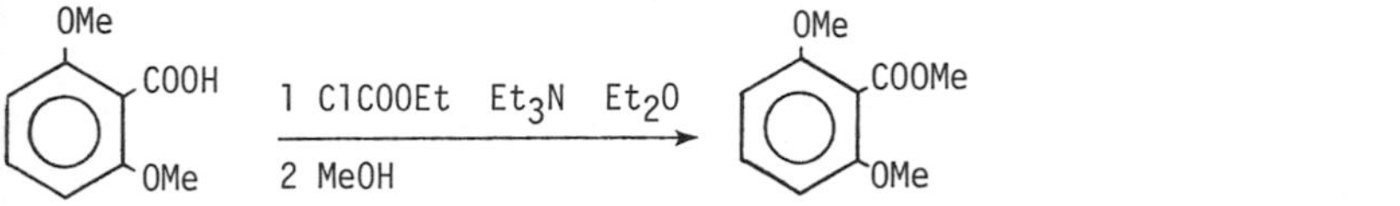

Rev Roumaine Chim (1966) 11 1237
(Chem Abs 66 65249)
JACS (1958) 80 5714
      (1969) 91 3931

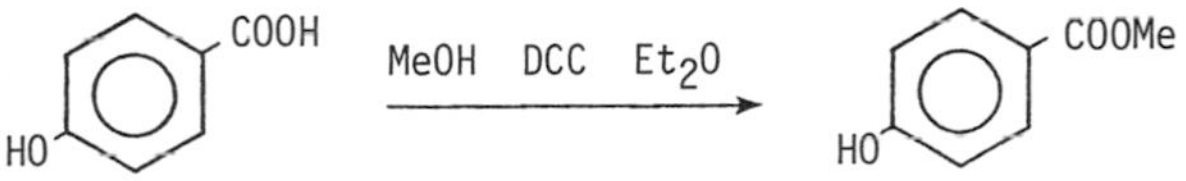

Tetrahedron (1965) 21 3531
Compt Rend (1963) 256 1804
           (1962) 255 945

45%

PhCH=CHCOOH $\xrightarrow[\text{MeONa}]{\text{MeOH   NN'-carbonyldiimidazole}}$ PhCH=CHCOOMe   83%

Ber (1962) 95 1284
JCS C (1969) 1610

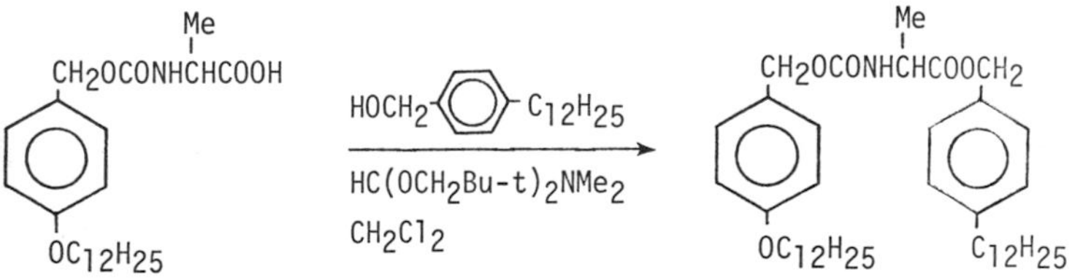

Helv (1965) 48 1746
JACS (1969) 91 5674

83%

$$(PhO)_2CHCOOH \xrightarrow{\text{EtOH} \quad Me_2C(OMe)_2} (PhO)_2CHCOOEt$$

Bull Soc Chim Fr (1964) 381
JOC (1959) 24 261

Further examples of the reaction RCOOH + ROH ⟶ RCOOR are included in
section 108 (Esters from Alcohols and Phenols) and section 30A (Protection
of Carboxylic Acids)

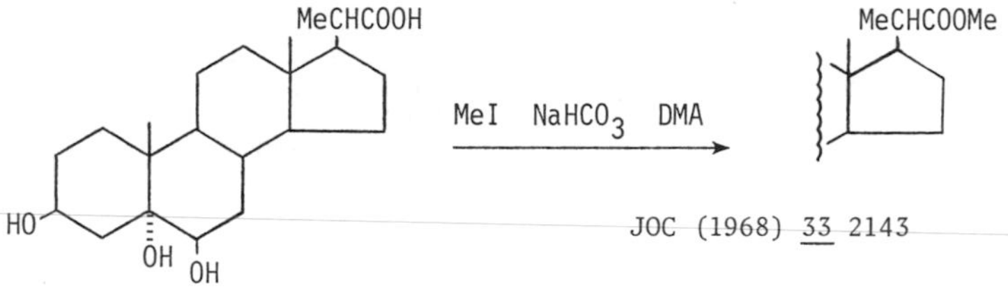

MeI   NaHCO$_3$ DMA

JOC (1968) 33 2143

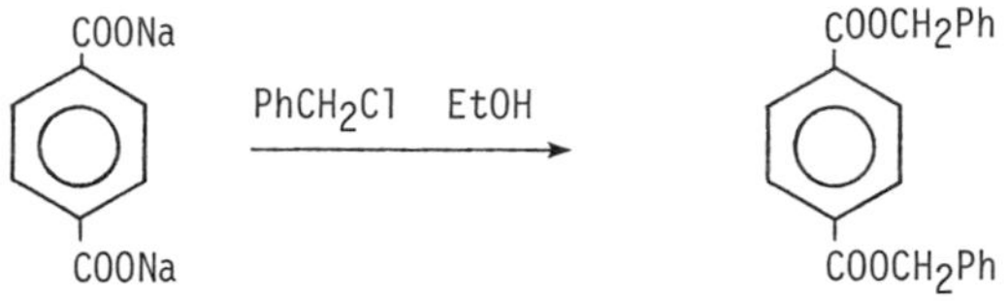

91%

PhCH$_2$Cl   EtOH

J Med Chem (1967) 10 706

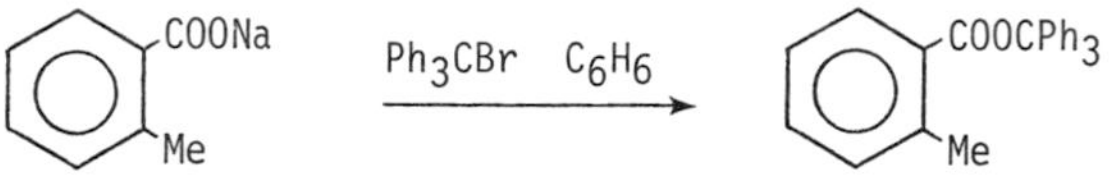

Ph$_3$CBr   C$_6$H$_6$

JOC (1962) 27 3595
JCS C (1966) 1191

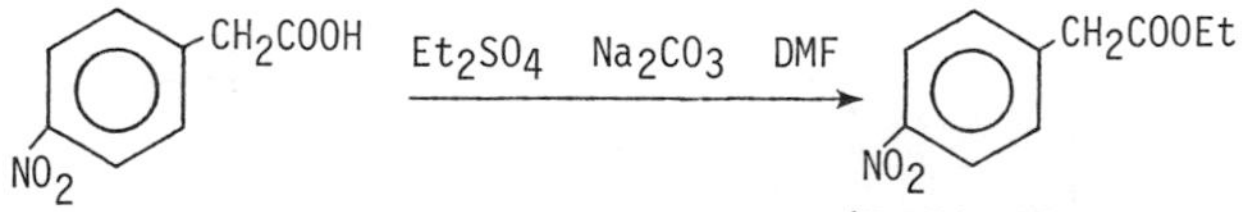

84%

Et$_2$SO$_4$ Na$_2$CO$_3$ DMF

(Applicable to hindered acids)

Monatsh (1968) 99 103
Helv (1943) 26 2283

88%

$C_6H_5(OH)$—CH=CHCOOH $\xrightarrow[\text{hexylamine \quad Me}_2\text{CO}]{\text{Et}_2\text{SO}_4 \quad \text{N-ethyldicyclo-}}$ $C_6H_5(OH)$—CH=CHCOOEt   92%

(Applicable to hindered acids)

JOC (1964) 29 2490
JACS (1969) 91 3544

PhCOOH $\xrightarrow[\text{xylene}]{C_6H_{13}Cl \quad Et_3N}$ PhCOOC$_6$H$_{13}$ JOC (1961) 26 5180   47%

$CH_2=CCH_2COOH$ (Me) $\xrightarrow[\text{2 AgNO}_3]{\text{1 NaOH \quad H}_2\text{O}}$ $CH_2=CCH_2COOAg$ (Me) $\xrightarrow[\text{Me}]{\text{MeI \quad Et}_2\text{O}}$ $CH_2=CCH_2COOMe$ (Me)   47%

JACS (1949) 71 3214
(1951) 73 5487
(1952) 74 3944

Bu$_3$CCOONa $\xrightarrow[]{\text{BuOSOCl \quad C}_6\text{H}_6}$ Bu$_3$CCOOBu   JACS (1947) 69 1046   82%

⌐COOH
└COOH $\xrightarrow[]{\text{HC(OEt)}_3}$ ⌐COOEt
└COOEt   (Applicable to hindered acids)   89%
Chem Ind (1965) 349

COOH
├—OH
HO—│
├—OH
COOH $\xrightarrow[]{\text{(BuO)}_3\text{B}}$ COOBu
├—OH
HO—│
├—OH
COOBu   98%

Zh Obshch Khim (1967) 37 1481
(Chem Abs 68 22157)

PhCOOH $\xrightarrow[]{\text{Me}_2\text{NCH(OR)}_2 \quad \text{C}_6\text{H}_6}$ PhCOOR   (R=Me, Et, CH$_2$Ph)   80-95%

Helv (1965) 48 1746

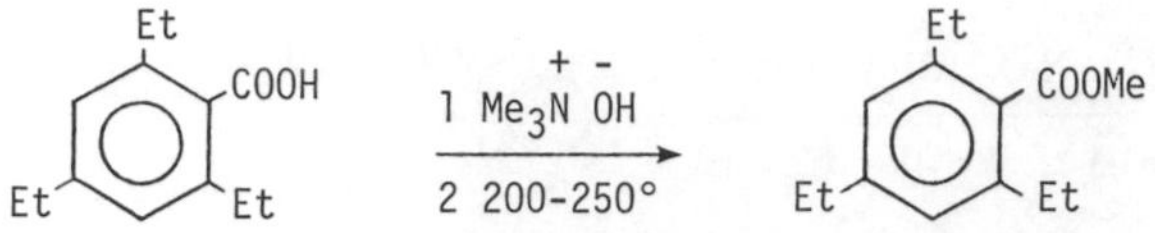

63-90%

JACS (1939) 61 1290
Z Physiol Chem (1936) 244 56

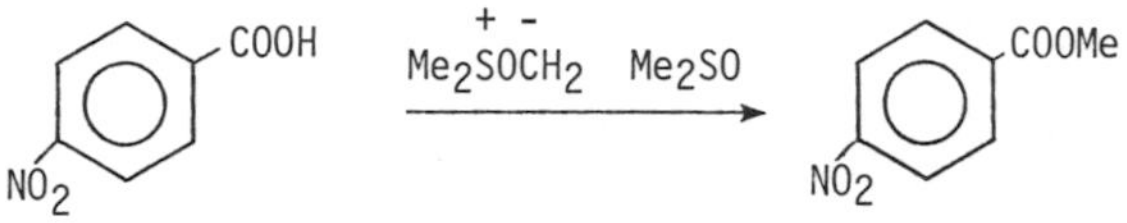

69%

Tetr Lett (1964) 867

$CH_2(COOH)_2$ →[$Me_2C=CH_2$ H_2SO_4] $CH_2(COOBu-t)_2$ 58-60%

Org Synth (1963) Coll Vol 4 261
JOC (1963) 28 1251

$PhCONHCH_2COOH$ →[$Et_3O^+ BF_4^-$ $NaHCO_3$ H_2O] $PhCONHCH_2COOEt$ 82%

Tetr Lett (1969) 1819
Ber (1956) 89 209 2060

$PhCOOH$ →[$(EtO)_3P$ $\overset{NCOOEt}{\overset{\|}{NCOOEt}}$ Et_2O] $PhCOOEt$ 95%

Bull Chem Soc Jap (1967) 40 2380

$PhCOOH$ →[CH_2N_2 Et_2O] $PhCOOMe$

JACS (1954) 76 4481

$PhCOOH \xrightarrow{Ph_2CN_2 \quad Et_2O} PhCOOCHPh_2$ 100%

Org Synth (1955) Coll Vol 3 351

For preparation of CH_2N_2 see Org Synth (1943) Coll Vol 2 165
                                      and Org React (1942) 1 50

"   "   " Me_2CN_2 and Pr_2CN_2 see JCS C (1966) 467

"   "   " $PhCHN_2$ " JACS (1964) 86 658

"   "   " anthryldiazomethane " Bull Chem Soc Jap (1967) 40 691

$C_{15}H_{31}COOH \xrightarrow[2 \; I_2]{1 \; Pb(OAc)_4 \quad tetrachloroethane} C_{15}H_{31}COOC_{15}H_{31}$ 54%

JOC (1963) 28 65

$(PrCH_2CO)_2O \xrightarrow{HgO \quad I_2 \quad BrCH_2CH_2Br} PrCH_2COOCH_2Pr$ 50%

JOC (1970) 35 3167

$BuCHCOCl$ over Et $\xrightarrow[\text{Pyr \quad hexane}]{\text{m-Chloroperbenzoic acid}}$ ~73%

JOC (1965) 30 3760
JACS (1950) 72 67

$Ph_2CHCOOH \xrightarrow{Electrolysis \quad HOAc} Ph_2CHOAc$ 37%

JCS (1952) 3624

Section 108   Esters from Alcohols and Phenols

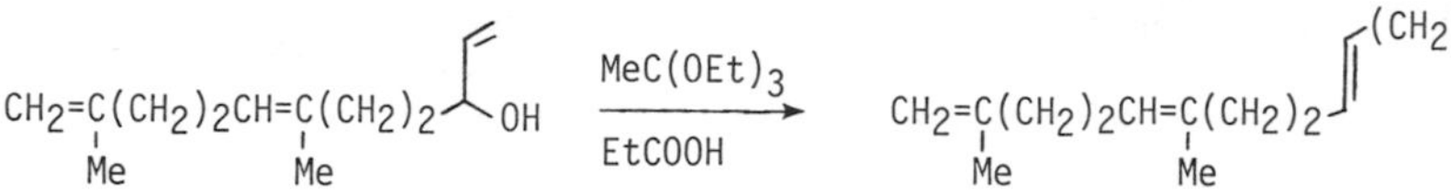

$$CH_2=C(CH_2)_2CH=C(CH_2)_2 \diagup OH \xrightarrow[\text{EtCOOH}]{MeC(OEt)_3} CH_2=C(CH_2)_2CH=C(CH_2)_2 \diagup (CH_2)_2COOEt \quad >91\%$$

with Me under each C.

JACS (1970) 92 741
Chem Comm (1970) 1512 1513

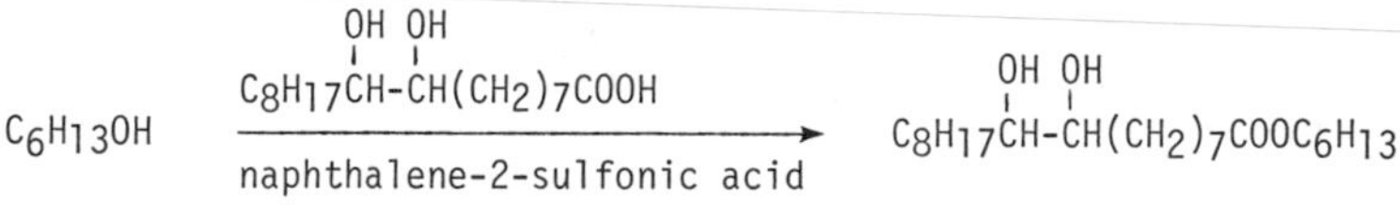

$$C_6H_{13}OH \xrightarrow[\text{naphthalene-2-sulfonic acid}]{C_8H_{17}\overset{OH}{CH}-\overset{OH}{CH}(CH_2)_7COOH} C_8H_{17}\overset{OH}{CH}-\overset{OH}{CH}(CH_2)_7COOC_6H_{13}$$

JACS (1945) 67 902
Rec Trav Chim (1951) 70 277

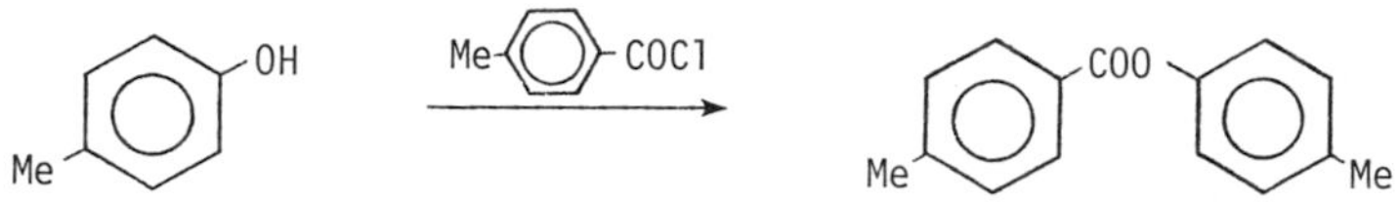

Annalen (1969) 726 216

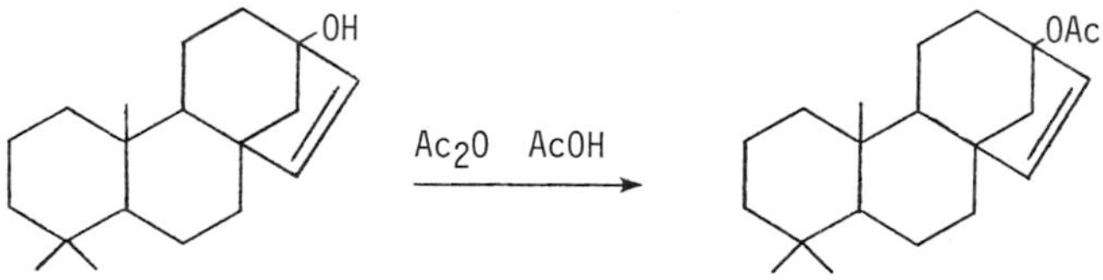

Ac$_2$O   AcOH

JCS C (1969) 2115

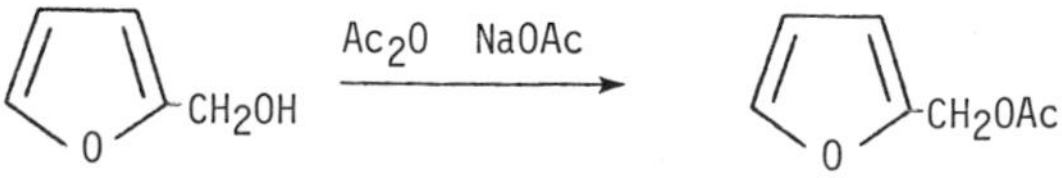

Ac$_2$O   NaOAc   87-93%

Org Synth (1941) Coll Vol 1 285
JOC (1965) 30 3480

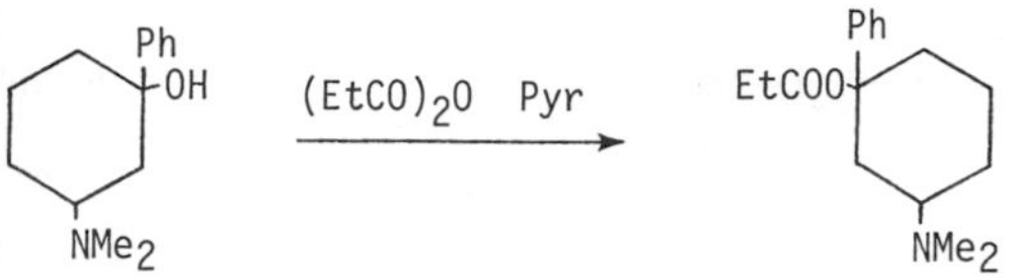
Ph
OH
NMe2
(EtCO)2O Pyr
Ph
EtCOO
NMe2

60%

J Med Chem (1970) 13 125

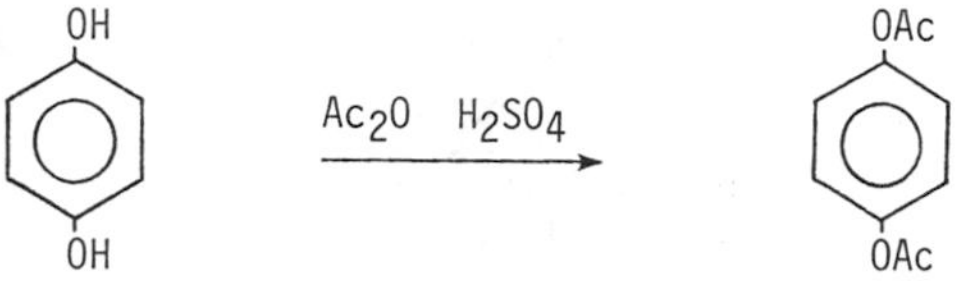
OH
OH
Ac2O H2SO4
OAc
OAc

96-98%

Org Synth (1955) Coll Vol 3 452

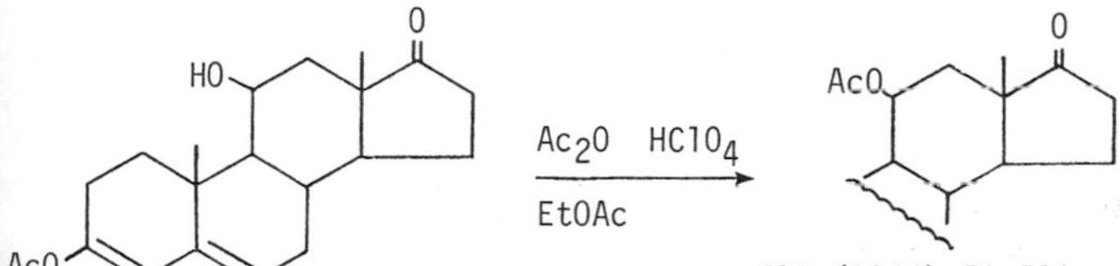
O
HO
AcO
Ac2O HClO4
EtOAc
O
AcO

JOC (1966) 31 324

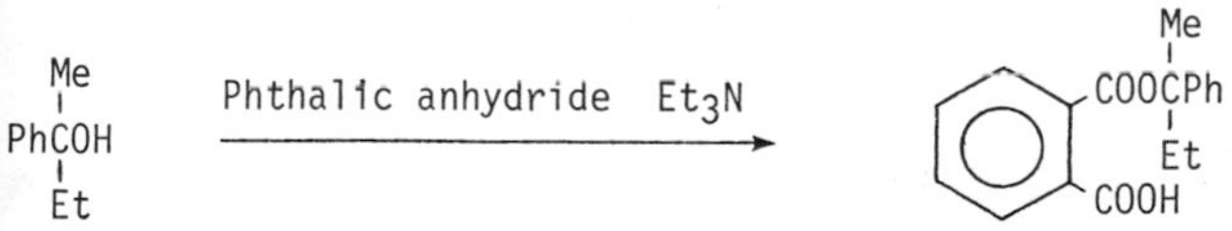
Me
PhCOH
Et
Phthalic anhydride Et3N
Me
COOCPh
Et
COOH

JCS (1957) 3148

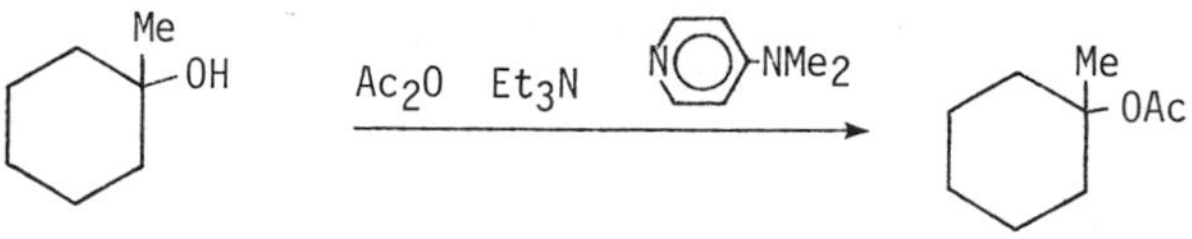
Me
OH
Ac2O Et3N
N
NMe2
Me
OAc

86%

Angew (1969) 81 1001
(Internat Ed 8 981)

t-BuOH $\xrightarrow{\text{EtCOCl} \quad \text{PhNMe}_2 \quad \text{Et}_2\text{O}}$ EtCOOBu-t

JACS (1943) 65 986
Helv (1944) 27 513
JCS (1952) 4883

$$t\text{-BuOH} \quad \xrightarrow[\text{Et}_2\text{O}]{\text{MeCOCl} \quad \text{tetramethylurea}} \quad t\text{-BuOAc} \qquad 50\text{-}60\%$$

Tetr Lett (1967) 3267

$$\text{Et}_3\text{COH} \quad \xrightarrow[\text{2 PhCOCl}]{\text{1 BuLi} \quad \text{THF}} \quad \text{PhCOOCEt}_3 \qquad 94\%$$

JOC (1970) <u>35</u> 1198

Esters from alcohols, Ph_3CNa and $(\text{RCO})_2\text{O}$ JOC (1963) <u>28</u> 582

| " | " | " | EtMgBr | " | " | JACS (1936) <u>58</u> 1384
JCS (1963) 3578 |

"   "   "   EtMgBr   "   "   JACS (1936) <u>58</u> 1384
                                                            JCS (1963) 3578

"   "   "   K   "   "   JACS (1961) <u>83</u> 423

"   "   " CaC_2 "   "   Monatsh (1966) <u>97</u> 62

"   "   "   Mg   and RCOCl   Org Synth (1955) Coll Vol 3 144

"   "   "   NaH   "   "   Chem Comm (1967) 259

$$\text{C}_{12}\text{H}_{25}\text{OH} \quad \xrightarrow[\text{TsCl} \quad \text{Pyr}]{\text{3,5-Dinitrobenzoic acid}} \quad$$

82%

(Applicable to hindered alcohols)
JACS (1955) <u>77</u> 6214
       (1963) <u>85</u> 2446

$$\text{PhOH} \quad \xrightarrow{\text{PhCOOH} \quad (\text{CF}_3\text{CO})_2\text{O}} \quad \text{PhCOOPh} \qquad 70\%$$

(Applicable to hindered alcohols and phenols)
Tetrahedron (1965) <u>21</u> 3531
Tetr Lett (1964) 1285
JCS (1965) 4594
For selective esterification of 2ry-OH in the presence of 1ry-OH
see JOC (1961) <u>26</u> 177

JACS (1969) <u>91</u> 3931
      (1958) <u>80</u> 5714

PhOH $\xrightarrow{\text{PhCOOH DCC Et}_2\text{O}}$ PhCOOPh

Tetrahedron (1965) 21 3531
Compt Rend (1963) 256 1804
             (1962) 255 945

PhCH$_2$OH $\xrightarrow{\text{PhCOOH N,N'-carbonyl diimidazole}}$ PhCOOCH$_2$Ph   89%

Ber (1962) 95 1284
JCS C (1969) 1610

Further examples of the reaction ROH + RCOOH ⟶ RCOOR are included in
section 107 (Esters from Carboxylic Acids, Acid Halides and Anhydrides)
and section 45A (Protection of Alcohols and Phenols)

JACS (1954) 76 4915
      (1945) 67 740

91%

$\text{Me}_2\text{C=CH(CH}_2)_2\underset{\text{Me}}{\text{C}}\text{=CH(CH}_2)_2\underset{\text{Me}}{\overset{\text{OH}}{\text{C}}}\text{CH=CH}_2$ $\xrightarrow[\text{Ac}_2\text{O}]{\text{HCOOH}}$ $\text{Me}_2\text{C=CH(CH}_2)_2\underset{\text{Me}}{\text{C}}\text{=CH(CH}_2)_2\underset{\text{Me}}{\overset{\text{HCOO}}{\text{C}}}\text{CH=CH}_2$

Bull Soc Chim Fr (1965) 484
Rec Trav Chim (1964) 83 1287 1294

Chem Comm (1965) 413
JACS (1958) 80 2906
Coll Czech (1961) 26 1723

79%

$$C_8H_{17}OH \xrightarrow[\text{}]{\text{HCOF \quad Et}_3\text{N \quad Et}_2\text{O}} HCOOC_8H_{17} \qquad 73\%$$

JACS (1960) <u>82</u> 2380

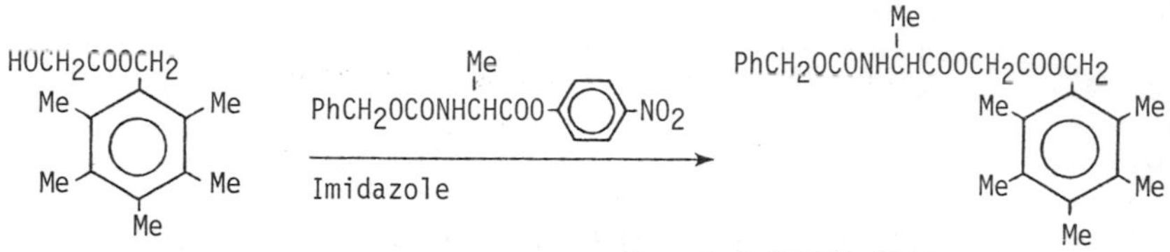

Chem Ind (1967) 1960

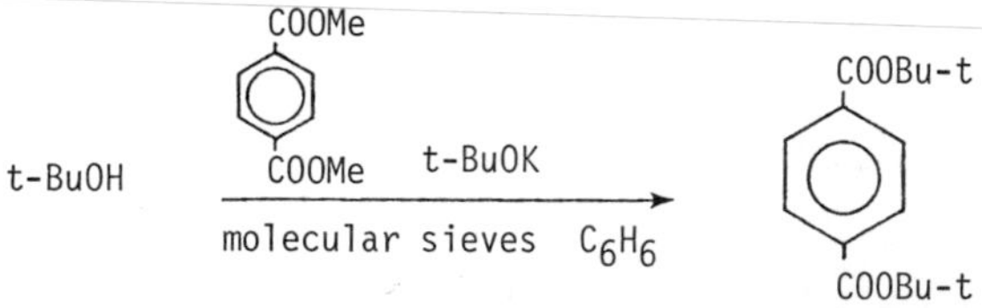

Chem Ind (1966) 1622

$$Me(CH_2)_7CH=CH(CH_2)_7CH_2OH \xrightarrow[\text{}]{\text{EtOAc \quad MeONa}} Me(CH_2)_7CH=CH(CH_2)_7CH_2OAc$$

Lipids (1967) <u>2</u> 437
JACS (1956) <u>78</u> 6347

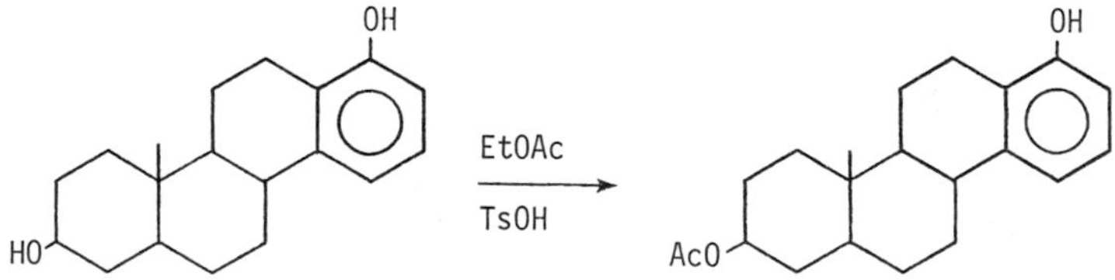

JACS (1956) <u>78</u> 6322

78%

$$\text{CH}_2=\overset{\overset{\displaystyle Me}{|}}{\text{COAc}} \quad \text{TsOH}$$

59%

J Med Chem (1969) 12 563

EtOH
$$\xrightarrow{\overset{\displaystyle + \quad -}{\text{MeCO SbF}_6} \quad \text{MeCN}}$$
EtOAc

(Applicable to hindered alcohols)
JACS (1962) 84 2733
Gazz (1967) 97 442
(Chem Abs 67 117049)

$$\xrightarrow{\text{CH}_2=\text{CO} \quad \text{C}_6\text{H}_6}$$

95%

(Applicable to hindered alcohols)
J Med Chem (1969) 12 432
JOC (1968) 33 3695
Org React (1946) 3 108

For preparation of $CH_2=CO$ see

t-BuOH
$$\xrightarrow{\overset{\displaystyle \text{Me}_2\text{C}\underset{\text{COO}}{\overset{\text{COO}}{\diagup}}\text{C}=\text{CMe}_2 \quad \text{K}_2\text{CO}_3}{}}$$
Me₂CHCOOBu-t

95%

Angew (1963) 75 841
(Internat Ed 2 608)

$$\xrightarrow{\text{MeCN}}$$

~91%

JACS (1969) 91 2162

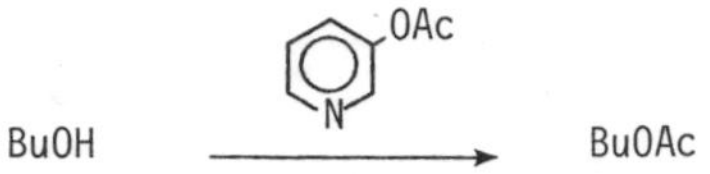
OAc
N
BuOH
BuOAc

93%

Bull Chem Soc Jap (1964) 37 864
Chem Ind (1965) 1498

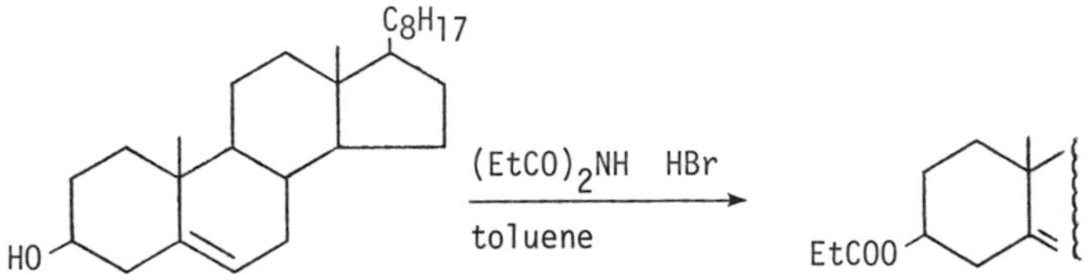
C8H17
HO
(EtCO)2NH HBr
toluene
EtCOO

85%

Roczniki Chem (1952) 26 692
(Chem Abs 49 2464)

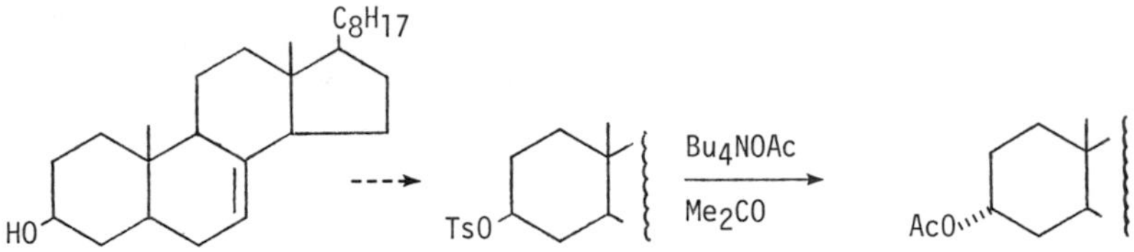
C8H17
HO
TsO
Bu4NOAc
Me2CO
AcO

JCS (1969) 1605

$$i\text{-}PrCH_2CH_2OH \xrightarrow{\text{t-Butylhypochlorite} \quad CCl_4} i\text{-}PrCH_2COOCH_2CH_2Pr\text{-}i$$

89%

Helv (1953) 36 1763

$$PrCH_2OH \xrightarrow{Na_2Cr_2O_7 \quad H_2SO_4 \quad H_2O} PrCOOCH_2Pr$$

41-47%

Org Synth (1941) Coll Vol 1 138

Section 109   Esters from Aldehydes

JOC (1949) 14 1013

$Me_2C=CH(CH_2)_2\overset{\overset{\displaystyle Me}{|}}{C}=CHCHO$ $\xrightarrow[\text{HOAc \quad MeOH}]{\text{NaCN \quad MnO}_2}$ $Me_2C=CH(CH_2)_2\overset{\overset{\displaystyle Me}{|}}{C}=CHCOOMe$

(Applicable to unsaturated and aromatic aldehydes only)
JACS (1968) 90 5616
Ber (1970) 103 3774

PrCHO $--\rightarrow$ $PrCH(OEt)_2$ $\xrightarrow{\text{MeCOO}_2\text{H \quad H}_2\text{SO}_4}$ PrCOOEt

JOC (1960) 25 1699

PhCHO $\xrightarrow{\text{MeOH \quad (NH}_4)_2\text{S}_2\text{O}_8 \quad \text{H}_2\text{SO}_4}$ PhCOOMe   100%

JOC (1968) 33 2525

$C_7H_{15}CHO$ $--\rightarrow$ $\xrightarrow[\text{t-BuOH}]{\text{h}\nu \quad \text{Me}_2\text{CO}}$ $C_7H_{15}COOEt$

Tetr Lett (1963) 2189

$C_5H_{11}CH_2CHO$ $\xrightarrow[\text{2 NaI \quad Me}_2\text{CO}]{\text{1 SO}_2\text{Cl}_2 \quad \text{CH}_2\text{Cl}_2}$ $C_5H_{11}\overset{\overset{\displaystyle}{|}}{C}HCHO$ \quad $\xrightarrow[\text{Ag}^+ \quad \text{Et}_2\text{O}]{\text{EtOH}}$ $C_5H_{11}CH_2COOEt$
$\qquad\qquad\qquad\qquad\qquad\qquad\qquad\qquad\qquad\quad I$

Tetr Lett (1968) 4415
JACS (1954) 76 2695

PhCHCHO $\xrightarrow{\text{NaN}_3 \quad \text{HCl} \quad \text{EtOH}}$ PhCHCOOEt   51%
| |
Me   Me   JACS (1954) $\underline{76}$ 4564

PrCHO $\xrightarrow{\text{Al(OEt)}_3}$ PrCOOCH$_2$Pr   81%

                                    JACS (1947) $\underline{69}$ 2605

PhCHO $\xrightarrow{\text{NaH} \quad \text{C}_6\text{H}_6}$ PhCOOCH$_2$Ph   92%

                                    JACS (1946) $\underline{68}$ 2647

For examples of the reaction RCHO ⟶ HCOOR see page 84 (Section 34, Alcohols and Phenols from Aldehydes)

Esters may also be prepared by oxidation of aldehydes to carboxylic acids followed by esterification. See section 19 (Carboxylic Acids and Acid Halides from Aldehydes)

Section 110   Esters from Alkyls, Methylenes and Aryls
○○○

No examples of the reaction RR ⟶ RCOOR' or R'COOR (R=alkyl, aryl etc.) occur in the literature. For the reaction RH ⟶ RCOOR' or R'COOR see section 116 (Esters from Hydrides)

Section 111   Esters from Amides
              °°°°°°°°°°°°°°°°°°°°

$PhCONH_2 \xrightarrow[105°]{MeOH \ BF_3} PhCOOMe$ 100%

Chem Comm (1969) 414

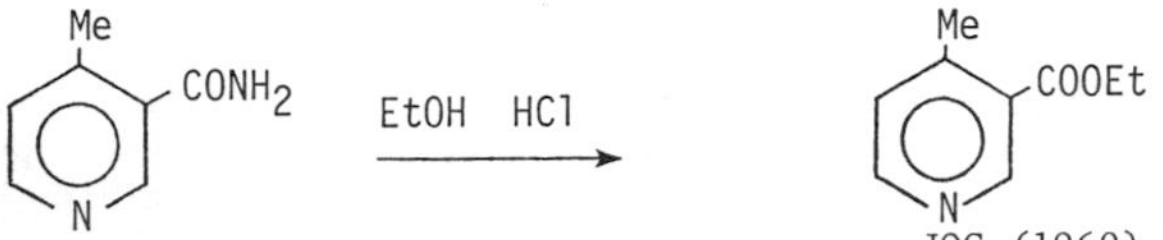

$\xrightarrow{EtOH \ HCl}$

85%

JOC (1960) 25 560
Arch Pharm (1957) 290 218
JOC (1970) 35 125

$PhCONH_2 \xrightarrow[170°]{C_8H_{17}Br \ H_2O} PhCOOC_8H_{17}$ 79%

JOC (1969) 34 3204

$MeCONHBu \xrightarrow[2 \ CCl_4 \ 77°]{1 \ HNO_3 \ Ac_2O} MeCOOBu$ 75%

JACS (1955) 77 6011
JOC (1969) 34 3834

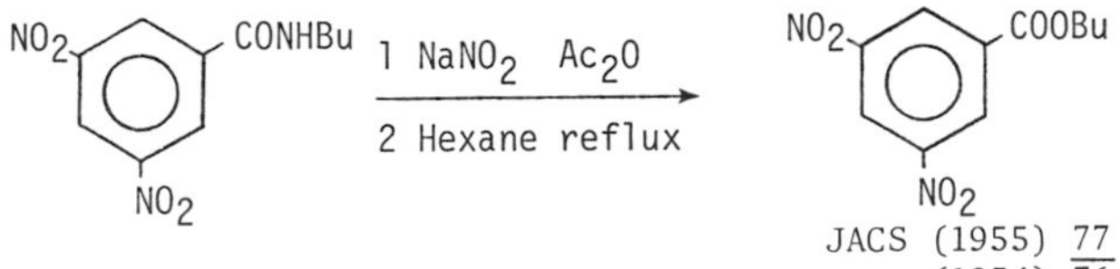

83%

JACS (1955) 77 6008 6011
       (1954) 76 4497
Tetr Lett (1965) 2627

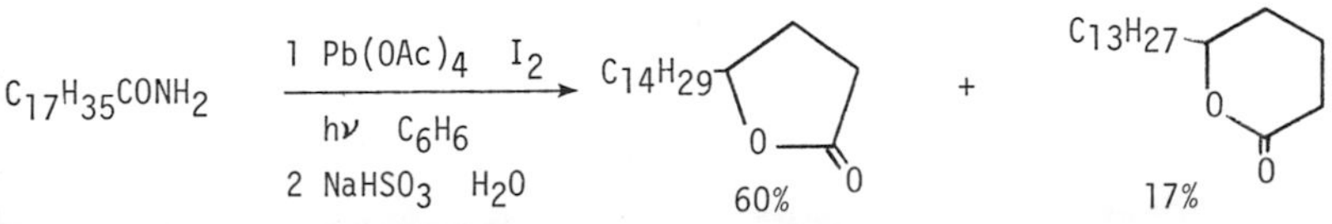

JCS (1965) 181

Section 112 Esters from Amines

$BuNH_2$ —(Ac_2O HOAc)→ BuNHAc —(1 $NaNO_2$ Ac_2O; 2 Hexane reflux)→ BuOAc 83%

JACS (1955) 77 6008 6011
(1954) 76 4497
JOC (1969) 34 3834
Chem Pharm Bull (1960) 8 266

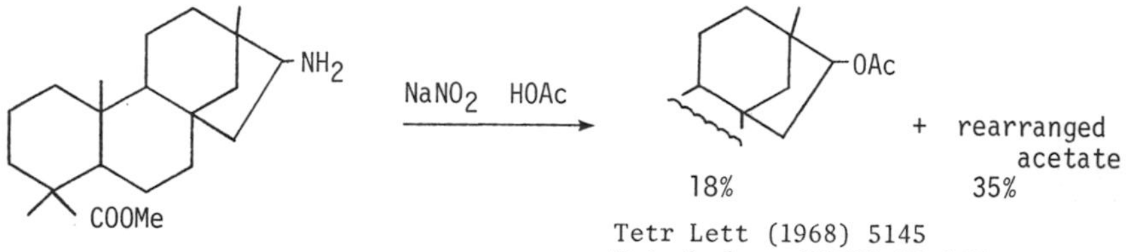
NH2
COOMe
NaNO2 HOAc
OAc
+ rearranged acetate
18%
35%

Tetr Lett (1968) 5145
Tetrahedron (1970) 26 147

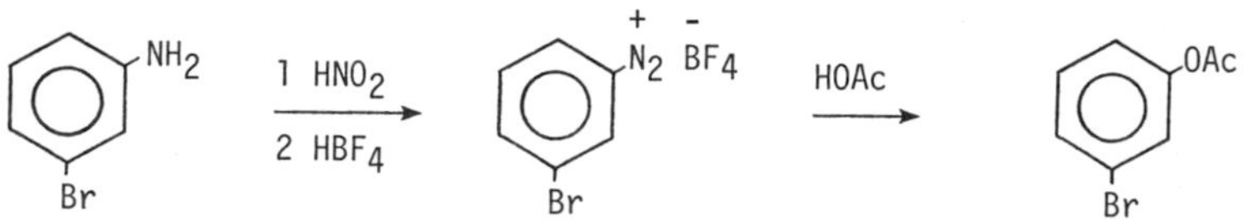
NH2
Br
1 HNO2
2 HBF4
N2 BF4
HOAc
OAc

JACS (1939) 61 143
(1933) 55 4954

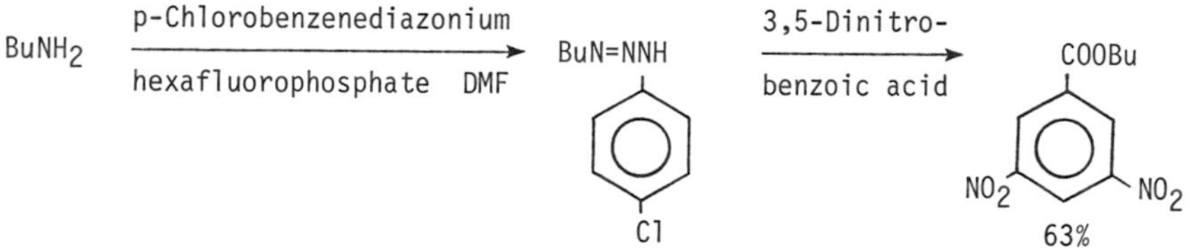
BuNH2
p-Chlorobenzenediazonium
hexafluorophosphate DMF
BuN=NNH
Cl
3,5-Dinitro-
benzoic acid
COOBu
NO2
63%

Tetr Lett (1961) 758

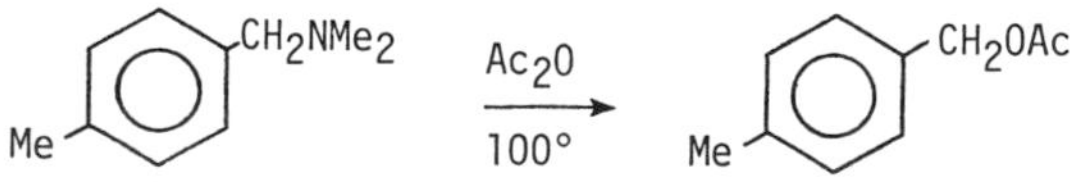
CH2NMe2
Me
Ac2O
100°
CH2OAc

Bull Soc Chim Fr (1914) 15 162

Section 113   Esters from Esters
             oooooooooooooooooooo

PhCH$_2$COOBu-t $\xrightarrow[\text{monoglyme}]{\text{BuBr   NaH}}$ PhCHCOOBu-t   66%
$\qquad\qquad\qquad\qquad\qquad\qquad\quad$ |
$\qquad\qquad\qquad\qquad\qquad\qquad\quad$ Bu   JOC (1965) 30 2937
$\qquad\qquad\qquad\qquad\qquad\qquad\qquad\qquad\qquad\qquad$ (1964) 29 2990

Me$_2$CHCOOEt $\xrightarrow[\text{2 EtI}]{\text{1 Ph}_3\text{CNa   Et}_2\text{O}}$ Me$_2$CCOOEt   58%
$\qquad\qquad\qquad\qquad\qquad\qquad\qquad\qquad\quad$ |
$\qquad\qquad\qquad\qquad\qquad\qquad\qquad\qquad\quad$ Et   JACS (1940) 62 2457

MeCH$_2$COOBu-t $\xrightarrow[\text{2 PhBr}]{\text{1 NaNH}_2\ \ \text{NH}_3}$ MeCHCOOBu-t   38%
$\qquad\qquad\qquad\qquad\qquad\qquad\qquad\qquad\quad$ |
$\qquad\qquad\qquad\qquad\qquad\qquad\qquad\qquad\quad$ Ph   JACS (1959) 81 1627

C$_{16}$H$_{33}$CH$_2$COOMe $\xrightarrow[\text{di-t-butyl peroxide}]{\substack{\text{Me}\\|\\ \text{i-Pr(CH}_2)_3\text{CHCH=CH}_2}}$ C$_{16}$H$_{33}$CHCOOMe   42%
$\qquad\qquad\qquad\qquad\qquad\qquad\qquad\qquad\qquad\quad$ i-Pr(CH$_2$)$_3$CH(CH$_2$)$_2$
$\qquad\qquad\qquad\qquad\qquad\qquad\qquad\qquad\qquad\qquad\qquad$ |
$\qquad\qquad\qquad\qquad\qquad\qquad\qquad\qquad\qquad\qquad\qquad$ Me
$\qquad\qquad\qquad\qquad\qquad\qquad$ J Amer Oil Chem Soc (1968) 45 453
$\qquad\qquad\qquad\qquad\qquad\qquad$ (Chem Abs 70 3210)

—COOPh $\xrightarrow[\text{2 h}\nu \text{ MeOH}]{\text{1 Me}_2\overset{+}{\text{S}}\overset{-}{\text{O}}\text{CH}_2\text{ THF}}$ —CH$_2$COOMe

$\qquad\qquad\qquad\qquad\qquad\qquad\qquad\qquad$ JACS (1964) 86 1640

Further examples of the alkylation and homologation of carboxylic acids
and esters are included in section 17 (Carboxylic Acids, Acid Halides
and Anhydrides from Carboxylic Acids and Acid Halides) and section 107
(Esters from Carboxylic Acids, Acid Halides and Anhydrides)

t-BuOK   molecular sieves
─────────────────────────→   > 90%
t-BuOH C_6H_6

Chem Ind (1966) 1622

$C_{17}H_{35}COOEt$ $\xrightarrow[\text{MeOH}]{\text{Ion exch resin (basic)}}$ $C_{17}H_{35}COOMe$ 80-90%

JOC (1969) 34 2032

$CH_2=CHCOOMe$ $\xrightarrow{C_{10}H_{21}OH \quad H_2SO_4}$ $CH_2=CHCOOC_{10}H_{21}$ 83%

JACS (1944) 66 1203
Org Synth (1955) Coll Vol 3 146 605

RCOOBu $\xrightarrow{\text{HBr} \quad Et_2O}$ RCOOEt

J Pharm Sci (1969) 58 949

PhCOOBu-t $\xrightarrow[\text{reflux}]{\text{MeOH}}$ PhCOOMe   62%

JACS (1941) 63 3382

$\xrightarrow{PhCH_2NMe_2 \quad 170°}$ 65%

JACS (1952) 74 547

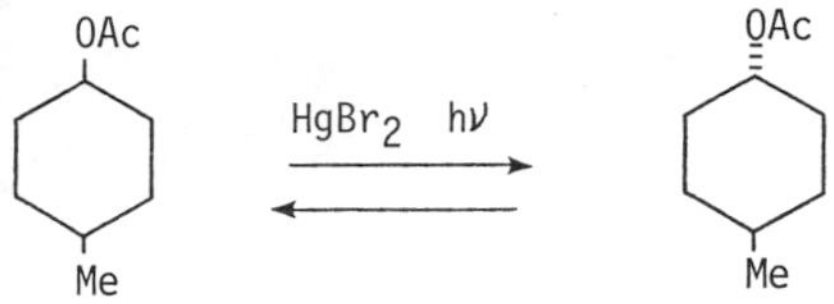

$$HgBr_2 \quad h\nu$$

JACS (1970) 92 1094

$$PhCH(COOEt)_2 \xrightarrow{470°} PhCH_2COOEt$$

JOC (1964) 29 1249

36%

$$EtCH(COOEt)_2 \xrightarrow{NaCN \quad Me_2SO} EtCH_2COOEt$$

Tetr Lett (1967) 215

80%

Section 114   Esters from Ethers
                ○○○○○○○○○○○○○○○○○○○

$$PhCH_2OBu\text{-}t \xrightarrow{O_3 \quad Freon \; 11} PhCOOBu\text{-}t$$

JACS (1968) 90 6777
Annalen (1929) 476 233

62%

$$PrCH_2OBu \xrightarrow{Trichloroisocyanuric\ acid \quad H_2O} PrCOOBu$$

Tetr Lett (1968) 5819

100%

$$PrCH_2OBu \xrightarrow{RuO_4 \quad CCl_4} PrCOOBu$$

JACS (1958) 80 6682

100%

$RCH_2OC_{15}H_{31}$ $\xrightarrow{\text{CrO}_3 \quad \text{HOAc} \quad \text{CH}_2\text{Cl}_2}$ $RCOOC_{15}H_{31}$ 48% (R=H)
9% (R=Me)

Chem Comm (1966) 752

$\xrightarrow{\text{CrO}_3 \quad \text{HOAc}}$ 50%

Carbohydrate Res (1970) 12 147

$CH_2=CH(CH_2)_3CH=CHCH_2OPh$ $\xrightarrow[\text{HOAc}]{\text{PdCl}_2(\text{Ph}_3\text{P})_2}$ $CH_2=CH(CH_2)_3CH=CHCH_2OAc$

Chem Comm (1970) 1392

$\begin{matrix} C_{17}H_{35}COOCH_2 \\ C_{17}H_{35}COOCH \\ Ph_3COCH_2 \end{matrix}$ $\xrightarrow{\text{C}_{15}\text{H}_{31}\text{COCl} \quad \text{AgClO}_4 \quad \text{MeCN}}$ $\begin{matrix} C_{17}H_{35}COOCH_2 \\ C_{17}H_{35}COOCH \\ C_{15}H_{31}COOCH_2 \end{matrix}$ 70%

Ber (1961) 94 812
JCS (1950) 1992

Bu_2O $\xrightarrow{\text{TsOAc} \quad \text{MeCN}}$ $BuOAc$ 50%

JACS (1968) 90 3878

$\xrightarrow[\text{Ac}_2\text{O}]{\text{BF}_3\cdot\text{Et}_2\text{O}}$ 93%

JOC (1965) 30 1734
Chem Rev (1954) 54 615

$$\text{BuOBu-t} \xrightarrow{\text{HOAc TsOH}} \text{BuOAc}$$

Tetr Lett (1970) 4269   94%

Section 115   Esters from Halides and Sulfonates
ooooooooooooooooooooooooooooooooooooooo

$$\text{PrBr} \xrightarrow[\substack{\text{2 MeCOCH}_2\text{CH}_2\text{COOEt} \\ \text{C}_6\text{H}_6 \\ \text{3 SOCl}_2 \\ \text{4 185-195°}}]{\text{1 Mg Et}_2\text{O}} \underset{\text{Me}}{\text{PrC=CHCH}_2\text{COOEt}} \xrightarrow[\text{EtOH}]{\text{H}_2 \quad \text{PtO}_2} \underset{\text{Me}}{\text{PrCHCH}_2\text{CH}_2\text{COOEt}}$$

JACS (1944) 66 1764

$$\text{BuBr} \xrightarrow[\text{2 MeCH=CHCOOBu-s}]{\text{1 Mg Et}_2\text{O}} \underset{\text{Me}}{\text{BuCHCH}_2\text{COOBu-s}}$$

~60%

Org Synth (1961) 41 60

$$\text{PhBr} \xrightarrow[\substack{\text{2 CdCl}_2 \\ \text{3 BrCH}_2\text{COOEt}}]{\text{1 Mg THF}} \text{PhCH}_2\text{COOEt}$$

53%

JOC (1968) 33 1675

$$\text{PhBr} \xrightarrow[\substack{\text{2 CuBr} \\ \text{3 N}_2\text{CHCOOEt}}]{\text{1 Li Et}_2\text{O}} \text{PhCH}_2\text{COOEt}$$

51%

Chem Comm (1969) 515

73%

JACS (1969) 91 4304

$$\underset{Me}{\overset{Me}{EtCHCOOEt}} \quad NaH$$

BuBr $\xrightarrow[\text{diglyme}]{}$ $\underset{\overset{|}{Et}}{\overset{Me}{Bu\overset{|}{C}COOEt}}$ 38%

JOC (1964) $\underline{29}$ 2990

Further examples of the alkylation of esters with halides are included in section 113 (Esters from Esters)

PhBr $\xrightarrow{\quad CH_3COOEt \quad NaNH_2 \quad NH_3 \quad}$ PhCH$_2$COOEt   42%

JACS (1959) $\underline{81}$ 1627

EtI $\overset{CH_2(COOEt)_2}{\dashrightarrow}$ EtCH(COOEt)$_2$ $\xrightarrow{\quad NaCN \quad Me_2SO \quad}$ EtCH$_2$COOEt

Tetr Lett (1967) 215

$C_{10}H_{21}Br$ $\xrightarrow[\text{EtONa \quad EtOH}]{MeCOCH_2COOEt}$ $\underset{MeCO}{C_{10}H_{21}\overset{|}{C}HCOOEt}$ $\xrightarrow[\text{EtOH}]{EtONa}$ $C_{10}H_{21}CH_2COOEt$ 84%

JOC (1962) $\underline{27}$ 622

$\xrightarrow[\text{2 (EtO)}_2CO]{\text{1 Mg \quad Et}_2O}$ 68-73%

Org Synth (1943) Coll Vol 2 282

$\xrightarrow[\text{2 C(OEt)}_4]{\text{1 Mg \quad Et}_2O}$ Acid $\dashrightarrow$

Ber (1905) $\underline{38}$ 561

PhBr $\xrightarrow[\text{2 ClCOOEt}]{\text{1 Mg   Et}_2\text{O}}$ PhCOOEt   75%

Ber (1903) <u>36</u> 3087

$C_7H_{15}I$ $\xrightarrow[\text{t-BuOH}]{\text{Ni(CO)}_4 \quad \text{t-BuOK}}$ $C_7H_{15}COOBu\text{-t}$ 66%

JACS (1969) <u>91</u> 1233
(1963) <u>85</u> 2779

RBr $\xrightarrow[\text{MeOH}]{\text{Ni(CO)}_4 \quad \text{MeONa}}$ RCOOMe   88% (R=Ph)
95% (R= PhCH=CH)

JACS (1969) <u>91</u> 1233

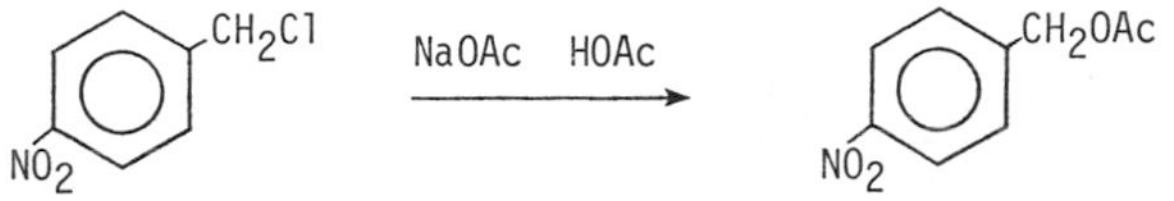

Org Synth (1955) Coll Vol 3 650
Ber (1956) <u>89</u> 1732

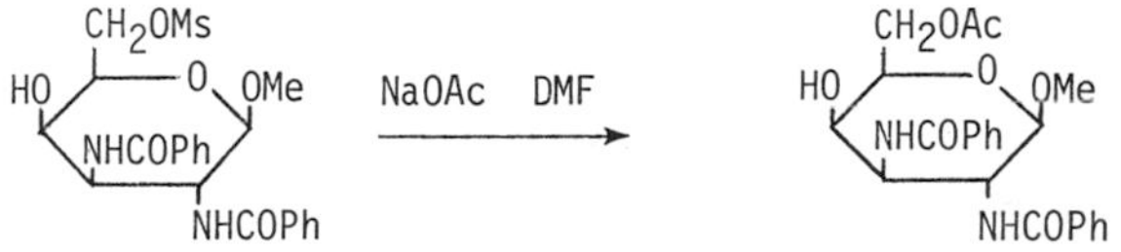

Ber (1970) <u>103</u> 37

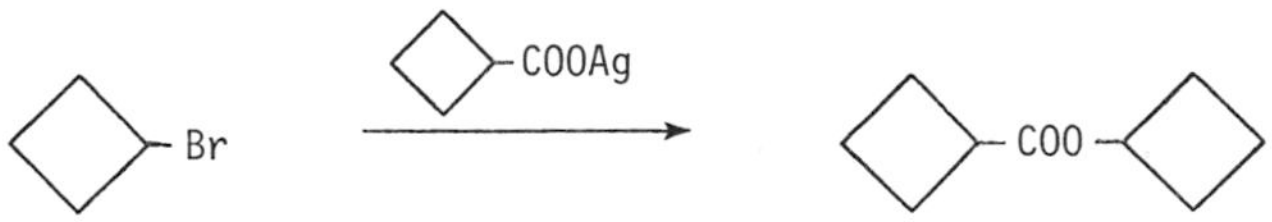

JACS (1951) <u>73</u> 5487
(1949) <u>71</u> 3214

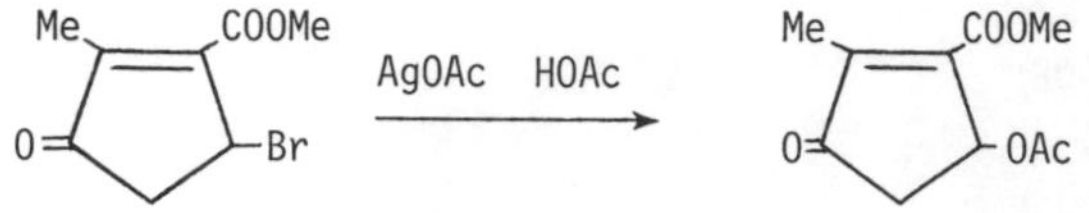

Tetrahedron (1968) 24 5421
Ber (1964) 97 443

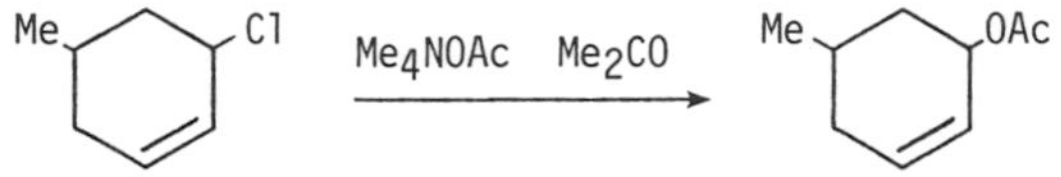

JACS (1955) 77 4042
     (1967) 89 2758

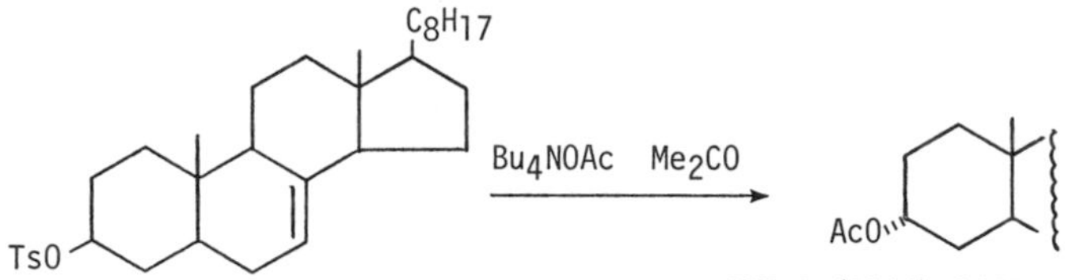

76%

JCS C (1969) 1605

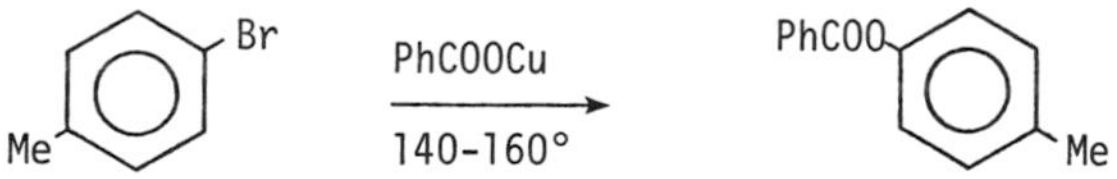

JACS (1966) 88 4521

79%

BuI $\xrightarrow{\text{MeCOO}_2\text{H HOAc xylene}}$ BuOAc

JOC (1969) 34 3974

78%

Section 116   Esters from Hydrides (RH)

This section contains examples of the reaction RH $\longrightarrow$ RCOOR' or R'COOR
(R=alkyl, allyl, vinyl, aryl etc.) and ArH $\longrightarrow$ Ar-X-COOR (X=alkyl chain)

$CH_2=CH(CH_2)_2COOEt$ $AlCl_3$

81%

JOC (1953) 18 1499
Tetrahedron (1967) 23 2481

PhH $\xrightarrow{ClCH_2CH_2COOEt \ \ AlCl_3}$ $PhCH_2CH_2COOEt$ 68%

Dokl (1945) 50 257
(Chem Abs 43 4638)

PhH $\xrightarrow{ClCOOPh \ \ AlCl_3}$ PhCOOPh   64%

JOC (1957) 22 325

$PhCH=CH_2$ $\xrightarrow[\text{MeONa \ \ MeOH}]{\text{Electrolysis \ \ CO (70 kg/cm}^2)}$ $PhCH=CHCOOMe$

Bull Chem Soc Jap (1965) 38 2122
JACS (1965) 87 3525

$PhCH=CH_2$ $\xrightarrow[\text{2 CO \ \ Et}_3\text{N \ \ MeOH}]{\text{1 PdCl}_2}$ $PhCH=CHCOOMe$ 41%

JOC (1969) 34 738

44%

JACS (1953) 75 3843

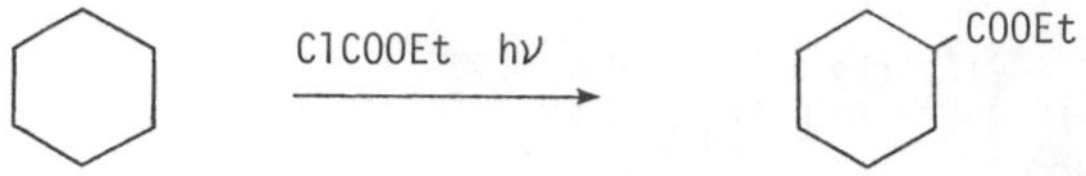
ClCOOEt hν
COOEt

15%

Bull Chem Soc Jap (1966) 39 2463

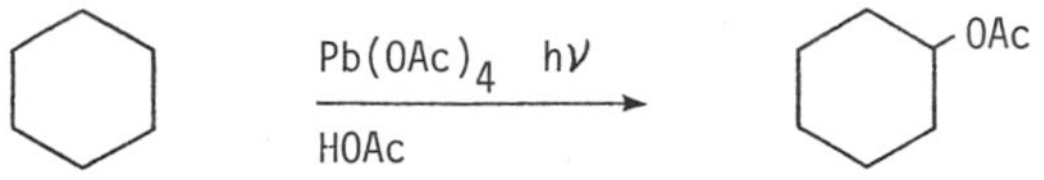
Pb(OAc)4 hν
HOAc
OAc

24%

Nature (1966) 209 395

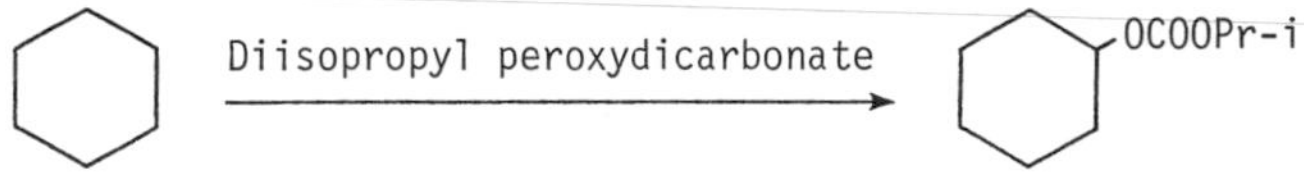
Diisopropyl peroxydicarbonate
OCOOPr-i

JCS (1965) 1932

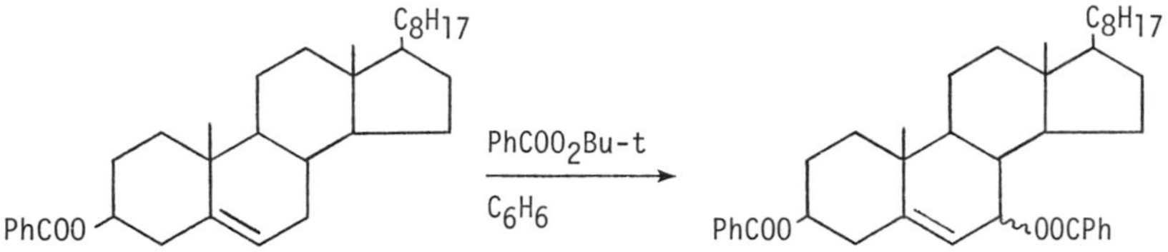
C8H17
PhCOO
PhCOO2Bu-t
C6H6
C8H17
PhCOO
OOCPh

Proc Chem Soc (1962) 63
JACS (1958) 80 756

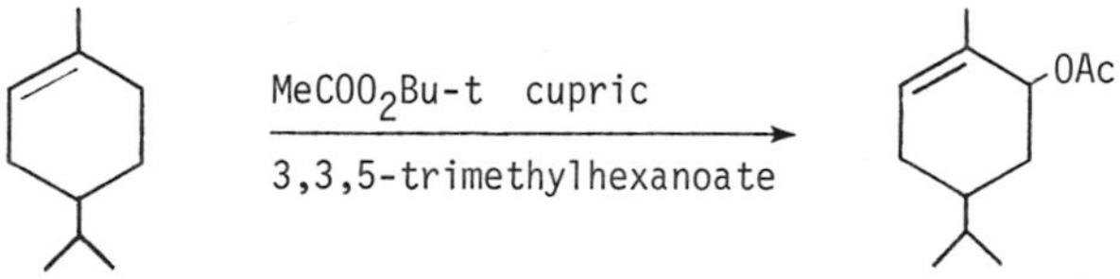
MeCOO2Bu-t cupric
3,3,5-trimethylhexanoate
OAc

65%

JACS (1962) 84 4969 774

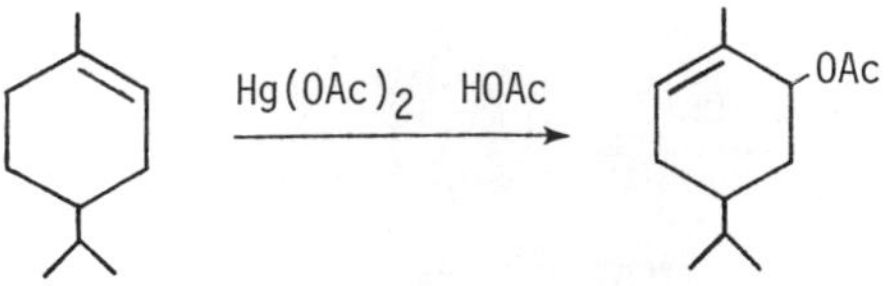
Hg(OAc)2 HOAc
OAc

Annalen (1953) 581 59

Tetrahedron (1964) 20 1017
Org React (1949) 5 331

EtCH=CHCH$_3$ $\xrightarrow{\text{Pd(OAc)}_2}$ EtCHCH=CH$_2$
                                                    |
                                                   OAc

JACS (1966) 88 2054

C$_5$H$_{11}$CH$_2$CH=CH$_2$ $\xrightarrow[\text{HOAc}]{\text{Electrolysis Et}_4\text{NTs}}$ C$_5$H$_{11}$CHCH=CH$_2$
                                                                                                          |
                                                                                                         OAc

Tetr Lett (1968) 6207

PhCH$_3$ $\xrightarrow[\text{(PhCOO)}_2\text{Cu   RCOOH}]{\text{Di-t-butyl peroxide}}$ RCOOCH$_2$Ph   (R=Ph, Me)

Proc Chem Soc (1962) 63

JOC (1966) 31 2033   90%

JACS (1968) 90 1082

JOC (1969) 34 3302   86%

$$\text{PhCH}_3 \xrightarrow[\text{I}_2]{\text{Benzoyl peroxide}} \text{PhCOO-C}_6\text{H}_4\text{-Me}$$

Bull Chem Soc Jap (1969) $\underline{42}$ 2733
Rec Trav Chim (1927) $\underline{46}$ 54

Section 117　Esters from Ketones

$$\text{Ph}_2\text{CO} \dashrightarrow \text{Ph}_2\text{CN}_2 \xrightarrow[\text{EtOH}]{\text{Ni(CO)}_4 \quad \text{CO}} \text{Ph}_2\text{CHCOOEt} \qquad 74\%$$

Ber (1960) $\underline{93}$ 1840

$$\text{Ph}_2\text{CO} \xrightarrow{\text{N}_2\text{H}_4 \quad \text{EtOH}} \text{Ph}_2\text{C=NNH}_2 \xrightarrow{\text{Pb(OAc)}_4 \quad \text{CH}_2\text{Cl}_2} \text{Ph}_2\text{CHOAc} \qquad 43\%$$

JCS $\underline{\text{C}}$ (1970) 1033

$$\text{PhCOMe} \xrightarrow{\text{Ph}_3\text{SnH} \quad \text{MeCOCl} \quad \text{C}_6\text{H}_6} \underset{\overset{|}{\text{Me}}}{\text{PhCHOAc}} \qquad 100\%$$

JACS (1966) $\underline{88}$ 1833 4970

$$\underset{\overset{|}{\text{Me}}}{\text{EtCHCOPr}} \xrightarrow{\text{Br}_2} \underset{\overset{|}{\text{Me}}}{\overset{\overset{\text{Br}}{|}}{\text{EtCCOPr}}} \xrightarrow[\text{Et}_2\text{O}]{\text{MeONa}} \underset{\overset{|}{\text{Me}}}{\overset{\overset{\text{Pr}}{|}}{\text{EtCCOOMe}}} \qquad <34\%$$

JACS (1942) $\underline{64}$ 300
Org React (1960) $\underline{11}$ 261

$$\text{Ph}_2\text{CHCOMe} \xrightarrow{\text{Cl}_2} \underset{\overset{|}{\text{Cl}}}{\text{Ph}_2\text{CCOMe}} \xrightarrow[\text{EtOH}]{\text{EtONa}} \text{Ph}_2\text{CHCH}_2\text{COOEt}$$

Org React (1960) $\underline{11}$ 261

$$\xrightarrow{\text{KOCl} \quad \text{MeOH}}$$

80%

JACS (1944) <u>66</u> 208

$$\xrightarrow{\begin{array}{c} 1 \ \ \text{i-PrCH}_2\text{ONO} \quad \text{HCl} \\ \hline 2 \ \ \text{NaOH} \quad \text{Me}_2\text{SO}_4 \quad \text{H}_2\text{O} \end{array}}$$

Compt Rend (1942) <u>214</u> 113

$$\text{Et}_2\text{CO} \quad \xrightarrow{\begin{array}{c} \text{H}_2\text{O}_2 \quad (\text{CF}_3\text{CO})_2\text{O} \\ \hline \text{Na}_2\text{HPO}_4 \quad \text{CH}_2\text{Cl}_2 \end{array}} \quad \text{EtCOOEt}$$

78%

JACS (1955) <u>77</u> 2287

$$\text{i-PrCH}_2\text{COMe} \quad \xrightarrow{\begin{array}{c} \text{HOOCCH=CHCO}_2\text{OH} \\ \hline \text{CH}_2\text{Cl}_2 \end{array}} \quad \text{i-PrCH}_2\text{OAc}$$

72%

Tetrahedron (1962) <u>17</u> 31

$$\xrightarrow{\text{PhCOO}_2\text{H} \quad \text{CHCl}_3}$$

67%

JACS (1949) <u>71</u> 14
Org React (1957) <u>9</u> 73
For use of m-chloroperbenzoic acid see JACS (1967) <u>89</u> 4530
and JOC (1964) <u>29</u> 2914

$$\text{C}_6\text{H}_{13}\text{COMe} \quad \xrightarrow{\begin{array}{c} \text{H}_2\text{O}_2 \quad \text{BF}_3\cdot\text{Et}_2\text{O} \end{array}} \quad \text{C}_6\text{H}_{13}\text{OAc}$$

62%

JOC (1962) <u>27</u> 24

$$C_5H_{11}COMe \quad \xrightarrow[\substack{PhNO_2}]{\substack{OOH \\ | \\ (CF_3)_2COH \quad CH_2Cl_2}} \quad C_5H_{11}OAc \qquad\qquad 81\%$$

Tetr Lett (1970) 2741

Esters may also be prepared by conversion of ketones into carboxylic acids
followed by esterification. See section 27 (Carboxylic Acids from Ketones)

Section 118   Esters from Nitriles
∘∘∘∘∘∘∘∘∘∘∘∘∘∘∘∘∘∘∘∘∘∘∘∘∘

$$PhCH_2CN \quad \xrightarrow{C_7H_{15}OH \quad TsOH} \quad PhCH_2COOC_7H_{15} \qquad\qquad 70\%$$

JOC (1958) 23 1225
(1960) 25 560

$$PhCH_2CN \quad \xrightarrow{EtOH \quad H_2O \quad H_2SO_4} \quad PhCH_2COOEt \qquad\qquad 83-87\%$$

Org Synth (1941) Coll Vol 1 270

$$PhCH_2CN \quad \xrightarrow{HOCH_2CH_2OH} \quad PhCH_2COOCH_2CH_2OH$$

JCS (1963) 2417

$$CH_2=CHCN \quad \xrightarrow{Ph_3P=CHCOOEt \quad C_6H_6} \quad CH_2=CHCOOEt$$

Tetr Lett (1967) 2407

Esters may also be prepared by hydrolysis of nitriles to carboxylic acids
followed by esterification. See section 28 (Carboxylic Acids from Nitriles)

Section 119 Esters from Olefins

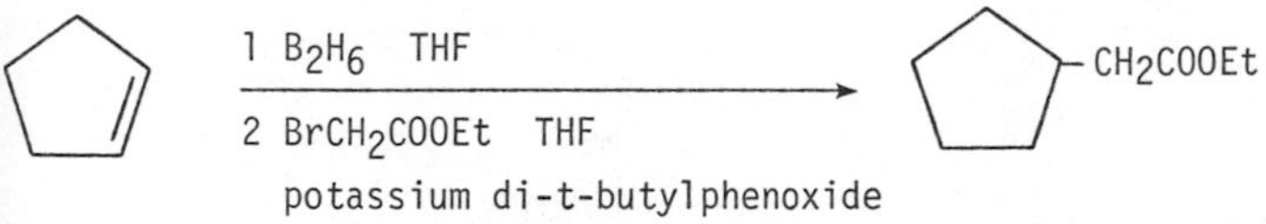
1 B_2H_6 THF
2 $BrCH_2COOEt$ THF
potassium di-t-butylphenoxide
CH_2COOEt

75%

JACS (1969) 91 6855 2146
(1968) 90 818 1911

$BuCH{=}CH_2$ —(1 B_2H_6 THF; 2 $N_2CHCOOEt$)→ $Bu(CH_2)_3COOEt$ 83%

JACS (1968) 90 6891

$C_5H_{11}CH{=}CH_2$ —(1 B_2H_6 THF; 2 $Me_2\overset{+}{S}\overset{-}{C}HCOOEt$)→ $C_5H_{11}(CH_2)_3COOEt$ 45%

JACS (1967) 89 6804

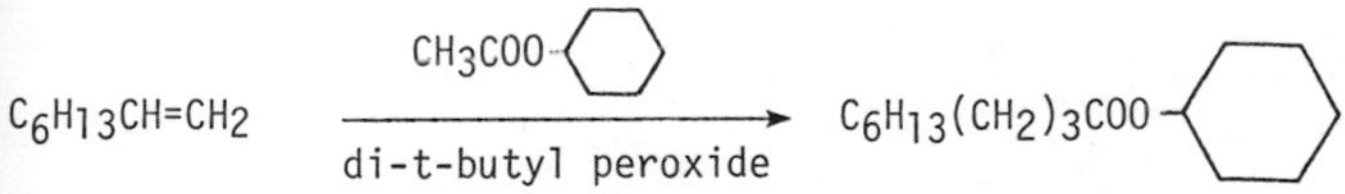
$C_6H_{13}CH{=}CH_2$
CH_3COO
di-t-butyl peroxide
$C_6H_{13}(CH_2)_3COO$

59%

JCS (1965) 1918

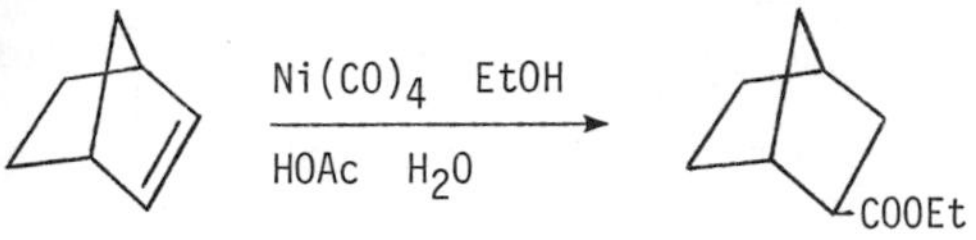
$Ni(CO)_4$ EtOH
HOAc H_2O
COOEt

72%

JCS (1963) 410

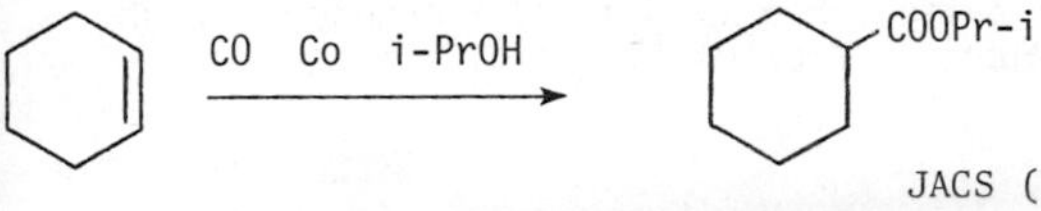
CO Co i-PrOH
COOPr-i

12%

JACS (1952) 74 4496

$$C_{12}H_{25}CH=CH_2 \xrightarrow[\text{MeOH}]{\text{CO} \quad H_2PtCl_6\text{-}SnCl_2} C_{12}H_{25}(CH_2)_2COOMe \qquad 65\%$$

JOC (1970) <u>35</u> 2846

$$\xrightarrow[\text{MeOH}]{\text{CO} \quad (Ph_3P)_2PdCl_2}$$

Angew (1968) <u>80</u> 352
(Internat Ed <u>7</u> 329)

$$\xrightarrow{\text{HCOOH} \quad HClO_4}$$

69%

JACS (1953) <u>75</u> 6212
Org Synth (1965) <u>45</u> 74

$$\xrightarrow{\text{HOAc} \quad h\nu}$$

JACS (1967) <u>89</u> 5199

$$\begin{array}{c} \text{COOEt} \\ | \\ (CH_2)_8CH=CH_2 \end{array} \xrightarrow[\text{2 EtOH} \quad 150\text{-}155°]{\text{1 O}_3 \quad \text{EtOH}} \begin{array}{c} \text{COOEt} \\ | \\ (CH_2)_8COOEt \end{array} \qquad 50\%$$

Can J Chem (1965) <u>43</u> 319

Esters may also be prepared by conversion of olefins into carboxylic acids
or alcohols followed by esterification. See section 29 (Carboxylic Acids
from Olefins) and section 44 (Alcohols from Olefins)

Section 120   Esters from Miscellaneous Compounds
               ○○

$$\xrightarrow[\text{MeOH}]{\text{H}_2 \quad \text{Pd-SrCO}_3}$$

JCS (1936) 192
JACS (1962) 84 4951

$$\text{PhCH=CHCOOCH}_2\text{Ph} \xrightarrow{\text{H}_2 \quad (\text{Ph}_3\text{P})_3\text{RhCl}} \text{PhCH}_2\text{CH}_2\text{COOCH}_2\text{Ph} \qquad 65\%$$

Compt Rend (1966) 263 251

$$\text{Me}_2\text{C=CHCOOEt} \xrightarrow{\text{Ph}_3\text{SnH} \quad h\nu \quad \text{MeOH}} \text{Me}_2\text{CHCH}_2\text{COOEt} \qquad 40\%$$

Tetr Lett (1967) 4805

$$\text{PhCH=CHCOOMe} \xrightarrow[\text{NCOOK Pyr HOAc}]{\overset{\text{NCOOK}}{\overset{\|}{}}} \text{PhCH}_2\text{CH}_2\text{COOMe} \qquad 96\%$$

JOC (1965) 30 3985

$$\text{PhCH=CHOAc} \xrightarrow[\text{NCOOK dioxane HOAc}]{\overset{\text{NCOOK}}{\overset{\|}{}}} \text{PhCH}_2\text{CH}_2\text{OAc} \qquad 50\%$$

JOC (1965) 30 3985

$$\xrightarrow[\text{HOAc}]{\text{H}_2 \quad \text{Pd} \quad \text{H}_2\text{SO}_4}$$

Ber (1941) 74B 315
Org React (1953) 7 263

ClCH$_2$COOPr-i $\xrightarrow{\text{Mg} \quad \text{i-PrOH}}$ MeCOOPr-i   63%

Proc Chem Soc (1963) 219

PhCH=CHCHO $\xrightarrow[\text{2 NaOH   MeOH}]{\text{1 PhNH}_2 \quad \text{NaCN}}$ PhCH$_2$CH$_2$$\overset{\text{OMe}}{\underset{|}{\text{C}}}$=NPh $\dashrightarrow$ PhCH$_2$CH$_2$COOMe

Chem Comm (1967) 1290
Chem Ind (1967) 583

Chapter 9

PREPARATION
OF
ETHERS
AND EPOXIDES

Section 121   Ethers and Epoxides from Acetylenes

No examples

Section 122   Ethers from Carboxylic Acids

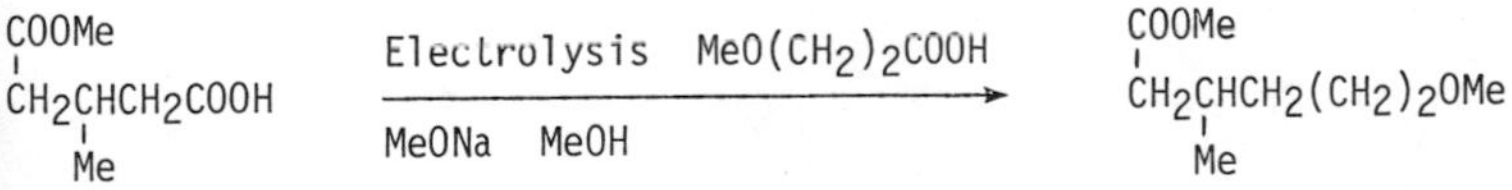

COOMe
|
CH₂CHCH₂COOH → Electrolysis   MeO(CH₂)₂COOH → COOMe
| |
Me   MeONa   MeOH   CH₂CHCH₂(CH₂)₂OMe
                                                  |
                                                  Me

JCS (1956) 1620
Advances in Org Chem (1960) 1 1

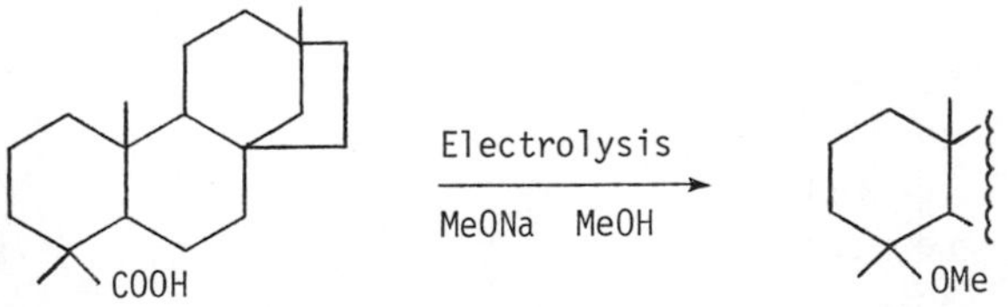

Electrolysis

MeONa   MeOH

JOC (1962) 27 4689
        (1960) 25 136
JACS (1960) 82 2645

Section 123   Ethers from Alcohols and Phenols

For the preparation of allyl, t-butyl, methoxymethyl, tetrahydropyranyl
and triphenylmethyl ethers see section 45A (Protection of Alcohols and
Phenols)

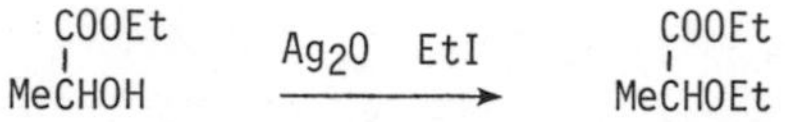

COOEt   COOEt
| Ag$_2$O EtI |
MeCHOH   ————————→   MeCHOEt

JACS (1932) 54 3732
JCS (1959) 2594

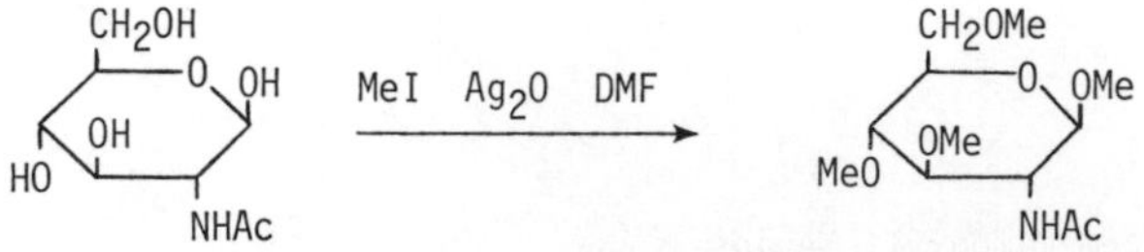

Angew (1955) 67 32
JCS C (1969) 2372
Can J Chem (1963) 41 1801

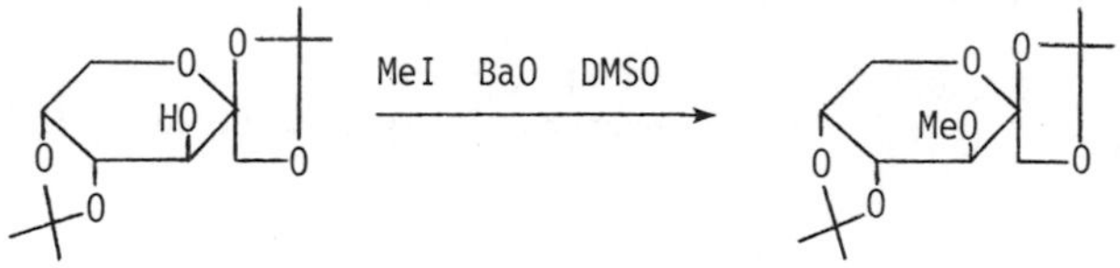

59%

JCS (1967) 2681

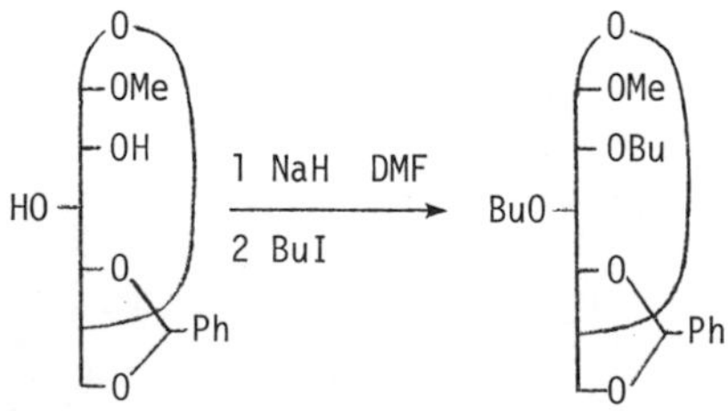

97%

Carbohydrate Res (1966) 2 167
Can J Chem (1966) 44 1591

$$\underset{\text{MeOCH}_2\text{CH}_2\text{OMe}}{\overset{\text{MeI \quad NaH}}{\longrightarrow}}$$

91%

Can J Chem (1966) 44 1591

$$\underset{\text{Et}}{\text{Me}_2\text{COH}} \quad \xrightarrow[\text{2 EtBr}]{\text{1 RMgBr \quad HMPA}} \quad \underset{\text{Et}}{\text{Me}_2\text{COEt}}$$

40%

Compt Rend (1968) C 266 1178

$$\xrightarrow[\text{2 EtI}]{\text{1 K \quad toluene}}$$

~41%

JOC (1961) 26 4553

$$\xrightarrow[\text{NaNH}_2 \quad \text{Et}_2\text{O}]{\text{Ph}-\bigcirc-\text{CH}_2\text{Cl}}$$

41%

JOC (1958) 23 1700

$$\underset{\overset{|}{\text{OCH}_2}}{\overset{\overset{|}{\text{OCH}_2}}{\text{PhCH}\quad \text{CHOH}}} \quad \xrightarrow[\text{2 C}_{18}\text{H}_{37}\text{OTs}]{\text{1 K \quad C}_6\text{H}_6} \quad \underset{\overset{|}{\text{OCH}_2}}{\overset{\overset{|}{\text{OCH}_2}}{\text{PhCH}\quad \text{CHOC}_{18}\text{H}_{37}}}$$

99%

JOC (1959) 24 409
        (1966) 31 498

$$\xrightarrow{\text{MeOH \quad H}_2\text{SO}_4}$$

Tetrahedron (1961) 16 85
J Med Chem (1965) 8 57

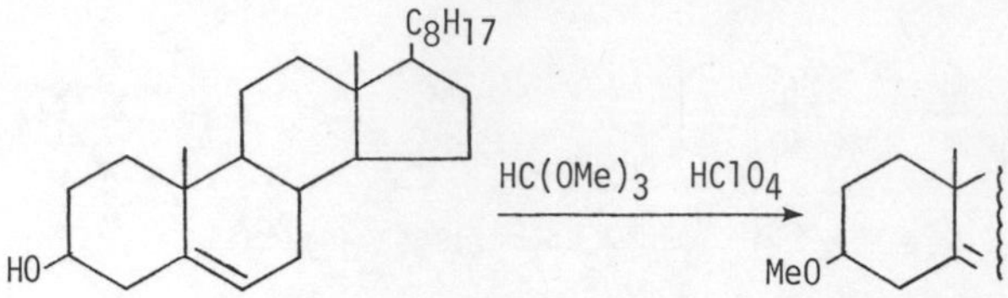

$$HC(OMe)_3 \quad HClO_4$$

52%

Steroids (1966) 8 495

$$C_8H_{17}OH \xrightarrow{\quad CH_2N_2 \quad HBF_4 \quad Et_2O \quad} C_8H_{17}OMe$$

84%

Tetrahedron (1959) 6 36
Annalen (1963) 662 38
(1964) 677 55

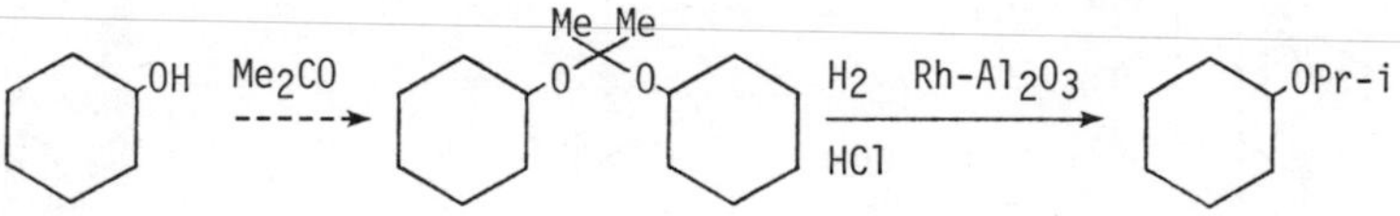

JOC (1961) 26 1026

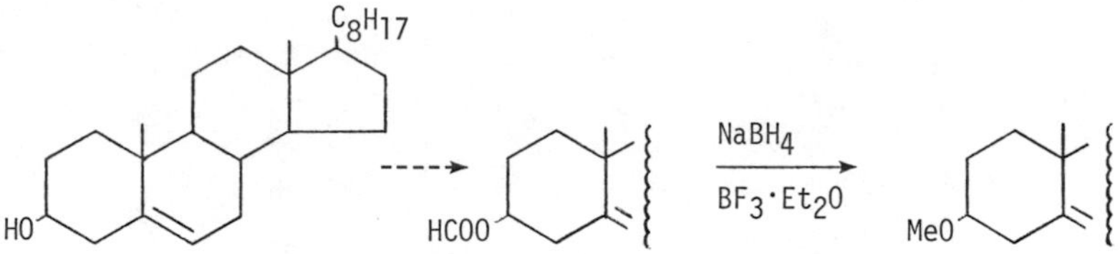

$$\xrightarrow[\quad BF_3 \cdot Et_2O \quad]{\quad NaBH_4 \quad}$$

JOC (1961) 26 1685
(1962) 27 2127

$$PhOH \xrightarrow[\quad 2 \quad C_5H_{11}Br \quad]{\quad 1 \quad NaNH_2 \quad NH_3 \quad} PhOC_5H_{11}$$

45%

JACS (1935) 57 510

$$PhOH \xrightarrow[\quad 150-190° \quad]{\quad PhCH_2Cl \quad Et_3N \quad} PhOCH_2Ph$$

46%

JOC (1961) 26 5180

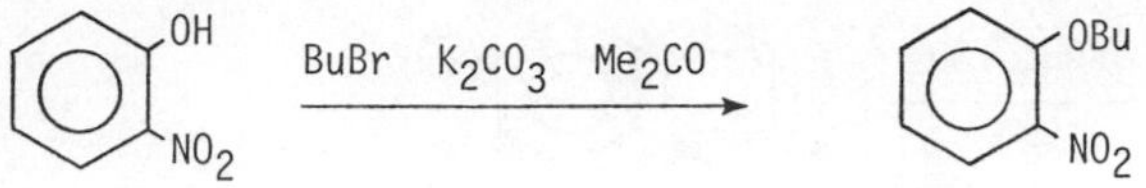

Org Synth (1955) Coll Vol 3 140 418
Tetr Lett (1964) 1431
JACS (1950) 72 3396

75-80%

PhOH

$\xrightarrow{\text{BuCHOTs} \quad \text{KOH} \quad \text{DMF}}$

PhOCHBu (with Me on the CH)

JACS (1969) 91 1376

38%

PhOH $\xrightarrow{\text{Me}_2\text{SO}_4 \quad \text{NaOH} \quad \text{H}_2\text{O}}$ PhOMe

Org Synth (1941) Coll Vol 1 58

72-75%

PhOH $\xrightarrow{\text{PrBr} \quad \text{NaOH} \quad \text{H}_2\text{O}}$ PhOPr

JACS (1947) 69 2451

63%

$\xrightarrow{\text{CH}_2\text{N}_2 \quad \text{Et}_2\text{O} \quad \text{MeOH}}$

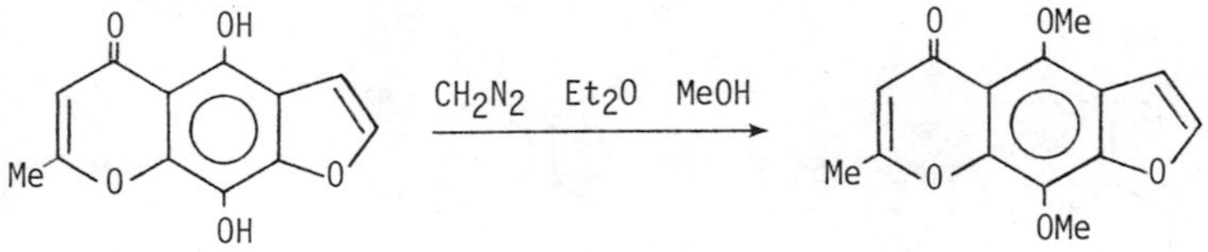

JACS (1950) 72 3396
       (1962) 84 2972

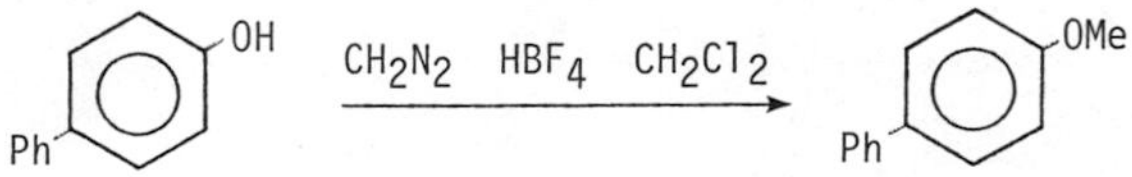

$\xrightarrow{\text{CH}_2\text{N}_2 \quad \text{HBF}_4 \quad \text{CH}_2\text{Cl}_2}$

Tetrahedron (1959) 6 36
Annalen (1963) 662 38

41%

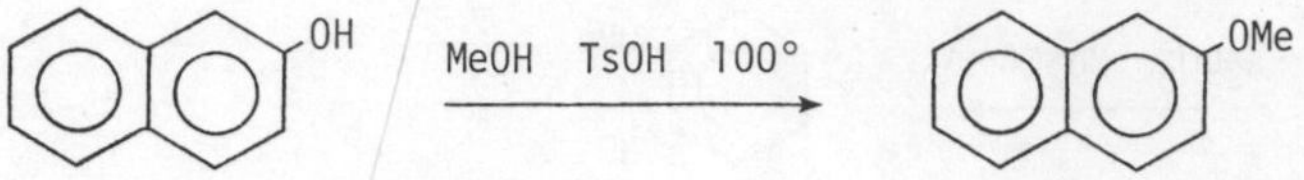

MeOH   TsOH   100°

70%

JOC (1960) 25 832

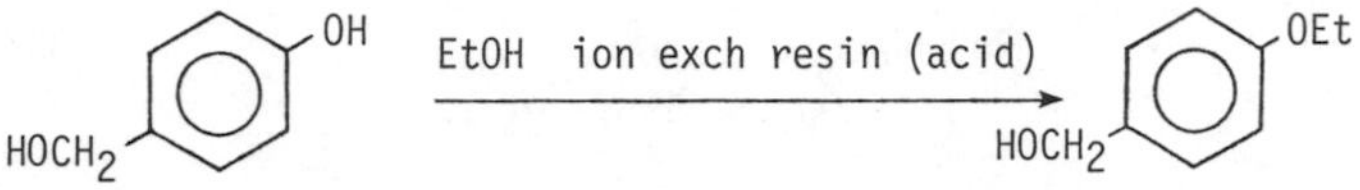

EtOH   ion exch resin (acid)

37%

JOC (1962) 27 2662

Me$_2$SOCH$_2$ (+ -)

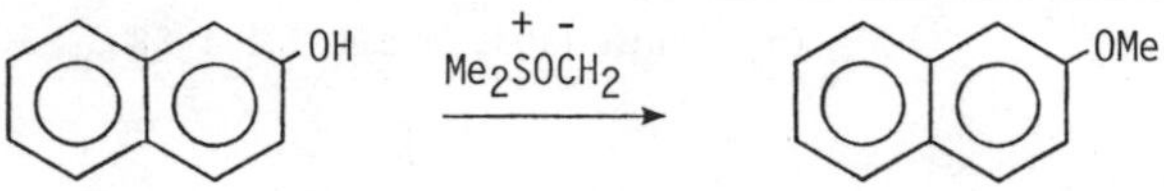

77%

Tetr Lett (1964) 867

PhOH $\xrightarrow{\text{Me}_3\text{O}^+ \text{BF}_4^-}$ PhOMe

73%

J Prakt Chem (1937) 147 257

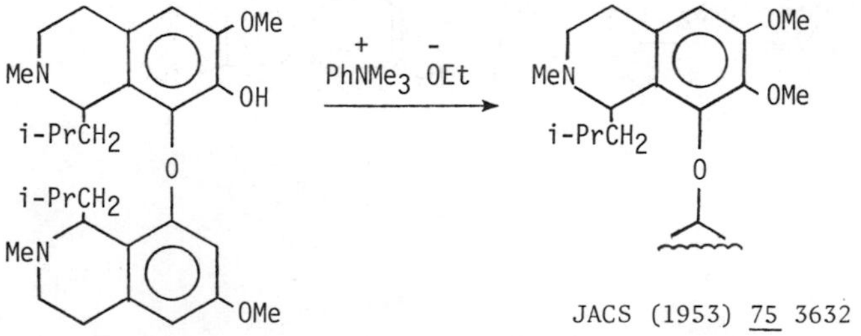

PhNMe$_3^+$ OEt$^-$

53%

JACS (1953) 75 3632

PhOH $\xrightarrow{\text{C(OEt)}_4}$ PhOEt

85%

Acta Chem Scand (1956) 10 1006

$\xrightarrow{\text{ROH   DCC}}$

87% (R=Me)
90% (R=CH$_2$Ph)

Ber (1962) 95 2997
Angew (1963) 75 377
(Internat Ed 2 218)

PhOH $\xrightarrow{\text{HC(OCH}_2\text{Ph)}_2\text{NMe}_2}$ PhOCH$_2$Ph   64%

Angew (1963) 75 296
(Internat Ed 2 211)

$\xrightarrow[\text{dioxane}]{\text{C}_2\text{H}_2 \text{ KOH}}$

$\xrightarrow{\text{H}_2 \text{ Ni}}$

Zh Obshch Khim (1964) 34 3424
(Chem Abs 62 2727)

$\xrightarrow[\text{Cu}_2\text{O   DMA}]{\text{Me}-\bigcirc-\text{Br}}$

81%

JCS (1965) 4953
Chem Rev (1946) 38 405

PhOH $\xrightarrow[\text{NaOH   H}_2\text{O}]{\text{Di-p-tolyliodonium bromide}}$

86%

JCS (1963) 4578

C$_7$H$_{15}$OH $\xrightarrow{\text{Pb(OAc)}_4 \text{ C}_6\text{H}_6}$ C$_3$H$_7$

37%

Tetr Lett (1963) 2091
Synthesis (1970) 209

JACS (1964) 86 3905 1528

66%

JOC (1963) 28 1388

Section 124   Ethers and Epoxides from Aldehydes

Pr_2NCH_2CHO $\dashrightarrow$ $Pr_2NCH_2CH(OMe)_2$ $\xrightarrow[\text{xylene}]{\text{PhMgBr}}$ $Pr_2NCH_2\underset{\underset{Ph}{|}}{C}HOMe$ 65%

JACS (1951) 73 4893

$PrCHO$ $\dashrightarrow$ $PrCH(OEt)_2$ $\xrightarrow[\text{Et}_2\text{O}]{\text{LiAlH}_4\ \ \text{AlCl}_3}$ $PrCH_2OEt$ 49%

JACS (1962) 84 2371
JOC (1958) 23 1088

$i\text{-}PrCH_2CHO$ $\dashrightarrow$ $i\text{-}PrCH_2CH(OEt)_2$ $\xrightarrow[\text{C}_6\text{H}_6]{\text{(i-Bu)}_2\text{AlH}}$ $i\text{-}PrCH_2CH_2OEt$ 83%

Izv (1959) 2255
(Chem Abs 54 10837)

$PhCHO$ $\dashrightarrow$ $PhCH(OEt)_2$ $\xrightarrow{\text{Et}_3\text{SiH}\ \ \text{ZnCl}_2}$ $PhCH_2OEt$

Compt Rend (1962) 254 1814

i-PrCHO $\xrightarrow[\text{MeOH}]{\text{H}_2 \quad \text{PtO}_2 \quad \text{HCl}}$ i-PrCH$_2$OMe

JCS (1963) 5598

PhCHO $\xrightarrow[\text{Co}_2\text{CO}_8]{\text{H}_2 \quad (3200 \text{ psi}) \quad \text{CO}}$ PhCH$_2$OCH$_2$Ph   43%

JACS (1950) $\underline{72}$ 4375

$\xrightarrow{(\text{Me}_2\text{N})_3\text{P}}$ 50-55%

Org Synth (1966) $\underline{46}$ 31
JACS (1963) $\underline{85}$ 1884

$\xrightarrow{\overset{+-}{\text{Ph}_2\text{SCHR}} \quad \text{THF}}$ 72% (R=Ph)
40% (R=Bu)

JACS (1964) $\underline{86}$ 918

PhCHO $\xrightarrow{\overset{+ \quad -}{\text{Me}_3\text{SO I}} \quad \text{NaH} \quad \text{Me}_2\text{SO}}$ PhCHCH$_2$ (epoxide)   56%

JACS (1965) $\underline{87}$ 1353

$\xrightarrow[\text{Me}_2\text{SO}]{\overset{+ \quad -}{\text{PhSOMe}} \quad \text{BF}_4 \\ \underset{}{|} \\ \text{NMe}_2 \quad \text{NaH}}$ 74%

JACS (1970) $\underline{92}$ 6594

$$C_{11}H_{23}CHO \xrightarrow{CH_2Br_2 \ Li \ THF} C_{11}H_{23}\overset{O}{\overset{/\backslash}{CH}}CH_2 \qquad 35\%$$

Chem Comm (1969) 1047

Some of the methods listed in section 132 (Ethers and Epoxides from Ketones) may also be applied to the preparation of ethers and epoxides from aldehydes

Section 125   Ethers and Epoxides from Alkyls, Methylenes and Aryls

No examples of the preparation of ethers and epoxides by replacement of alkyl, methylene and aryl groups occur in the literature. For the conversion RH ⟶ ROR' (R=alkyl, aryl etc.) see section 131 (Ethers from Hydrides)

Section 126   Ethers and Epoxides from Amides

No examples

Section 127   Ethers from Amines

$$PhNH_2 \dashrightarrow PhN_2^+ Cl^- \xrightarrow{MeOH} PhOMe \qquad 93\%$$

JACS (1955) 77 1745
(1958) 80 6072
Org React (1944) 2 262

Section 128   Ethers from Esters

$$\xrightarrow[\text{MeOH}]{PhCH_2Cl \ NaHCO_3}$$

48%

JCS (1948) 376
Ber (1943) 76 466

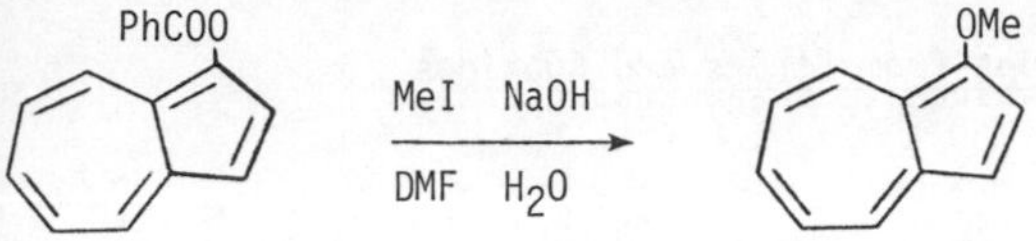
PhCOO
MeI NaOH
DMF H_2O
OMe

81%

JOC (1964) 29 2805

HCOOBu $\xrightarrow[MeOCH_2CH_2OMe]{PhCH_2Br\ \ NaH}$ $BuOCH_2Ph$ 43%

Tetr Lett (1965) 1713

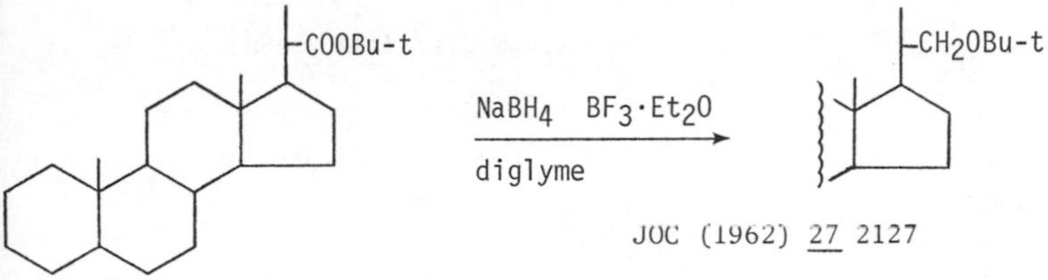
COOBu-t
$NaBH_4$ $BF_3 \cdot Et_2O$
diglyme
CH_2OBu-t

76%

JOC (1962) 27 2127

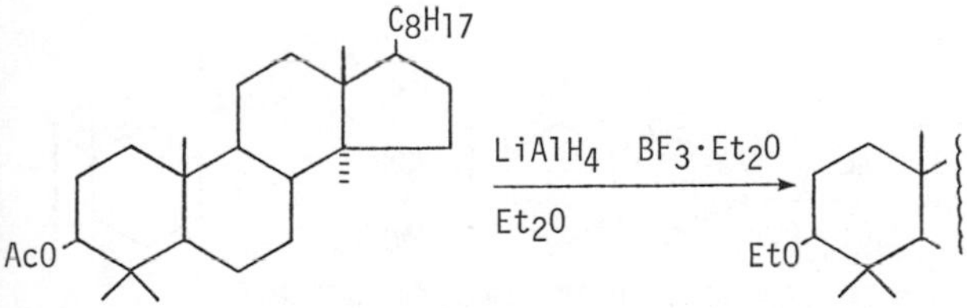
C_8H_{17}
AcO
$LiAlH_4$ $BF_3 \cdot Et_2O$
Et_2O
EtO

38%

JOC (1961) 26 4553 1685

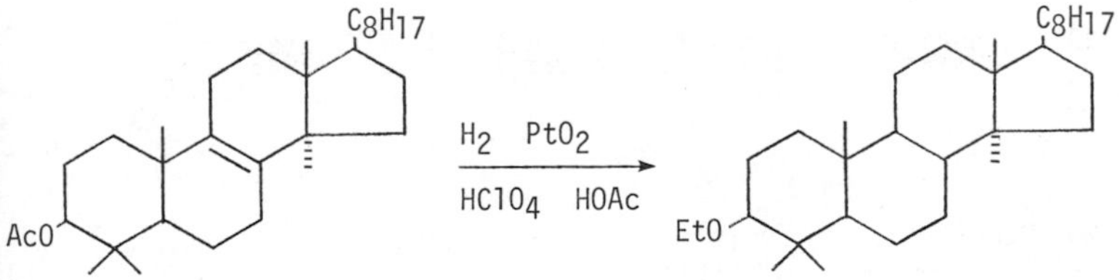
C_8H_{17}
AcO
H_2 PtO_2
$HClO_4$ HOAc
C_8H_{17}
EtO

24%

JOC (1964) 29 228
Chem Ind (1964) 975

Section 129 Ethers and Epoxides from Ethers and Epoxides

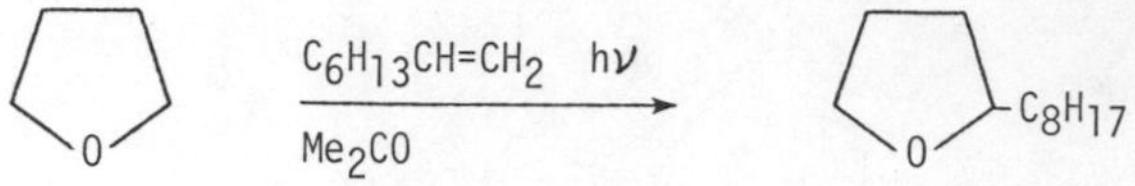

$C_6H_{13}CH=CH_2$ hν
Me_2CO
C_8H_{17}

JOC (1964) 29 2031
Tetrahedron (1967) 23 3193

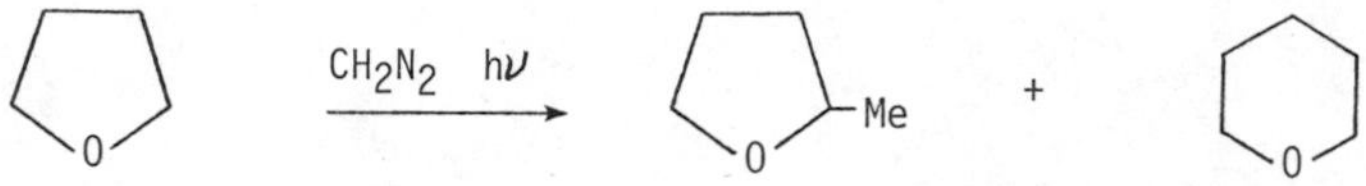

CH_2N_2 hν
Me
+

Chem Comm (1966) 454

$$Et_2O \underset{}{\overset{MeOSO_2F}{\rightleftarrows}} EtOMe$$

Chem Comm (1968) 1533

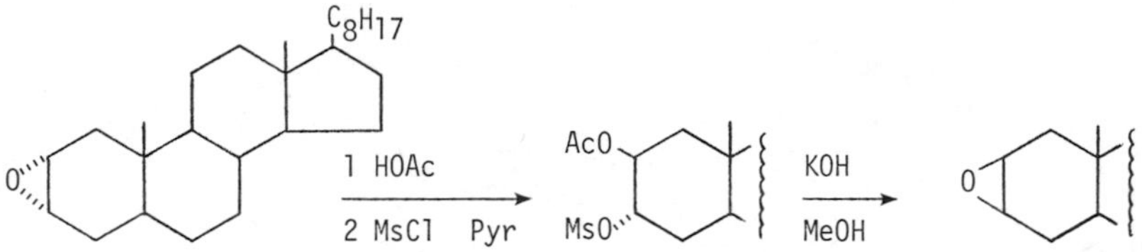

C_8H_{17}
1 HOAc
2 MsCl Pyr
AcO
MsO
KOH
MeOH

Helv (1949) 32 275

Section 130 Ethers and Epoxides from Halides and Sulfonates

$$PrBr \xrightarrow[PhPr\text{-}i]{Mg\quad MeCH(OEt)_2} PrCH(Me)OEt \qquad 73\%$$

Rec Trav Chim (1962) 81 238
JACS (1951) 73 4893

BuBr   ---→   BuLi $\xrightarrow{ClCH_2OBu}$ BuCH$_2$OBu   70%

Ber (1964) <u>97</u> 636

$\xrightarrow[\text{Et}_2\text{O}]{\text{Cyclohexanol}\quad\text{NaNH}_2}$ 44%

JOC (1958) <u>23</u> 1700

C$_{18}$H$_{37}$OTs $\xrightarrow{\text{t-BuOK}\quad\text{t-BuOH}}$ C$_{18}$H$_{37}$OBu-t   67%

JACS (1964) <u>86</u> 3072   99%

PhBr $\xrightarrow[\text{Me}_2\text{SO}]{\text{t-BuOK}\quad\text{t-BuOH}}$ PhOBu-t   42-46%

Org Synth (1965) <u>45</u> 89
JOC (1968) <u>33</u> 259

RBr $\xrightarrow[\text{2 PhCOO}_2\text{Bu-t}]{\text{1 Mg}\quad\text{Et}_2\text{O}}$ ROBu-t   74% (R=C$_8$H$_{17}$)
                                                            65% (R=Ph)
JACS (1959) <u>81</u> 4230
Org Synth (1963) <u>43</u> 55

$\xrightarrow[\text{collidine}]{\text{PhOH}\quad\text{Cu}_2\text{O}}$ 67%

JCS (1965) 4953
Chem Rev (1946) <u>38</u> 405

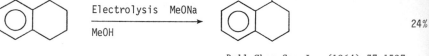

100%

Further examples of the preparation of ethers by the reaction
ROH + RHal ⟶ ROR are included in section 123 (Ethers from Alcohols and
Phenols)

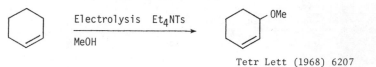

$$\text{i-PrBr} \xrightarrow[\substack{2 \; ClCH_2COMe \\ 3 \; KOH}]{1 \; Mg \quad Et_2O} \underset{\underset{Me}{|}}{\overset{\overset{O}{\diagup\!\!\backslash}}{\text{i-PrC-CH}_2}}$$

JACS (1946) 68 2339

Section 131   Ethers from Hydrides (RH)

This section lists examples of the reaction RH ⟶ ROR

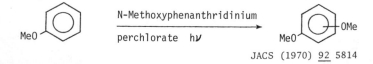

Tetr Lett (1968) 6207   21%

Electrolysis   MeONa

MeOH

Bull Chem Soc Jap (1964) 37 1597
Tetr Lett (1968) 2415   24%

N-Methoxyphenanthridinium

perchlorate hν

JACS (1970) 92 5814

Section 132   Ethers and Epoxides from Ketones
                ○○○○○○○○○○○○○○○○○○○○○○○○○○○○○○○○○○○○

MeCO
  | 1 MeMgI   Me₂COBu   40%
  Et   ──────────────────► |
             2 BuBr   HMPA   Et
                                          Compt Rend (1968) $\underline{C}$ 266 1178

              H₂   PtO₂   HCl
Bu₂CO   ──────────────────►   Bu₂CHOMe   54%
              MeOH
                                        JCS (1963) 5598
                                        Chem Comm (1967) 422

                                    H₂   Rh-Al₂O₃
Me₂CO --→   Me₂C(OBu)₂   ──────────────────►   Me₂CHOBu   <47%
                                    JOC (1961) $\underline{26}$ 1026

                                                                            <92%

                                        JACS (1962) $\underline{84}$ 2371
                                        JOC (1958) $\underline{23}$ 1088

                                                                            < 50%

                                Compt Rend (1962) $\underline{C}$ 254 1814

$$Ph_2CO \dashrightarrow Ph_2CN_2 \xrightarrow{\text{h}\nu \quad \text{MeOH}} Ph_2CHOMe$$

Annalen (1958) <u>614</u> 19

Some of the methods listed in section 124 (Ethers and Epoxides from Aldehydes) may also be applied to the preparation of ethers from ketones

$$PhCOBu\text{-}t \xrightarrow{\text{LiCHCl}_2} \underset{\underset{Bu\text{-}t}{|}}{Ph\overset{O}{\hat{C}}\text{-}CHCl} \xrightarrow{\text{BuLi}} \underset{\underset{Bu\text{-}t}{|}}{Ph\overset{O}{\hat{C}}\text{-}CHBu}$$

Tetr Lett (1969) 2181

Me$_3$S$^+$ I$^-$ NaH

THF

JACS (1965) <u>87</u> 1353

77%

Me$_3$SO$^+$ I$^-$ NaH

Me$_2$SO

JACS (1965) <u>87</u> 1353
(1970) <u>92</u> 6594

72%

$$Me_2CHSOCMe_2^- \; Na^+$$
$$\underset{\text{Me}_2\text{SO}}{\overset{\overset{\|}{\text{NTs}}}{}}$$

JACS (1970) <u>92</u> 5753

63%

$$C_6H_{13}COMe \xrightarrow{\text{CH}_2\text{Br}_2 \quad \text{Li} \quad \text{THF}} C_6H_{13}\underset{\underset{Me}{|}}{\overset{O}{\hat{C}}\text{-}CH_2}$$

52%

Chem Comm (1969) 1047

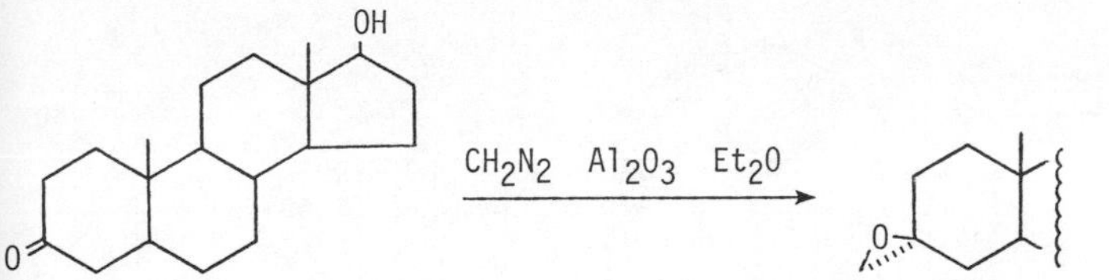

CH₂N₂ Al₂O₃ Et₂O

50%

Tetr Lett (1969) 305
Org React (1954) 8 364

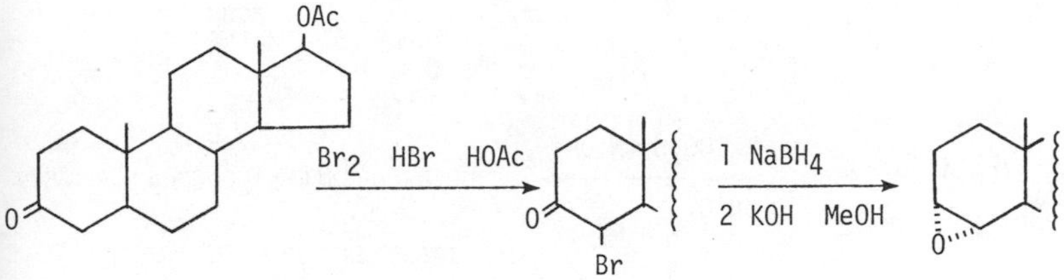

Br₂ HBr HOAc

1 NaBH₄

2 KOH MeOH

20%

JACS (1953) 75 1704

Section 133   Ethers and Epoxides from Nitriles
∘∘∘∘∘∘∘∘∘∘∘∘∘∘∘∘∘∘∘∘∘∘∘∘∘∘∘∘∘∘∘∘∘∘∘∘∘∘

No examples

Section 134   Ethers and Epoxides from Olefins
∘∘∘∘∘∘∘∘∘∘∘∘∘∘∘∘∘∘∘∘∘∘∘∘∘∘∘∘∘∘∘∘∘∘∘∘

$$\underset{\underset{Me}{|}}{C}H=C(CH_2)_2\underset{\underset{Me}{|}}{C}H=C(CH_2)_2CH=CMe_2$$

with COOMe

1 Hg(OAc)₂

MeOH

2 KOH

3 NaBH₄

$$\underset{\underset{Me}{|}}{C}H=C(CH_2)_2\underset{\underset{Me}{|}}{C}H=C(CH_2)_2CH_2\underset{\underset{OMe}{|}}{C}Me_2$$

with COOMe

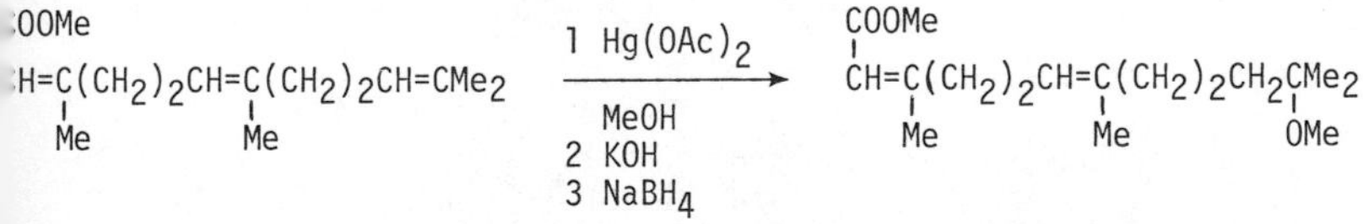

74%

J Med Chem (1969) 12 191
JACS (1969) 91 5646

$$\text{(cyclohexene with Et)} \xrightarrow{\text{h}\nu \quad \text{MeOH}} \text{(cyclohexane with Et, OMe)} \qquad 60\%$$

JACS (1967) <u>89</u> 5199
Acc of Chem Res (1969) <u>2</u> 33

Review: Epoxidation and Hydroxylation of Ethylenic Compounds with
      Organic Peracids
                           Org React (1953) <u>7</u> 378

$$HOCH_2CH_2\underset{Me}{CH}(CH_2)_2CH{=}CMe_2 \xrightarrow[\text{CH}_2\text{Cl}_2]{\text{MeCOO}_2\text{H} \quad \text{NaOAc}} HOCH_2CH_2\underset{Me}{CH}(CH_2)_2\overset{O}{\overset{/\backslash}{CH}}CMe_2 \qquad 87\%$$

Helv (1965) <u>48</u> 182
JACS (1960) <u>82</u> 4328
Org Synth (1963) Coll Vol 4 860

$$C_{10}H_{21}CH{=}CH_2 \xrightarrow{\text{CF}_3\text{COO}_2\text{H} \quad \text{Na}_2\text{CO}_3 \quad \text{CH}_2\text{Cl}_2} C_{10}H_{21}\overset{O}{\overset{/\backslash}{CH}}CH_2 \qquad 90\%$$

JACS (1955) <u>77</u> 89

$$C_6H_{13}CH{=}CH_2 \xrightarrow{\text{HOOCCH}{=}\text{CHCOO}_2\text{H} \quad \text{CH}_2\text{Cl}_2} C_6H_{13}\overset{O}{\overset{/\backslash}{CH}}CH_2 \qquad 80\%$$

Tetrahedron (1962) <u>17</u> 31

$$C_8H_{17}CH{=}CH(CH_2)_7COOH \xrightarrow{\text{PhCOO}_2\text{H} \quad \text{Me}_2\text{CO}} C_8H_{17}\overset{O}{\overset{/\backslash}{CH}}CH(CH_2)_7COOH \qquad 62\text{-}67\%$$

Org React (1953) <u>7</u> 378

For use of m-chloroperbenzoic acid see JOC (1970) <u>35</u> 251
                          and Tetr Lett (1965) 849
For preparation and use of monoperphthalic acid see Org React (1953) <u>7</u> 378
                          and JACS (1960) <u>82</u> 6373
For preparation of various aliphatic and aromatic peracids see
                                JOC (1962) <u>27</u> 1336

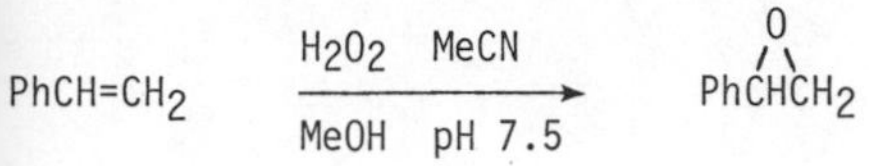

$$PhCH=CH_2 \xrightarrow[\text{MeOH \quad pH 7.5}]{H_2O_2 \quad MeCN} PhCHCH_2 \overset{O}{\triangle}$$

JOC (1961) 26 659
JCS C (1970) 731

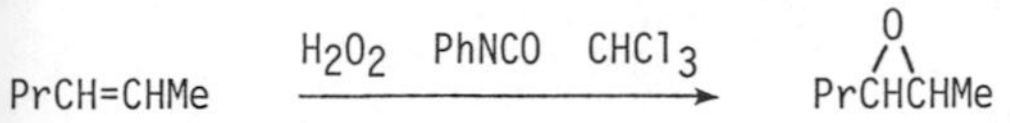

$$PrCH=CHMe \xrightarrow{H_2O_2 \quad PhNCO \quad CHCl_3} PrCHCHMe \overset{O}{\triangle} \qquad 39\%$$

Tetr Lett (1970) 2029

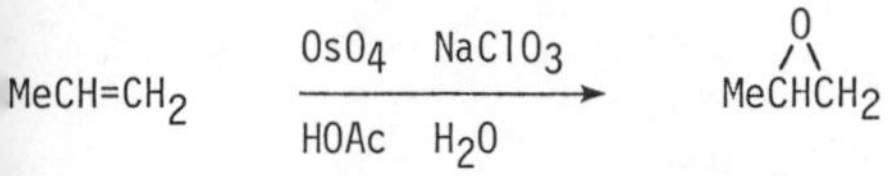

$$MeCH=CH_2 \xrightarrow[\text{HOAc \quad H_2O}]{OsO_4 \quad NaClO_3} MeCHCH_2 \overset{O}{\triangle}$$

Chem Comm (1968) 1610

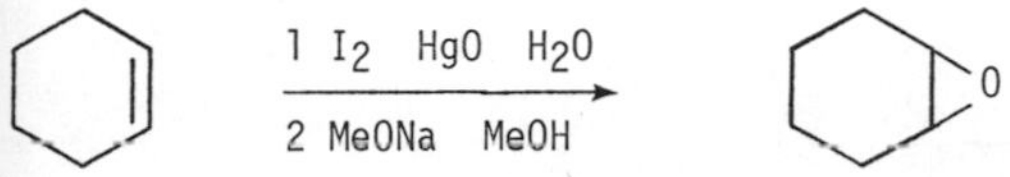

$$\xrightarrow[\text{2 MeONa \quad MeOH}]{1 \; I_2 \quad HgO \quad H_2O}$$

Tetrahedron (1964) 20 1017

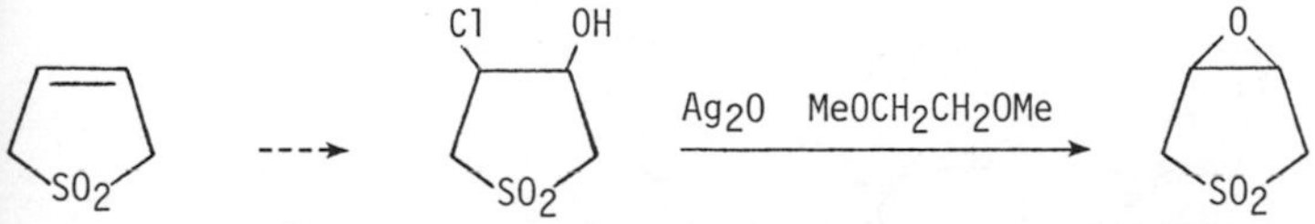

$$\xrightarrow{Ag_2O \quad MeOCH_2CH_2OMe}$$

JOC (1967) 32 3888

$$\xrightarrow[\text{CCl}_4]{AcOBr} \xrightarrow[\text{MeOH}]{NaOH}$$

27%

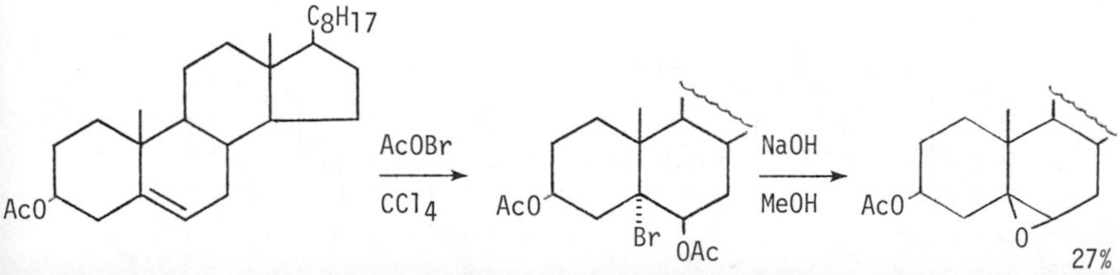

JACS (1959) 81 2826

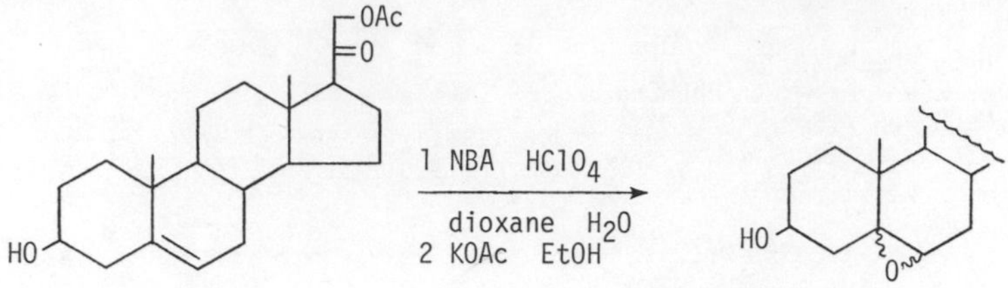

JCS (1956) 4417
JACS (1959) 81 2195

Section 135   Ethers and Epoxides from Miscellaneous Compounds
○○

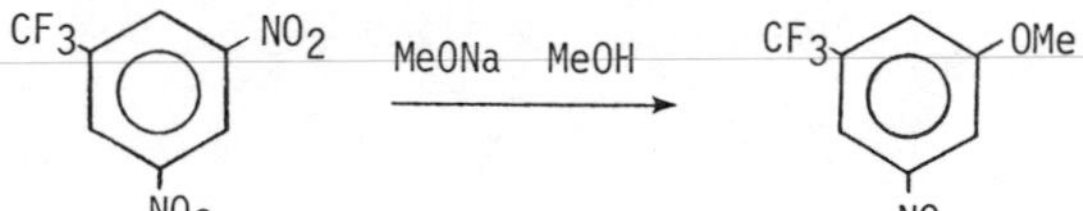

91%

JCS (1951) 2013

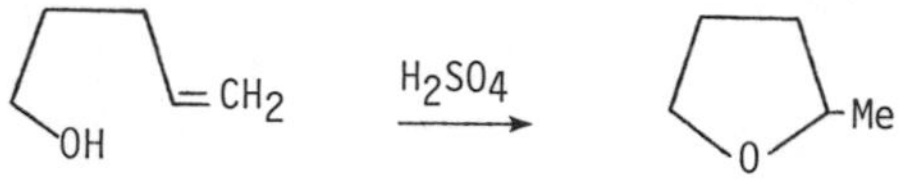

Bull Soc Chim Fr (1944) 11 365

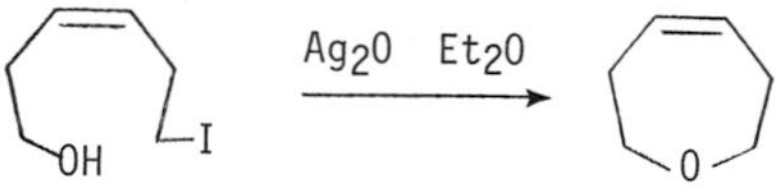

JACS (1958) 80 3132

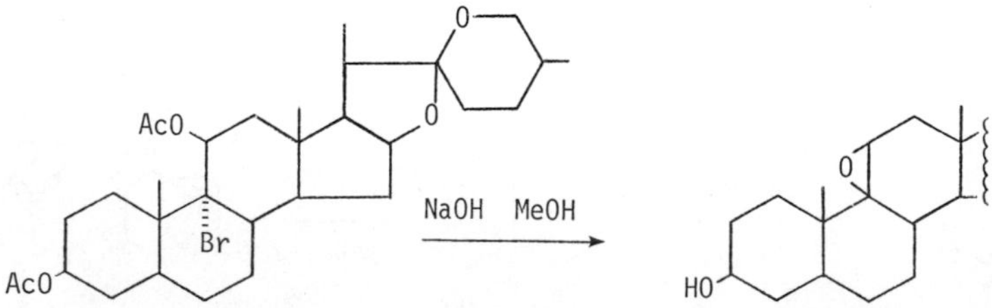

62%

JACS (1959) 81 2826

Chapter 10   PREPARATION
OF
HALIDES
AND SULFONATES

Section 136   Halides from Acetylenes

RC≡CH $\xrightarrow{\text{HBr}}$ RCH=CHBr $\xrightarrow[\text{dioxane}]{\overset{\displaystyle\text{NCOOK}}{\underset{\displaystyle\text{NCOOK HOAc}}{\|}}}$ RCH_2CH_2Br

JOC (1965) 30 3985
JACS (1936) 58 1806

Section 137   Halides from Carboxylic Acids

$\overset{\text{COOMe}}{\underset{}{|}}$
$(CH_2)_8COOH$ $\xrightarrow[\text{MeONa   MeOH}]{\text{Electrolysis   } F(CH_2)_9COOH}$ $\overset{\text{COOMe}}{\underset{}{|}}$ $(CH_2)_{17}F$

JACS (1956) 78 2255
Advances in Org Chem (1960) 1 1

i-Pr(CH$_2$)$_3$$\overset{}{\underset{\text{Me}}{\text{CHCH}_2\text{COOH}}}$ $\xrightarrow[\text{2 Br}_2\text{   CCl}_4]{\text{1 NH}_3\text{   AgNO}_3\text{   H}_2\text{O}}$ i-Pr(CH$_2$)$_3$$\overset{}{\underset{\text{Me}}{\text{CHCH}_2\text{Br}}}$ 55%

JCS (1953) 132
Chem Rev (1956) 56 219

RCOOH $\xrightarrow{\text{Br}_2\text{   HgO   CCl}_4}$ RBr

46% (R=C$_{11}$H$_{23}$)
<83% (R=Ph)
JOC (1965) 30 415

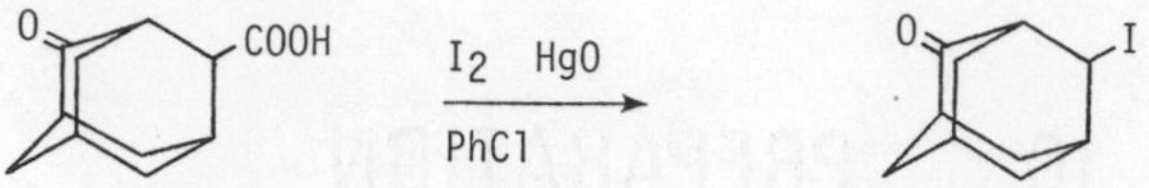

Ber (1968) 101 2010

38%

$C_8H_{17}COONa \xrightarrow{\text{F}_2 \quad \text{H}_2\text{O}} C_8H_{17}F$ ~50%

JOC (1969) 34 2446

$C_{17}H_{35}COOH \xrightarrow{\text{I}_2 \quad \text{benzoyl peroxide}} C_{17}H_{35}I$

J Amer Oil Chem Soc (1969) 46 615

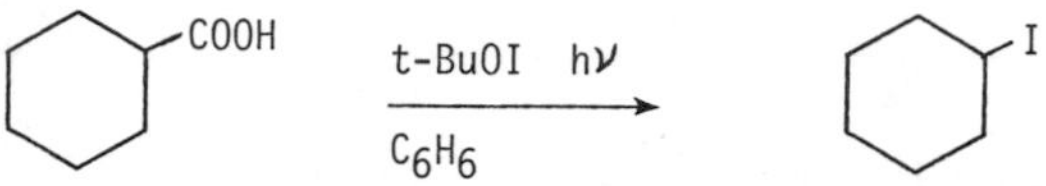

JCS (1965) 2438   70%

$RCOOH \xrightarrow[\text{CCl}_4]{\text{I}_2 \quad \text{Pb(OAc)}_4 \quad h\nu} RI$

100% (R=C_5H_{11})
61% (R=Ph)

JCS (1965) 2438
JOC (1963) 28 65

$C_{15}H_{31}COOH \xrightarrow[\text{HOAc}]{\text{Pb}_3\text{O}_4 \quad \text{Ac}_2\text{O}} (C_{15}H_{31}COO)_4Pb \xrightarrow[\text{C}_2\text{H}_2\text{Cl}_4]{\text{I}_2} C_{15}H_{31}I$ 72%

JOC (1963) 28 65
Annalen (1970) 735 47

BuCOOH $\xrightarrow{\text{Pb(OAc)}_4 \quad \text{LiCl} \quad \text{C}_6\text{H}_6}$ BuCl   96%

JOC (1965) <u>30</u> 3265

$\xrightarrow[\text{2 Pyr·Br}_2]{\text{1 Hg(OAc)}_2 \quad \text{N-methylpyrrolidone}}$

JOC (1969) <u>34</u> 1904

RCOOH $\xrightarrow{\text{SOCl}_2}$ RCOCl $\xrightarrow[\text{CH}_2\text{Cl}_2]{\text{RhCl(Ph}_3\text{P)}_3}$ RCl   86% (R=PhCH$_2$)

61% (R=Ph)

JACS (1968) <u>90</u> 99 94
(1966) <u>88</u> 3452

Section 138   Halides and Sulfonates from Alcohols and Phenols
oooooooooooooooooooooooooooooooooooooooooooooooooo

C$_{12}$H$_{25}$OH $\xrightarrow{\text{HBr} \quad \text{H}_2\text{SO}_4}$ C$_{12}$H$_{25}$Br   91%

Org Synth (1941) Coll Vol 1 29
JCS (1953) 132

HO(CH$_2$)$_6$OH $\xrightarrow{\text{KI} \quad \text{polyphosphoric acid}}$ I(CH$_2$)$_6$I   83-85%

Org Synth (1963) Coll Vol 4 323

Et$_3$COH $\xrightarrow{\text{HCl}}$ Et$_3$CCl   94%

JOC (1966) <u>31</u> 1090

EtCHOH → HCl MeCN → EtCHOC=NH·HCl → 130° → EtCHCl   56%
| | | |
Me   Me   Me   Me

JACS (1955) 77 2341

HCl CCl₃CN CHCl₃   85-90%

Tetr Lett (1970) 2517

SOCl₂ Pyr   91%

JCS (1942) 684
    (1943) 99
    (1950) 3650
Tetr Lett (1970) 2931
        (1971)   87

PhCH₂CH₂OH → 1 SOCl₂ Pyr / 2 SOBr₂ Pyr·HBr → PhCH₂CH₂Br   91%

Chem Ind (1954) 931
JOC (1961) 26 3645

C₁₆H₃₃OH → I₂ P₄ → C₁₆H₃₃I   85%

Org Synth (1943) Coll Vol 2 322

CH₂=CCH₂CH₂C≡CCH₂OH → PBr₃ LiBr Et₂O / collidine → CH₂=CCH₂CH₂C≡CCH₂Br   87%
| |
Me   Me

Chem Comm (1969) 611
JCS (1950) 3650

Et_2CHOH $\xrightarrow[\text{Et}_2\text{O}]{\text{PBr}_3 \text{ or PBr}_5}$ Et_2CHBr

JOC (1961) 26 3645
    (1959) 24 143
J Lipid Res (1966) 7 453

22-47%

$\underset{(CH_2)_9OH}{\overset{COOH}{|}}$ $\xrightarrow[\text{2 H}_2\text{O}]{\text{1 PCl}_5}$ $\underset{(CH_2)_9Cl}{\overset{COOH}{|}}$

JCS (1936) 1605
    (1970) 1124

$C_7H_{15}OH$ $\xrightarrow[\text{2 I}_2 \quad \text{CH}_2\text{Cl}_2]{1 \quad \text{(catechol PCl)} \quad \text{Pyr} \quad \text{Et}_2\text{O}}$ $C_7H_{15}I$

JOC (1967) 32 4160
    (1968) 33 300

77%

$BuOH$ $\xrightarrow{\text{(PhO)}_3\text{P} \quad \text{MeI}}$ BuI

JCS (1953) 2224
    (1954) 2281

90%

$\xrightarrow{\text{Ph}_3\text{P}\cdot\text{I}_2 \quad \text{DMF}}$

59%

JACS (1964) 86 2093

$MeCH=C=CHCH_2OH$ $\xrightarrow{\text{Ph}_3\text{P}\cdot\text{Br}_2 \quad \text{Pyr}}$ $MeCH=C=CHCH_2Br$

JCS (1967) 2260
JACS (1969) 91 7405
JOC (1965) 30 3469

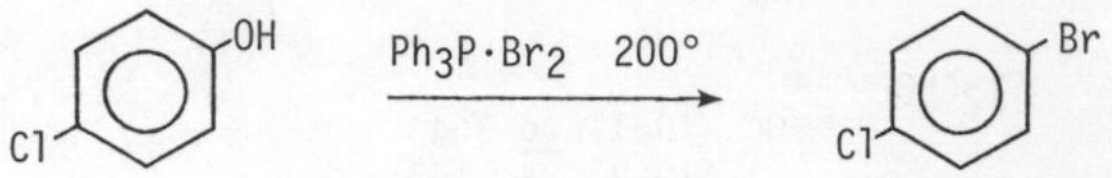

$Ph_3P \cdot Br_2$ 200°

90%

JACS (1964) $\underline{86}$ 964

$PhCH_2OH$ $\xrightarrow{Ph_3P \cdot BrCN}$ $PhCH_2Br$

Annalen (1959) $\underline{626}$ 26

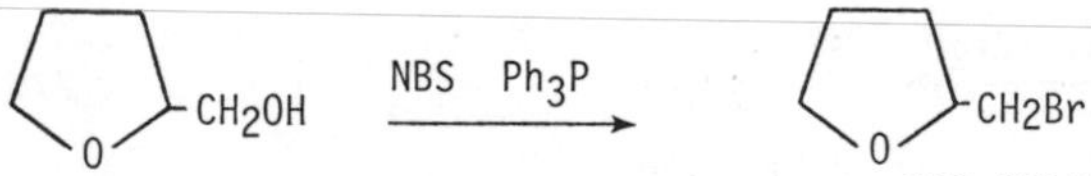

NBS Ph_3P

92%

JOC (1969) $\underline{34}$ 212

$\underset{Me}{PrCHOH}$ $\xrightarrow{Ph_3P \cdot F_2}$ $\underset{Me}{PrCHF}$

52%

Chem Pharm Bull (1968) $\underline{16}$ 1009 1784

$C_8H_{17}OH$ $\xrightarrow{R_3P \quad CCl_4}$ $C_8H_{17}Cl$ (R=Ph or C_8H_{17})   94%

Can J Chem (1968) $\underline{46}$ 86
Chem Comm (1968) $1\overline{35}8$
Chem Ind (1966) 900

$C_5H_{11}OH$ $\xrightarrow{CBr_4 \quad R_3P \quad Et_2O}$ $C_5H_{11}Br$ (R=Ph or C_8H_{17})   97%

Can J Chem (1968) $\underline{46}$ 86

$$\text{[OH-O-O}\diagup\quad \xrightarrow{\quad CCl_4 \quad (Me_2N)_3P \quad}\quad \text{[Cl-O-O}\diagup \qquad 70\%$$

Chem Comm (1968) 1350

$$ROH \xrightarrow{\quad COCl_2 \quad} ROCOCl \xrightarrow{\quad Ph_3P \quad} RCl$$

65% (R=Et)
67% (R=Ph)

Annalen (1966) <u>698</u> 106

$$\xrightarrow{\quad COCl_2 \quad} ClCOO\text{—} \xrightarrow[Me_2CO]{\quad NaX \quad} X\text{—}$$

67% (X=Cl)
49% (X=I)

JOC (1967) <u>32</u> 2633

$$\underset{Me}{tCHOH} \xrightarrow[\quad 2\ TlF_2 \quad]{\quad 1\ COCl_2 \quad} \underset{Me}{EtCHOCOF} \xrightarrow{\quad Pyr \quad} \underset{Me}{EtCHF} \qquad 64\%$$

JACS (1955) <u>77</u> 3099

$$\xrightarrow[\quad CH_2Cl_2 \quad]{\quad Et_2NCF_2CHClF \quad} \qquad 35\%$$

Tetr Lett (1962) 1065

$$\xrightarrow[\quad C_2H_2Cl_4 \quad]{\quad \overset{+}{Me_2N}=CHCl\ \overset{-}{Cl} \quad} \qquad \sim 70\%$$

Chem Comm (1967) 1152

$$\underset{\underset{Me}{|}}{EtCHOH} \xrightarrow{\overset{\overset{Cl}{|}}{Cl_2C=CNEt_2}} \underset{\underset{Me}{|}}{EtCHCl} \qquad 69\%$$

JACS (1960) <u>82</u> 909

$$C_{10}H_{21}OH \xrightarrow[\substack{2 \ HF \\ 3 \ KF_2 \quad 250-300°}]{1 \ Ph_2NCN \quad t\text{-BuOK}} C_{10}H_{21}F \qquad 68\%$$

Can J Chem (1965) <u>43</u> 3173

$$\underset{\underset{Me}{|}}{C_6H_{13}CHOH} \xrightarrow[\substack{2 \ Ph_2CHCl}]{1 \ Naphthyl \ isothiocyanate \quad t\text{-BuOK}} \underset{\underset{Me}{|}}{C_6H_{13}CHCl} \qquad 47\%$$

JCS (1953) 3572

$$MeOH \xrightarrow{I_2 \quad B_2H_6} MeI \qquad 70\%$$

Chem Ind (1964) 1582
(1965) 223

$$\underset{\underset{Me}{|}}{EtCHOH} \xrightarrow[\substack{Et_2O}]{\overset{\overset{Br}{|}}{C_6H_{13}CHCH_2CH_2COBr}} \underset{\underset{Br}{|}}{C_6H_{13}CHCH_2CH_2COOCHEt} \xrightarrow{175-180°} \underset{\underset{Me}{|}}{EtCHBr}$$

40%

JACS (1956) <u>78</u> 4967

$$\underset{\underset{Me}{|}}{(MeO)_2CHCH_2CH_2C=CHCH_2OH} \xrightarrow[\substack{2 \ TsCl \quad LiCl}]{1 \ MeLi \quad HMPA \quad Et_2O} \underset{\underset{Me}{|}}{(MeO)_2CHCH_2CH_2C=CHCH_2Cl}$$

Tetr Lett (1969) 1393

$$\overset{CH_2OH}{\diagup} \xrightarrow{\text{TsCl \quad Pyr}} \overset{CH_2OTs}{\diagup} \xrightarrow{\text{NaI \quad Me}_2\text{CO}} \overset{CH_2I}{\diagup} \quad \sim 39\%$$

JACS (1969) 91 4771

Further examples of the preparation of halides from sulfonates are included in section 145 (Halides and Sulfonates from Halides and Sulfonates)

$$C_{16}H_{33}OH \xrightarrow{\text{TsCl \quad Pyr}} C_{16}H_{33}OTs$$

JACS (1933) 55 345
      (1970) 92 553

$$\text{MeO}\!-\!\!\bigcirc\!\!-\!\text{OH} \xrightarrow{\text{MsCl \quad NaOH \quad H}_2\text{O}} \text{MeO}\!-\!\!\bigcirc\!\!-\!\text{OMs}$$

JCS (1955) 522

$$PhCH_2OH \xrightarrow[\text{2 TsCl}]{\text{1 NaH \quad Et}_2\text{O}} PhCH_2OTs \qquad\qquad 80\%$$

JACS (1953) 75 3443
      (1960) 82 3982

$$\overset{OH}{\underset{Me}{\square}} \xrightarrow{\text{MsCl \quad Et}_3\text{N \quad CH}_2\text{Cl}_2} \overset{OMs}{\underset{Me}{\square}}$$

JOC (1970) 35 3195

$$\overset{PhCHOH}{\underset{Me}{|}} \xrightarrow[\text{2 m-Chloroperbenzoic acid}]{\text{1 \quad Me}\!-\!\bigcirc\!-\text{SOCl}} \overset{PhCHOTs}{\underset{Me}{|}} \qquad 53\%$$

Tetr Lett (1969) 2705

Section 139   Halides from Aldehydes

$$PhCHO \xrightarrow[\text{2 HCOOEt}]{\text{1 BuMgBr}} \underset{\text{Bu}}{PhCHBr}$$

Zh Obshch Khim (1955) <u>25</u> 505
(Chem Abs <u>50</u> 3356)   56%

$$i\text{-}PrCHO \xrightarrow[\text{CCl}_4]{\text{Di-t-butyl peroxide}} i\text{-}PrCl$$

JACS (1965) <u>87</u> 2194   45%

Section 140   Halides from Alkyls

The conversion RR' $\longrightarrow$ RHal (R,R'=alkyl or aryl) is included here. For
the reaction RH $\longrightarrow$ RHal see section 146 (Halides from Hydrides)

$$\xrightarrow{S_2Cl_2 \quad SO_2Cl_2 \quad Fe}$$

JACS (1947) <u>69</u> 3146   45%

Section 141   Halides from Amides

$$\underset{\text{Me}}{Bu\overset{|}{C}H}(CH_2)_3\underset{\text{Me}}{\overset{|}{C}H}(CH_2)_3NHCOPh \xrightarrow{PBr_3} \underset{\text{Me}}{Bu\overset{|}{C}H}(CH_2)_3\underset{\text{Me}}{\overset{|}{C}H}(CH_2)_3Br$$

Z Physiol Chem (1942) <u>273</u> 225
Org Synth (1941) Coll Vol 1 428

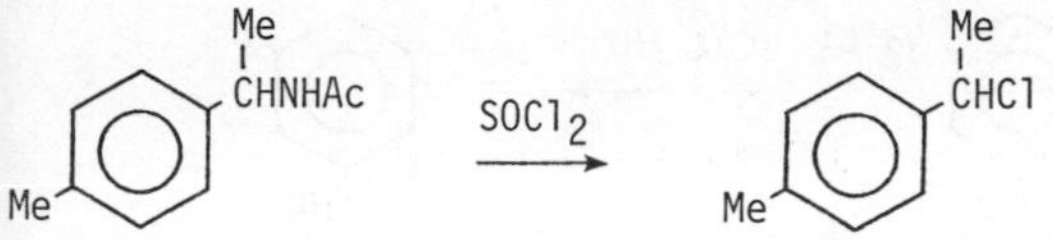

JACS (1962) 84 769

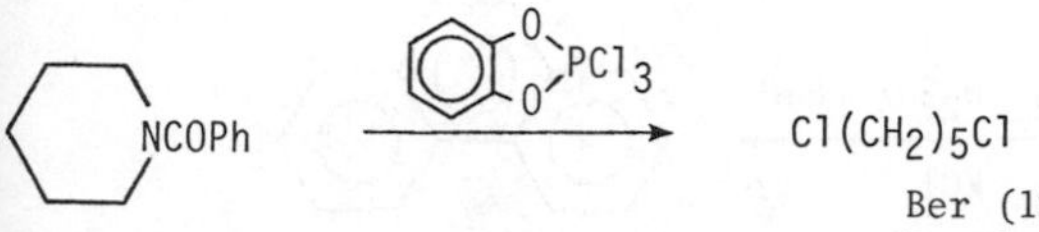

Cl(CH$_2$)$_5$Cl   55%

Ber (1963) 96 1387
Org Synth (1941) Coll Vol 1 428

Section 142   Halides from Amines
             °°°°°°°°°°°°°°°°°°°°°°

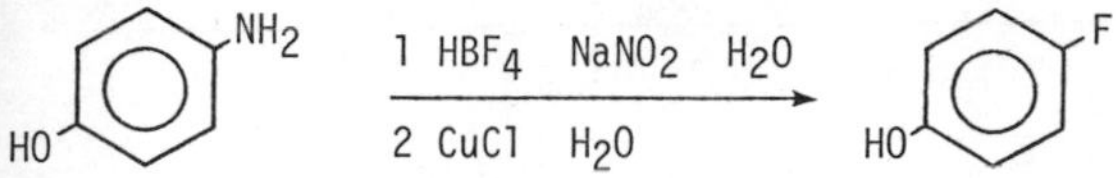

71%

JACS (1956) 78 6037
Chem Ind (1961) 179
Org Synth (1944) Coll Vol 2 295
Org React (1949) 5 193

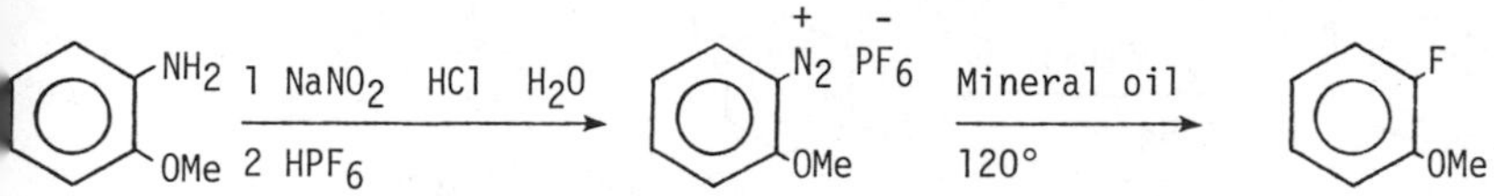

JOC (1961) 26 5149
    (1963) 28 568

PhNH$_2$ $\xrightarrow{\text{CuX}_2 \quad \text{NO} \quad \text{MeCN}}$ PhX

73% (X=Cl)
65% (X=Br)

Rec Trav Chim (1966) 85 857
JACS (1947) 69 1221

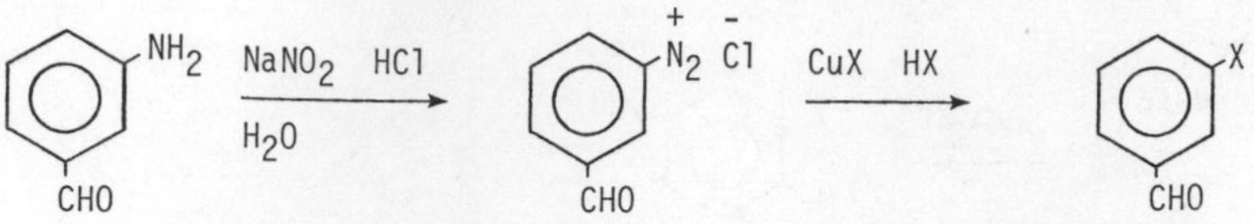

~75% (X=Cl)
~65% (X=Br)
Org Synth (1944) Coll Vol 2 130

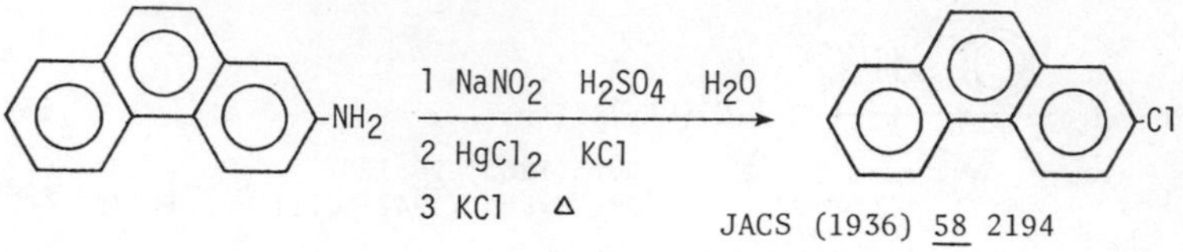

1 NaNO$_2$ H$_2$SO$_4$ H$_2$O
2 HgCl$_2$ KCl
3 KCl Δ

JACS (1936) 58 2194

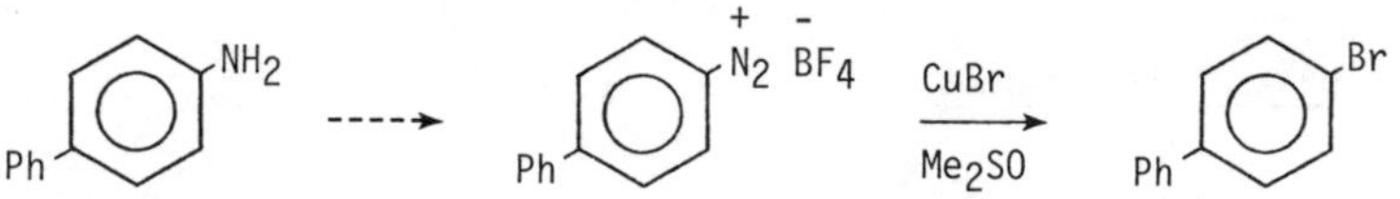

~97%

Org Prep and Procedures (1969) 1 221

$$PhNH_2 \xrightarrow{NaNO_2 \quad HCl \quad H_2O} PhN_2^+ Cl^- \xrightarrow{KI \quad H_2O} PhI \qquad 74\text{-}76\%$$

Org Synth (1944) Coll Vol 2 351
Chem Ind (1962) 1760

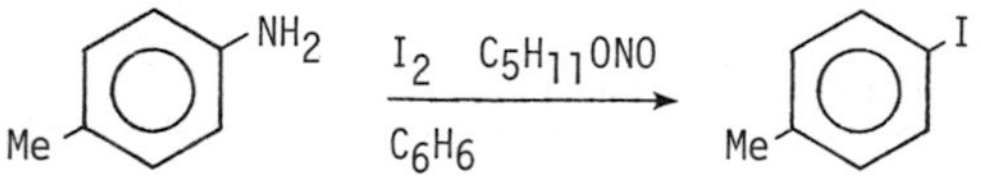

40%

JOC (1968) 33 1636

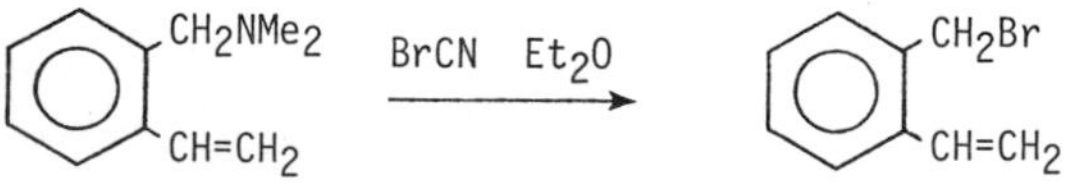

50%

Org React (1953) 7 198
Ber (1910) 43 3209

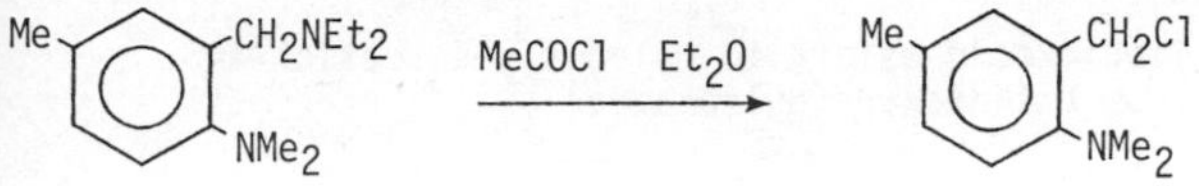

Ber (1960) $\underline{93}$ 1310 1305

BuNH$_2$ $\xrightarrow[\text{2 HBr Et}_2\text{O}]{\begin{array}{c}\text{1 p-Chlorobenzenediazonium}\\ \text{hexafluorophosphate}\end{array}}$ BuBr

Tetr Lett (1961) 758

Bu$_3$N $\xrightarrow{\text{ClCOOPh CH}_2\text{Cl}_2}$ BuCl   (Not isolated)

JCS (1967) 2015

Pr$_2$NPh $\xrightarrow{\text{HBr 156°}}$ PrBr

JOC (1963) $\underline{28}$ 3144

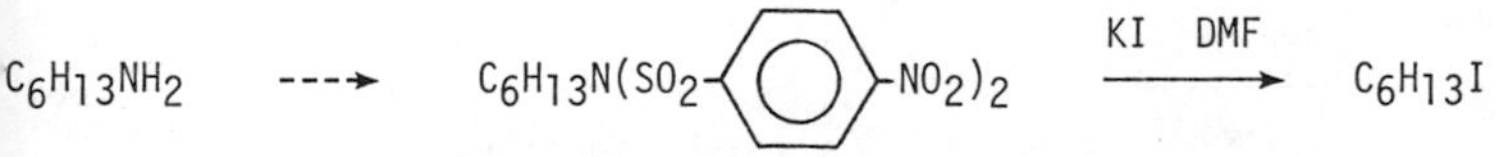

C$_6$H$_{13}$NH$_2$ $\dashrightarrow$ C$_6$H$_{13}$N(SO$_2$⟨○⟩-NO$_2$)$_2$ $\xrightarrow{\text{KI DMF}}$ C$_6$H$_{13}$I   79%

JACS (1969) $\underline{91}$ 2384

I$_2$ N-methylmorpholine

CHCl$_3$

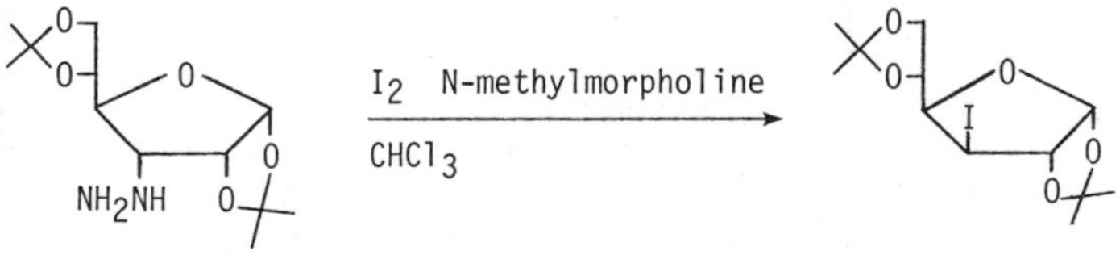

72%

JCS $\underline{C}$ (1967) 252

Halides may also be prepared from amines via amide intermediates. See section 141 (Halides from Amides)

Section 143   Halides from Esters
∘∘∘∘∘∘∘∘∘∘∘∘∘∘∘∘∘∘∘∘∘

For the conversion of sulfonic acid esters to halides see section 145
(Halides and Sulfonates from Halides and Sulfonates)

PhCH$_2$COOBu $\xrightarrow{\text{Br}_2 \quad \text{P}_4}$ BuBr

Arch Pharm (1953) <u>286</u> 108
JCS <u>B</u> (1967) 1067   80%

PhCOOBu $\xrightarrow{\text{PCl}_3}$ BuCl

Ber (1963) <u>96</u> 1387   73%

$\xrightarrow{\text{AlCl}_3 \quad \text{Et}_2\text{O}}$

JCS (1957) 2071   97%
Zesz Nauk Pol Slask (1964) <u>24</u> 193
(Chem Abs <u>63</u> 11338)
JACS (1940) <u>62</u> 1432

$\xrightarrow[\text{190-210°}]{\text{Pyr·HCl}}$ BuCl

Monatsh (1952) <u>83</u> 1398   66%

R'COOR $\xrightarrow{\text{LiI} \quad \text{DMF}}$ RI   (Not isolated)

JCS (1965) 6655

$\xrightarrow{\text{PhMgI} \quad \text{Bu}_2\text{O}}$ BuI

JACS (1942) <u>64</u> 1450   54%

Section 144   Halides from Ethers
              °°°°°°°°°°°°°°°°°°°°°°°

$EtCH(CH_2)_{10}OMe$ $\xrightarrow{\text{KI \quad polyphosphoric acid}}$ $EtCH(CH_2)_{10}I$ ~97%
   | |
   Me   Me

JACS (1965) 87 5452

$MeO(CH_2)_{20}OMe$ $\xrightarrow[180°]{\text{HI \quad H}_2\text{O}}$ $I(CH_2)_{20}I$

Ber (1942) 75B 1715
JCS (1957) 463
Chem Rev (1954) 54 615

$\xrightarrow{\text{HBr \quad H}_2\text{O}}$ 81%

Helv (1946) 29 1204

$\xrightarrow{\text{PhCOCl \quad ZnCl}_2 \quad \text{CHCl}_3}$

84%

J Prakt Chem (1938) 151 61
JCS (1954) 2819

Pr_2O $\xrightarrow{\text{MeCOI}}$ PrI 45%

JACS (1933) 55 378

$PhCH_2OMe$ $\xrightarrow{\text{Cl}_2\text{CHOMe \quad ZnCl}_2}$ $PhCH_2Cl$ 88%

Ber (1959) 92 83

Et_2O $\xrightarrow{\text{SOCl}_2 \quad \text{SnCl}_4}$ EtCl

Ber (1936) 69B 1036

ROBu $\xrightarrow{\text{BBr}_3}$ BuBr

77% (R=Bu)
76% (R=Ph)

JACS (1942) 64 1128
Tetrahedron (1968) 24 2289

$(C_5H_{11})_2O$ $\xrightarrow{\text{Ph}_3\text{P·Br}_2 \quad \text{PhCN}}$ $C_5H_{11}Br$ 78%

JACS (1964) 86 5037

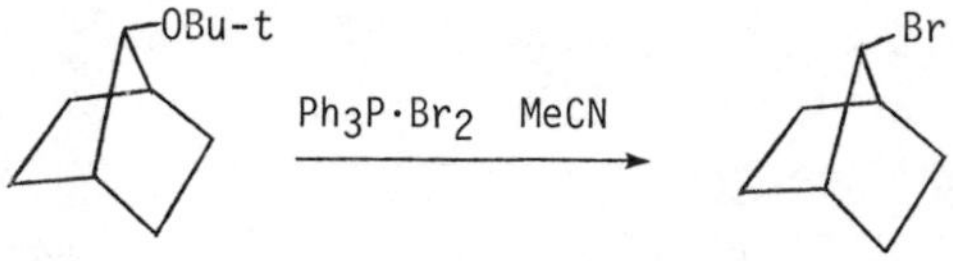

$\xrightarrow{\text{Ph}_3\text{P·Br}_2 \quad \text{MeCN}}$ 83%

Chem Ind (1969) 200

$(n\text{-Bu})_2O$ $\xrightarrow{\text{ICl} \quad \text{LiBH}_4}$ s-BuI

Chem Ind (1965) 223
          (1964) 1582
Tetr Lett (1967) 4131

Et_2O $\xrightarrow{\text{Pyr·BH}_2\text{I} \quad \text{C}_6\text{H}_6}$ EtI

JACS (1968) 90 6260

Section 145   <u>Halides and Sulfonates from Halides and Sulfonates</u>

$Ph(CH_2)_4Cl$ $\xrightarrow[\text{2 } TsO(CH_2)_3Cl]{\text{1 Mg } Et_2O}$ $Ph(CH_2)_7Cl$ 24%

JACS (1928) <u>50</u> 1491
(1946) <u>68</u> 1101

$\underset{\underset{Et}{|}}{Me_2CCl}$ $\xrightarrow{CH_2=CH_2 \quad AlCl_3}$ $\underset{\underset{Et}{|}}{Me_2CCH_2CH_2Cl}$ 25%

JACS (1945) <u>67</u> 1152

$C_{10}H_{21}I$ $\xrightarrow[\text{2 MeI NaI DMF}]{\text{1 } PhCH_2SLi \quad THF}$ $C_{10}H_{21}CH_2I$

Tetr Lett (1968) 5787

$CH_2=CHCH_2Cl$ $\xrightarrow{CH_2N_2 \quad Cu}$ $CH_2=CHCH_2CH_2Cl$ + $\underset{\underset{CH_2}{\diagup\diagdown}}{CH_2-CH_2CH_2Cl}$

Ber (1966) <u>99</u> 2855

$C_{11}H_{23}Cl$ $\xrightarrow{KF \quad HOCH_2CH_2OH}$ $C_{11}H_{23}F$ 48%

JACS (1957) <u>79</u> 2311

$C_6H_{13}Br$ $\xrightarrow{KF \quad HOCH_2CH_2OH}$ $C_6H_{13}F$ 40-45%

Org Synth (1963) Coll Vol 4 525

$C_{12}H_{25}Br$ $\xrightarrow[\text{2 FClO}_3]{\text{1 Li   THF   pet ether}}$ $C_{12}H_{25}F$ 39%

Ber (1969) <u>102</u> 1944

$\xrightarrow{\text{HgF}_2 \quad \text{CHCl}_3}$ 50%

JACS (1948) <u>70</u> 2310
Org React (1944) <u>2</u> 49

$C_6H_{13}OTs$ $\xrightarrow{\text{KF   diethylene glycol}}$ $C_6H_{13}F$ 90%

Chem Ind (1958) 157
Can J Chem (1956) <u>34</u> 757
JACS (1955) <u>77</u> 4899

$\xrightarrow{\text{Bu}_4\text{NF} \quad \text{MeCN}}$ 69%

Carbohydrate Res (1967) <u>5</u> 292

$C_8H_{17}Br$ $\xrightarrow{\text{ClF}_2\text{CCOOAg}}$ $C_8H_{17}Cl$ 89%

Tetr Lett (1970) 3447

$HC{\equiv}CCH_2CH_2OTs$ $\xrightarrow{\text{LiCl   EtOCH}_2\text{CH}_2\text{OH}}$ $HC{\equiv}CCH_2CH_2Cl$ 90%

JCS (1950) 3650

$$BuOTs \xrightarrow{\text{Pyr·HX} \quad 200°} BuX$$

89% (X=Cl)
66% (X=Br)

Monatsh (1952) 83 1398
JOC (1961) 26 2883

$$HOCH_2C\equiv CCH_2Cl \xrightarrow{\text{NaBr} \quad \text{MeOH}} HOCH_2C\equiv CCH_2Br \qquad 62\%$$

JACS (1955) 77 165

$$PrI \xrightarrow[\text{2 Br}_2]{\text{1 Mg} \quad \text{Et}_2O} PrBr \qquad \text{JACS (1919) } 41 \text{ 287} \qquad 30\text{-}40\%$$

$$\underset{\underset{Me}{|}}{C_6H_{13}CHOTs} \xrightarrow{\text{NaBr} \quad \text{Me}_2SO} \underset{\underset{Me}{|}}{C_6H_{13}CHBr} \qquad 52\%$$

JOC (1961) 26 3645
(1962) 27 624
(1970) 35 2803

$$\underset{\underset{Me}{|}}{\overset{\overset{COOH}{|}}{(CH_2)_8CHBr}} \xrightarrow{\text{NaI} \quad \text{Me}_2CO} \underset{\underset{Me}{|}}{\overset{\overset{COOH}{|}}{(CH_2)_8CHI}}$$

JCS (1936) 1605
JACS (1968) 90 6225

$$\xrightarrow{\text{KI} \quad \text{DMF}}$$

82%

JOC (1969) 34 3519

$$\underset{\text{Br}}{\text{(cyclohexenyl)}} \quad \xrightarrow[\substack{2\ HgBr_2 \\ 3\ I_2\ Pyr}]{1\ Mg\ Et_2O} \quad \underset{\text{I}}{\text{(cyclohexenyl)}}$$

JACS (1969) <u>91</u> 5774
(1939) <u>61</u> 1585

13%

$$HC\equiv CCH_2CH_2OTs \quad \xrightarrow{NaI\ Me_2CO} \quad HC\equiv CCH_2CH_2I$$

JCS (1950) 3650

64%

$$C_7H_{15}I \quad \xrightarrow{AgOTs\ MeCN} \quad C_7H_{15}OTs$$

JACS (1959) <u>81</u> 4113
JOC (1962) <u>27</u> 2365

$$PhBr \quad \xrightarrow[2\ FClO_3]{1\ Li\ Et_2O} \quad PhF$$

Ber (1969) <u>102</u> 1944

42%

$$PhI \quad \xrightarrow{h\nu\ CCl_4} \quad PhCl$$

JOC (1970) <u>35</u> 528

76%

$$PhBr \quad \dashrightarrow \quad PhMgBr \quad \xrightarrow{Ph_3PCl_2\ Et_2O} \quad PhCl$$

JOC (1967) <u>32</u> 3710

$$PhI \quad \xrightarrow[2\ Br_2]{1\ Mg\ Et_2O} \quad PhBr$$

JACS (1919) <u>41</u> 287

30-40%

PhBr $\xrightarrow[\text{2 } I_2]{\text{1 Mg Et}_2\text{0}}$ PhI   90%

JACS (1919) 41 287
     (1945) 67 1479
JOC (1938) 3 55

PhBr $\xrightarrow[\text{2 ICN}]{\text{1 Mg Et}_2\text{0}}$ PhI   Annales de Chimie (1915) 4 28   64%

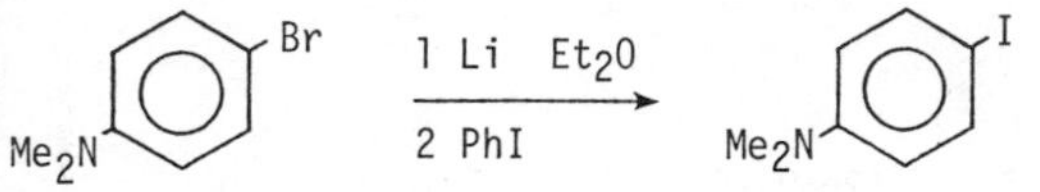

$\xrightarrow[\text{2 PhI}]{\text{1 Li Et}_2\text{0}}$ 51%

JACS (1950) 72 2767

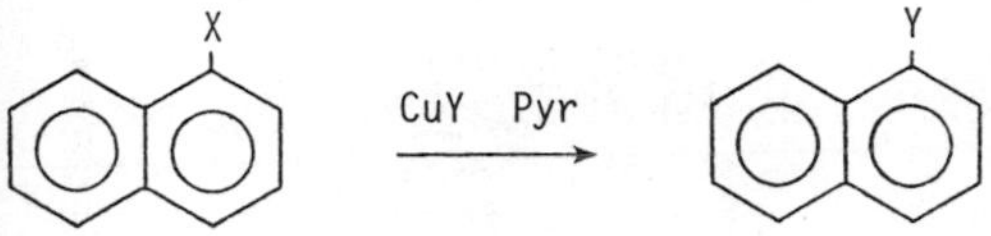

$\xrightarrow{\text{CuY Pyr}}$

X=Br	Y=Cl	84%
I	Cl	100%
I	Br	100%
Br	I	30%

JCS (1964) 1097 1108
Proc Chem Soc (1962) 113

Section 146   Halides from Hydrides (RH)

PhH $\xrightarrow[\text{BF}_3 \text{ or BBr}_3]{\text{FCH}_2\text{CH}_2\text{X}}$ PhCH$_2$CH$_2$X   50-94%

(X=Cl, Br or I)

JOC (1964) 29 2317

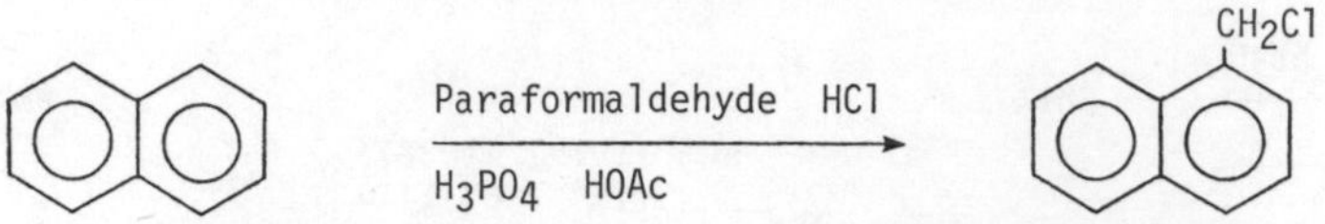

Paraformaldehyde HCl
———————————————————→
H$_3$PO$_4$ HOAc

74-77%

Org Synth (1955) Coll Vol 3 195
Org React (1942) 1 63

XeF$_2$ HF CCl$_4$
PhH ———————————————→ PhF   68%

JACS (1969) 91 1563
JOC (1970) 35 723 4020
For aromatic fluorination with CF$_3$OF see JACS (1970) 92 7494

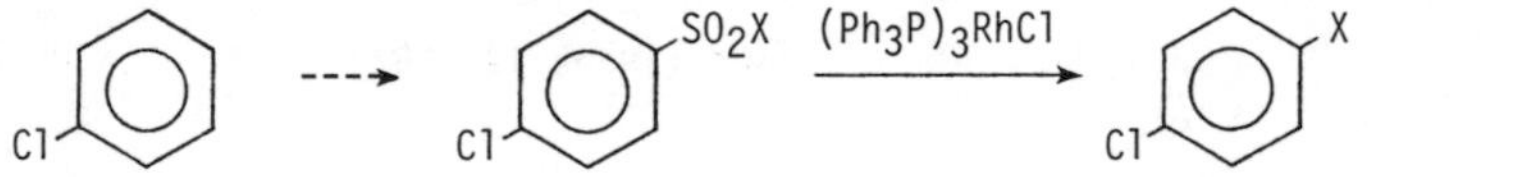

(X=F, Cl or Br)

JOC (1970) 35 1895
Monatsh (1915) 36 719

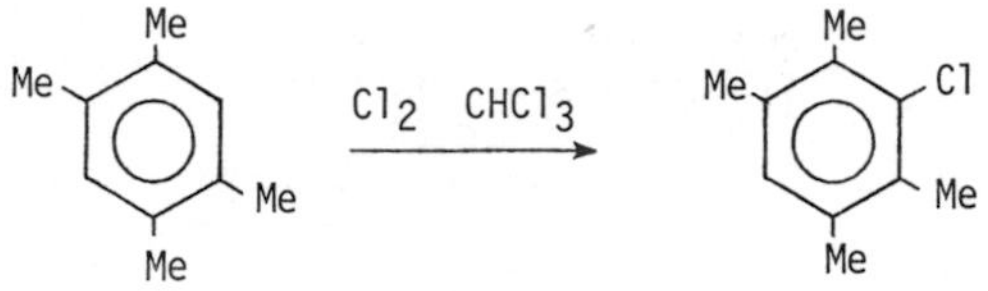

Cl$_2$ CHCl$_3$
——————————→

61%

JACS (1936) 58 1

For aromatic chlorination with TiCl$_4$ and CF$_3$CO$_2$H see Tetr Lett (1970) 2611
 "   "   "   " trichlorocyanuric acid see JOC (1970) 35 719

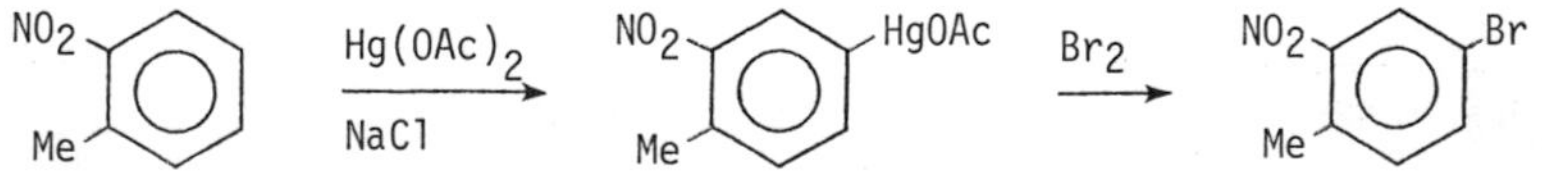

Hg(OAc)$_2$
——————————→
NaCl

Br$_2$
——————→

JCS (1926) 637

Zh Org Khim (1969) 5 387
(Chem Abs 70 105613)

Org Synth (1955) Coll Vol 3 138

For aromatic bromination with Tl(OAc)$_3$ and Br$_2$ see Tetr Lett (1969) 1623
"   "   "   " NBS   " JOC (1965) 30 304
                                    and JACS (1958) 80 4327
"   "   "   " CBr$_4$ see JACS (1932) 54 2025
"   "   "   " CuBr$_2$ " JACS (1964) 86 427
"   "   "   " 1,3-dibromo-5,5-dimethylhydantoin see
                                    An Argentina (1950) 38 181 188
                                    (Chem Abs 45 2873)

22%

Org React (1954) 8 258
JACS (1950) 72 2767

86-87%

Org Synth (1932) Coll Vol 1 323

85-91%

Org Synth (1963) Coll Vol 4 547

For aromatic iodination with I$_2$ and oleum see JCS C (1970) 1480
"   "   "   " I$_2$ " KIO$_3$ " JACS (1943) 65 1273
"   "   "   " I$_2$ " Ag$_2$SO$_4$ " JCS (1952) 150
"   "   "   " I$_2$ " electrolysis see Tetr Lett (1968) 1831
"   "   "   " AlI$_3$ " CuCl$_2$ see JOC (1970) 35 3436
"   "   "   " (CF$_3$COO)$_3$Tl and KI see Tetr Lett (1969) 2427

$$PrCH_2CH=CHMe \xrightarrow[\text{CCl}_4]{\text{NBS \ dibenzoyl peroxide}} PrCHCH=CHMe \qquad 58\text{-}64\%$$
$$\underset{Br}{|}$$

Org Synth (1963) Coll Vol 4 108

$$\xrightarrow[\text{C}_6\text{H}_6]{\text{NBS \ dibenzoyl peroxide}}$$

71-79%

Org Synth (1963) Coll Vol 4 921

$$\xrightarrow{\text{Br}_2 \ h\nu}$$

Org Synth (1963) Coll Vol 4 984

For reviews of allylic and benzylic halogenation see Chem Rev (1948) 43 271
and Angew (1959) 71 349

The following reagents may also be used for allylic/benzylic halogenation:
N-Chloro-N-cyclohexylbenzenesulfonamide Annalen (1967) 703 34

SO_2Cl_2	JACS (1939) 61 2142
	and JCS (1951) 1851
PCl_5	JOC (1969) 34 3655
$PhICl_2$	JOC (1964) 29 3692
1,2-Dibromotetrachloroethane	Chem Ind (1963) 1954
$BrCCl_3$	JACS (1960) 82 391
CBr_4	JACS (1932) 54 2025
$Ph_2C=NBr$	JOC (1967) 32 223
CCl_3SO_2Br	JOC (1965) 30 38
1,3-Dibromo-5,5-dimethylhydantoin	Helv (1958) 41 70
t-BuOI	JACS (1968) 90 808

$$\xrightarrow{\text{CF}_3\text{OF} \ h\nu}$$

JACS (1970) 92 7494

JOC (1964) 29 3692

For chlorination of saturated compounds with SO_2Cl_2 see JCS (1951) 1851
"   "   "   "   "   " CCl_3SO_2Cl " JACS (1960) 82 5246
"   "   "   "   "   "   NCS   " JOC (1953) 18 649

30%

JOC (1953) 18 649
JCS (1952) 2240

For bromination of saturated compounds with Br_2 see
                                        Rec Trav Chim (1964) 83 67
"   "   "   "   "   " CCl_3SO_2Br see
                                        JOC (1965) 30 38
"   "   "   "   "   " $Ph_2C=NBr$ see
                                        JOC (1967) 32 223

$EtCH_2Me$ $\xrightarrow[CCl_4]{\text{t-BuOCl}\ \ HgI_2\ \ h\nu}$ EtCHMe   35%
                                              |
                                              I   JACS (1968) 90 808

Section 147   Halides from Ketones
               oooooooooooooooooooooo

64%

Tetr Lett (1965) 525

Section 148   Halides from Nitriles
               oooooooooooooooooooooo

No examples

Section 149   Halides from Olefins
                oooooooooooooooooooooo

The conversion of olefins into saturated halides is considered in this
section. For allylic halogenation see section 146 (Halides from Hydrides)

$C_6H_{13}CH=CH_2$ $\xrightarrow[130°]{CH_3Cl \quad di-t-butyl \ peroxide}$ $C_6H_{13}(CH_2)_3Cl$ 23%

Chem Comm (1966) 258

$\xrightarrow{HF}$

Org React (1944) 2 49
JOC (1970) 35 4020   60%

$\xrightarrow{HCl \quad HOAc}$ 97%

Helv (1955) 38 1587
JOC (1966) 31 1090
JACS (1957) 79 456

$C_6H_{13}CH=CH_2$ $\xrightarrow[\text{2 N-Chloropiperidine}]{\text{1 } B_2H_6 \quad THF}$ $C_6H_{13}CH_2CH_2Cl$ + $C_6H_{13}\underset{Cl}{CHMe}$

JOC (1965) 30 4313

$BuCH=CH_2$ $\xrightarrow[\substack{\text{2 } Br_2 \\ \text{3 MeONa}}]{\text{1 } B_2H_6 \quad THF}$ $BuCH_2CH_2Br$ 93%

JACS (1970) 92 6660 7212

$C_7H_{15}CH=CH_2$ $\xrightarrow[\substack{\text{2 } Hg(OAc)_2 \\ \text{3 } Br_2 \quad CCl_4}]{\text{1 } B_2H_6 \quad THF}$ $C_7H_{15}CH_2CH_2Br$ 69%

JACS (1970) 92 3221
Chem Comm (1970) 372

$PrCH=CH_2$ $\xrightarrow{\text{HBr\quad ascaridole\quad pentane}}$ $PrCH_2CH_2Br$ 81%

        $\xrightarrow{\text{HBr\quad HOAc}}$ $PrCHMe$ >84%
                                        |
                                        Br

JACS (1934) <u>56</u> 1642
Org React (1963) <u>13</u> 150

$AcO(CH_2)_9CH=CH_2$ $\xrightarrow[C_6H_6]{\text{HBr\quad dibenzoyl peroxide}}$ $AcO(CH_2)_9CH_2CH_2Br$ 82%

JOC (1946) <u>11</u> 281
JACS (1946) <u>68</u> 1101

$MeCOCH_2CH=CH_2$ $\xrightarrow{\text{HBr\quad h}\nu\text{\quad pentane}}$ $MeCOCH_2CH_2CH_2Br$

Tetrahedron (1969) <u>25</u> 5149

$PhC=CH_2$ $\xrightarrow[\text{2\quad}I_2\text{\quad NaOH\quad MeOH}]{\text{1\quad}B_2H_6\text{\quad THF}}$ $PhCHCH_2I$ 63%
 |
Me   Me

JACS (1968) <u>90</u> 5038

$\xrightarrow{\text{KI\quad H}_3PO_4}$ 88-90%

Org Synth (1963) Coll Vol 4 543

$\xrightarrow{\text{HI\quad C}_6H_6}$ 83%

JCS (1957) 463

$$\underset{\underset{Et}{|}}{BuC}=CH_2 \quad \xrightarrow[2\ I_2]{1\ Et_2AlH} \quad \underset{\underset{Et}{|}}{BuCHCH_2I}$$

Annalen (1954) <u>589</u> 91

Section 150   <u>Halides from Miscellaneous Compounds</u>

$$PhCH=CHBr \quad \xrightarrow[]{\underset{\underset{NCOOK}{\parallel}}{NCOOK}\ dioxane\ HOAc}} \quad PhCH_2CH_2Br$$

PhCH=CHBr $\xrightarrow{\underset{NCOOK}{\overset{NCOOK}{\parallel}}\ dioxane\ HOAc}$ $PhCH_2CH_2Br$ 22%

JOC (1965) <u>30</u> 3985

$PhNO_2$ $\xrightarrow{SOCl_2\ \ 190\text{-}200°}$ PhCl   100%

Monatsh (1915) <u>36</u> 723

$\xrightarrow{HCl\ \ h\nu\ \ CHCl_3}$

Tetr Lett (1969) 4603   ~90%

PhNO $\xrightarrow[HCl\ \ H_2O]{NH_2OH\ \ CuCl_2}$ PhCl   15-20%

JCS (1949) S181

$PhCH_2CH_2SO_2Cl$ $\xrightarrow{200\text{-}220°}$ $PhCH_2CH_2Cl$ 59%

Zh Obshch Khim (1953) <u>23</u> 204
(Chem Abs <u>48</u> 2568)

$\xrightarrow{SOCl_2\ \ 180°}$

Monatsh (1915) <u>36</u> 719
JOC (1970) <u>35</u> 1895

Chapter 11   PREPARATION

OF

HYDRIDES

This chapter lists hydrogenolysis and related reactions by which functional groups are replaced by hydrogen, e.g. $RCH_2X \longrightarrow RCH_2-H$ or $R-H$

Section 151   Hydrides from Acetylenes

No examples of the reaction $RC{\equiv}CR \longrightarrow RH$ occur in the literature. For the hydrogenation of acetylenes see section 61 (Alkyls and Methylenes from Acetylenes)

Section 152   Hydrides from Carboxylic Acids

This section lists examples of the decarboxylation of acids, $RCOOH \longrightarrow RH$ (R=alkyl, aryl, vinyl etc). For the conversion $RCOOH \longrightarrow RCH_3$ see section 62 (Alkyls from Carboxylic Acids)

$$C_{17}H_{35}COOH \xrightarrow{\text{Ni} \quad 350°} C_{17}H_{36}$$

JOC (1944) $\underline{9}$ 319

MeONa
Δ

64%

JACS (1934) $\underline{56}$ 715

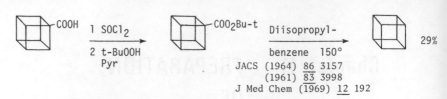

$$\text{[cubane]}-COOH \xrightarrow[\substack{2 \ t\text{-BuOOH} \\ Pyr}]{1 \ SOCl_2} \text{[cubane]}-CO_2Bu\text{-}t \xrightarrow[\substack{benzene \ 150°}]{Diisopropyl-} \text{[cubane]} \quad 29\%$$

JACS (1964) 86 3157
     (1961) 83 3998
J Med Chem (1969) 12 192

$$\text{[cyclohexylidene]}=CHCOOH \underset{}{\overset{200°}{\rightleftharpoons}} \text{[cyclohexenyl]}-CH_2COOH \xrightarrow{240°} \text{[cyclohexylidene]}=CH_2 \quad {\sim}60\%$$

JCS (1930) 1603
JACS (1950) 72 4359

CH=CHCOOH
[phenyl ring, OAc] $\xrightarrow{\text{Cu   quinoline}}$ CH=CH$_2$ [phenyl ring, OAc]   37%

JACS (1950) 72 1200
Org Synth (1963) Coll Vol 4 857
Chem Comm (1967) 96

$$Ph(C{\equiv}C)_3COOH \xrightarrow[Me_2CO]{\text{Copper tetrammine sulfate}} Ph(C{\equiv}C)_3H \quad 79\%$$

Ber (1964) 97 2586

[furan with Me and COOH] $\xrightarrow{\text{Cu   quinoline}}$ [furan with Me]   83-89%

Org Synth (1963) Coll Vol 4 628

[MeO, MeO, MeO substituted polycyclic aromatic with COOH and OMe] $\xrightarrow{\text{CuSO}_4 \quad \text{quinoline}}$ [MeO, MeO, MeO substituted polycyclic aromatic with OMe]

JACS (1951) 73 1414

Section 153   Hydrides from Alcohols and Phenols

This section lists examples of the hydrogenolysis of alcohols and phenols, ROH $\longrightarrow$ RH and RR'CHOH $\longrightarrow$ RH (R=alkyl or aryl, R'=H or aryl). For the conversion ROH $\longrightarrow$ RR' (R'=alkyl or aryl) see section 63 (Alkyls and Aryls from Alcohols)

$$\begin{array}{c} \text{Me} \\ | \\ \text{i-PrCCH}_2\text{OH} \\ | \\ \text{Me} \end{array} \quad \xrightarrow[300°]{\text{H}_2 \text{ (965 atmos)} \quad \text{Co-Al}_2\text{O}_3} \quad \begin{array}{c} \text{Me} \\ | \\ \text{i-PrCMe} \\ | \\ \text{Me} \end{array} \quad \quad 20\%$$

JACS (1948) 70 3793
     (1951) 73 553
     (1933) 55 1293

$$\text{Me}_3\text{COH} \quad \xrightarrow{\text{H}_2 \quad \text{Pt} \quad \text{CF}_3\text{COOH}} \quad \text{Me}_3\text{CH}$$

JOC (1964) 29 2325

$$\begin{array}{c} \text{C}_5\text{H}_{11}\text{CHOH} \\ | \\ \text{HOCH}_2\text{CH(CHCH}_2)_5\text{(CHOH)}_3\text{CHMe} \\ | \qquad\qquad | \\ \text{OH} \qquad\quad \text{OH} \\ \qquad\quad \text{HOCH}_2\text{CHCH(CH}_2)_9 \\ \qquad\qquad | \\ \qquad\qquad \text{Me} \end{array} \quad \xrightarrow[\text{2 LiAlH}_4]{\text{1 P}_4 \quad \text{HI} \quad \text{H}_2\text{O}} \quad \begin{array}{c} \text{C}_5\text{H}_{11}\text{CH}_2 \\ | \\ \text{MeCH(CH}_2)_{13}\text{CHMe} \\ \text{MeCHCH}_2\text{(CH}_2)_9 \\ | \\ \text{Me} \end{array} \quad 13\%$$

JACS (1962) 84 2170
JCS (1959) 1044
Tetrahedron (1970) 26 2199

JCS C (1967) 136
Chem Ind (1963) 1354
Org React (1953) 7 263

$$\begin{array}{c} \text{PhCHMe} \\ | \\ \text{OH} \end{array} \quad \xrightarrow{\text{Na} \quad \text{NH}_3} \quad \text{PhCH}_2\text{Me}$$

JCS (1945) 809
    (1949) 2531

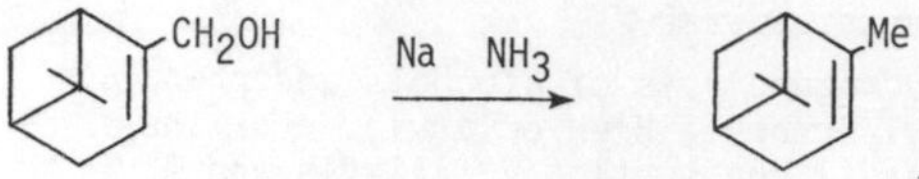
CH2OH
Na NH3
Me

85%

Ber (1956) 89 1549
JCS (1957) 1969
(1945) 809

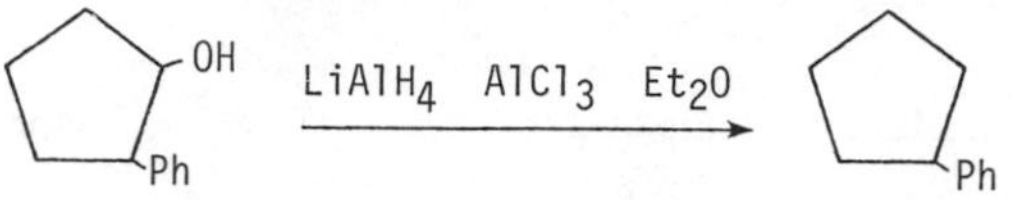
OH
Ph
LiAlH4 AlCl3 Et2O
Ph

78%

JOC (1964) 29 121

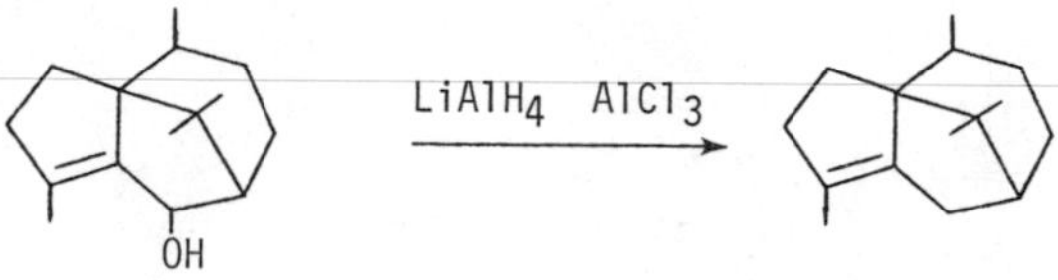
OH
LiAlH4 AlCl3

Tetr Lett (1967) 2447

For reduction of benzylic alcohols see JCS (1957) 3755

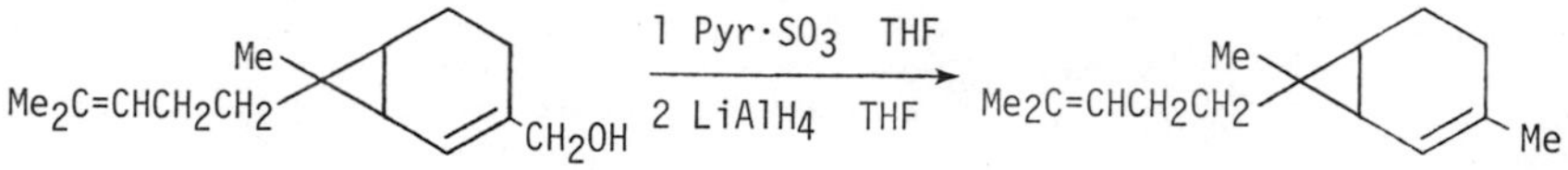
Me
Me2C=CHCH2CH2
CH2OH
1 Pyr·SO3 THF
2 LiAlH4 THF
Me
Me2C=CHCH2CH2
Me

Tetr Lett (1969) 1837
JOC (1969) 34 3667

86%

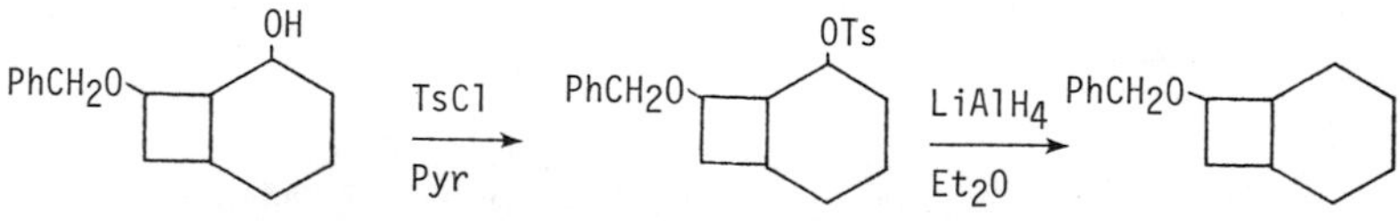
OH
PhCH2O
TsCl
Pyr
OTs
PhCH2O
LiAlH4
Et2O
PhCH2O

79%

JACS (1970) 92 553
(1955) 77 1820

Further examples of the reduction of sulfonates are included in section 160 (Hydrides from Halides and Sulfonates)

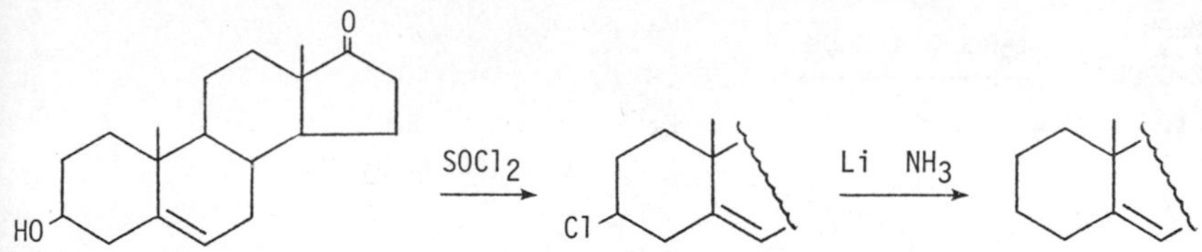

JOC (1968) 33 1196

For further examples of the reaction ROH ⟶ RHal and RHal ⟶ RH see
section 138 (Halides and Sulfonates from Alcohols and Phenols) and
section 160 (Hydrides from Halides and Sulfonates)

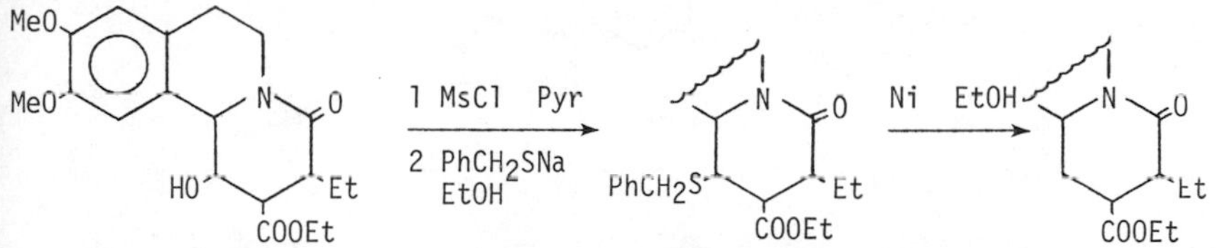

Helv (1959) 42 2431
JACS (1953) 75 384
      (1963) 85 173

77%

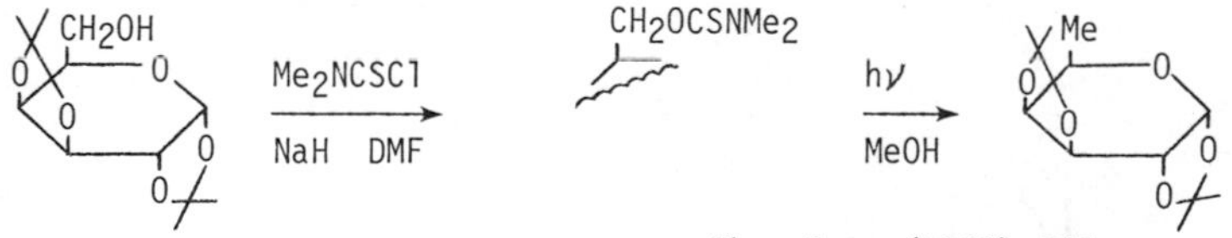

Chem Comm (1968) 323
Carbohydrate Res (1967) 4 115

~25%

$C_{17}H_{35}CH_2OH$ $\xrightarrow[250°]{H_2 \text{ (100-200 atmos)   Ni}}$ $C_{17}H_{36}$

JACS (1933) 55 1293

PhCHCH$_2$OH
  |
  Me
$\xrightarrow{Ni \quad EtOH}$
PhCH$_2$
  |
  Me

JACS (1957) 79 1696

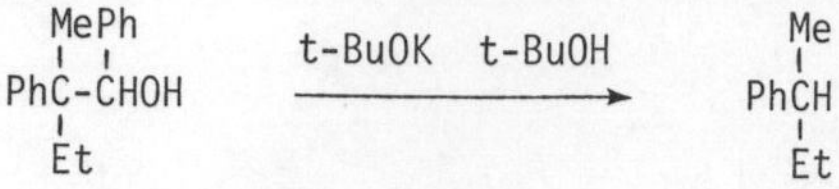
MePh
PhC-CHOH
Et
t-BuOK t-BuOH
Me
PhCH
Et

JACS (1969) 91 1009

Alcohols (ROH) may also be converted into hydrides (RH) via ester or ether intermediates. See section 158 (Hydrides from Esters) and section 159 (Hydrides from Ethers)

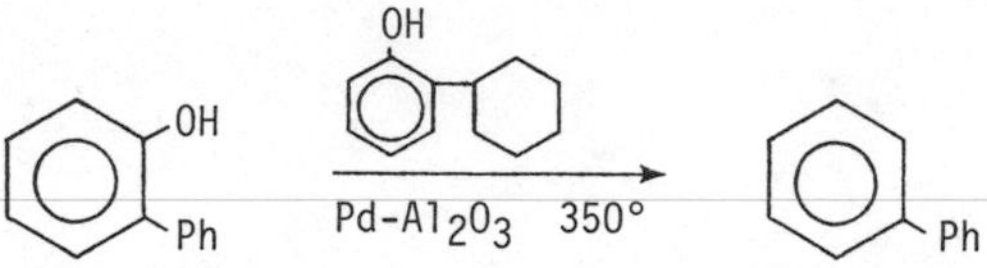
OH
OH
Ph
Pd-Al2O3 350°
Ph

93%

Tetr Lett (1969) 1577

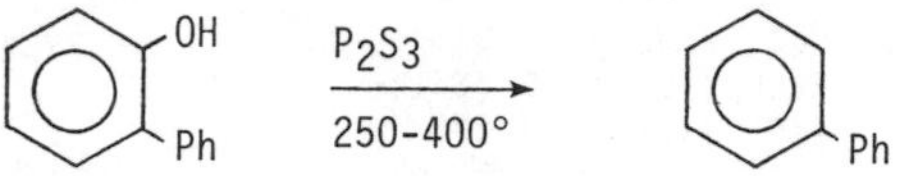
OH
Ph
P2S3
250-400°
Ph

18%

JOC (1961) 26 2528

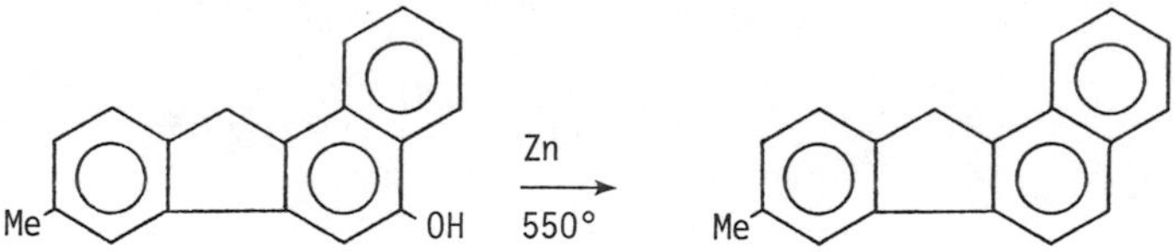
Me
OH
Zn
550°
Me

24%

JACS (1951) 73 3439

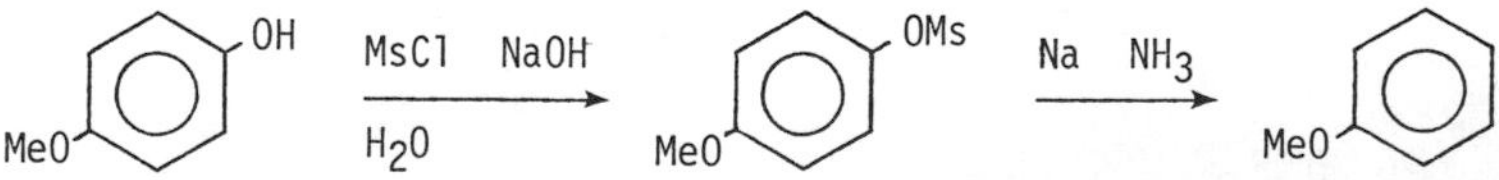
OH
MeO
MsCl NaOH
H2O
OMs
MeO
Na NH3
MeO

JCS (1955) 522

Further examples of the reduction of sulfonates are included in section 160 (Hydrides from Halides and Sulfonates)

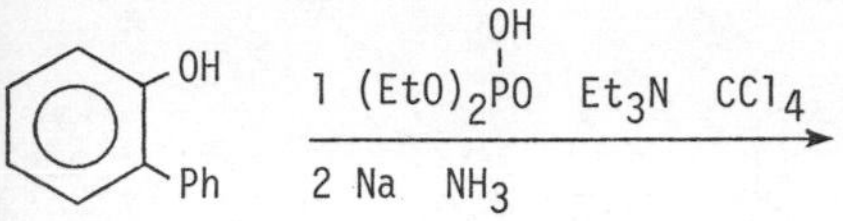
OH
Ph
OH
1 $(EtO)_2PO$ Et_3N CCl_4
2 Na NH_3
Ph

48%

JOC (1958) 23 131
J Med Chem (1965) 8 409
JCS (1955) 522

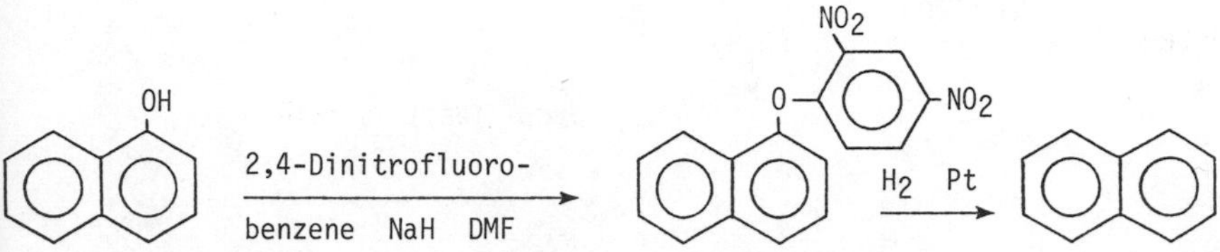
OH
2,4-Dinitrofluoro-
benzene NaH DMF
NO_2
O
NO_2
H_2 Pt

JOC (1964) 29 3124

40%

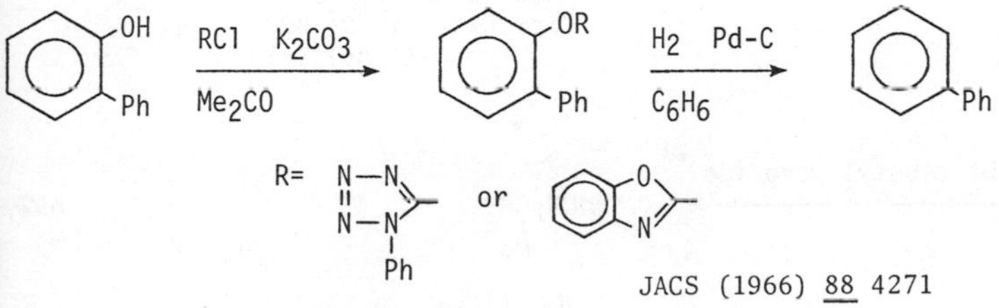
OH
Ph
RCl K_2CO_3
Me_2CO
OR
Ph
H_2 Pd-C
C_6H_6
Ph
R=
N—N
N—N
Ph
or
O
N

70-76%

JACS (1966) 88 4271

Section 154 Hydrides from Aldehydes

This section lists examples of the decarbonylation of aldehydes, RCHO ⟶ RH. For the conversion RCHO ⟶ RMe see section 64 (Alkyls from Aldehydes)

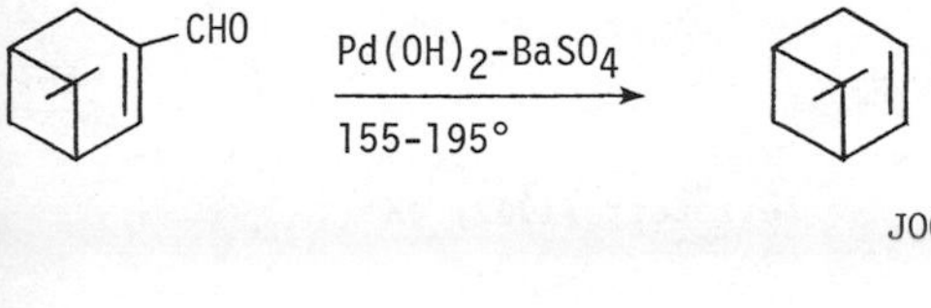
CHO
$Pd(OH)_2-BaSO_4$
155-195°

76%

JOC (1959) 24 1369

$$\xrightarrow[210°]{\text{Pd-C}}$$

80%

JOC (1960) 25 2215

$$\underset{\text{Me}}{\text{PhCHCHO}} \xrightarrow{\text{Ni   EtOH}} \underset{\text{Me}}{\text{PhCH}_2}$$

JACS (1957) 79 1696

$$\text{RCHO} \xrightarrow{\text{RhCl(Ph}_3\text{P)}_3} \text{RH} \qquad \text{R=C}_6\text{H}_{13}, \text{ PhCH=}\underset{\text{Et}}{\text{C}} \text{ or } \text{Cl-}$$

JACS (1968) 90 99
Tetr Lett (1970) 823
        (1968) 1899

$$\underset{\text{Me}}{\text{PhCH}_2\text{CHCHO}} \xrightarrow{\text{Di-t-butyl peroxide}} \underset{\text{Me}}{\text{PhCH}_2\text{CH}_2}$$

69%

JOC (1964) 29 1663
JACS (1941) 63 226

$$\xrightarrow{\text{h}\nu \text{ EtOH}}$$

90%

Proc Chem Soc (1963) 114

$$\underset{\text{Et}}{\text{BuCHCHO}} \xrightarrow[140-145°]{\text{h}\nu \text{ PhCH}_2\text{SSCH}_2\text{Ph}} \underset{\text{Et}}{\text{BuCH}_2}$$

68%

Tetr Lett (1962) 43

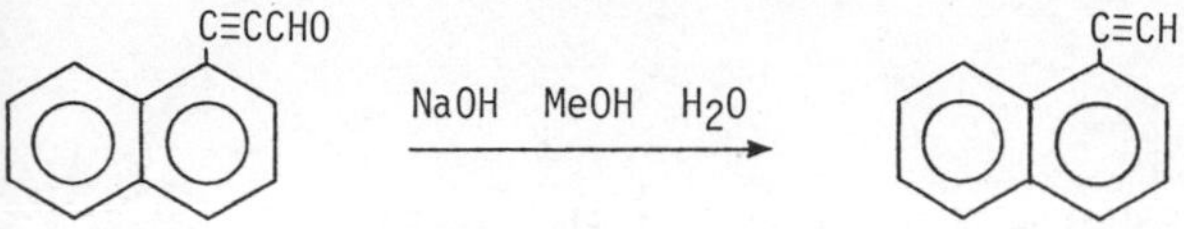

JCS C (1969) 2173

Section 155　Hydrides from Alkyls

This section lists examples of the conversion RR' ⟶ RH (R,R'=alkyl, aryl etc.)

$Ph_2CHCHPh_2$　$\xrightarrow[\text{methylcyclohexane}]{H_2 \quad Cu\text{-}Cr \text{ oxide}}$　Ph_2CH_2　　　　　　　90%

JACS (1932) 54 1668

$t\text{-}BuCH_2Pr\text{-}i$　$\xrightarrow{AlCl_3 \quad C_6H_6}$　$t\text{-}BuH$

JACS (1935) 57 2415

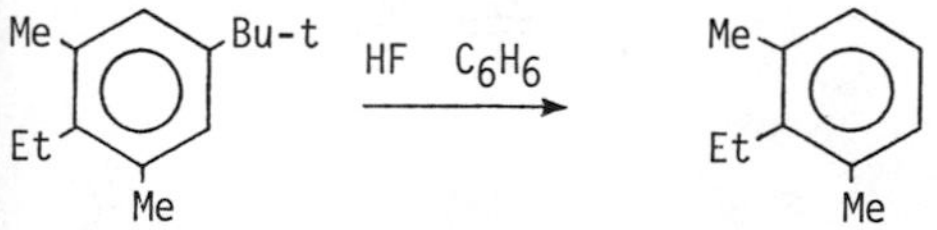

60%

JACS (1954) 76 4952
(1937) 59 1417

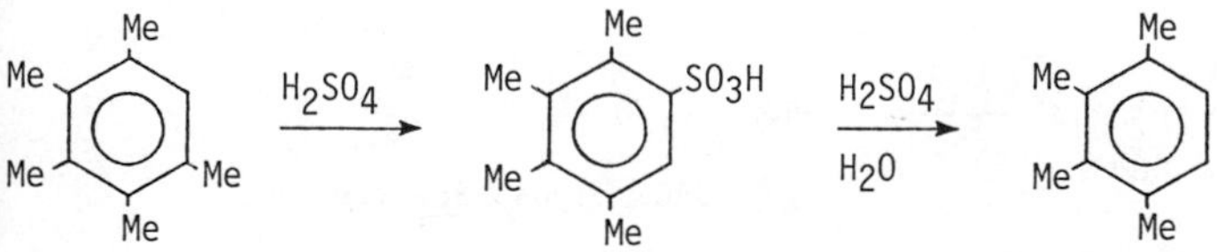

88%

Org React (1942) 1 370

JCS $\underline{C}$ (1968) 2915

The conversion Ar-Alkyl $\longrightarrow$ ArH may also be achieved via intermediate carboxylic acids ArCOOH. See section 20 (Carboxylic Acids from Alkyls and Aryls) and section 152 (Hydrides from Carboxylic Acids)

Section 156   Hydrides from Amides

This section lists examples of the conversions $RCONH_2 \longrightarrow RH$ and $RNHCOR' \longrightarrow RH$. For the conversions $RCONR'_2 \longrightarrow RR''$ and $RNHCOR' \longrightarrow RR''$ (R''=alkyl or aryl) see section 66 (Alkyls and Aryls from Amides)

$\dfrac{H_3PO_4}{190\text{-}210°}$ 95%

JACS (1951) $\underline{73}$ 436

$Ph_2CHCONH_2$ $\dfrac{BuLi \quad THF \quad hexane}{}$ Ph_2CH_2 86%

JACS (1969) $\underline{91}$ 7774

$PhC{\equiv}CCONH_2$ $\dfrac{NaNH_2 \quad NH_3 \quad Et_2O}{}$ $PhC{\equiv}CH$ 90%

JCS (1963) 4402

PhNHAc $\dashrightarrow$ PhNAc $\dfrac{Et_2O}{25°}$ PhH
                          |
                         NO

JACS (1964) $\underline{86}$ 3180

Amides may also be converted into hydrides via intermediate amines or carboxylic acids. See section 157 (Hydrides from Amines) and section 152 (Hydrides from Carboxylic Acids)

Section 157　　Hydrides from Amines

This section lists examples of the conversions $RNH_2 \longrightarrow RH$, $RCH_2NH_2 \longrightarrow RH$
and $R'NR_3^+ X^- \longrightarrow R'H$. For the conversion $RNH_2 \longrightarrow RR'$ (R'=alkyl or aryl)
see section 67 (Alkyls and Aryls from Amines)

Review: Replacement of the Aromatic Primary Amino Group by Hydrogen
　　　　　　　　　　　　　　　　　Org React (1944) 2 262

JOC (1963) 28 568

~75%

80%

JACS (1952) 74 3074
Org Synth (1963) Coll Vol 4 947
(1955) Coll Vol 3 295

For reduction of diazonium salt with $NaBH_4$ see JACS (1961) 83 1251
　　"　　　　"　　　"　　　"　　　"　　"　　HCHO derivatives see
　　　　　　　　　　　　　　　　　　　　　　Angew (1958) 70 211
　　"　　　　"　　　"　　　"　　　"　　"　　H_3PO_2, HCHO, EtOH, Zn or N_2H_4 see
　　　　　　　　　　　　　　　　　　　　　　Org React (1944) 2 262
For selective diazotization of $ArNH_2$ in the presence of $AliphNH_2$ see
　　　　　　　　　　　　　　　　　　JACS (1949) 71 2137

Tetr Lett (1969) 1577

Aromatic amines (RNH_2) may also be converted into hydrides (RH) via
N-acetyl derivatives. See section 156 (Hydrides from Amides)

JACS (1964) 86 1152

$$BuNH_2 \xrightarrow{\ HNF_2\ } BuH$$

JACS (1963) 85 97
(1964) 86 2233

61%

$$\underset{\underset{Et}{|}}{\overset{\overset{Me}{|}}{PhCNHNH_2}} \xrightarrow{\ KIO_4\ \ KOH\ \ H_2O\ } \underset{\underset{Et}{|}}{\overset{\overset{Me}{|}}{PhCH}}$$

JACS (1963) 85 1108

68%

$$PhCH_2CH_2NH_2 \xrightarrow{\ Ni\ \ EtOH\ } PhMe \ + \ PhCH_2Me$$

JACS (1957) 79 1696

JACS (1951) 73 4122
Org React (1953) 7 263

99%

Section 158   Hydrides from Esters

This section lists examples of the hydrogenolysis of esters, R'COOR ⟶ RH.
For the conversion R'COOR ⟶ R'R" or RR" (R"=alkyl) see section 68
(Alkyls and Aryls from Esters). The reduction of esters of sulfonic acids
(e.g. ROTs ⟶ RH) is included in section 160 (Hydrides from Halides and
Sulfonates)

$$\underset{\underset{COOH}{|}}{\overset{\overset{Me}{|}}{PhCOAc}} \xrightarrow{\ H_2\ \ Pd\text{-}C\ \ EtOAc\ } \underset{\underset{COOH}{|}}{\overset{\overset{Me}{|}}{PhCH}}$$

JCS C (1967) 136
Org React (1953) 7 263

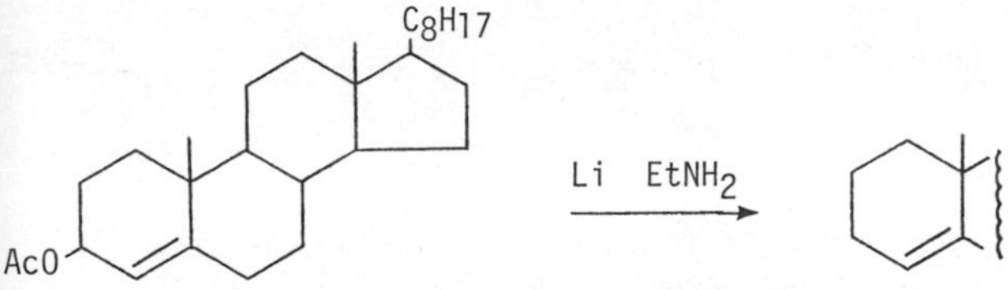
C_8H_{17}
AcO
Li EtNH2

94%

JCS (1957) 1969
JOC (1968) 33 435

Section 159 Hydrides from Ethers

This section lists examples of the hydrogenolysis of ethers, ROR' ⟶ RH. For the conversion ROR' ⟶ RR" (R"=alkyl or aryl) see section 69 (Alkyls and Aryls from Ethers)

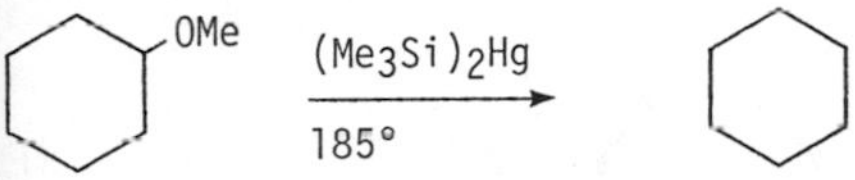
OMe
(Me3Si)2Hg
185°

58-76%

JCS C (1967) 2188

$PhCH_2OEt$ $\xrightarrow{(i\text{-}Bu)_2AlH}$ PhMe 92%

Izv (1959) 2255
(Chem Abs 54 10837)

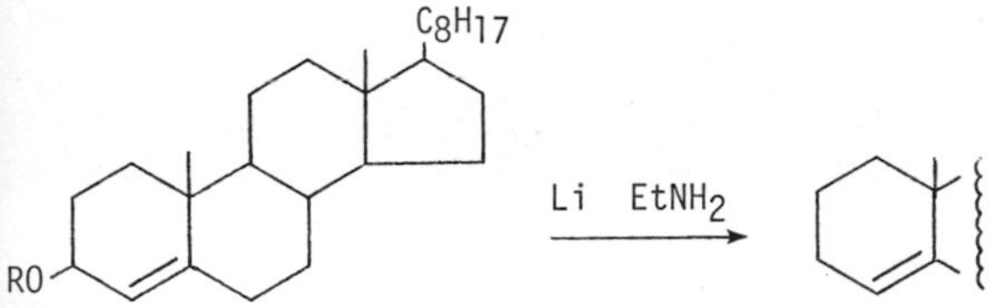
C_8H_{17}
RO
Li EtNH2

R=Me or $PhCH_2$

JCS (1957) 1969

$PhCH{=}CHCH_2OPh$ $\xrightarrow{LiAlH_4\ \ NiCl_2\ \ THF}$ $PhCH{=}CHMe$

JACS (1957) 79 5463

PhCH$_2$OBu $\xrightarrow{\text{H}_2 \quad \text{Ni}}$ PhMe   Org React (1953) 7 263   ~93%
                                                    Chem Ind (1963) 1354

Further examples of the hydrogenolysis of allyl and benzyl ethers are
included in section 39 (Alcohols and Phenols from Ethers and Epoxides)
and section 45A (Protection of Alcohols and Phenols)

Section 160   Hydrides from Halides and Sulfonates
oooooooooooooooooooooooooooooooooooooooooo

$\xrightarrow[\text{MeOH}]{\text{H}_2 \quad \text{Pd} \quad \text{Et}_3\text{N}}$

JOC (1969) 34 3519   81%

$\xrightarrow[\text{MeOH}]{\text{H}_2 \quad \text{Ni} \quad \text{NaOH}}$

JACS (1950) 72 561   37%
Helv (1946) 29 378 360

PrF $\xrightarrow[190°]{\text{H}_2 \quad \text{Pd-C}}$ PrH   J Phys Chem (1956) 60 1454   100%

$\xrightarrow[\text{HOAc}]{\text{H}_2 \quad \text{Pd-C} \quad \text{NaOAc}}$

JCS C (1969) 2600

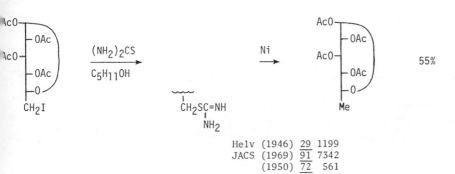

N_2H_4 Pd-C   EtOH   88%

Chem Ind (1959) 1348

H_2 Ni   KOH
—————————→   90%
MeOH

Ber (1958) 91 1376

H_2 Ni   EtOH
—————————→   90%

JCS (1949) S178
Aust J Chem (1963) 16 647

$(NH_2)_2CS$ Ni
——————→   ——→   55%
$C_5H_{11}OH$

$CH_2SC=NH$
        NH_2

Helv (1946) 29 1199
JACS (1969) 91 7342
     (1950) 72 561

EtSK   t-BuOH   Ni   EtOH
——————————→   ——————→

JACS (1953) 75 384

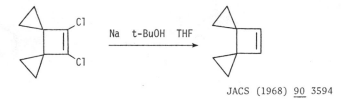

JOC (1944) 9 1

COOH
|
(CH₂)₉CH₂Br

$$\xrightarrow[\text{or Zn HOAc}]{\text{Ni-Al NaOH MeOH}}$$
or Zn-Cu EtOH

COOH
|
(CH₂)₉Me   100%

Bull Soc Chim Fr (1967) 2018

C₁₆H₃₃I $\xrightarrow{\text{Zn HOAc}}$ C₁₆H₃₄   85%

Org Synth (1943) Coll Vol 2 320

$\xrightarrow{\text{Li t-BuOH THF}}$ 94%

(Aliphatic halides are also reduced)

Tetr Lett (1962)   449
              (1968) 1575
JOC (1964) 29 160

$\xrightarrow{\text{Na t-BuOH THF}}$ 80%

JACS (1968) 90 3594

$$\text{i-Pr(CH}_2)_3\overset{\text{Et}}{\underset{\text{Me}}{\text{CCl}}} \xrightarrow{\text{Li NH}_3} \text{i-Pr(CH}_2)_3\overset{\text{Et}}{\underset{\text{Me}}{\text{CH}}}$$

Rec Trav Chim (1964) 83 367
JACS (1951) 73 3329
Chem Comm (1969) 138

$$\text{33\% (X=F)}$$
$$\text{83\% (X=Cl)}$$

Proc Chem Soc (1963) 219

$$\text{PhX} \xrightarrow{\text{Mg i-PrOH decalin}} \text{PhH}$$

89% (X=Cl or Br)
95% (X=I)

Proc Chem Soc (1963) 219

$$\text{PrCHBr} \atop \text{Me} \quad \xrightarrow[\text{2 } H_2SO_4 \ H_2O]{\text{1 Mg Bu}_2O} \quad \text{PrCH}_2 \atop \text{Me}$$

50-53%

Org Synth (1943) Coll Vol 2 478

$$C_8H_{17}Br \xrightarrow{\text{BuLi hexane}} C_8H_{18} \ + \ C_8H_{17}Bu$$

JOC (1963) 28 280

$$C_8H_{17}Cl \xrightarrow{\text{NaBH}_4 \ \text{Me}_2SO} C_8H_{18}$$

42%

JOC (1969) 34 3923
Tetr Lett (1969) 3495

$$\text{PrOTs} \xrightarrow{\text{NaBH}_4 \ \text{Me}_2SO} \text{PrH}$$

90%

JOC (1969) 34 3923

$$\left. \begin{array}{l} C_8H_{17}Br \\ C_6H_{13}CHCH_2Br \\ \phantom{C_6H_{13}CH}\!Br \end{array} \right\} \xrightarrow{\text{LiAlH}_4} C_8H_{18}$$

~70%

JACS (1948) 70 3664 3738
Org React (1951) 6 469

$$\text{PhBr} \xrightarrow{\text{LiAlH}_4 \ \text{THF}} \text{PhH}$$

95%

JOC (1969) 34 3918

$$\xrightarrow[\text{2 } H_2 \quad Pt \quad HOAc]{\text{1 } LiAlH_4 \quad Et_2O}$$

Tetrahedron (1964) $\underline{20}$ 2903
Helv (1949) $\underline{32}$ 1371
JACS (1970) $\underline{92}$ 553

CN
|
$(CH_2)_3CH_2Br$ $\xrightarrow{Bu_3SnH}$ CN
|
$(CH_2)_3Me$

JOC (1969) $\underline{34}$ 2014
Synthesis (1970) 499

60%

$\xrightarrow[150°]{Ph_3SnH}$

JOC (1960) $\underline{25}$ 2203
    (1969) $\underline{34}$ 2014
    (1959) $\underline{24}$ 294

75%

BuBr $\xrightarrow[Me_2SO \quad H_2O]{Cr^{2+} \quad NH_2CH_2CH_2NH_2}$ BuH

JACS (1966) $\underline{88}$ 4094
Angew (1968) $\underline{80}$ 271
(Internat Ed $\underline{7}$ 247)

100%

$\xrightarrow[DMF \quad H_2O]{Cr(ClO_4)_2 \quad NH_2CH_2CH_2NH_2}$

(Vinyl halides are also reduced)
Tetrahedron (1968) $\underline{24}$ 3503

100%

$\xrightarrow[collidine]{PhCH_2ONa \quad Cu_2O}$

JCS $\underline{C}$ (1969) 308

95%

Ac$_2$O   Cu$_2$O   Pyr   64%

JCS (1964) 1112
Proc Chem Soc (1962) 113

PhCl $\xrightarrow{\text{h}\nu \quad \text{i-PrOH}}$ PhH   72%

Tetr Lett (1969) 1267

Electrolysis   Et$_4$NBr   94%

DMF

(Aromatic halides may also be selectively reduced in the
                              presence of aliphatic halides)
JOC (1970) 35 1232
Ber (1968) 101 4179

Section 161   Hydrides from Hydrides

This section lists examples of hydrocarbon epimerization. Many related
reactions are included in section 65 (Alkyls and Aryls from Alkyls and
Aryls)

hν HgBr$_2$
$\xrightarrow{\hspace{2cm}}$ 70%
cyclohexane

JACS (1970) 92 1094

Section 162   Hydrides from Ketones

This section lists examples of the conversion R$_2$CO $\longrightarrow$ RH. For the
conversion R$_2$CO $\longrightarrow$ R$_2$CH$_2$ or R$_2$CHR' see section 72 (Alkyls, Methylenes
and Aryls from Ketones)

(t-Bu)$_2$CO $\xrightarrow{\text{NaNH}_2 \quad \text{C}_6\text{H}_6}$ t-BuH   +   t-BuCONH$_2$

Compt Rend (1910) 150 661
Org React (1957) 9 1

Bull Acad Sci URSS (1941) 167
(Chem Abs 37 3749)

JACS (1946) 68 2176

$$PhC \equiv CCOPh \xrightarrow{\text{NaNH}_2 \quad \text{NH}_3 \quad \text{THF}} PhC \equiv CH \qquad\qquad 25\%$$

Bull Chem Soc Jap (1962) 35 1488

Section 163   Hydrides from Nitriles

This section lists examples of the conversion RCN $\longrightarrow$ RH. For the
conversion RCN $\longrightarrow$ RMe see section 73 (Alkyls from Nitriles)

$$PhO(CH_2)_3\underset{\underset{CN}{|}}{CHPh} \xrightarrow[\text{toluene}]{\text{Na} \quad \text{EtOH}} PhO(CH_2)_3CH_2Ph \qquad\qquad 89\%$$

JACS (1934) 56 1614
Gazz (1963) 93 525
(Chem Abs 59 8825)

$$RCN \xrightarrow{\text{Na} \quad \text{NH}_3} RH$$

35% ($R=C_{12}H_{25}$)
90% ($R=PhCH_2$)

JACS (1969) 91 2059

$$\underset{\underset{CH_2NMe_2}{|}}{\underset{Me_2NCHCH_2}{|}}{Ph_2CCN} \xrightarrow[\text{Et}_2\text{O} \quad \text{xylene}]{\text{EtMgBr}} \underset{\underset{CH_2NMe_2}{|}}{\underset{Me_2NCHCH_2}{|}}{Ph_2CH}$$

JACS (1952) 74 5793
Chem Comm (1970) 350

Tetr Lett (1968) 1975

JACS (1953) 75 3600

Section 164   Hydrides from Olefins

This section lists examples of the conversion $R_2C=CR_2 \longrightarrow R_2CH_2$. For the hydrogenation, dimerization and alkylation of olefins see section 74 (Alkyls, Methylenes and Aryls from Olefins)

$$Ph_2C=CH_2 \xrightarrow{\quad NH_2NHNa \quad Et_2O \quad} Ph_2CH_2 \qquad\qquad 97\%$$

Angew (1962) 74 650
(Internat Ed 1 456)

Section 165   Hydrides from Miscellaneous Compounds

This section lists examples of the replacement of miscellaneous functional groups by hydrogen (RX $\longrightarrow$ RH)

$$C_6H_{13}SH \xrightarrow[300°]{H_2 \text{ (100 atmos)} \quad MoS_2} C_6H_{14} \qquad\qquad 90\%$$

Coll Czech (1966) 31 2202

$$C_8H_{17}SH \xrightarrow{\quad (EtO)_3P \quad} C_8H_{18} \qquad\qquad 88\%$$

JACS (1956) 78 6414

$$Ph-X-Ph \xrightarrow{\text{Ni EtOH}} PhH$$

68% (X=S)
75% (X=SO)
65% (X=SO$_2$)

JACS (1943) 65 1013
For use of Ni$_2$B see JOC (1965) 30 1316

$$C_{10}H_{21}-X-C_{10}H_{21} \xrightarrow{\text{Li MeNH}_2} C_{10}H_{22} \qquad (X=S \text{ or } SO_2)$$

JACS (1960) 82 2872

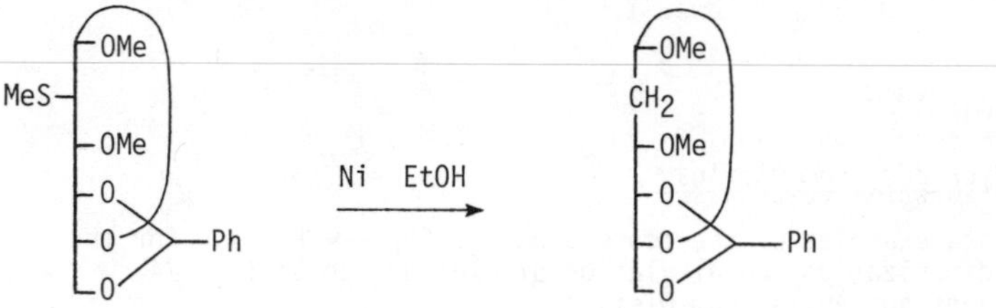

$$\xrightarrow{\text{Ni EtOH}}$$

95%

Helv (1946) 29 371
JACS (1953) 75 384
(1963) 85 173

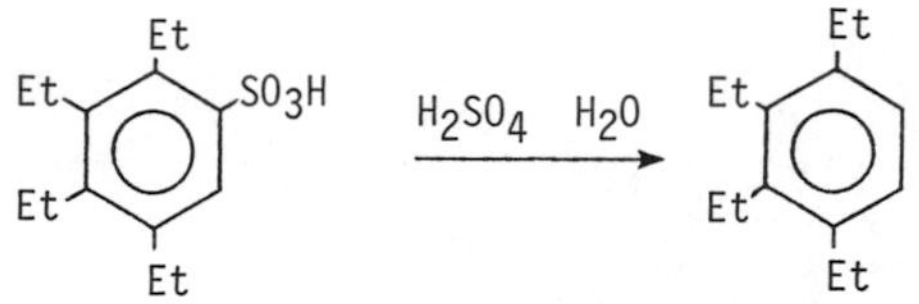

$$\xrightarrow{\text{H}_2\text{SO}_4 \quad \text{H}_2\text{O}}$$

91%

Org React (1942) 1 370

Chapter 12   PREPARATION
OF
KETONES

Section 166   <u>Ketones from Acetylenes</u>
∘∘∘∘∘∘∘∘∘∘∘∘∘∘∘∘∘∘∘∘∘∘∘∘∘

EtC≡CEt $\xrightarrow{\text{1 NaBH}_4 \quad \text{BF}_3 \cdot \text{Et}_2\text{O} \quad \text{diglyme}}{\text{2 H}_2\text{O}_2 \quad \text{H}_2\text{O}}$ EtCH$_2$COEt   62%

JACS (1961) <u>83</u> 3834
      (1967) <u>89</u> 5086
Org React (1963) <u>13</u> 1

JCS (1952) 4086
Helv (1945) <u>28</u> 1355
      (1943) <u>26</u> 680

75%

JCS (1954) 3257

80%

C$_6$H$_{13}$C≡CH $\xrightarrow{\text{HgO} \quad \text{BF}_3 \cdot \text{Et}_2\text{O}}{\text{HOCH}_2\text{CH}_2\text{OH}}$ C$_6$H$_{13}$CMe   ---→   C$_6$H$_{13}$COMe
                                                                     O O

Chem Comm (1967) 200
JACS (1934) <u>56</u> 1130
      (1936) <u>58</u>   80

$$EtC \equiv CEt \xrightarrow[\text{2 } H_2O]{\text{1 } ClSO_2CNO \quad CH_2Cl_2} EtCOCH_2Et \qquad 67\%$$

Tetr Lett (1970) 27

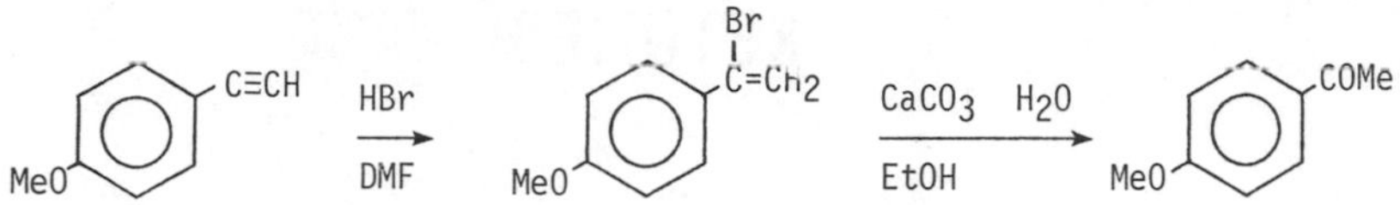

Helv (1964) 47 194

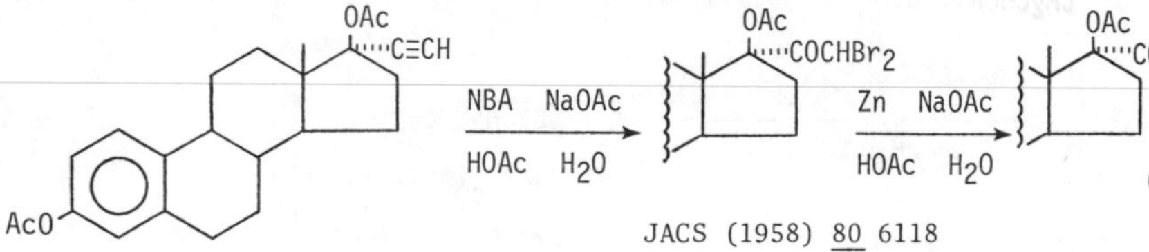

73%

JACS (1958) 80 6118

Section 167   Ketones from Carboxylic Acids, Acid Halides and Anhydrides

$$C_{17}H_{34}COOH \xrightarrow[\text{electrolysis}]{HOOCCH_2\overset{Me}{\underset{|}{C}}HCH_2\overset{Me}{\underset{|}{C}}HCH_2COMe} C_{17}H_{34}CH_2\overset{Me}{\underset{|}{C}}HCH_2\overset{Me}{\underset{|}{C}}HCH_2COMe$$

Acta Chem Scand (1956) 10 478
Advances in Org Chem (1960) 1 1

$$PhCH_2COOH \xrightarrow{MeLi \quad Et_2O} PhCH_2COMe \qquad 76\%$$

Acta Chem Scand (1952) 6 782
Org React (1970) 18 1

$$\text{JACS (1969) } \underline{91} \text{ 456}$$
$$\text{(1970) } \underline{92} \text{ 2590}$$

$$\text{EtCOONa} \xrightarrow[\text{}]{\text{EtMgBr Et}_2\text{O}} \text{Et}_2\text{CO}$$

Ber (1909) $\underline{42}$ 4500
JACS (1933) $\underline{55}$ 1258
JCS (1950) 2012

Review: The Synthesis of Ketones from Acid Halides and Organometallic
Compounds of Magnesium, Zinc and Cadmium

Org React (1954) $\underline{8}$ 28

$$\text{C}_{17}\text{H}_{35}\text{COCl} \xrightarrow[\text{Et}_2\text{O C}_6\text{H}_6]{\text{PhCH}_2\text{CH}_2\text{MgBr-CdCl}_2} \text{C}_{17}\text{H}_{35}\text{COCH}_2\text{CH}_2\text{Ph} \qquad 65\%$$

Org React (1954) $\underline{8}$ 28
JACS (1950) $\underline{72}$ 5333
For use of aryl cadmium reagents see   JACS (1945) $\underline{67}$ 740

$$\begin{array}{c}\text{COOEt}\\|\\(\text{CH}_2)_{10}\text{COCl}\end{array} \xrightarrow[\text{Et}_2\text{O C}_6\text{H}_6]{\text{C}_{18}\text{H}_{37}\text{MgBr-ZnCl}_2} \begin{array}{c}\text{COOEt}\\|\\(\text{CH}_2)_{10}\text{COC}_{18}\text{H}_{37}\end{array} \qquad \sim79\%$$

Org React (1954) $\underline{8}$ 28

$$\begin{array}{c}\text{COOEt}\\|\\(\text{CH}_2)_8\text{COCl}\end{array} \xrightarrow[\text{}]{\text{i-PrCH}_2\text{MgBr Et}_2\text{O}} \begin{array}{c}\text{COOEt}\\|\\(\text{CH}_2)_8\text{COCH}_2\text{Pr-i}\end{array}$$

Org React (1954) $\underline{8}$ 28
JOC (1961) $\underline{26}$ 1768
Tetr Lett (1970) 2523

$$\text{C}_5\text{H}_{11}\text{COCl} \xrightarrow[\text{}]{\text{Bu}_2\text{CuLi}} \text{C}_5\text{H}_{11}\text{COBu} \qquad 79\%$$

Tetr Lett (1970) 4647

$(PhCO)_2O$ $\xrightarrow{\text{t-BuMgCl-CdCl}_2 \ \ \text{Et}_2\text{O}}$ PhCOBu-t   40%

JOC (1941) $\underline{6}$ 462
(1948) $\underline{13}$ 592
Ber (1964) $\underline{97}$ 1649

$C_5H_{11}COOH$ $\dashrightarrow$ $C_5H_{11}CON$⟨imidazole⟩ $\xrightarrow[\text{Et}_2\text{O}]{\text{EtMgBr}}$ $C_5H_{11}COEt$ ~64%

Annalen (1962) $\underline{655}$ 90

PhCOCl $\xrightarrow[\text{HMPA}]{\text{Me}_3\text{SiCH}_2\text{Ph} \ \ \text{BuLi}}$ $PhCOCH_2Ph$

Tetr Lett (1970) 1137

PhCOCl $\xrightarrow{\text{Bu}_2\text{Hg} \ \ \text{AlBr}_3 \ \ \text{CH}_2\text{Cl}_2}$ PhCOBu   73%

J Organometallic Chem (1969) $\underline{17}$ P21

PrCOCl $\xrightarrow{(C_6H_{13})_3Al \ \ CH_2Cl_2}$ $PrCOC_6H_{13}$ 78%

Tenside (1967) $\underline{4}$ 167
(Chem Abs $\underline{67}$ 116507)

PhCOCl $\xdashrightarrow{\text{CH}_2\text{N}_2}$ $PhCOCHN_2$ $\xrightarrow{\text{p-Tolylcopper}}$ $PhCOCH_2$⟨C6H4⟩Me   ~31%

Chem Comm (1969) 515

$C_{10}H_{21}COCl$ $\xrightarrow[\text{C}_6\text{H}_6 \ \ \text{EtOH}]{\overset{\text{R}}{\underset{|}{\text{MgC(COOEt)}_2}}}$ $C_{10}H_{21}CO\overset{\text{R}}{\underset{|}{C}}(COOEt)_2$ $\xrightarrow[\substack{\text{EtCOOH} \\ \text{H}_2\text{O}}]{\text{H}_2\text{SO}_4}$ $C_{10}H_{21}COCH_2R$

JCS (1950) 322
Org Synth (1963) Coll Vol 4 285

For use of malonic tetrahydropyranyl esters see JCS (1952) 3945
    "   "   "   "   t-butyl   "   " JACS (1952) $\underline{74}$ 831
    "   "   "   "   benzyl   "   " JCS (1950) $\overline{325}$
               and JACS (1963) $\underline{85}$ 1409

$PhCOCl$ $\xrightarrow[\text{Ph}_3\text{CNa}\quad\text{Et}_2\text{O}]{\text{Et}_2\text{CHCOOEt}}$ $\underset{\underset{\text{COOEt}}{|}}{PhCOCEt_2}$ $\xrightarrow[\text{or HI}\quad\text{H}_2\text{O}]{\text{H}_2\text{SO}_4\quad\text{HOAc}}$ $PhCOCHEt_2$ 44%

JACS (1941) <u>63</u> 3163

$PhCOCl$ $\xrightarrow{\text{Naphthalene}\quad\text{AlCl}_3\quad\text{PhNO}_2}$

JCS (1949) S99
Chem Rev (1955) <u>55</u> 229

For further examples of the acylation of aromatic compounds see section
176 (Ketones from Hydrides)

$PhCOOH$ $\xrightarrow[250°]{\text{C}_8\text{H}_{17}\text{COOH}\quad\text{Fe}}$ $PhCOC_8H_{17}$ 70%

JOC (1963) <u>28</u> 879
Org Synth (1963) Coll Vol 4 854

$(C_5H_{11}CO)_2O$ $\xrightarrow{\text{BF}_3}$ $(C_5H_{11})_2CO$ 64%

JACS (1950) <u>72</u> 3294

$(PrCO)_2O$ $\xrightarrow{\text{(Me}_3\text{Si)}_2\quad\text{AlCl}_3}$ $PrCOMe$ 75%

Compt Rend (1965) <u>261</u> 1329

$PhCOCl$ $\xrightarrow{\text{Et}_3\text{SiH}\quad\text{RhCl}_3(\text{PBu}_2\text{Ph})_3}$ Ph_2CO

Chem Comm (1970) 1703

$PhCOCl$ $\xrightarrow[\text{2 H}_2\text{O}]{\text{1 MeCH=PPh}_3\quad\text{C}_6\text{H}_6}$ $PhCOCH_2Me$ < 85%

Tetr Lett (1960) (4) 7
Annalen (1964) <u>674</u> 11
Ber (1962) <u>95</u> 1513

PrCOOAg $\xrightarrow{\begin{array}{c}\text{1 Bu}_4\overset{+}{\text{P}}\ \overset{-}{\text{Br}}\quad\text{MeOH}\\\text{2 300°}\end{array}}$ PrCOBu   39%

JOC (1963) 28 1133

1 CH$_2$N$_2$
Et$_2$O
2 HBr
Et$_2$O

Zn   NaI
HOAc

Tetr Lett (1969) 51
JACS (1943) 65 1516   75%

1 CH$_2$=C(OMe)$_2$
2 HCl   H$_2$O   HOAc

JACS (1956) 78 6086

C$_7$H$_{15}$COCl $\xrightarrow{\begin{array}{c}\text{1 Me}_3\text{N}\quad\text{Et}_2\text{O}\\\text{2 H}_2\text{O}\end{array}}$ (C$_7$H$_{15}$)$_2$CO   95%

JACS (1947) 69 2444
Org Synth (1963) Coll Vol 4 555
Aust J Chem (1955) 8 506
For preparation of cyclic ketones from dicarboxylic acids see
JACS (1948) 70 34

(Me$_2$CHCO)$_2$O $\xrightarrow{\begin{array}{c}\text{Pyridine N-oxide}\\\text{PhCl}\end{array}}$ Me$_2$CO

Tetr Lett (1965) 233

KMnO$_4$
KOH   H$_2$O

NaBiO$_3$
H$_3$PO$_4$ H$_2$O

JOC (1964) 29 2914
JCS (1953) 2129 3580

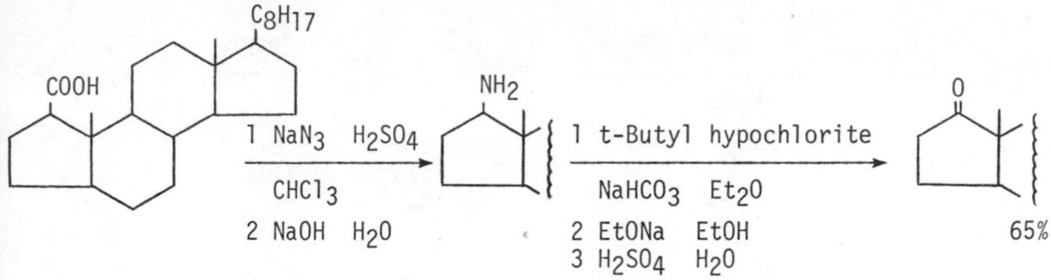

JOC (1965) $\underline{30}$ 3775

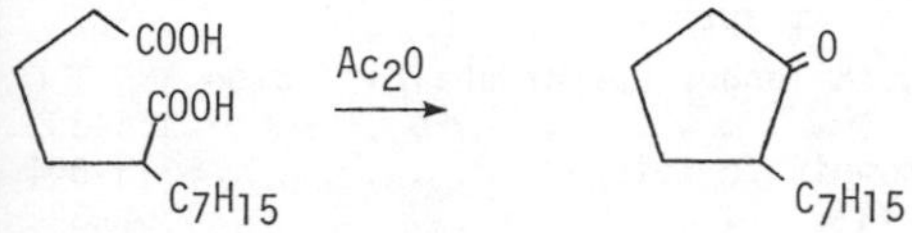

~80%

Compt Rend (1955) $\underline{240}$ 317
Helv (1945) $\underline{28}$ 1651

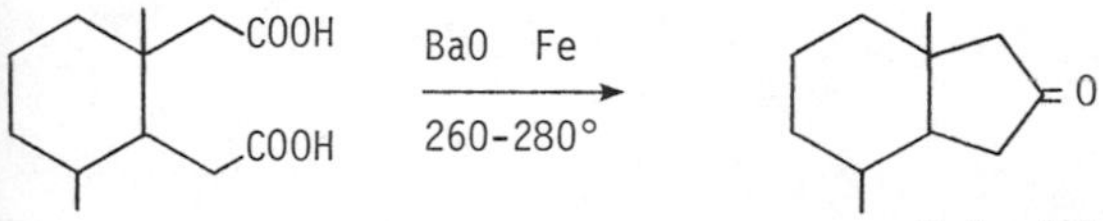

Helv (1947) $\underline{30}$ 2158
Org Synth (1941) Coll Vol 1 192

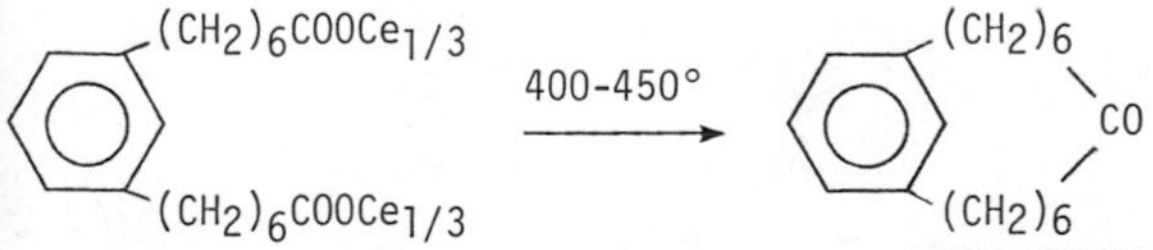

Helv (1932) $\underline{15}$ 1220 1459
      (1928) $\underline{11}$ 1174

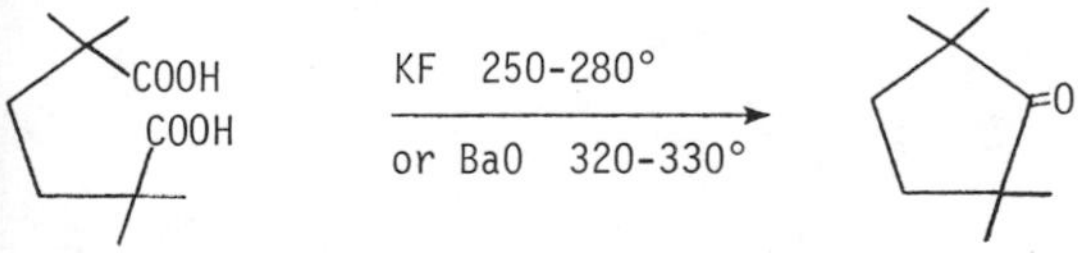

52-72%

JOC (1962) $\underline{27}$ 1034

$Bu_2C(COOH)_2$ $\xrightarrow[\text{2 KOH MeOH H}_2\text{O}]{\text{1 Pb(OAc)}_4 \text{ Pyr C}_6\text{H}_6}$ Bu_2CO 70%

Tetr Lett (1966) 6145
JOC (1964) $\underline{29}$ 2914

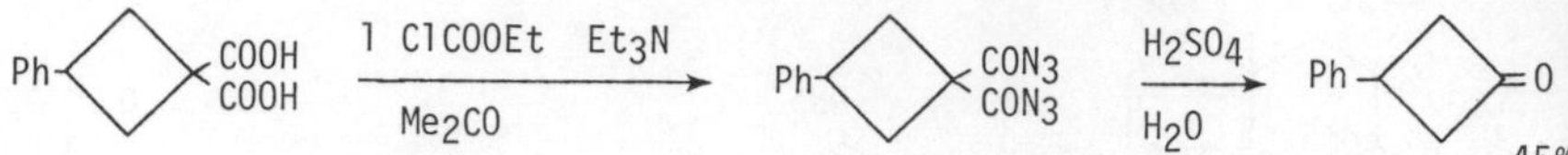

JOC (1962) <u>27</u> 1647
    (1964) <u>29</u> 2914

45%

Section 168   Ketones from Alcohols and Phenols

$Me_2CCH=CH_2$ (with OH)

$$\underset{CH_2=CMe \quad TsOH \quad MeOH}{\overset{OEt}{\longrightarrow}}$$

$Me_2C=CHCH_2CH_2COMe$ 41%

Helv (1967) <u>50</u> 2091

$\underset{\text{OH}}{Me}$ HOCHCH=CH_2   - - →   MeCOCH_2COOCHCH=CH_2 ($\overset{Me}{}$)

$$\overset{Ph_2O}{\underset{220°}{\longrightarrow}}$$

MeCH=CHCH_2CH_2COMe   76%
trans

Tetr Lett (1969) 3253

i-PrOH $\underset{BF_3}{\overset{MeCOCH_2COOEt}{\longrightarrow}}$ MeCOCHCOOEt ($\underset{i-Pr}{}$)   - - →   MeCOCH_2 ($\underset{i-Pr}{}$)

60-67%

Org Synth (1955) Coll Vol 3 405

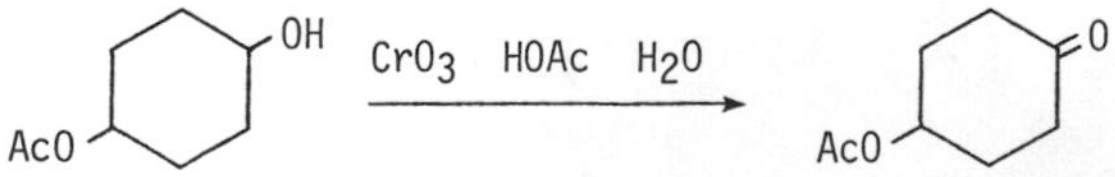

69%

JCS (1949) 615
    (1940) 10
For use of CrO_3-$Mn(NO_3)_2$ see JOC (1967) <u>32</u> 1098

$$\text{MeO}-\text{C}_6\text{H}_4-(\text{CH}_2)_3\text{CHCH}=\text{CH}_2 \quad \xrightarrow[\text{(2 phases)}]{\text{CrO}_3 \quad \text{Et}_2\text{O} \quad \text{H}_2\text{O}} \quad \text{MeO}-\text{C}_6\text{H}_4-(\text{CH}_2)_3\text{COCH}=\text{CH}_2$$

63%

JCS C (1966) 1972
Chem Pharm Bull (1964) 12 1184
JOC (1971) 36 387

$$\xrightarrow[\text{HOAc}]{\text{Na}_2\text{Cr}_2\text{O}_7}$$

96%

Org Synth (1963) Coll Vol 4 195
Annalen (1967) 707 203

$$\text{PhCHC}\equiv\text{CH} \quad \xrightarrow[\text{(Jones' reagent)}]{\text{CrO}_3 \quad \text{H}_2\text{SO}_4 \quad \text{Me}_2\text{CO} \quad \text{H}_2\text{O}} \quad \text{PhCOC}\equiv\text{CH}$$
$$\text{OH}$$

77%

JCS (1946) 39
     (1970) 2631
JOC (1960) 25 1434

$$\xrightarrow[\text{DMF}]{\text{CrO}_3 \quad \text{H}_2\text{SO}_4}$$

79%

Ber (1961) 94 729

$$\text{C}_6\text{H}_{13}\text{CHMe} \quad \xrightarrow{\text{CrO}_3\cdot\text{Pyr} \quad \text{CH}_2\text{Cl}_2} \quad \text{C}_6\text{H}_{13}\text{COMe}$$
$$\text{OH}$$

97%

Tetr Lett (1968) 3363
For use of CrO$_3$ in Pyr-H$_2$O see Tetrahedron (1962) 18 1351
For in situ preparation of CrO$_3$-Pyr see JOC (1970) 35 4000

$$\xrightarrow[\text{CCl}_4]{\text{t-Butyl chromate}}$$

Biochem J (1962) 84 195
Helv (1957) 40 487

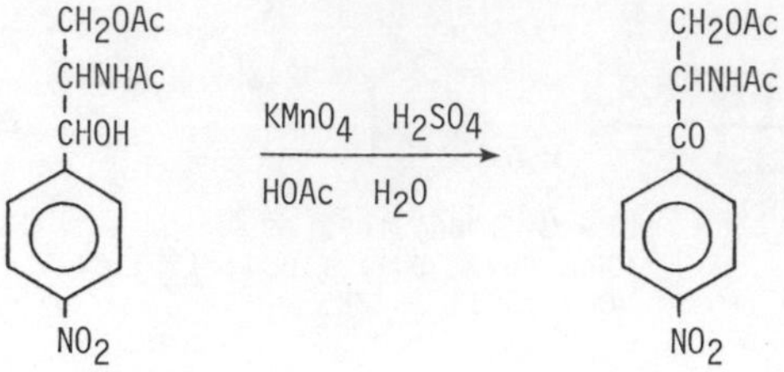

CH$_2$OAc
|
CHNHAc
|
CHOH

$\xrightarrow[\text{HOAc} \quad \text{H}_2\text{O}]{\text{KMnO}_4 \quad \text{H}_2\text{SO}_4}$

CH$_2$OAc
|
CHNHAc
|
CO

86%

NO$_2$ NO$_2$

Ber (1960) <u>93</u> 387
(1961) <u>94</u> 169

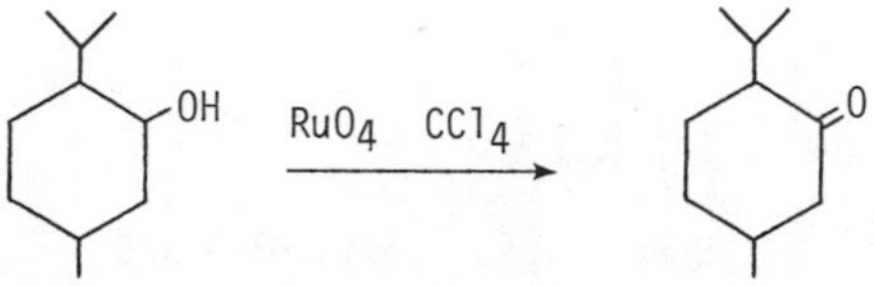

OH $\xrightarrow{\text{RuO}_4 \quad \text{CCl}_4}$ =O

96%

JACS (1958) <u>80</u> 6682

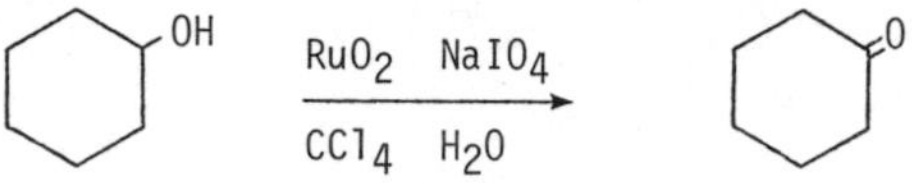

OH $\xrightarrow[\text{CCl}_4 \quad \text{H}_2\text{O}]{\text{RuO}_2 \quad \text{NaIO}_4}$ =O

79%

Tetr Lett (1967) 4729
(1970) 4233
Tetrahedron (1963) <u>19</u> 1959

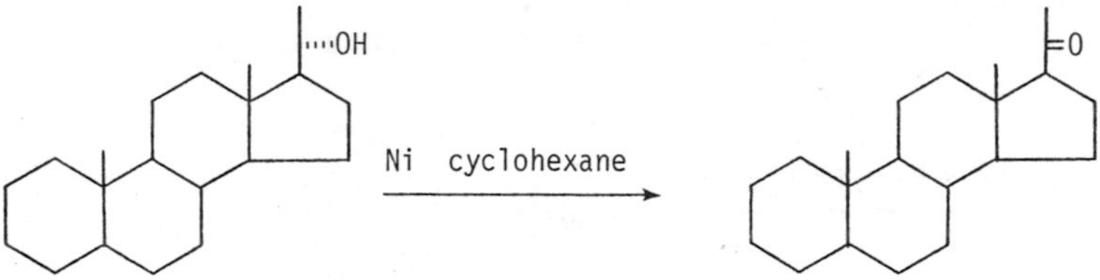

$\xrightarrow{\text{Ni} \quad \text{cyclohexane}}$

JCS <u>C</u> (1969) 968
JOC (1968) <u>33</u> 175 2814
(1958) <u>23</u> 899

BuCHMe $\xrightarrow{\text{Li} \quad \text{NH}_2\text{CH}_2\text{CH}_2\text{NH}_2}$ BuCOMe
|
OH

45%

JOC (1962) <u>27</u> 2662

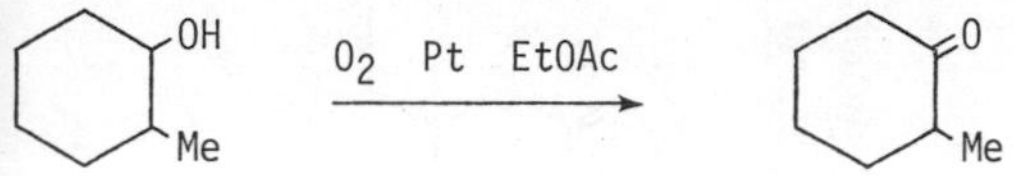
OH
Me
O_2 Pt EtOAc
O
Me

50%

JACS (1955) 77 190
Tetrahedron (1968) 24 6583
Angew (1957) 69 600
Advances in Carbohydrate Chem (1962) 17 169

PrCH(OH)Me $\xrightarrow{O_2 \quad h\nu \quad Ph_2CO}$ PrCOMe

Ber (1963) 96 509

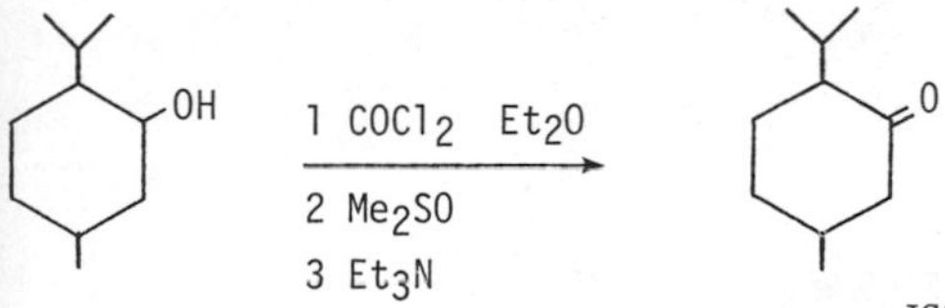
OH
1 $COCl_2$ Et_2O
2 Me_2SO
3 Et_3N
O

20%

JCS (1964) 1855

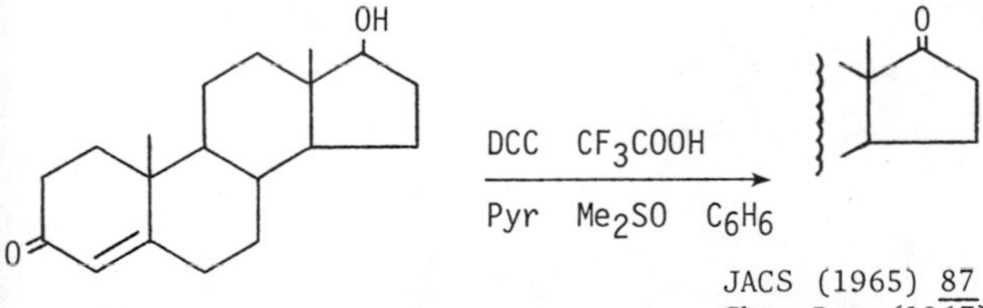
OH
O
DCC CF_3COOH
Pyr Me_2SO C_6H_6
O

92%

JACS (1965) 87 5670
Chem Rev (1967) 67 247
Can J Chem (1966) 44 2517

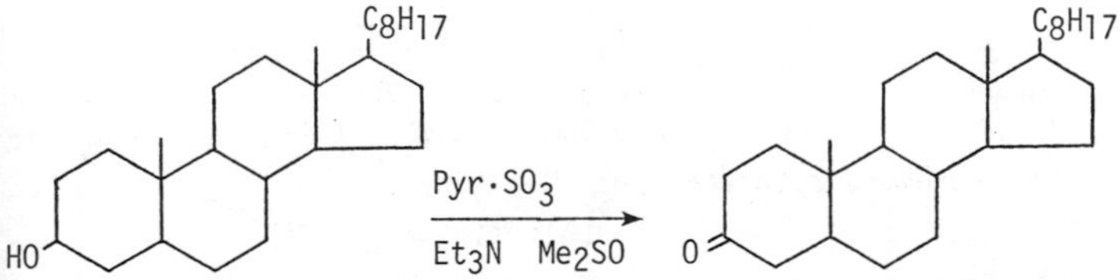
C_8H_{17}
HO
Pyr·SO_3
Et_3N Me_2SO
C_8H_{17}
O

83%

JACS (1967) 89 5505

Ac$_2$O   Me$_2$SO

30%

JACS (1967) $\underline{89}$ 2416
Carbohydrate Res (1966) $\underline{2}$ 251

Al(OPr-i)$_3$ Me$_2$CO

C$_6$H$_6$

80%

Org React (1951) $\underline{6}$ 207

t-BuOLi   EtCOMe

68%

JCS $\underline{C}$ (1969) 804

1 MeMgI

2 Me$_2$CO

Tetr Lett (1964) 3481

$$Ph(CH_2)_3\underset{\underset{OH}{|}}{C}HMe \xrightarrow{\text{NBA   Me}_2\text{CO   H}_2\text{O}} Ph(CH_2)_3COMe$$

59%

Chem Comm (1965) 5
JACS (1954) $\underline{76}$ 3682
Chem Rev (1963) $\underline{63}$ 21

For oxidation with NBS   see JCS (1959) 2594
                                and JOC (1957) $\underline{22}$ 1678

"   "   "   N-bromocaprolactam see JOC (1960) $\underline{25}$ 263

"   "   "   NCS   " Helv (1953) $\underline{36}$ 1763

"   "   "   Tribromoisocyanuric acid see
                                Bull Chem Soc Jap (1958) $\underline{31}$ 450

"   "   "   1-chlorobenzotriazole see JCS $\underline{C}$ (1969) 1474

$$\text{Br}_2 \quad \text{NaHCO}_3 \qquad \text{H}_2\text{O}$$

44%

JACS (1945) 67 312

$$\underset{\underset{\text{OH}}{|}}{\text{MeCH}_2\text{CMe}_2} \xrightarrow{\text{Br}_2} \underset{\underset{\text{BrBr}}{|\ |}}{\text{MeCHCMe}_2} \xrightarrow{\text{H}_2\text{O}} \text{MeCOCHMe}_2$$

55%

JACS (1933) 55 1136

$$\underset{\underset{\text{OH}}{|}}{\text{PhCHMe}} \xrightarrow[\text{Pyr} \quad \text{CCl}_4]{\text{t-Butyl hypochlorite}} \text{PhCOMe}$$

83%

Helv (1953) 36 1763
JACS (1955) 77 172

$$\underset{\underset{\text{OH}}{|}}{\text{C}_5\text{H}_{11}\text{CHMe}} \xrightarrow{\text{INO}_3 \quad \text{Pyr} \quad \text{CHCl}_3} \text{C}_5\text{H}_{11}\text{COMe}$$

65%

JCS C (1970) 676

$$\underset{\underset{\text{OH}}{|} \quad \underset{\text{OH}}{|}}{\text{MeCHCH}_2\text{CH}_2\text{CHMe}} \xrightarrow{\text{Pb(OAc)}_4 \quad \text{Pyr}} \text{MeCOCH}_2\text{CH}_2\text{COMe}$$

89%

Tetr Lett (1964) 3071
JACS (1960) 82 4956

$$\xrightarrow[\text{C}_6\text{H}_6]{\text{Ag}_2\text{CO}_3\text{-celite}}$$

99%

Compt Rend (1968) C 267 900
JACS (1955) 77 490

Tetr Lett (1967) 4193
(1968) 5685

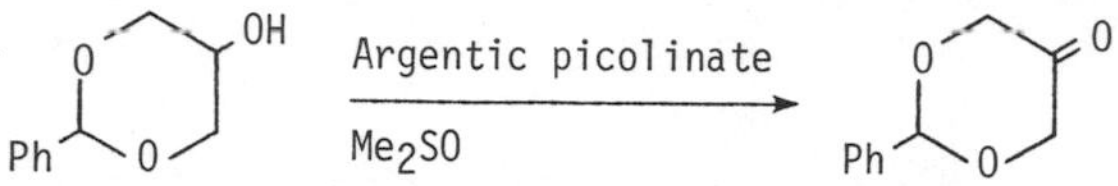

55%

Tetr Lett (1967) 415
Can J Chem (1969) 47 1649

EtCHMe $\xrightarrow[\text{O}_2]{\text{PdCl}_2 \quad \text{Cu(NO}_3)_2}$ EtCOMe
|
OH

JOC (1967) 32 2816

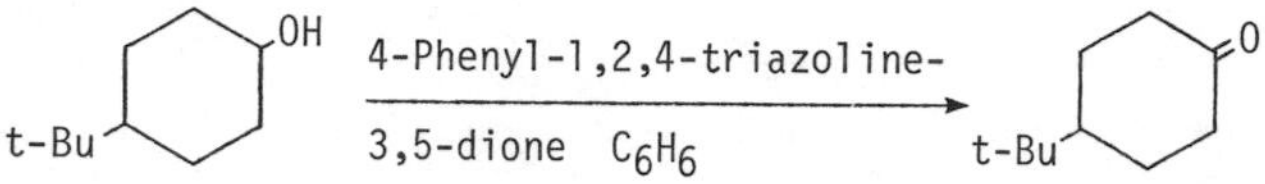

75%

Chem Comm (1966) 744

PhCHMe $\xrightarrow[]{\overset{\overset{\text{NCOOEt}}{\|}}{\text{NCOOEt}} \quad \text{C}_6\text{H}_6}$ PhCOMe
|
OH

87%

JOC (1967) 32 727
Bull Chem Soc Jap (1968) 41 1491

Me$_2$CHOH $\xrightarrow{\text{XeO}_3 \quad \text{H}_2\text{O}}$ Me$_2$CO

JACS (1964) 86 2078

$$\xrightarrow[\text{MeCN}]{\text{MnO}_2}$$

100%

Proc Chem Soc (1964) 110
Chem Comm (1966) 121

CH=CHC=CHCH=CHCHOH
OH   Me   Me

$$\xrightarrow[\text{Pet ether}]{\text{MnO}_2}$$

CH=CHC=CHCH=CHCOMe
OH   Me

(Applicable to allylic and benzylic alcohols only)
JCS (1952) 1094
JACS (1955) 77 4330
JOC (1959) 24 1051

$$\xrightarrow[\text{HMPA}]{\text{CrO}_3}$$

(Applicable to allylic and benzylic alcohols only)
Bull Soc Chim Fr (1969) 335

PhCHCH=CH$_2$ $\xrightarrow{\text{Chloranil   CCl}_4}$ PhCOCH=CH$_2$ 66%
   OH
                              (Applicable to allylic and benzylic alcohols only)
                                        JCS (1956) 3070

OH
CHEt

$$\xrightarrow{\text{DDQ   dioxane}}$$

COEt

(Applicable to allylic and benzylic alcohols only)
Acta Chem Scand (1961) 15 218
Tetr Lett (1960) (9) 14
Chem Pharm Bull (1970) 18 2343
Chem Rev (1967) 67 153

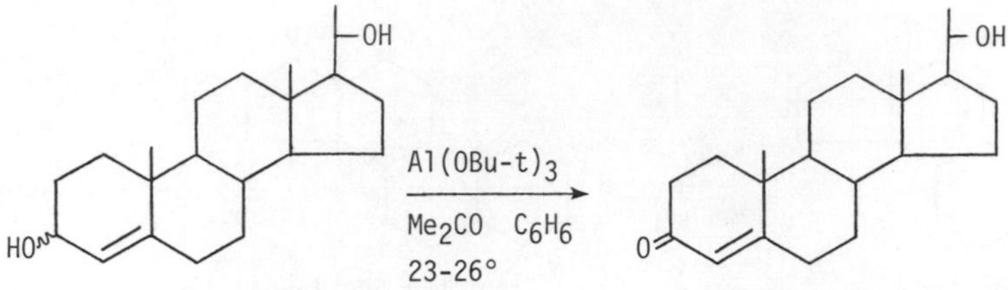
OH
HO
Al(OBu-t)3
Me2CO C6H6
23-26°
OH
O

>65%

(Method for selective oxidation of allylic alcohols)

Helv (1961) 44 179

PhCH(C_8H_{17})OH $\xrightarrow{N_2O_4 \quad CHCl_3}$ $PhCOC_8H_{17}$ 92%

(Applicable to benzylic alcohols only)

JCS (1957) 5087

PhCH(Me)OH $\xrightarrow{O_2 \quad h\nu \quad Me_2SO}$ PhCOMe 53%

(Applicable to benzylic alcohols only)

Bull Chem Soc Jap (1965) 38 1225

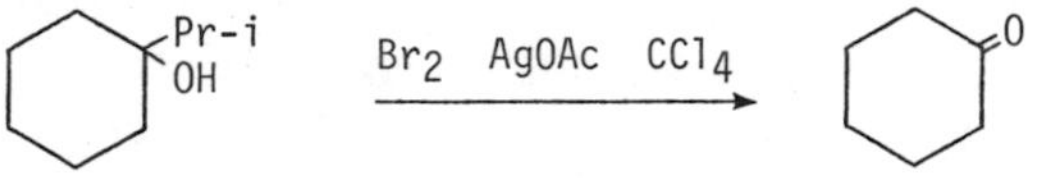
Pr-i
OH
Br2 AgOAc CCl4
O

38%

JACS (1965) 87 1807

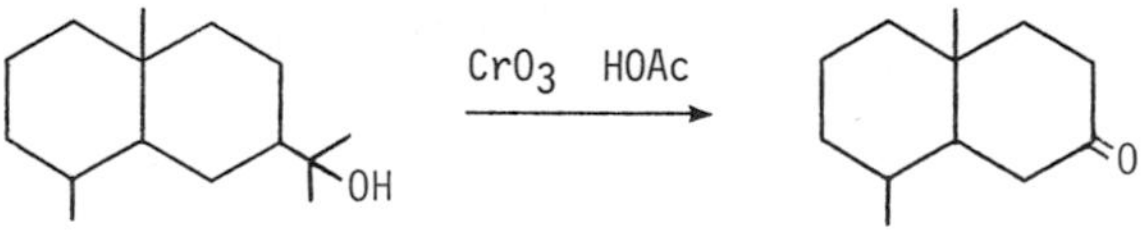
OH
CrO3 HOAc
O

Helv (1942) 25 1364

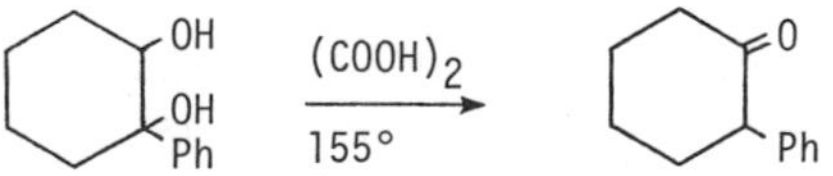
OH
OH
Ph
(COOH)2
155°
O
Ph

82%

Rec Trav Chim (1948) 67 489
JACS (1957) 79 147
Helv (1948) 31 1077

$$\underset{\substack{|\\Me}}{\overset{\substack{OH\ OH\\|\ \ |}}{PhC-CHMe}} \xrightarrow[Pyr]{Ac_2O} \underset{\substack{|\\Me}}{\overset{\substack{OH\ OAc\\|\ \ |}}{PhC-CHMe}} \xrightarrow[170°]{Zn} \underset{\substack{|\\Me}}{PhCHCOMe} \qquad \sim67\%$$

JOC (1970) <u>35</u> 660

$$\underset{\substack{OH\ OH\\|\ \ |}}{Me_2C-CMe_2} \xrightarrow{H_2SO_4\ \ H_2O} Me_3CCOMe \qquad\qquad 65\text{-}72\%$$

Org Synth (1941) Coll Vol 1 462
JACS (1960) <u>82</u> 4965

NaBiO$_3$ (or CrO$_3$)
────────────
HOAc

100%

JACS (1960) <u>82</u> 4965

Pb(OAc)$_4$ (or CrO$_3$)
────────────
HOAc

Helv (1959) <u>42</u> 1620

CrO$_3$
────────
HOAc

Helv (1938) <u>21</u> 546
JACS (1962) <u>84</u> 2241

$$\underset{\substack{OH\ OH\\|\ \ |}}{Me_2C-CMe_2} \xrightarrow{\text{Nickel peroxide}} Me_2CO \qquad\qquad 61\%$$

Chem Pharm Bull (1964) <u>12</u> 403

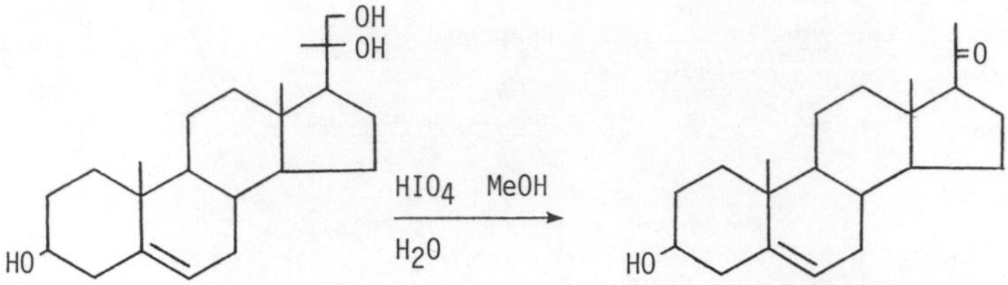
OH
OH
HO
HIO_4 MeOH
H_2O
O
HO

Helv (1941) 24 828
JACS (1956) 78 3087
Org React (1944) 2 341

$Me_2C(OH)-C(OH)Me_2$ $\xrightarrow{K_2S_2O_8\quad AgNO_3\quad H_2O}$ Me_2CO 100%

JACS (1954) 76 6345

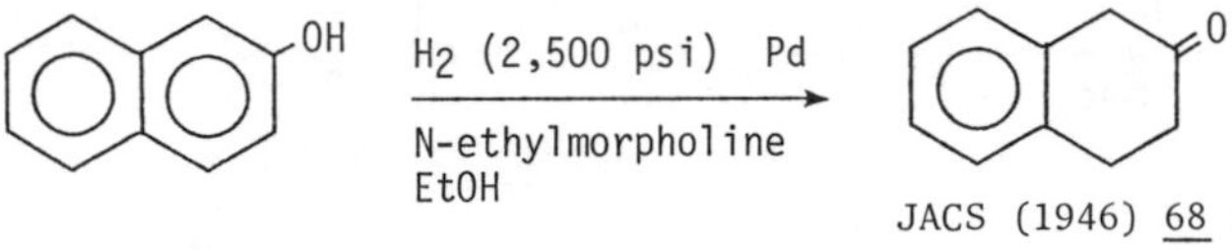
OH
H_2 (2,500 psi) Pd
N-ethylmorpholine
EtOH
O

40%

JACS (1946) 68 2172

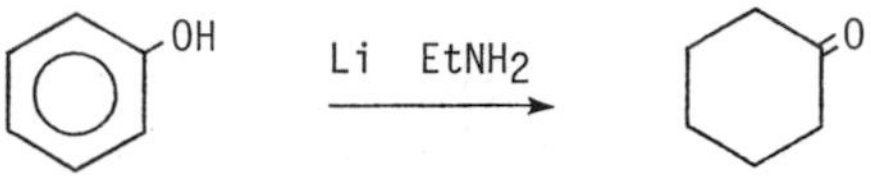
OH
Li $EtNH_2$
O

96%

JACS (1955) 77 6042

Section 169 Ketones from Aldehydes

$C_6H_{13}CHO$ $\xrightarrow[\text{glyme}]{Ph_3\overset{+}{P}CH(OMe)Me\ \overset{-}{Cl}\quad t\text{-}BuOK}$ $C_6H_{13}CH=C(OMe)Me$ $\xrightarrow[\text{MeOH}]{HCl}$ $C_6H_{13}CH_2COMe$ 57%

Tetr Lett (1964) 3323

$C_6H_{13}CHO$ $\xrightarrow[2\ I_2]{1\ Ph_3P=CHMe\quad THF}$ $C_6H_{13}CH_2COMe$

Tetr Lett (1970) 447

$$C_5H_{11}CHO \xrightarrow[\text{THF}]{\overset{\overset{-}{S}Me}{\underset{}{MeCPO(OEt)_2}} \overset{+}{Li}} C_5H_{11}CH=\overset{\overset{SMe}{|}}{C}Me \xrightarrow[H_2O]{HgCl_2 \quad MeCN} C_5H_{11}CH_2COMe \quad 50\%$$

JOC (1970) <u>35</u> 777

$$PhCHO \xrightarrow[\text{EtOH} \quad H_2O]{CH_2(CN)_2 \quad \text{glycine}} PhCH=C(CN)_2 \xrightarrow[\text{2 NaOH} \quad H_2O]{1 \quad CH_2N_2 \quad Et_2O} PhCOMe \quad 99\%$$

Tetr Lett (1963) 955

$$PhCHO \dashrightarrow PhC\overset{\displaystyle O-}{\underset{\displaystyle O-}{\overset{|}{H}}} \xrightarrow[\text{2 } H_2O]{1 \quad BuLi \quad Et_2O \quad cyclohexane} PhCOBu \quad 87\%$$

JOC (1965) <u>30</u> 226

$$PhCHO \xrightarrow{Cyclohexylamine} PhCH=N\text{—}\bigcirc \xrightarrow[Me_2SO]{NaH} Ph\overset{\overset{Me}{|}}{C}=N\text{—}\bigcirc \xrightarrow{Acid} PhCOMe$$

Can J Chem (1970) <u>48</u> 570

$$EtCHO \xrightarrow[\text{KCN}]{Me_2NH \cdot HCl} Et\overset{\overset{}{}}{C}H\underset{\underset{NMe_2}{|}}{C}N \xrightarrow[\text{2 } H_2O]{1 \quad EtBr \quad KNH_2 \quad NH_3} EtCOEt \quad <75\%$$

Bull Soc Chim Fr (1961) 1653
JACS (1960) <u>82</u> 1960
JOC (1961) <u>26</u> 4740

JOC (1951) <u>16</u> 221
Chem Comm (1967) 1258

$$\underset{\underset{Me}{|}}{PhCHCHO} \xrightarrow{Br_2} \underset{\underset{Me}{|}}{\overset{Br}{PhCCHO}} \xrightarrow{\overset{-}{CN}} \underset{\underset{Me}{|}}{Ph\overset{\overset{O}{\diagup\!\backslash}}{C\text{-}CHCN}} \xrightarrow[\text{2 MeMgBr}]{\text{1 } ZnBr_2 \quad Et_2O} \underset{\underset{Me}{|}}{PhCHCOMe} \quad 50\text{-}70\%$$

Tetr Lett (1970) 2947

$$PhCHO \xrightarrow{BuMgBr} \underset{\underset{Bu}{|}}{PhCHOMgBr} \xrightarrow[\substack{NCOOEt}]{\overset{NCOOEt}{\underset{\|}{}}} PhCOBu \qquad 50\%$$

Bull Chem Soc Jap (1968) 41 1491
JCS C (1966) 313

JCS (1943) 15
JACS (1963) 85 955
Org React (1949) 5 413
Chem Rev (1955) 55 283

Org React (1954) 8 364

JACS (1950) 72 2737

$$C_6H_{13}CHO \xrightarrow[\text{peroxide}]{C_6H_{13}CH=CH_2 \quad \text{diacetyl}} C_6H_{13}COC_8H_{17} \qquad 75\%$$

JOC (1949) 14 248
Aust J Chem (1967) 20 2033

PhCHO $\xrightarrow[\text{MeOH}\quad H_2O]{\text{HOAc}\quad\text{KOH}\quad\text{electrolysis}}$ PhCOMe   18%

Tetr Lett (1968) 1781

$Me_3CCHO \xrightarrow{AlCl_3} Me_2CHCOMe$ 100%

Ber (1936) 69B 2244

PhCHCHO
|
Me $\xrightarrow[\text{2 } H_2SO_4 \quad H_2O]{\text{1 } NH_2CONHNH_2}}$ PhCH$_2$COMe   65%

Zh Obshch Khim (1948) 18 2000
(Chem Abs 43 4632)

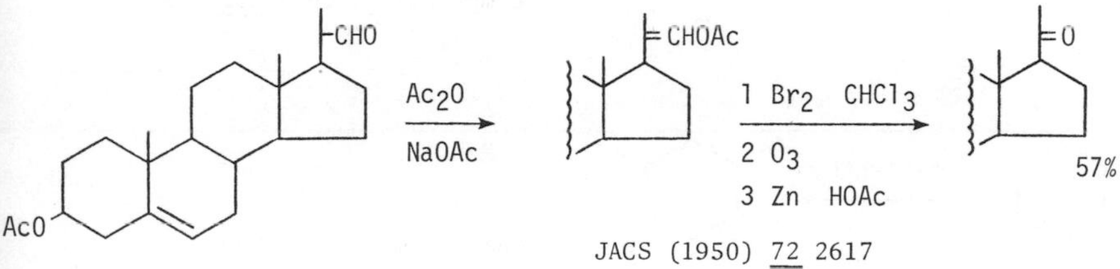

JACS (1950) 72 2617

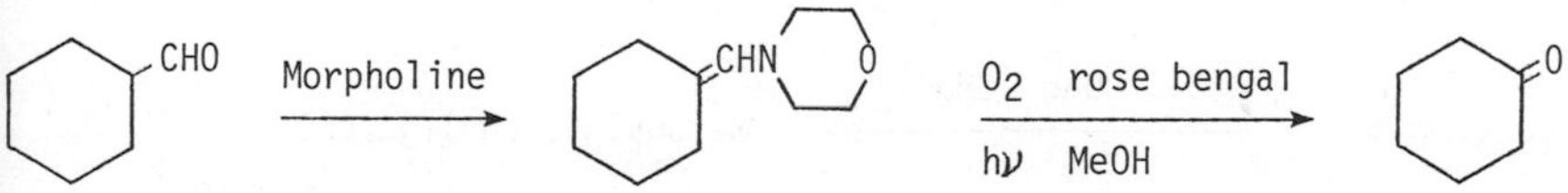

Tetr Lett (1968) 3271 3267
Chem Comm (1969) 314

For oxidation of enamines with Na$_2$Cr$_2$O$_7$ see JACS (1955) 77 1212 1216
   "   "   "   "   "   O$_3$ " JACS (1952) 74 3627

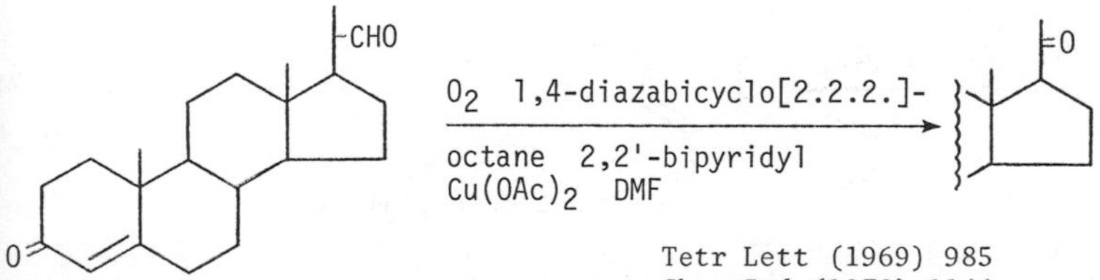

Tetr Lett (1969) 985
Chem Ind (1970) 1144
Ber (1967) 100 259

Section 170   Ketones from Alkyls and Methylenes
oooooooooooooooooooooooooooooooooooooooo

This section lists examples of the oxidation of methylenes (RCH$_2$R' $\longrightarrow$ RCOR'), the replacement of alkyl groups by ketonic groups and the degradation of alkyl groups to ketones (R$_2$CHR' $\longrightarrow$ R$_2$CO). For the acylation of hydrocarbons (RH $\longrightarrow$ RCOR') see section 176 (Ketones from Hydrides)

$$\xrightarrow[h\nu \quad CCl_4]{Cl_2 \quad NO \quad HCl}$$

Ber (1965) <u>98</u> 3493 3501

$$\xrightarrow{NOCl \quad h\nu \quad C_6H_6}$$

71%

JOC (1953) <u>18</u> 115
Compt Rend (1965) <u>260</u> 4514

$$\xrightarrow{h\nu \quad PhNO_2}$$

Chem Comm (1970) 1390

C$_{15}$H$_{31}$COOMe $\xrightarrow{CrO_3 \quad HOAc \quad CH_2Cl_2}$ Me(CH$_2$)$_n$CO(CH$_2$)$_{13-n}$COOMe

n=7 to 11

Chem Ind (1966) 2168
JACS (1952) <u>74</u> 3910

$$\xrightarrow{CrO_3 \cdot Pyr \quad CH_2Cl_2}$$

Ph   Ph

71

JOC (1969) <u>34</u> 3587

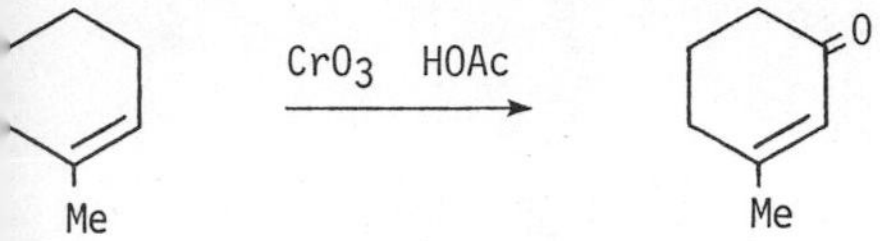
CrO3 HOAc
Me
O
Me

< 20%

JACS (1941) 63 758
(1949) 71 2226

For oxidation with Na_2CrO_4 in HOAc see JOC (1970) 35 192

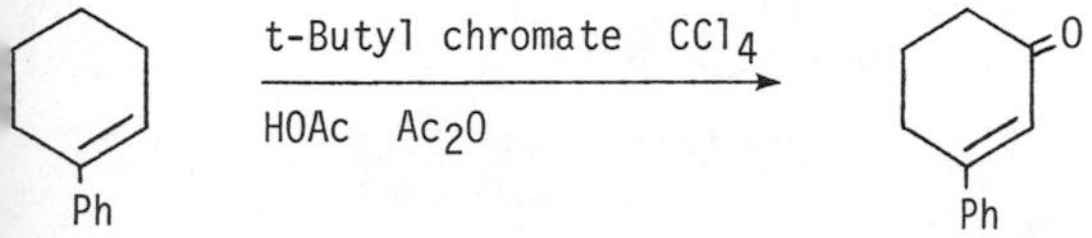
t-Butyl chromate CCl4
HOAc Ac2O
Ph
O
Ph

22%

JCS (1951) 516
Helv (1952) 35 284

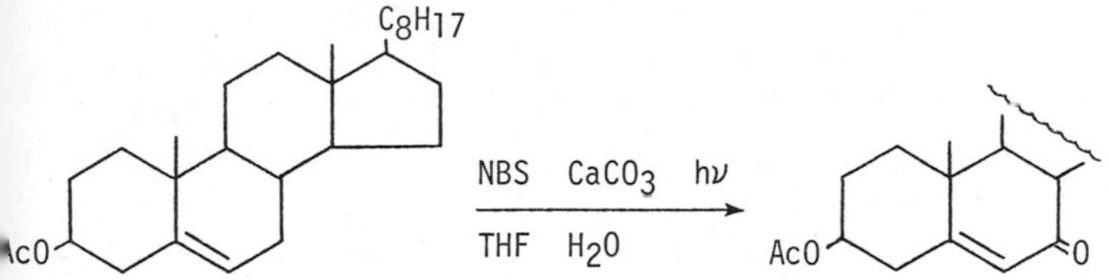
C8H17
AcO
NBS CaCO3 hν
THF H2O
AcO
O

81%

Chem Comm (1969) 1220

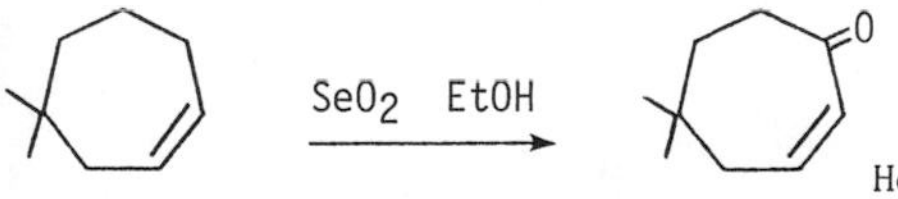
SeO2 EtOH
O

Helv (1940) 23 524 1477
Org React (1949) 5 331

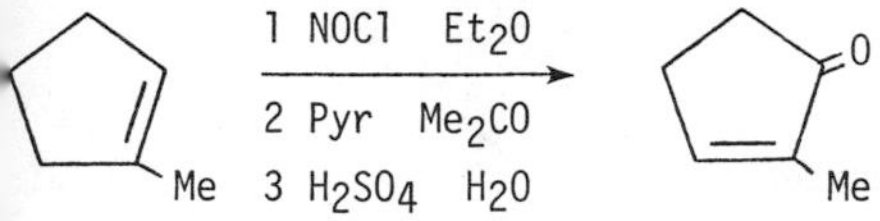
Me
1 NOCl Et2O
2 Pyr Me2CO
3 H2SO4 H2O
O
Me

Zh Org Khim (1965) 1 865
(Chem Abs 63 6873)
JCS (1951) 516

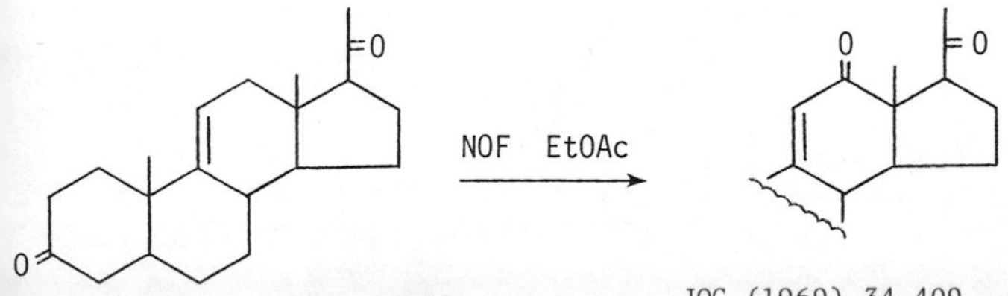
O
O
NOF EtOAc
O
O

30%

JOC (1969) 34 409

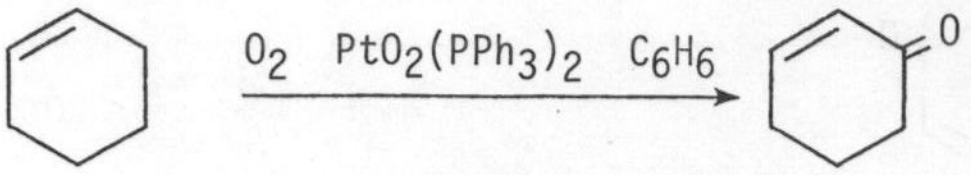
O2 PtO2(PPh3)2 C6H6
O

JACS (1967) 89 4809

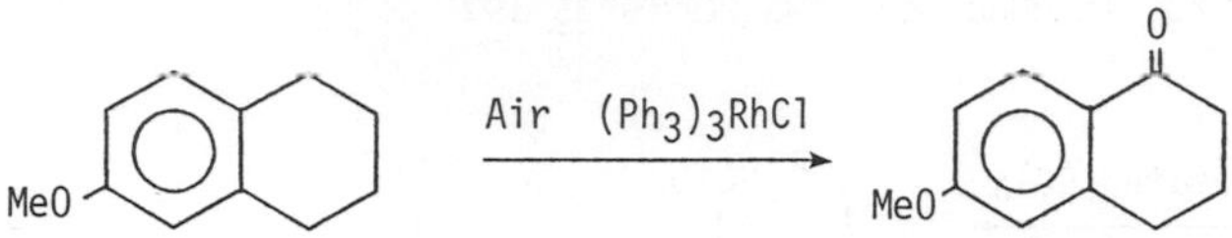
MeO
Air (Ph3)3RhCl
O
MeO

40%

Tetr Lett (1968) 2917
(1967) 3665

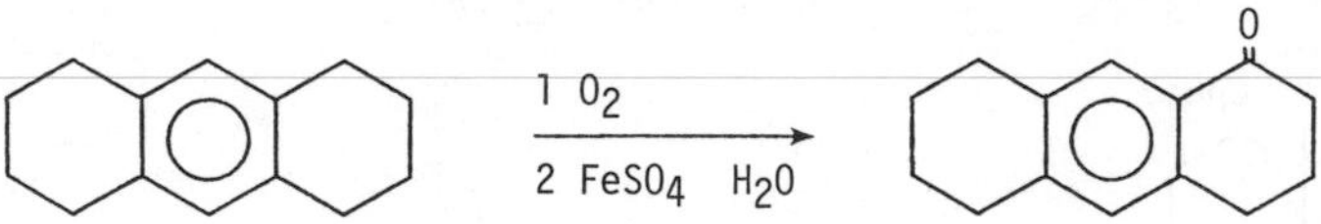
1 O2
2 FeSO4 H2O
O

Ber (1943) 76B 1130

$PhCH_2Me$ —Argentic picolinate / Me_2SO→ PhCOMe

Tetr Lett (1967) 415

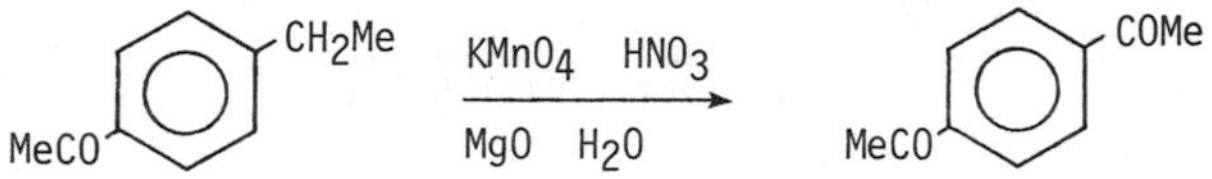
CH2Me
MeCO
KMnO4 HNO3
MgO H2O
COMe
MeCO

82%

JOC (1961) 26 4151

$PhCH_2Me$ —$(NH_4)_2S_2O_8$ $AgNO_3$ H_2O→ PhCOMe <73%

Org Prep and Procedures (1970) 2 207

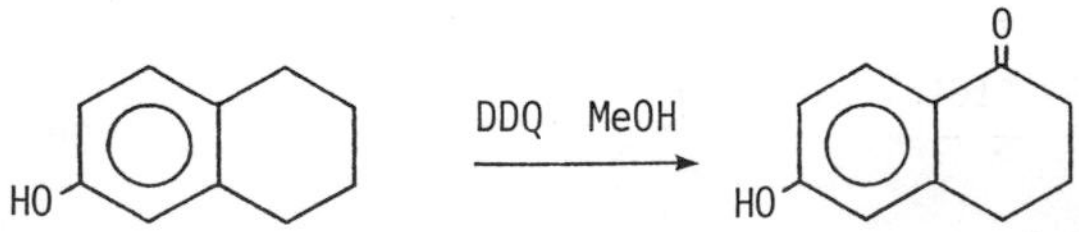
HO
DDQ MeOH
O
HO

78%

Chem Ind (1970) 158

$$\xrightarrow[\text{HNO}_3]{(\text{NH}_4)_2\text{Ce(NO}_3)_6}$$

76%

Tetr Lett (1966) 4493

$$\xrightarrow{\text{MeCOCl} \quad \text{AlCl}_3}$$

72%

JACS (1942) 64 2421
Helv (1960) 43 1473

$$\xrightarrow[\text{H}_2\text{O}]{\text{CrO}_3 \quad \text{HOAc}}$$

Helv (1935) 18 986

Section 171   Ketones from Amides
                ○○○○○○○○○○○○○○○○○○○○○○

PrCHCONMe₂ $\xrightarrow{\text{EtLi} \quad \text{Et}_2\text{O}}$ PrCHCOEt   80%
    | |
    Me   Me   JOC (1959) 24 701
                                                     JACS (1939) 61 232
                                                     Ber (1959) 92 2555

PhCHMe   Ni   200°
    | $\xrightarrow{\hspace{1.5cm}}$ PhCOMe
HCONH
                        Compt Rend (1947) 225 457

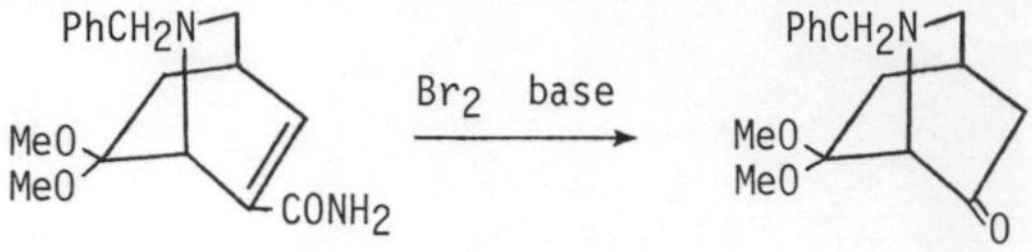
PhCH2N
MeO
MeO
CONH2
Br2 base
PhCH2N
MeO
MeO
O

JACS (1968) 90 2448
Org React (1946) 3 267

$(Me_2CH)_2NAc \xrightarrow{K_2S_2O_8 \quad K_2HPO_4 \quad H_2O} Me_2CO$

JOC (1964) 29 3632

Amides may also be converted into ketones via intermediate amines. See section 96 (Amines from Amides) and section 172 (Ketones from Amines)

Section 172 Ketones from Amines

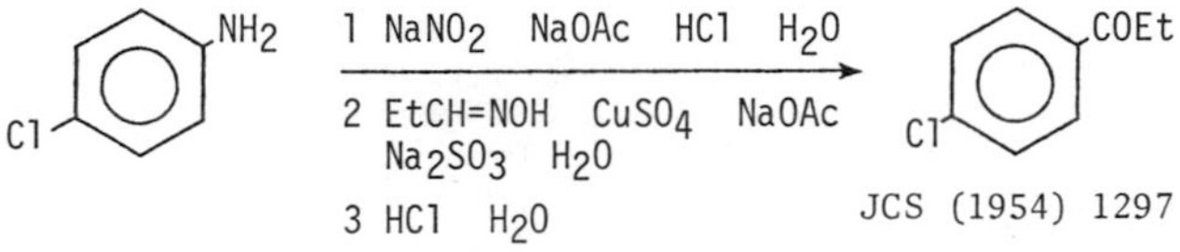
NH2
Cl
1 NaNO2 NaOAc HCl H2O
2 EtCH=NOH CuSO4 NaOAc
Na2SO3 H2O
3 HCl H2O
COEt
Cl

30%

JCS (1954) 1297

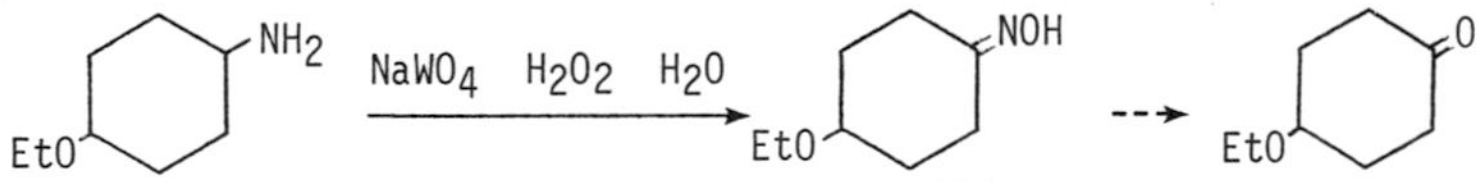
NH2
EtO
NaWO4 H2O2 H2O
NOH
EtO
O
EtO

87%

Ber (1960) 93 132

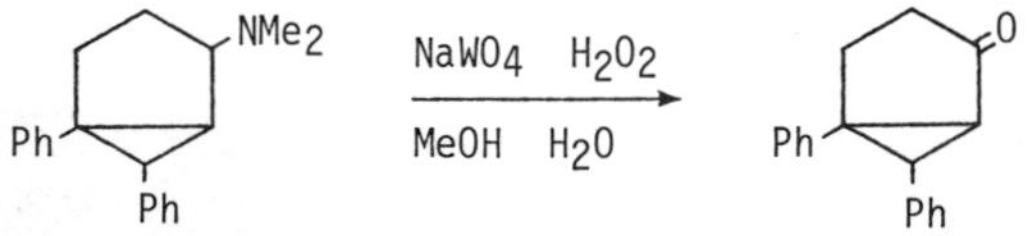
NMe2
Ph
Ph
NaWO4 H2O2
MeOH H2O
O
Ph
Ph

54%

JACS (1968) 90 4892

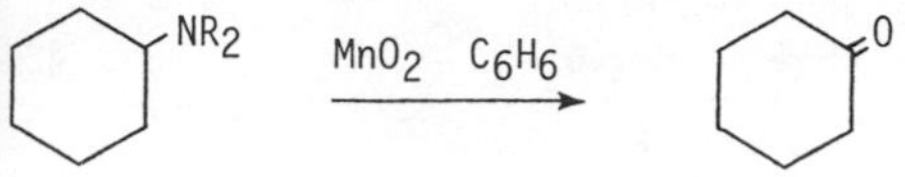
NR2
MnO2 C6H6
O

(R=H or Me)

JCS (1960) 3559
JCS C (1966) 995

$(Me_2CH)_2NH \xrightarrow{Ph_2CO\ \ h\nu\ \ C_6H_6} Me_2C{=}NCHMe_2 \xrightarrow[H_2O]{HCl} Me_2CO$ ~95%

JACS (1965) 87 2996

EtCH(Me)NHR $\xrightarrow[t\text{-}BuOH\ \ H_2O]{KMnO_4\ \ CaSO_4}$ EtCOMe

91% (R=H)
96% (R=EtCH(Me))

JOC (1967) 32 3129

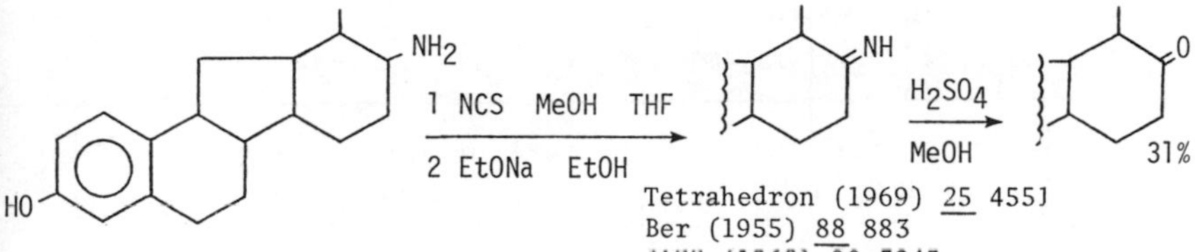
NH2
1 NCS MeOH THF
2 EtONa EtOH
HO
NH
H2SO4
MeOH
O

31%

Tetrahedron (1969) 25 4551
Ber (1955) 88 883
JACS (1968) 90 3245

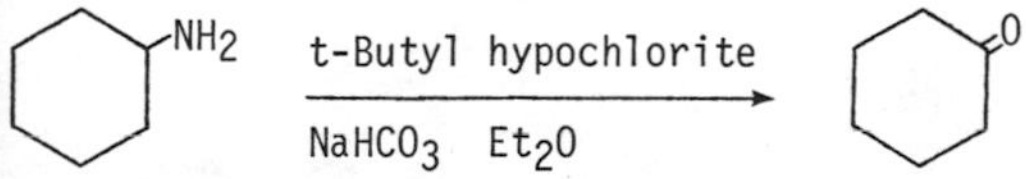
NH2
t-Butyl hypochlorite
NaHCO3 Et2O
O

73%

JACS (1954) 76 5554

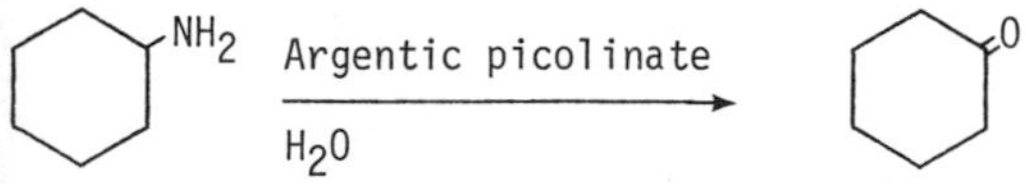
NH2
Argentic picolinate
H2O
O

41%

JCS (1965) 4962

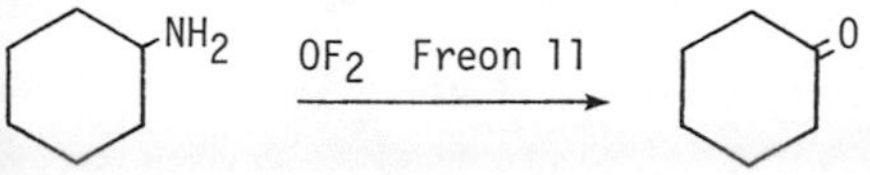
NH2
OF2 Freon 11
O

66%

JACS (1964) 86 1392

Me_2CHNH_2 $\xrightarrow{\begin{array}{l}\text{1 2,4-Dinitrobenzaldehyde Pyr}\\ \text{2 HCl } H_2O\end{array}}$ Me_2CO 46%

JCS (1954) 209

$(CH_2)_{11}$ $CHNH_2$ $\xrightarrow{\begin{array}{l}\text{1 3,5-Di-t-butyl-1,2-benzoquinone}\\ \text{(or 3,5-dinitromesitylglyoxal)}\\ \text{2 } (COOH)_2\end{array}}$ $(CH_2)_{11}$ CO 86-97%

JACS (1969) 91 1429

$\overset{Me}{\underset{|}{C_6H_{13}CHNH_2}}$ $\xrightarrow{\text{2,4-Dinitrochlorobenzene}}$ $\overset{Me}{\underset{|}{C_6H_{13}CHNH}}$ $\xrightarrow[H_2O]{CrO_3 \ H_2SO_4}$ $C_6H_{13}COMe$ 21%

JOC (1962) 27 452

$\xrightarrow{H_2O_2 \ H_2O}$ $\xrightarrow{\begin{array}{l}\text{1 } (CF_3CO)_2O\\ \text{2 NaOH } H_2O\end{array}}$

57%

Tetrahedron (1967) 23 4681
JACS (1968) 90 5622

Section 173   <u>Ketones from Esters</u>
oooooooooooooooooooo

$C_8H_{17}CH=CH(CH_2)_7COOMe$ $\xrightarrow[\text{xylene}]{\text{EtONa}}$ $\overset{\displaystyle C_8H_{17}CH=CH(CH_2)_6\underset{|}{CH}COOMe}{C_8H_{17}CH=CH(CH_2)_7CO}$

$\downarrow{\begin{array}{l}\text{KOH}\\ \text{EtOH}\end{array}}$

$[C_8H_{17}CH=CH(CH_2)_7]_2CO$

Aust J Chem (1955) 8 506

$$\text{i-PrCH}_2\text{COOEt} \quad \xrightarrow{\text{NaH}} \quad \underset{\underset{\text{i-Pr}}{|}}{\text{i-PrCH}_2\text{COCHCOOEt}} \quad \dashrightarrow \quad (\text{i-PrCH}_2)_2\text{CO}$$

52%   JACS (1946) $\underline{68}$ 2647

$$\text{Me}_2\text{CHCOOEt} \quad \xrightarrow[\text{2 PhCOCl}]{\text{1 Ph}_3\text{CNa}\quad\text{Et}_2\text{O}} \quad \underset{\underset{\text{COPh}}{|}}{\text{Me}_2\text{CCOOEt}} \quad \xrightarrow[\text{H}_2\text{O}]{\text{H}_2\text{SO}_4\quad\text{HOAc}} \quad \text{Me}_2\text{CHCOPh}$$

JACS (1941) $\underline{63}$ 3163 3156

$$\text{C}_7\text{H}_{15}\text{COOEt} \quad \xrightarrow[\text{hexane}]{\text{t-BuLi}\quad\text{THF}} \quad \underset{\underset{\text{C}_6\text{H}_{13}}{|}}{\text{C}_7\text{H}_{15}\text{COCHCOOEt}} \quad \xrightarrow[\text{H}_2\text{O}]{\text{LiOH}} \quad (\text{C}_7\text{H}_{15})_2\text{CO}$$

Dokl (1964) $\underline{155}$ 1352
(Chem Abs $\underline{61}$ 1750)

For the preparation of ketones by alkylation of β-ketoesters with halides
followed by hydrolysis see section 175 (Ketones from Halides and Sulfonates)

$$\text{C}_5\text{H}_{11}\text{COOEt} \quad \xrightarrow[\text{Me}_2\text{SO}]{\overset{+\quad-}{\text{MeSOCH}_2}} \quad \text{C}_5\text{H}_{11}\text{COCH}_2\text{SOMe}$$

1 MeI   NaH   Me$_2$SO
$\xrightarrow{\text{2 Al-Hg}\quad\text{H}_2\text{O}\quad\text{THF}}$ C$_5$H$_{11}$COCH$_2$Me 62%

$\xrightarrow{\text{Al-Hg}\quad\text{H}_2\text{O}\quad\text{THF}}$ C$_5$H$_{11}$COMe 100%

JOC (1966) $\underline{31}$ 2355
JACS (1965) $\underline{87}$ 1345

$$\text{C}_{17}\text{H}_{35}\text{COOMe} \quad \xrightarrow[\text{Me}_2\text{SO}\quad\text{THF}]{\overset{-\quad+}{\text{MeSO}_2\text{CH}_2}\text{ Na}} \quad \text{C}_{17}\text{H}_{35}\text{COCH}_2\text{SO}_2\text{Me} \quad \xrightarrow[\text{H}_2\text{O}\quad\text{THF}]{\text{Al-Hg}} \quad \text{C}_{17}\text{H}_{35}\text{COMe} \quad 67\%$$

JOC (1968) $\underline{33}$ 61

$$\xrightarrow[\text{THF}]{[\text{LiCH}_2\text{SONAr}]^{-}\text{ Li}^{+}}$$

84%

JACS (1968) $\underline{90}$ 5548

PhCH$_2$COOEt $\xrightarrow[\text{2 HCl H}_2\text{O}]{\overset{\overset{\text{OEt}}{|}}{\text{1 CH}_2=\text{CNMe}_2}}$ PhCH$_2$COMe   54%

Chem Pharm Bull (1969) <u>17</u> 2314

PhCOOEt $\xrightarrow[\text{2 NaOH MeOH}]{\text{1 Ph}_3\text{P=CH}_2 \text{ Et}_2\text{O}}$ PhCOMe   41%

Ber (1962) <u>95</u> 1513

Me$_2$CHOAc $\xrightarrow{\text{Br}_2 \text{ H}_2\text{O pH 4.6}}$ Me$_2$CO

JACS (1967) <u>89</u> 3555

$\xrightarrow[\text{Me}_2\text{CO}]{\text{CrO}_3 \text{ H}_2\text{SO}_4}$ 83-87%

Org Synth (1962) <u>42</u> 79

PhCH$_2$CH$_2$$\overset{\overset{}{}}{\underset{\underset{\text{Me}}{|}}{\text{C}}}$(COOEt)$_2$ $\xrightarrow[\text{2 NaNO}_2 \text{ HOAc}]{\text{1 N}_2\text{H}_4}$ PhCH$_2$CH$_2$$\underset{\underset{\text{Me}}{|}}{\text{C}}$(CON$_3$)$_2$ $\xrightarrow[\text{2 Acid}]{\text{1 ROH}}$ PhCH$_2$CH$_2$COMe

Org React (1946) <u>3</u> 337   28%
JOC (1962) <u>27</u> 1647

Section 174   <u>Ketones from Ethers and Epoxides</u>

(Me$_2$CH)$_2$O $\xrightarrow[\text{BuOH CuCl}]{\text{t-Butyl perbenzoate}}$ Me$_2$C(OBu)$_2$ $\xrightarrow{\text{Acid}}$ Me$_2$CO

Angew (1961) <u>73</u> 65
Tetrahedron (1961) <u>13</u> 241

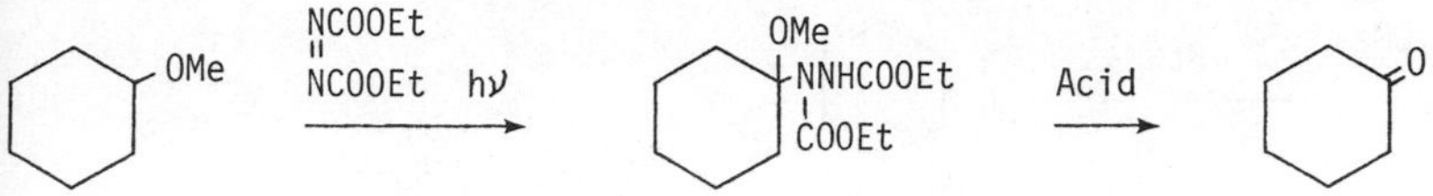

Chem Comm (1965) 259

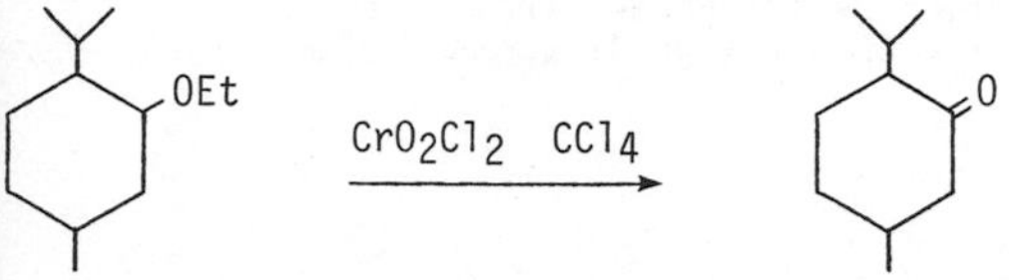

J Prakt Chem (1938) 151 61   87%

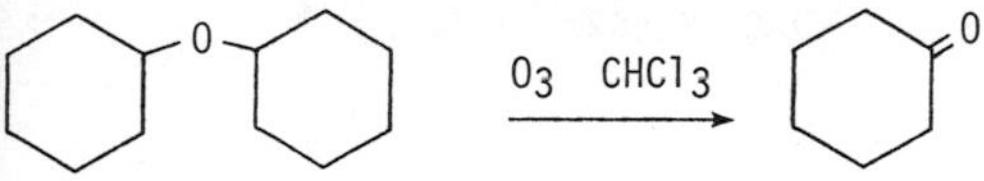

JOC (1962) 27 2392   25%

Me_2CHOCH_2Ph $\xrightarrow{\text{Br}_2 \quad \text{H}_2\text{O}}$ Me_2CO + $PhCHO$

                                        45%   55%

                        JACS (1967) 89 3550

$\underset{\underset{Me}{|}}{EtCHOMe}$ $\xrightarrow[\text{MeOH}]{\text{Electrolysis \quad MeONa}}$ $\underset{\underset{Me}{|}}{EtC(OMe)_2}$ $\dashrightarrow$ $EtCOMe$

                        JACS (1969) 91 2803

$Me_2CHOCPh_3$ $\xrightarrow{\sim 228°}$ Me_2CO

                        JACS (1930) 52 753
                             (1924) 46 2580

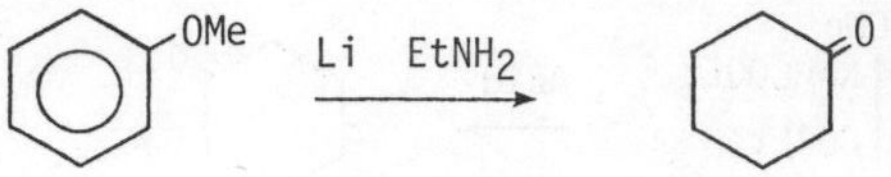

40%

JACS (1955) <u>77</u> 6042
JCS (1942) 689
Org Synth (1963) Coll Vol 4 903

Some of the methods listed in section 54 (Aldehydes from Ethers and
Epoxides) may also be applied to the preparation of ketones from ethers

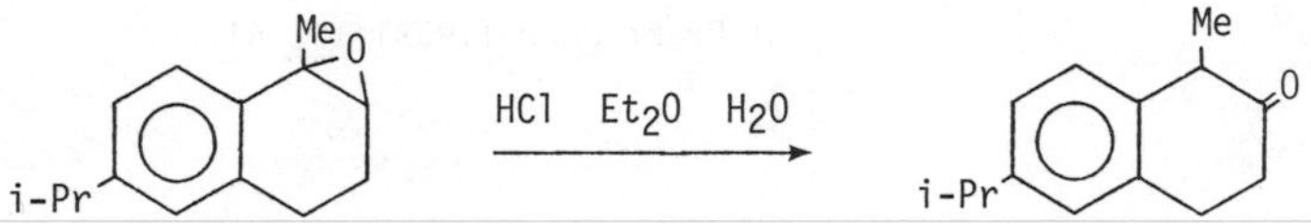

79%

JACS (1962) <u>84</u> 284

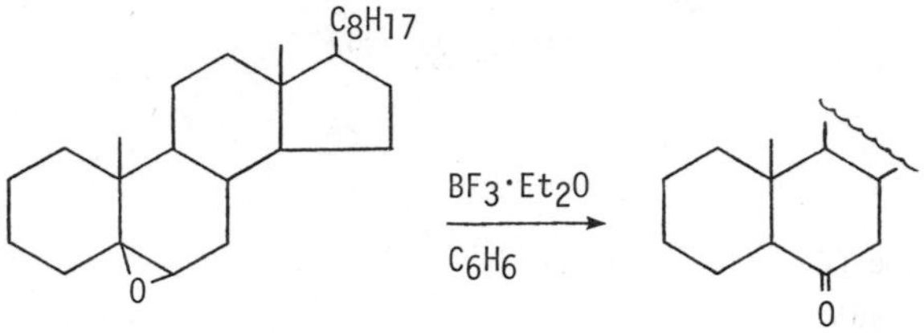

JCS (1957) 4596 4765
Helv (1953) <u>36</u> 398

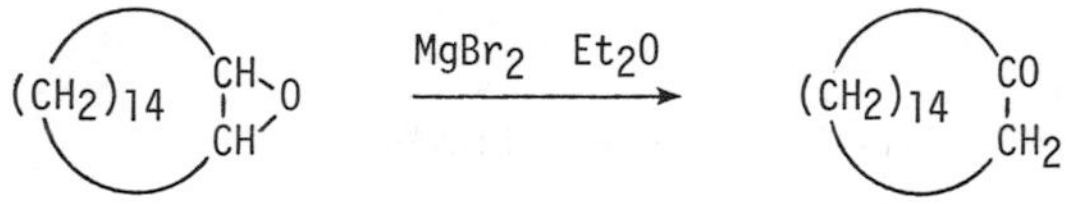

75%

Helv (1948) <u>31</u> 1077
Chem Rev (1959) <u>59</u> 737

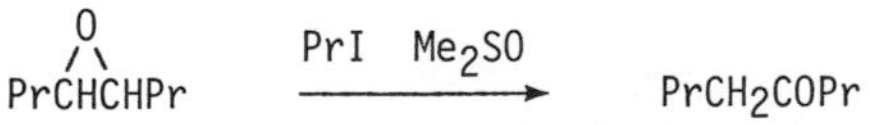

100%

Chem Comm (1968) 227

Bu$_3$PO LiClO$_4$ C$_6$H$_6$ 80%

JACS (1968) 90 4193
     (1965) 87 1405

MeCHCHMe $\xrightarrow{\text{Co}_2(\text{CO})_8 \quad \text{MeOH}}$ MeCH$_2$COMe   77%

JOC (1962) 27 2706

PhC-CH$_2$ $\xrightarrow[\text{MeOH H}_2\text{O}]{\text{H}_2\text{O}_2 \quad \text{KOH}}$ PhCOMe   85-90%
    |
    Me

JACS (1957) 79 503

Section 175   Ketones from Halides and Sulfonates
∘∘

BuBr $\xrightarrow[\text{EtOH}]{\text{MeCOCH}_2\text{COOEt} \quad \text{EtONa}}$ MeCOCHCOOEt $\xrightarrow{\text{NaOH H}_2\text{O}}$ MeCOCH$_2$Bu   ∼40%
                                              |
                                              Bu

                                          Org Synth (1941) Coll Vol 1 248 351
For cleavage of β-ketoesters with CaI$_2$ see   JOC (1966) 31 3267
    "   "   "   "   " LiI   " Org Synth (1965) 45 7
    "   "   "   "   by pyrolysis see JOC (1957) 22 1189

BuBr $\xrightarrow[\text{THF}]{\overset{-\quad -\quad +}{\text{CH}_2\text{COCHCOOMe} \; 2\text{Na}}}$ BuCH$_2$COCH$_2$COOMe $\xrightarrow[\text{2 BuLi}\atop\text{3 PhCH}_2\text{Cl}]{\text{1 NaH}}$ BuCHCOCH$_2$COOMe
                                                                            |
                                                                            CH$_2$Ph
                                                    ⋮ Hydrolysis   ⋮ Hydrolysis
                                            BuCH$_2$COMe   BuCHCOMe
                                                                                              |
                                                                                              CH$_2$Ph

JACS (1970) 92 6702

$C_8H_{17}Br$ $\xrightarrow[\text{NaH \quad t-BuOH}]{CH_2(COOBu\text{-}t)_2}$ $C_8H_{17}CH(COOBu\text{-}t)_2$ $\xrightarrow[\text{2 } C_7H_{15}COCl]{1 \text{ NaH \quad } C_6H_6}$ $C_8H_{17}\underset{\underset{COC_7H_{15}}{|}}{C}(COOBu\text{-}t)_2$

$\downarrow$ TsOH   HOAc

$C_8H_{17}CH_2COC_7H_{15}$

JACS (1952) 74 831   46%
JCS (1952) 3945
   (1950) 325

BuI $\xrightarrow[\text{EtOH}]{MeCOCH_2COMe \quad K_2CO_3}$ $BuCH_2COMe$ 60%

JOC (1965) 30 3321

BuBr $\xrightarrow[\underset{Cl-\text{(pyran)}}{2}]{1 \text{ Mg \quad } Et_2O}$ Bu-(tetrahydropyran) $\xrightarrow[\text{340-350°}]{Pt\text{-}C}$ $BuCO(CH_2)_3Me$ 71%

Izv (1964) 747
(Chem Abs 61 3057)

BuBr $\xrightarrow[\text{2 PhCN}]{1 \text{ Mg \quad } Et_2O}$ PhCOBu

JACS (1957) 79 881

Further examples of the reaction RMgX + R'CN → RCOR' are included in section 178 (Ketones from Nitriles)

$\xrightarrow[\text{2 } CdCl_2 \\ \text{3 MeCOCl \quad } C_6H_6]{1 \text{ Mg \quad EtBr \quad } Et_2O}$

JCS (1955) 3986
Org React (1954) 8 28

BuBr $\xrightarrow[\text{2 } Ac_2O]{1 \text{ Mg \quad } Et_2O}$ BuCOMe   79%

JACS (1945) 67 154
JOC (1948) 13 592

Further examples of the preparation of ketones by reaction of organometallic derivatives of Li, Mg, Cd and Zn with carboxylic acids, acid halides and anhydrides are included in section 167 (Ketones from Carboxylic Acids, Acid Halides and Anhydrides)

i-PrCl $\xrightarrow[\text{3 CO}_2]{\substack{\text{1 Li Et}_2\text{O} \\ \text{2 Reflux}}}$ (i-PrCH$_2$CH$_2$)$_2$CO

JACS (1953) $\underline{75}$ 1771

PhBr $\xrightarrow[\text{2 CO}_2]{\text{1 Li Et}_2\text{O}}$ Ph$_2$CO

JACS (1933) $\underline{55}$ 1258
(1955) $\underline{77}$ 2806   70%

i-PrCH$_2$CH$_2$Br $\xrightarrow[\text{2 EtCONMe}_2]{\text{1 Li Et}_2\text{O}}$ i-PrCH$_2$CH$_2$COEt   75%

JOC (1959) $\underline{24}$ 701

$\xrightarrow[\text{2 Me}_2\text{NCOOEt}]{\text{1 EtLi Et}_2\text{O}}$ 75%

Tetr Lett (1970) 5219

BuBr $\xrightarrow[\substack{\text{2 CNCMe}_2 \\ \quad\text{CH}_2\text{Bu-t}}]{\text{1 Li}}$ $\underset{\text{CH}_2\text{Bu-t}}{\overset{\text{Li}}{\text{BuC=NCMe}_2}}$ $\xrightarrow[\text{2 (COOH)}_2]{\text{1 EtBr}}$ BuCOEt   87%

JACS (1970) $\underline{92}$ 6675

BuX $\xrightarrow[\text{2 PrCHO}]{\text{1Mg}}$ $\underset{\overset{|}{\text{Pr}}}{\text{BuCHOMgX}}$ $\xrightarrow{\substack{\text{NCOOEt} \\ \| \\ \text{NCOOEt}}}$ BuCOPr   66%

Bull Chem Soc Jap (1968) $\underline{41}$ 1491

i-PrBr $\xrightarrow[\text{2 MeCH=CHCOMe}]{\text{1 Mg Et}_2\text{O}}$ MeCHCH$_2$COMe
                                                  |
                                                i-Pr

JACS (1951) <u>73</u> 2721

For promotion of 1:4 addition of Grignard
                    reagents by CuX$_2$ see JACS (1941) <u>63</u> 2308
                                    and JACS (1965) <u>87</u> 82
                                    and JOC (1966) <u>31</u> 3128

$\xrightarrow[\text{2 2-Chlorocyclohexanone}]{\text{1 Mg Et}_2\text{O}}$

48%

JOC (1959) <u>24</u> 843
     (1947) <u>12</u> 737

PhBr $\xrightarrow[\substack{\text{2 CuBr}\\\text{3 N}_2\text{CHCOPh}}]{\text{1 Li Et}_2\text{O}}$ PhCH$_2$COPh

35%

Chem Comm (1969) 515

$\xrightarrow{\text{1 Mg THF}}$

                                    Me
                                    |
                              CH$_2$CHCOMe

82%

3 MeLi
4 (COOH)$_2$

JACS (1969) <u>91</u> 5887

BuBr $\xrightarrow{\text{1 Mg Et}_2\text{O}}$ BuCOCH$_2$CH$_2$Ph

71%

3 (COOH)$_2$ H$_2$O

JACS (1970) <u>92</u> 1084

PhCl $\dashrightarrow$ PhMgCl $\dashrightarrow$ PhHgCl $\xrightarrow{\text{CO RhCl}_3\text{ MeCN}}$ Ph$_2$CO

<48%

JACS (1968) <u>90</u> 5546

BuBr   - - →   BuHgBr $\xrightarrow[\text{2 Ph}_3\text{P \quad C}_6\text{H}_6]{\text{1 Co}_2\text{(CO)}_8 \quad \text{THF}}$ Bu$_2$CO   <42%

JACS (1969) <u>91</u> 3037

$\xrightarrow[\substack{\text{2 9-Borabicyclo[3.3.1]-}\\ \text{nonane}\\ \text{3 MeSO}_3\text{H}\\ \text{4 PhCOCH}_2\text{Br \quad t-BuOK \quad THF}}]{\text{1 Li \quad Et}_2\text{O}}$

Me—C$_6$H$_4$—Br → Me—C$_6$H$_4$—CH$_2$COPh   95%

JACS (1969) <u>91</u> 4304 6852

PhCH$_2$Br $\xrightarrow{\text{Li[PhCOFe(CO)}_4\text{] \quad C}_6\text{H}_6}$ PhCH$_2$COPh   67%

JOC (1970) <u>35</u> 4183
Trans New York Acad Sci (1965) <u>27</u> 724

C$_6$H$_{13}$Br $\xrightarrow[\substack{\text{2 Ph}_3\text{PCMe}_2 \text{ Cl \quad THF}\\ \quad\quad \text{COOEt}}]{\text{1 Mg}}$ C$_6$H$_{13}$COPr-i   45%

Tetr Lett (1969) 23

PhCH$_2$Br —

   1   C$_5$H$_5$N$^+$CH=CMe—O$^-$ DMF $\xrightarrow{}$ 2 Zn   HOAc   H$_2$O   PhCH$_2$CH$_2$COMe   85%

   1   C$_5$H$_5$N^{+-}NCHCOCHN^{-+}NC$_5$H$_5$ DMA $\xrightarrow{}$ 2 Zn   HOAc   H$_2$O   (PhCH$_2$CH$_2$)$_2$CO   40%

Aust J Chem (1967) <u>20</u> 2441

BuBr   - - →   BuC(COOH)$_2$ $\xrightarrow[\text{Pyr \quad C}_6\text{H}_6]{\text{Pb(OAc)}_4}$ BuCOEt
      |
      Et

Tetr Lett (1966) 6145

BuBr --→ BuOCHCH=CHMe $\xrightarrow[\text{Me}_2\text{NCH}_2\text{CH}_2\text{NMe}_2]{\text{PrLI}}$ BuCHCH$_2$COMe   ~30%

with Me under BuOCHCH=CHMe and Me under BuCHCH$_2$COMe

Tetr Lett (1969) 821

i-PrI

1 MeCH(S,S ring)   BuLi   THF
2 HgCl$_2$
  →   i-PrCOMe   84%

1 CH$_2$(S,S ring)   BuLi   THF
2 BuLi   i-PrI
3 HgCl$_2$
  →   (i-Pr)$_2$CO   70%

Angew (1965) <u>77</u> 1134
(Internat Ed <u>4</u> 1075)
Tetr Lett (1969) 173
JOC (1968) <u>33</u> 298
Synthesis (1969) 17

(cyclopentene-Cl) $\xrightarrow[\text{2 H}_2\text{SO}_4\ \text{H}_2\text{O}]{\text{1 Na}_2\text{Cr}_2\text{O}_7\ \text{H}_2\text{O}}$ (cyclopentenone)

Ber (1956) <u>89</u> 1732

C$_6$H$_{13}$CHMe (with I below) $\xrightarrow{\text{MgO} \quad \text{Me}_2\text{SO}}$ C$_6$H$_{13}$COMe

JCS (1964) 520
JOC (1959) <u>24</u> 1792
    (1960) <u>25</u> 670
Chem Rev (1967) <u>67</u> 247   32%

(cyclopentane-Br) $\xrightarrow{\text{Me}_3\overset{+}{N}\overset{-}{O} \quad \text{CHCl}_3}$ (cyclopentanone)

Ber (1961) <u>94</u> 1360   50%

C$_6$H$_{13}$CHMe (with Cl below) $\xrightarrow[\text{2 O}_2\ \text{Et}_2\text{O}]{\text{1 Mg} \quad \text{Et}_2\text{O}}$ C$_6$H$_{13}$CHMe (with OOH below) $\xrightarrow{\text{NaOH}}$ C$_6$H$_{13}$COMe

91%

JACS (1955) <u>77</u> 6032
Ber (1960) <u>93</u> 2151

Tetr Lett (1970) 2679

Ber (1963) 96 1899
    (1961) 94 1987

$Cl(CH_2)_6Br$

1 LiCH(S(CH_2)_3S)   THF

2 BuLi

$(CH_2)_6$ CO   68%

JOC (1968) 33 300

Ph_2CBr_2

1 Morpholine

2 HCl H_2O

Ph_2CO 63%

Bull Soc Chim Fr (1965) 3544
Org Synth (1941) Coll Vol 1 95
JACS (1966) 88 3515

Section 176   Ketones from Hydrides (RH)
oooooooooooooooooooooooooooo

This section lists examples of the replacement of hydrogen by ketonic
groups e.g. RH ⟶ RCOMe (R=alkyl, aryl, vinyl etc.). For the oxidation
of methylenes R_2CH_2 ⟶ R_2CO, and the degradation of alkyls R_2CHR' ⟶
R_2CO, see section 170 (Ketones from Alkyls and Methylenes)

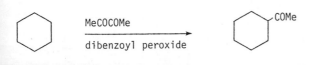

MeCOCOMe

dibenzoyl peroxide

COMe

JACS (1968) 90 3588

$$Me_2CHCH_3 \xrightarrow{\quad MeCOCl \quad AlCl_3 \quad} Me_2CHCH_2COMe$$

Ber (1936) <u>69B</u> 2244

$$BuCH=CH_2 \xrightarrow{\quad MeCOOH \quad (CF_3CO)_2O \quad} BuCH=CHCOMe \qquad 22\%$$

JCS (1953) 3628

Review: The Friedel-Crafts Acylation Reaction and its Application to
Polycyclic Aromatic Hydrocarbons Chem Rev (1955) <u>55</u> 229

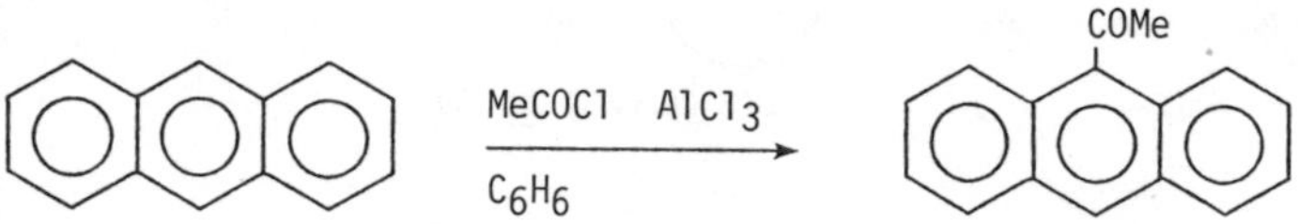

MeCOCl   AlCl$_3$

C$_6$H$_6$

57-60%

Org Synth (1963) Coll Vol 4 8
(1955) Coll Vol 3 23

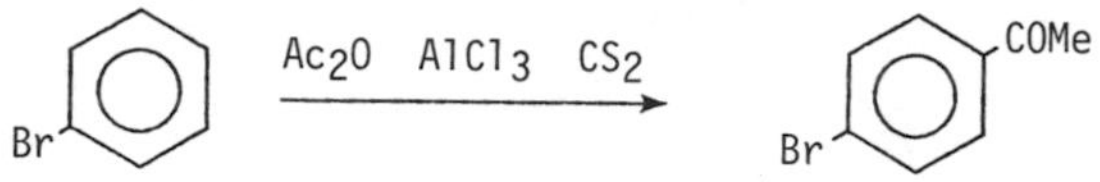

Ac$_2$O   AlCl$_3$ CS$_2$

69-79%

Org Synth (1941) Coll Vol 1 109

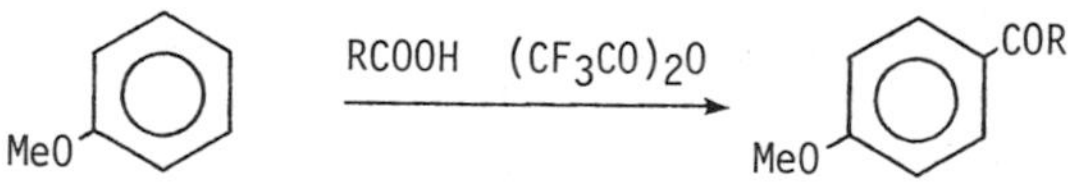

RCOOH   (CF$_3$CO)$_2$O

78% (R=Me)
56% (R=Ph)

JCS (1951) 718

$$PhH \xrightarrow{\quad \overset{+}{R}C\overset{-}{O} \; SbF_6 \quad} PhCOR$$

81-93% (R=Et)
86-93% (R=Ph)

JACS (1962) <u>84</u> 2733

JOC (1970) 35 2351

83%

JOC (1952) 17 1281
JCS (1932) 642

70%

Section 177   Ketones from Ketones

Direct alkylation and arylation of ketones with
                        halides, olefins and alcohols . . . page 419-421
Alkylation and arylation of ketones by indirect methods 421-426
Homologation and ring expansion of ketones 427-429
Transposition of carbonyl groups 429-431
Degradation and ring contraction of ketones 432-433

All reactions of enones forming <u>saturated</u> ketones (e.g. RCH=CHCOR ⟶
RCH$_2$-CHCOR) are listed in section 180 (Ketones from Miscellaneous Compounds)
    |
    R'

t-BuCOCH$_2$Me $\xrightarrow[\text{2 BuI}]{\text{1 NaH THF}}$ t-BuCOCHMe   57%
                                                  |
                                              Bu

Bull Soc Chim Fr (1968) 4990
Tetrahedron (1968) 24 6583

JACS (1958) 80 5220

40%

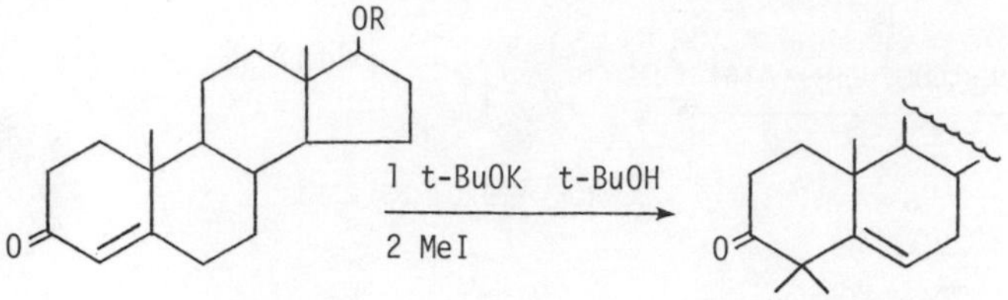

JCS (1956) 4490
    (1960) 67

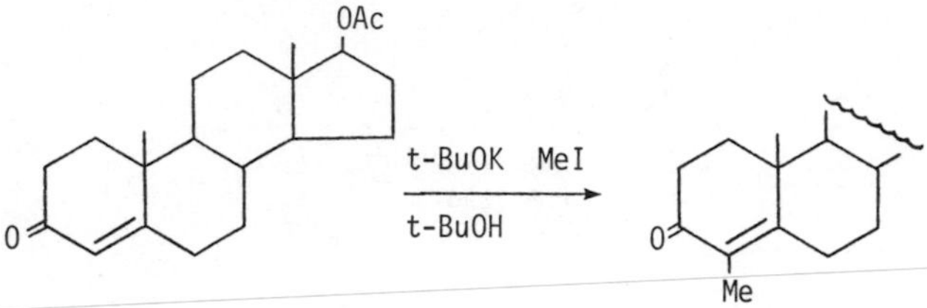

44%

JACS (1960) 82 2847
     (1962) 84 3402

$$MeCOCH_3 \xrightarrow[\text{Et}_2O \quad C_6H_6]{\overset{\overset{\displaystyle Et}{|}}{Me_2CONa} \quad PhCH_2Br} MeCOC(CH_2Ph)_3$$

Bull Soc Chim Fr (1961) 836
                  (1956) 1392

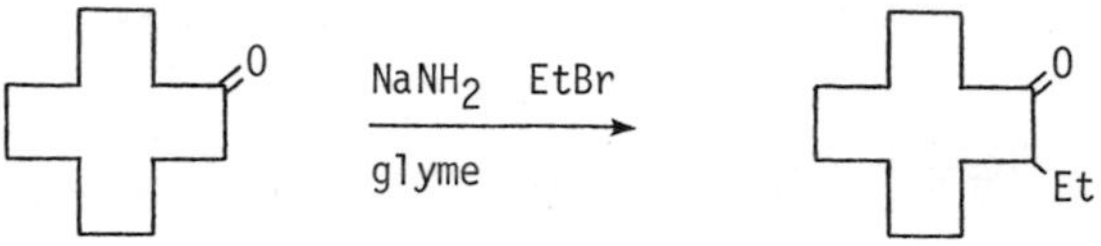

45%

JACS (1969) 91 1264
     (1953) 75 369
Org Synth (1955) Coll Vol 3 44

$$MeCOCH_2Et \xrightarrow[]{NaNH_2 \quad PhBr \quad NH_3} \underset{\underset{\displaystyle Ph}{|}}{MeCOCHEt}$$

65%

JACS (1959) 81 1169

$$PhCOCH_2Et \xrightarrow[\text{2 EtBr}]{\text{1 Ph}_3CNa \quad Et_2O} \underset{\underset{\displaystyle Et}{|}}{PhCOCHEt}$$

62%

JACS (1957) 79 881
     (1959) 81 1745

i-PrCOCHMe$_2$ $\xrightarrow[\text{2 Me}_2\text{SO}_4 \text{ or MeI}]{\text{1 BuMgBr   HMPA}}$ i-PrCOCMe$_2$
                                                              |
                                                              Me

Compt Rend (1966) <u>C</u> <u>263</u> 488

1 (Me$_3$Si)$_2$NNa
2 MeI

Chem Comm (1969) 1498

PhCOCH$_3$ $\xrightarrow[\text{di-t-butyl peroxide}]{C_6H_{13}CH=CH_2}$ PhCOCH$_2$C$_8$H$_{17}$ 10%

JCS (1965) 1918

PhCOCH$_3$ $\xrightarrow[\text{xylene}]{\text{PhCH}_2\text{OH   PhCH}_2\text{OLi}}$ PhCOCH$_2$CH$_2$Ph   70%

JACS (1956) <u>78</u> 4950

HCOOEt   MeONa
C$_6$H$_6$

MeI   K$_2$CO$_3$
Me$_2$CO

NaOH
H$_2$O

JCS (1954) 1373
JACS (1947) <u>69</u> 1361
     (1957) <u>79</u> 6313

32%

CO(OMe)$_2$
MeONa

1 MeI   MeOH
2 HCl   H$_2$O
    HOAc

83%

Tetrahedron (1957) <u>1</u> 49

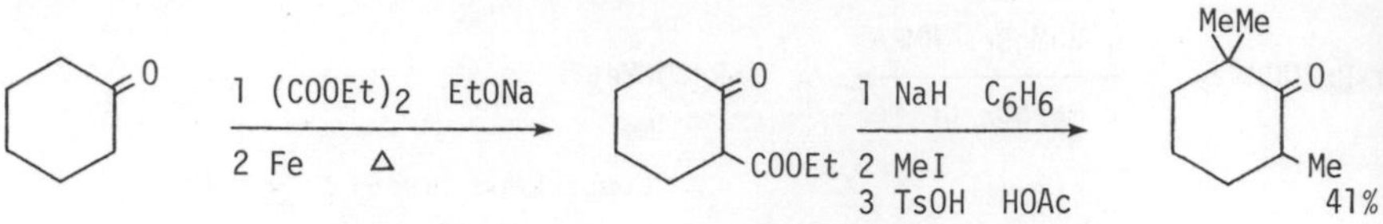

JACS (1958) 80 4072
Org Synth (1943) Coll Vol 2 531

For the preparation of β-ketoesters from ketones via enamines see
JACS (1963) 85 207

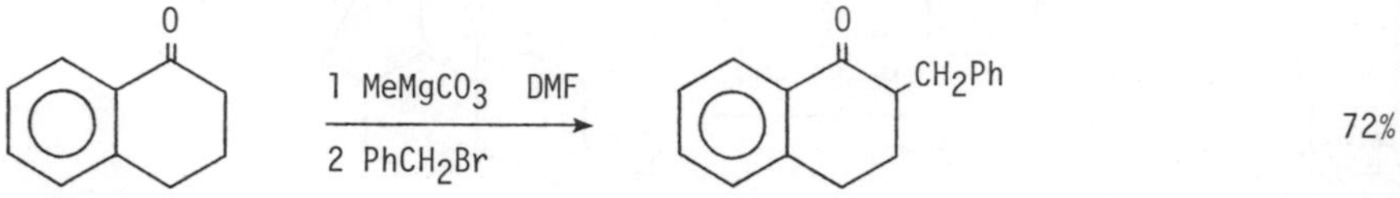

JACS (1959) 81 2598

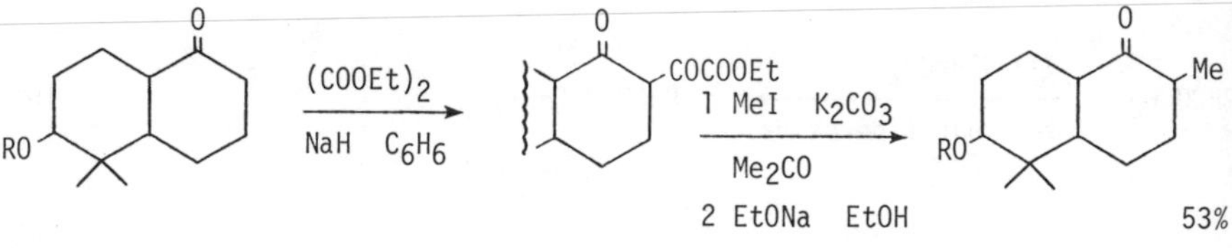

JACS (1958) 80 1967 5220

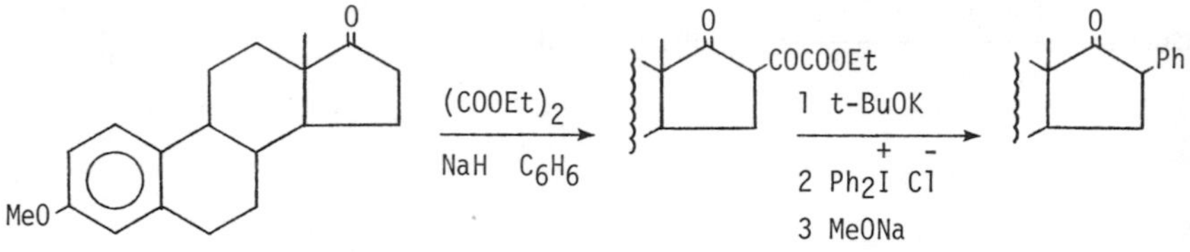

J Med Chem (1967) 10 106

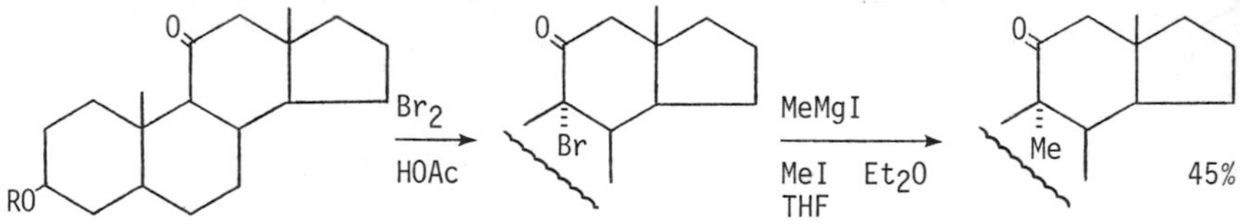

JCS (1964) 1161
JOC (1961) 26 2426
Tetr Lett (1969) 2269

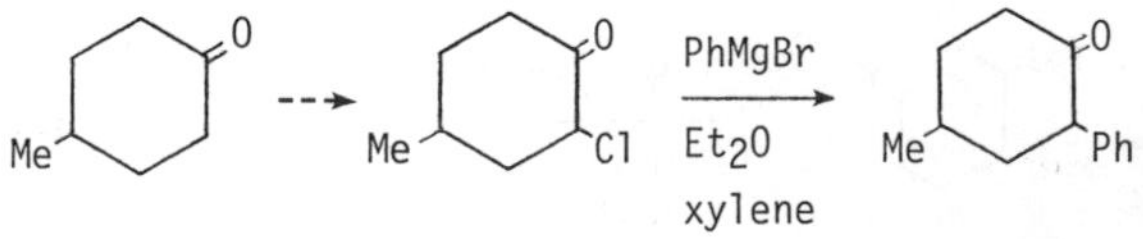

JOC (1947) 12 737
    (1959) 24 843
JCS (1957) 4089

i-PrCOCHMe$_2$ --→ i-PrCOCMe$_2$ $\xrightarrow{\text{i-Pr}_2\text{CuLi}}$ i-PrCOCMe$_2$
                        | |
                        Br   i-Pr

Tetr Lett (1971) 177

PhCOCH$_3$ --→ PhCOCH$_2$Br $\xrightarrow[\text{THF}]{\text{Et}_3\text{B}\quad\text{t-BuOK}}$ PhCOCH$_2$Et

JACS (1968) <u>90</u> 6218
(1969) <u>91</u> 4304 6852

C$_6$H$_{13}$COCH$_2$C$_5$H$_{11}$ --→ C$_6$H$_{13}$COCHC$_5$H$_{11}$ $\xrightarrow[\text{2 MeI}]{\text{1 Zn}\quad\text{Me}_2\text{SO}\quad\text{C}_6\text{H}_6}$ C$_6$H$_{13}$COCHC$_5$H$_{11}$
                                          | |
                                          Br   Me

JACS (1967) <u>89</u> 5727

$\xrightarrow[\text{CCl}_4]{\text{Ac}_2\text{O}}$

$\xrightarrow[\substack{\text{MeOCH}_2\text{CH}_2\text{OMe}\\ \text{2 MeI}}]{\text{1 MeLi}\quad\text{Et}_2\text{O}}$

JOC (1965) <u>30</u> 2502
Tetr Lett (1971) 105

$\xrightarrow[\text{HClO}_4]{\text{(PhCO)}_2\text{O}}$

$\xrightarrow[\text{2 CH}_2\text{I}_2\quad\text{Zn-Cu}]{\text{1 MeLi}}$

65-75%

JOC (1969) <u>34</u> 1962

$\xrightarrow[\text{C}_6\text{H}_6]{\text{Pyrrolidine}}$

$\xrightarrow[\text{2 H}_2\text{SO}_4\quad\text{H}_2\text{O}]{\text{1 BuI}\quad\text{toluene}}$

44%

JACS (1963) <u>85</u> 207
Advances in Org Chem (1963) <u>4</u> 1

For preparation of enamines from unreactive ketones
see JCS (1965) 5142
and JOC (1967) <u>32</u> 213

C5H11COCH3 $\xrightarrow{\text{Cyclohexylamine}}$ C5H11CCH3
                                                          ‖
                                                          N—⬡

$\xrightarrow[\text{2 BuI}\ \ \text{3 Acid}]{\text{1 EtMgBr}}$ C5H11COCH2Bu

JACS (1963) 85 2178

⬡=O $\xrightarrow[\text{2 MeI}\ \ \text{K}_2\text{CO}_3]{\text{1 HCHO}\ \ \text{Me}_2\text{NH·HCl}}$ [ring with O and CH2NMe3⁺ I⁻] + −

$\xrightarrow[\substack{\text{pentyl}\\ \text{borane}}]{\text{Tricyclo-}}$ [ring with O and CH2—cyclopentyl] 85%

JACS (1968) 90 4166

PhCH2COMe $\xrightarrow[\text{Me}_2\text{NH·HCl}]{\text{HCHO}}$ PhCHCOMe
                                                                  |
                                                              CH2NMe2·HCl

$\xrightarrow[\text{Ni}\ \ \text{EtOH}]{\text{H}_2\ (60\text{-}100\ \text{atmos})}$ PhCHCOMe   56%
                                                                          |
                                                                          Me

JACS (1953) 75 1128

[naphthalenone structure with O] $\xrightarrow[\text{EtOH}\ \ \text{H}_2\text{O}]{\text{PhCHO}\ \ \text{NaOH}}$ [structure with O and =CHPh] $\xrightarrow[\text{EtOH}]{\text{H}_2\ \ \text{Pd-C}}$ [structure with O and CH2Ph] 91%

Org Prep and Procedures (1970) 2 37

[decalone structure with O] $\xrightarrow[\text{C}_6\text{H}_6]{\text{HCOOEt}\ \ \text{MeONa}}$ [structure with O and =CHOH] $\xrightarrow[\text{HCl}\ \ \text{MeOH}]{\text{H}_2\ \ \text{Pd-C}}$ [structure with O and Me] 70%

JOC (1965) 30 2502

[naphthalenone structure with O] $\xrightarrow[\substack{\text{C}_6\text{H}_6\\ \text{2 PhCOCl}\ \ \text{Pyr}}]{\text{1 HCOOEt}\ \ \text{MeONa}}$ [structure with O and =CHOCOPh] $\xrightarrow[\substack{\text{Et}_3\text{N}\\ \text{i-PrOH}}]{\text{H}_2\ \ \text{PtO}_2}$ [structure with O and Me] 60-78%

Synthesis (1970) 476
Org React (1954) 8 119

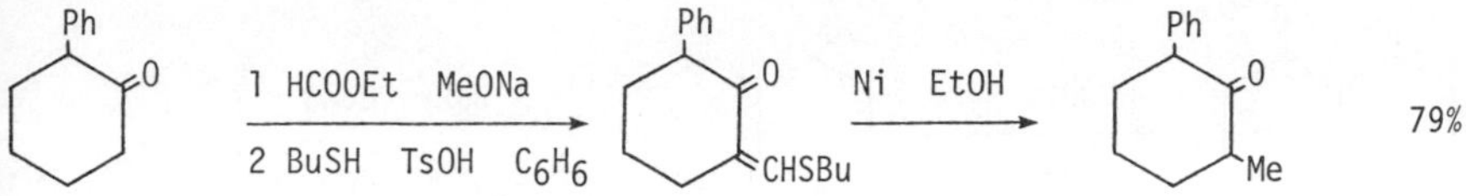

Chem Ind (1960) 1534

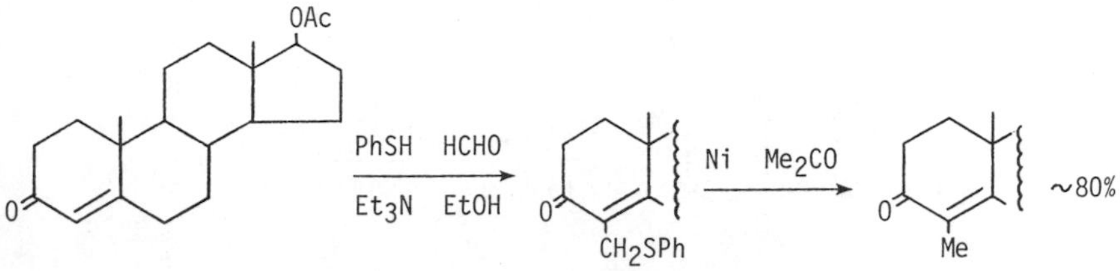

JCS (1962) 1091

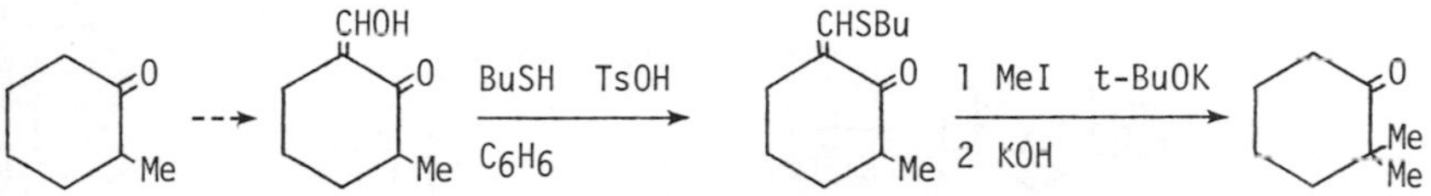

JOC (1962) 27 1615 1620
Tetrahedron (1968) 24 3095   74%

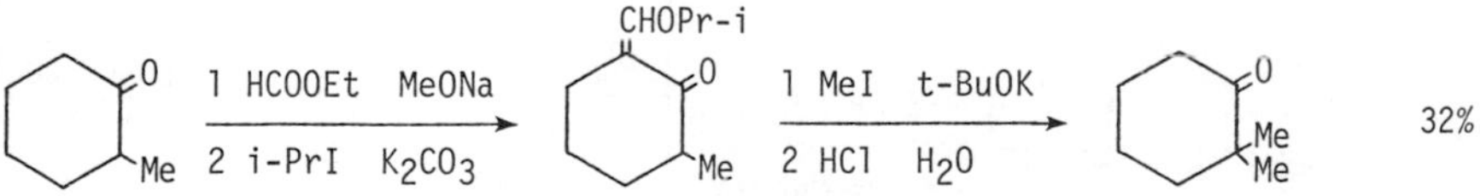

JACS (1947) 69 1361   32%

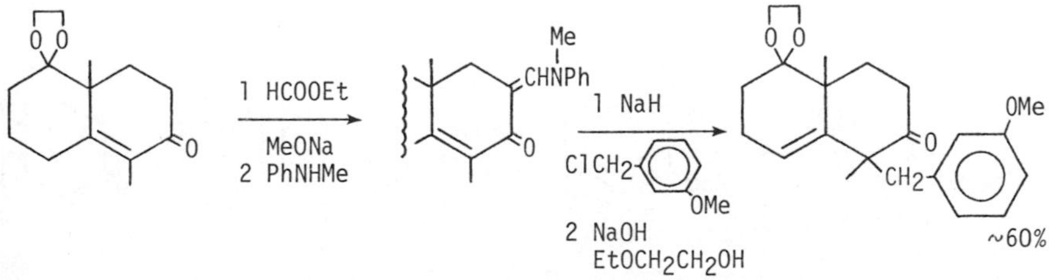

Tetrahedron (1969) 25 4011
JOC (1970) 35 468

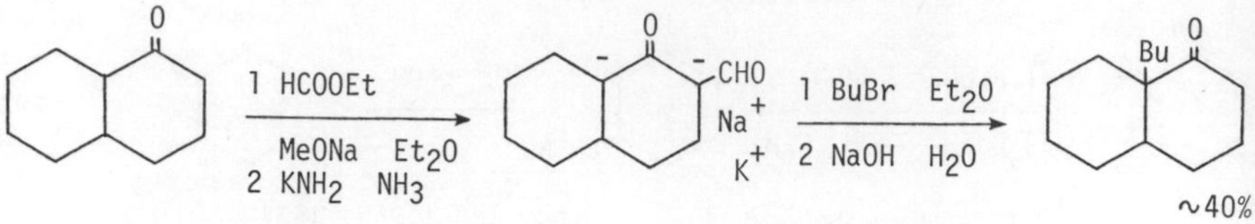

JACS (1965) **87** 82

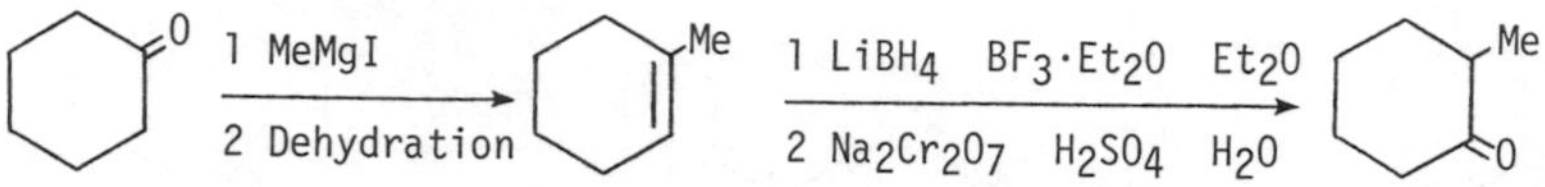

JACS (1961) **83** 2951

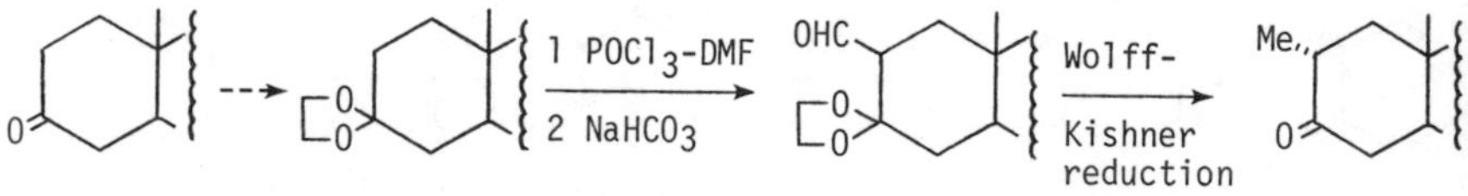

Tetr Lett (1964) 2161

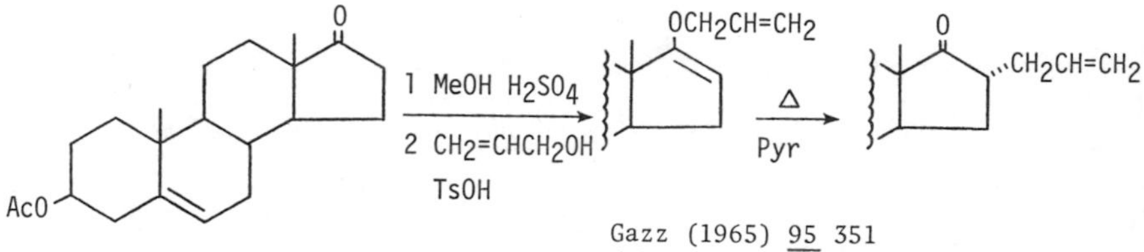

Gazz (1965) **95** 351
(Chem Abs **63** 11647)

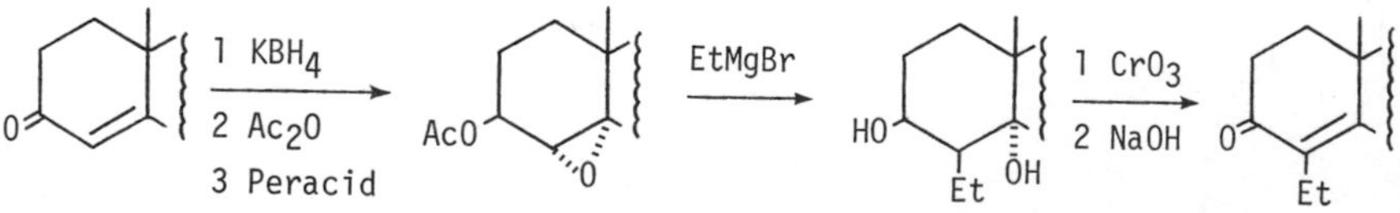

Bull Soc Chim Fr (1964) 321

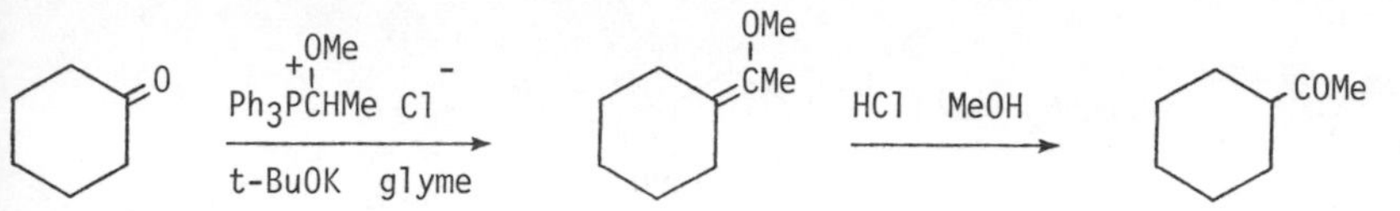

Tetr Lett (1964) 3323

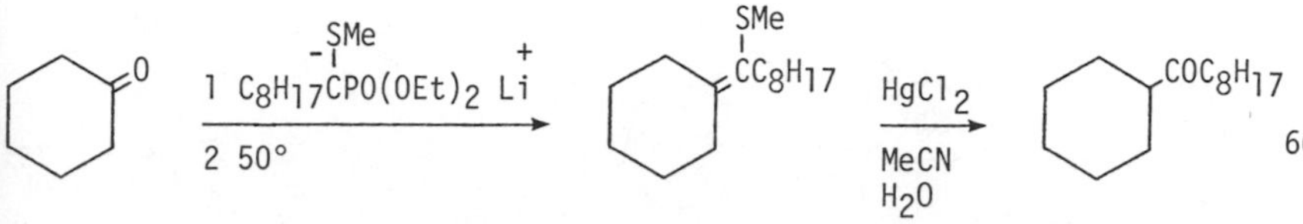

JOC (1970) 35 777

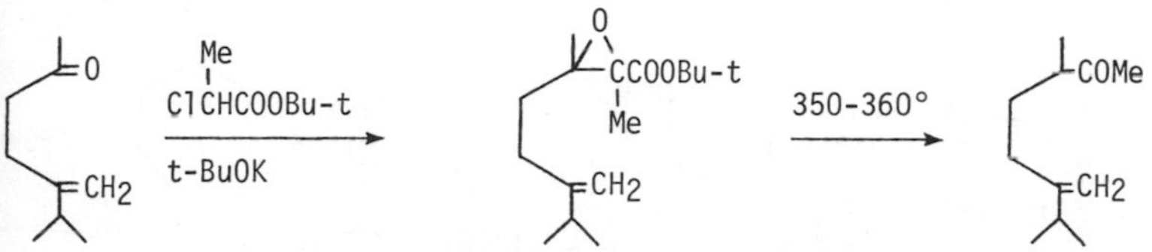

JACS (1963) 85 955
Org React (1949) 5 413
Chem Rev (1955) 55 283

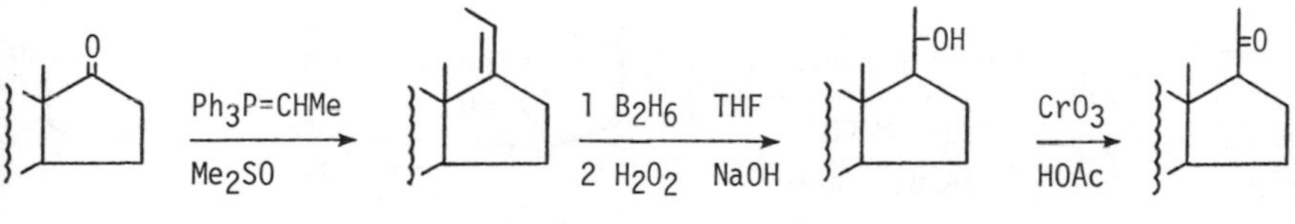

JOC (1966) 31 24

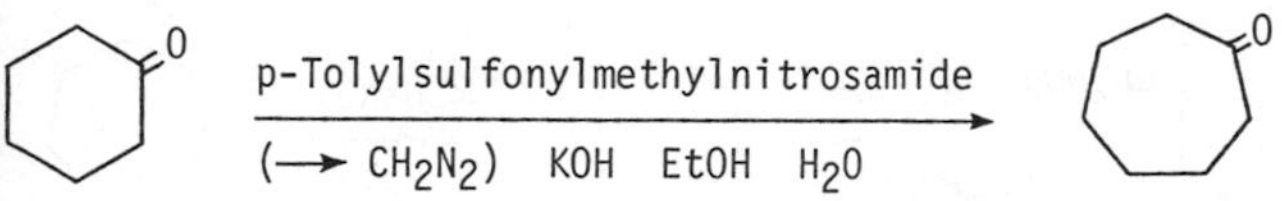

Org Synth (1963) Coll Vol 4 225
Org React (1954) 8 364
JACS (1966) 88 3515

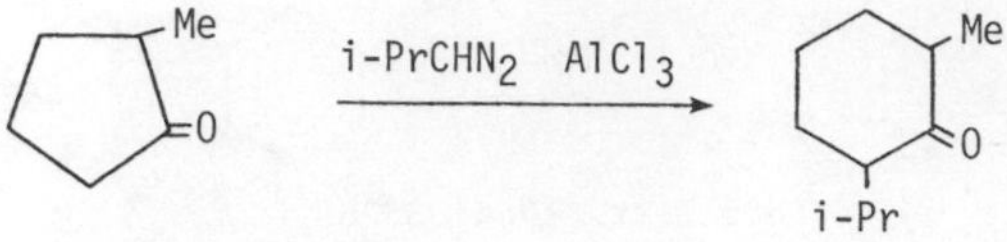

i-PrCHN$_2$ AlCl$_3$

Tetr Lett (1962) 775

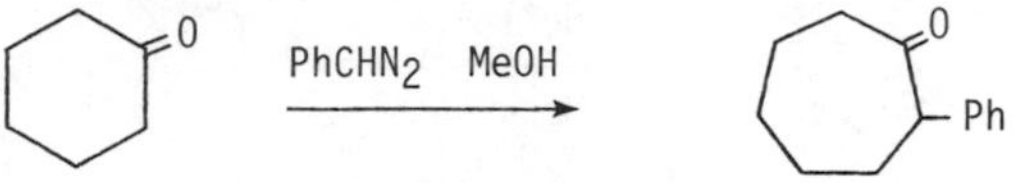

PhCHN$_2$ MeOH

76%

Org React (1954) $\underline{8}$ 364

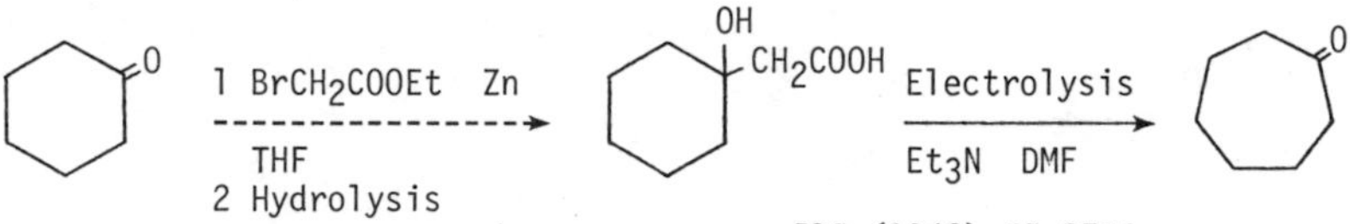

1 BrCH$_2$COOEt   Zn

THF
2 Hydrolysis

Electrolysis
Et$_3$N   DMF

<55%

JOC (1968) $\underline{33}$ 2704

1 HCN   piperidine

2 Ac$_2$O   MeCOCl

3 LiAlH$_4$ Et$_2$O

NaNO$_2$
HOAc
H$_2$O

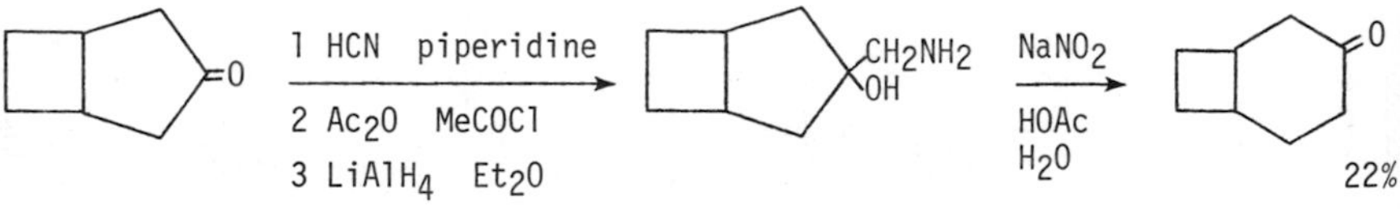

22%

JOC (1964) $\underline{29}$ 2914
JACS (1952) $\overline{74}$ 2278
JCS $\underline{C}$ (1970) $\overline{1454}$

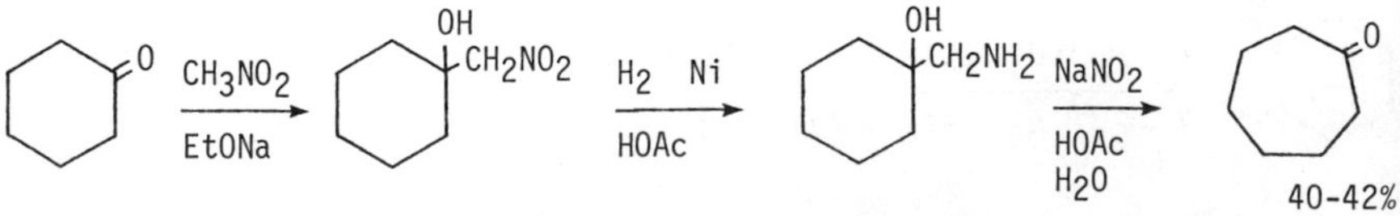

CH$_3$NO$_2$
EtONa

H$_2$ Ni
HOAc

NaNO$_2$
HOAc
H$_2$O

40-42%

Org Synth (1963) Coll Vol 4 221

JACS (1965) 87 1353

1 HC≡CCOOMe
2 HCl H₂O
3 H₂ Pd-Al₂O₃
4 NaOH
JOC (1967) 32 926

$$EtCOCH_2Me \xrightarrow{HClO_4 \quad H_2O} EtCH_2COMe \qquad \sim 63\%$$

JOC (1960) 25 1252
     (1969) 34 806

JOC (1963) 28 2626

11%

1 NaBH₄
2 Ac₂O
  Pyr
3 O₃

Zn
HOAc

JCS C (1970) 244

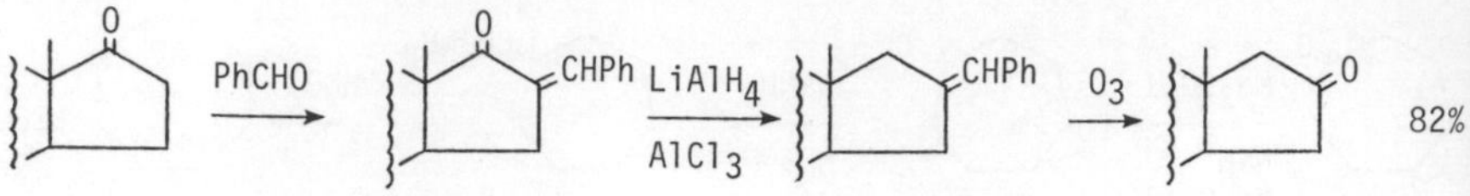

Chem Comm (1967) 898
Compt Rend (1967) C 265 929

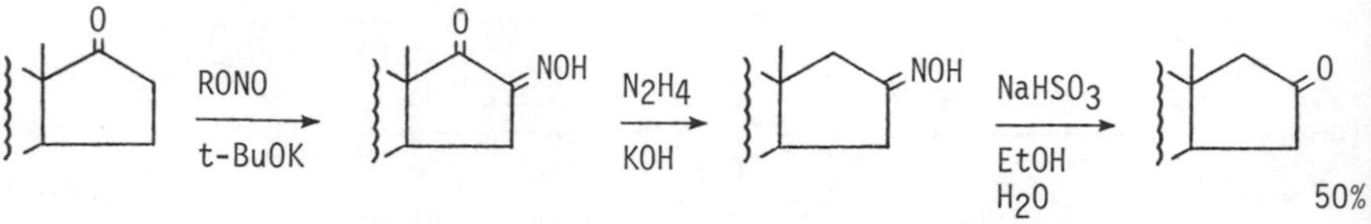

82%

Chem Comm (1968) 1350

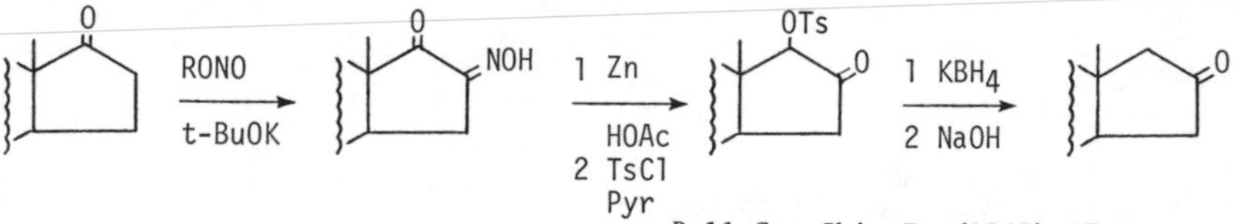

50%

Bull Soc Chim Fr (1965) 67

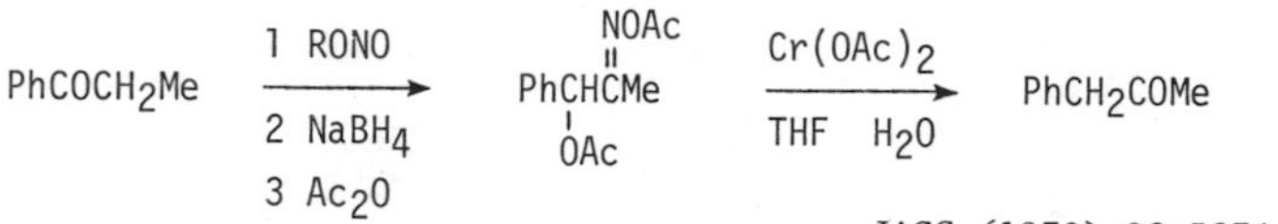

$$\text{PhCOCH}_2\text{Me} \xrightarrow[\substack{2 \text{ NaBH}_4 \\ 3 \text{ Ac}_2\text{O}}]{1 \text{ RONO}} \overset{\overset{\text{NOAc}}{\|}}{\underset{\underset{\text{OAc}}{|}}{\text{PhCHCMe}}} \xrightarrow[\text{THF} \quad \text{H}_2\text{O}]{\text{Cr(OAc)}_2} \text{PhCH}_2\text{COMe}$$

JACS (1970) 92 5276

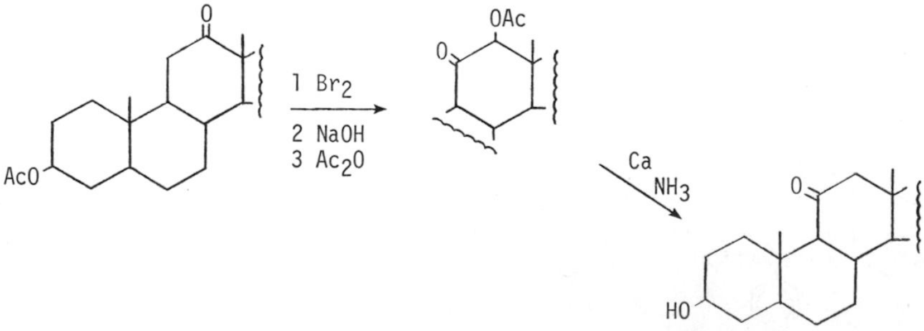

JCS (1956) 4330 4344

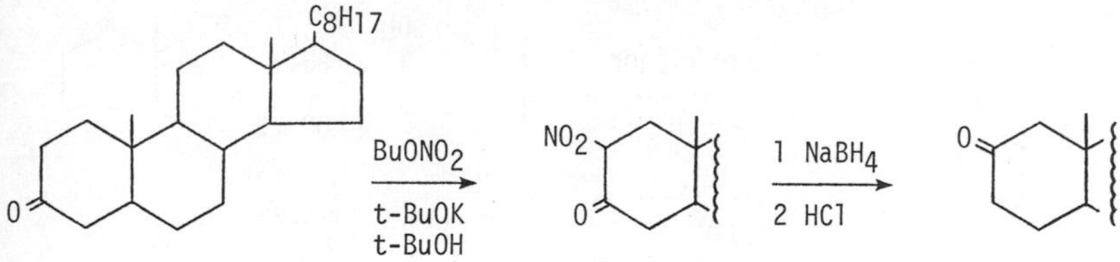

JOC (1968) 33 1733

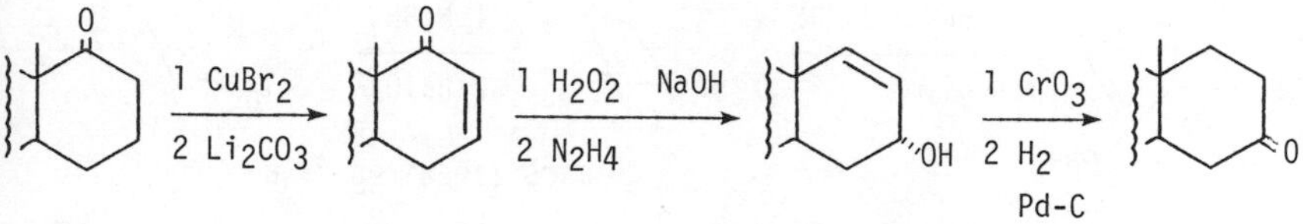

JCS C (1970) 1454

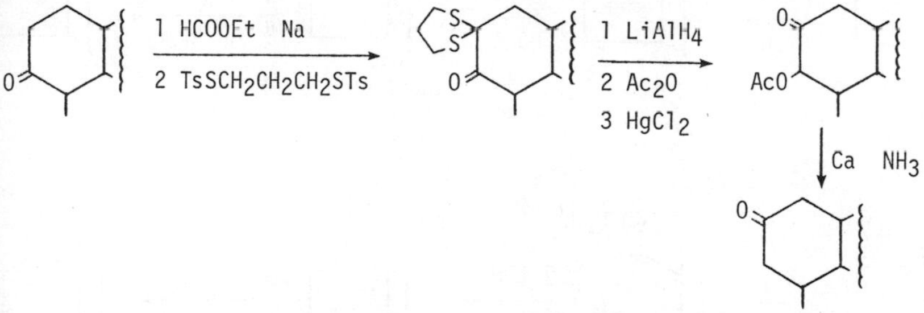

JOC (1969) 34 4188

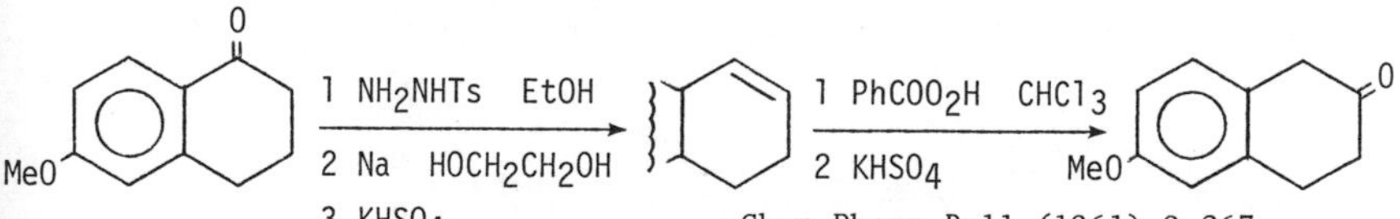

Chem Pharm Bull (1961) 9 267

PrCOCH₂Pr $\xrightarrow{\text{h}\nu \quad \text{t-BuOH}}$ PrCOMe

Tetr Lett (1968) 5385
Chem Comm (1969) 204

O_2 t-BuOK

t-BuOH THF

Base

60-70°

87%

Chem Ind (1966) 25

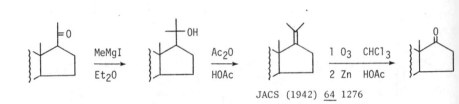

O_2 hν t-BuOH

monoglyme

1 $NaBH_4$

2 $NaIO_4$

JACS (1968) 90 2448

MeMgI

Et_2O

Ac_2O

HOAc

1 O_3 $CHCl_3$

2 Zn HOAc

JACS (1942) 64 1276

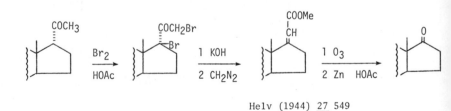

Br_2

HOAc

1 KOH

2 CH_2N_2

1 O_3

2 Zn HOAc

Helv (1944) 27 549

Ac_2O $HClO_4$

CCl_4

1 O_3

2 H_2 Pd-BaSO$_4$

35%

JCS (1962) 1572

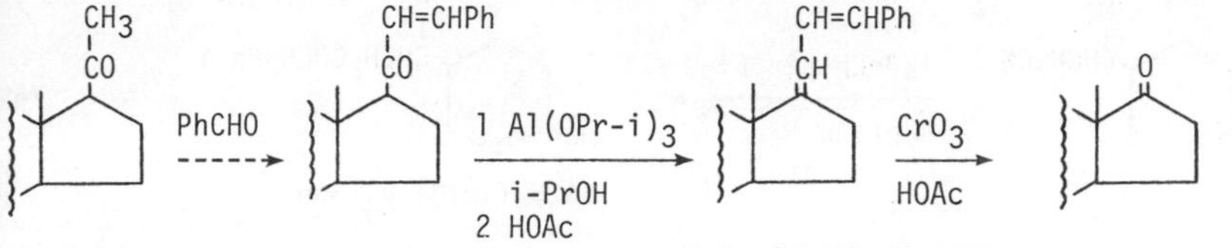

Helv (1949) 32 1795

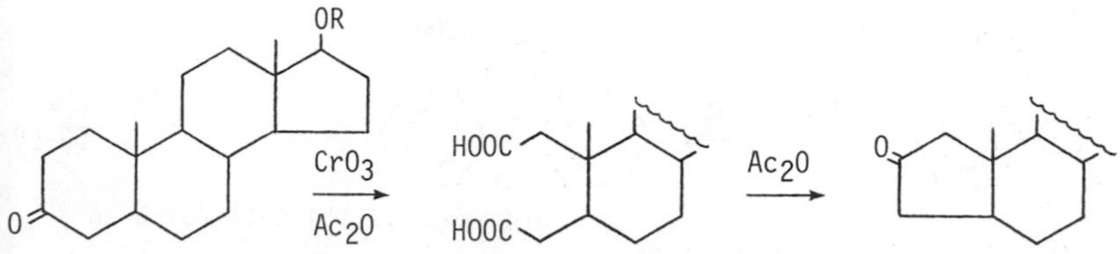

77%

Bull Soc Chim Fr (1958) 1573

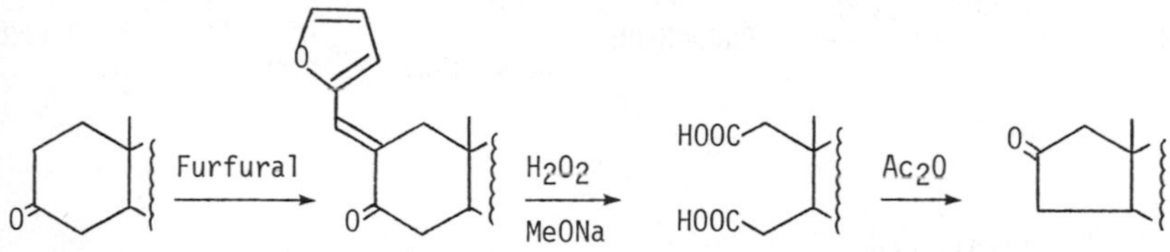

Bull Soc Chim Fr (1958) 1573

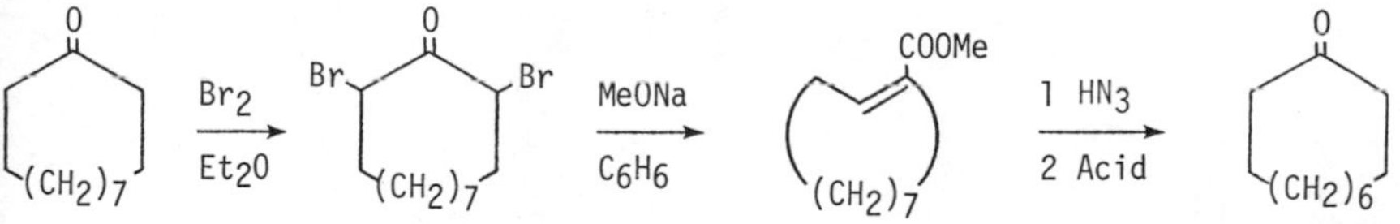

85-90%

JOC (1968) 33 2157

Section 178   Ketones from Nitriles
ooooooooooooooooooooooo

PhCN $\xrightarrow[\text{2 BuI}]{\text{1 MeCH}_2\text{Li   HMPA}}$ PhCOCHMe   70%
                                                    Bu

Compt Rend (1967) C 265 245
Bull Soc Chim Fr (1968) 4990

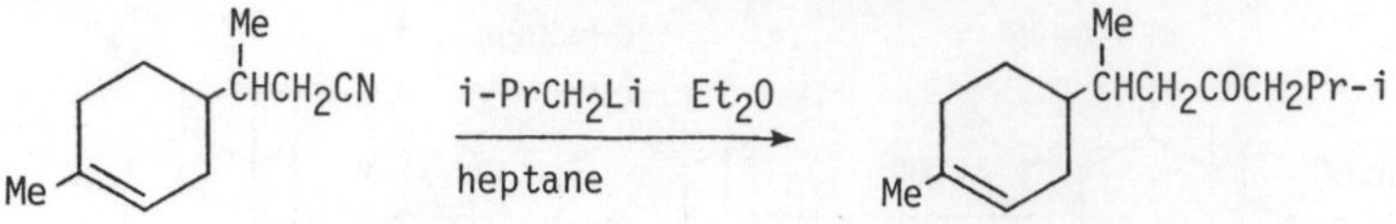

JACS (1970) <u>92</u> 336

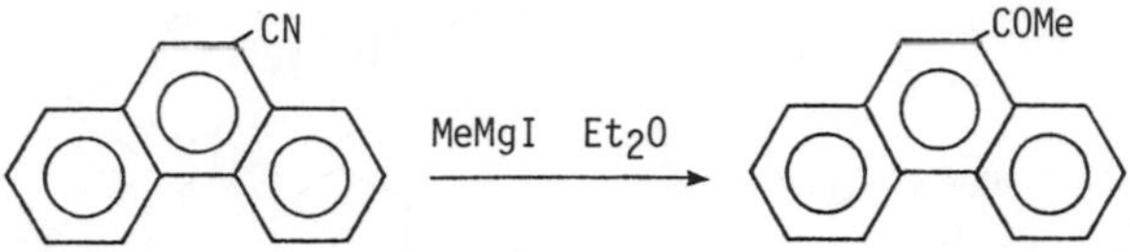

76%

Org Synth (1955) Coll Vol 3 26

52-59%

$$\text{BuCN} \xrightarrow{\text{Bu}_3\text{P=CHPh} \quad \text{Et}_2\text{O}} \text{BuCOCH}_2\text{Ph}$$

JACS (1967) <u>89</u> 7009

73%

$$\text{C}_{11}\text{H}_{23}\text{CN} \xrightarrow{\text{Et}_3\text{Al} \quad \text{C}_6\text{H}_6} \text{C}_{11}\text{H}_{23}\text{COEt}$$

Ber (1964) <u>97</u> 2661

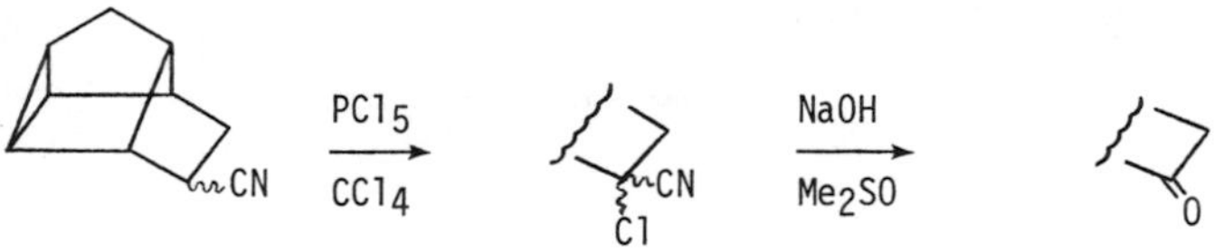

89%

Tetr Lett (1967) 437

47%

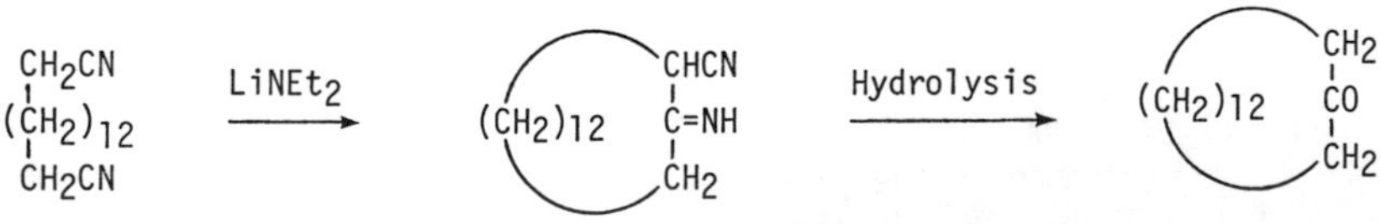

Annalen (1933) <u>504</u> 94
(1934) <u>513</u> 43

Section 179   Ketones from Olefins
             °°°°°°°°°°°°°°°°°°°°°°°

EtCH=CH$_2$ $\xrightarrow{\begin{array}{l}1\ B_2H_6\quad THF\\[6pt]2\ BrCH_2COMe\quad potassium\ 2,6\text{-}\\ \quad di\text{-}t\text{-}butylphenoxide\end{array}}$ Et(CH$_2$)$_3$COMe   84%

JACS (1969) 91 6852 2147

$\xrightarrow{\begin{array}{l}1\ B_2H_6\quad THF\\ 2\ N_2CHCOMe\\ 3\ NaOH\quad H_2O\end{array}}$ —CH$_2$COMe   67%

JACS (1968) 90 5936

$\xrightarrow{\begin{array}{l}1\ B_2H_6\quad THF\\ 2\ MeCH=CHCOMe\\ \quad air\quad i\text{-}PrOH\end{array}}$

Me
|
CHCH$_2$COMe   98%

JACS (1970) 92 714

$\xrightarrow{\begin{array}{l}1\ B_2H_6\quad THF\\ 2\ BuC\equiv CBr\\ 2\ NaOH\quad H_2O_2\quad H_2O\end{array}}$ —COCH$_2$Bu   79%

JACS (1967) 89 5086

C$_6$H$_{13}$CH=CH$_2$ $\xrightarrow[\ \ \ \ ]{B_2H_6}$ (C$_6$H$_{13}$CH$_2$CH$_2$)$_3$B $\xrightarrow{\begin{array}{l}1\ NaCN\quad diglyme\\ 2\ (CF_3CO)_2O\\ 3\ NaOH\quad H_2O_2\quad H_2O\end{array}}$ (C$_6$H$_{13}$CH$_2$CH$_2$)$_2$CO

~95%

Chem Comm (1970) 1529

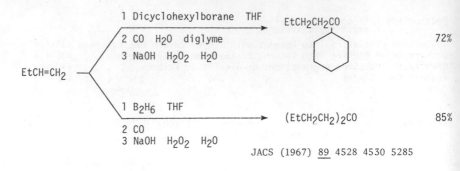

EtCH=CH₂

1 Dicyclohexylborane   THF
2 CO   H₂O   diglyme
3 NaOH   H₂O₂   H₂O

EtCH₂CH₂CO ⬡   72%

1 B₂H₆   THF
2 CO
3 NaOH   H₂O₂   H₂O

(EtCH₂CH₂)₂CO   85%

JACS (1967) 89 4528 4530 5285

⬠(with double bond) → (MeCOCl   AlCl₃) / cyclohexane → ⬠—COMe   49%

Ber (1936) 69B 1820

C₆H₁₃CH=CH₂   →   MeCH₂COMe / di-t-butyl peroxide   →   C₆H₁₃CH₂CH₂CHCOMe (Me)   ~30%

JCS (1965) 1918

(bicyclic =CH₂)   →   MeCHO / dibenzoyl peroxide   →   (bicyclic CH₂COMe)   30%

Aust J Chem (1967) 20 2033
JOC (1964) 29 245
    (1949) 14 248

(cyclohexene-Me)   →   1 LiBH₄   BF₃·Et₂O / 2 Na₂CrO₃   H₂SO₄   H₂O   →   (cyclohexanone with Me)   78%

JACS (1961) 83 2951
JOC (1962) 27 2938
Org React (1963) 13 1

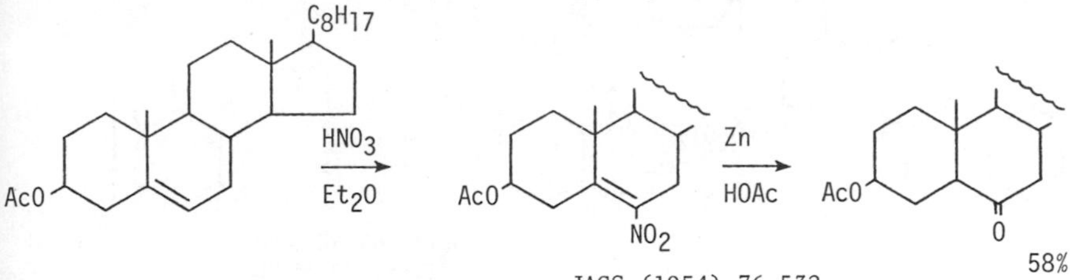

JACS (1954) 76 532

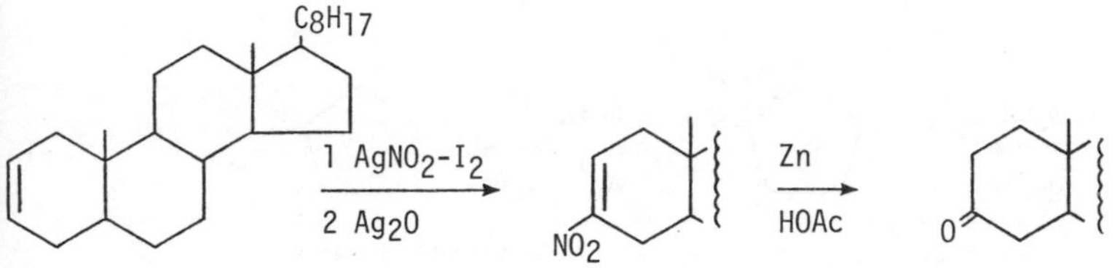

JOC (1969) 34 2628

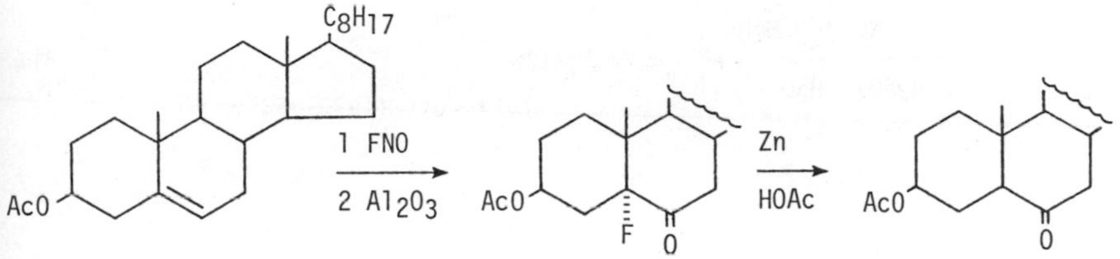

Chem Ind (1965) 1929

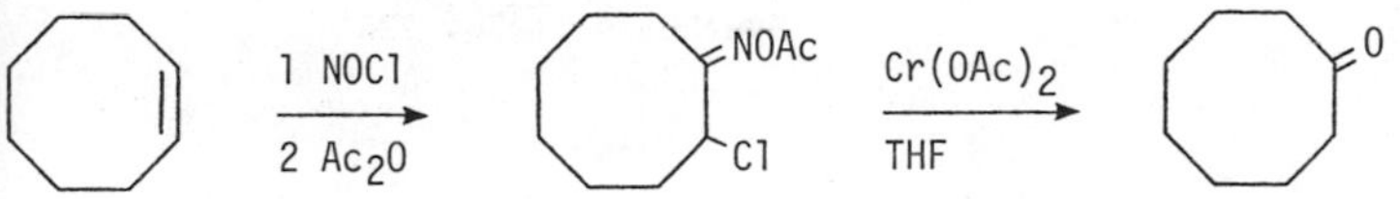

JACS (1970) 92 5276
Zh Org Khim (1966) 2 2178
(Chem Abs 66 75730)

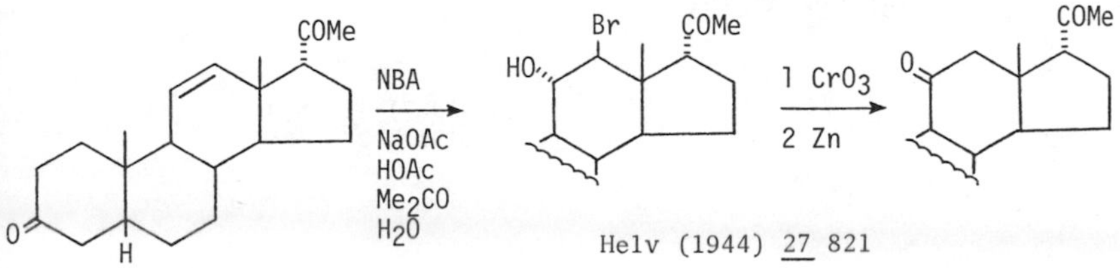

Helv (1944) 27 821

$$\xrightarrow[\text{CHCl}_3]{\text{CrO}_2\text{Cl}_2}$$

$$\xrightarrow[\text{2 Zn}]{\text{1 CrO}_3}$$

JACS (1956) 78 3749
JOC (1968) 33 3970

$$\xrightarrow[\text{MeCN}]{\text{N}_3\text{CN}}$$

$$\xrightarrow[\text{Ag}^+ \quad \text{H}_2\text{O}]{\text{Acid}}$$

91%

JACS (1964) 86 4506

$$\underset{\text{Me}}{\text{PhC=CH}_2} \xrightarrow[\text{2 H}_2\text{SO}_4 \quad \text{H}_2\text{O}]{\text{1 Tl(NO}_3)_3 \quad \text{MeOH}} \text{PhCH}_2\text{COMe}$$ 81%

Tetr Lett (1970) 5275

$$\text{C}_6\text{H}_{13}\text{CH=CH}_2 \xrightarrow[\text{PrOH}]{\text{O}_2 \quad \text{PdCl}_2 \quad \text{CuCl}_2\cdot 2\text{H}_2\text{O}} \text{C}_6\text{H}_{13}\text{COMe}$$ 68%

JOC (1969) 34 3949
Synthesis (1970) 225

$$\xrightarrow[\text{BF}_3\cdot\text{Et}_2\text{O} \quad \text{CH}_2\text{Cl}_2]{(\text{CF}_3\text{CO})_2\text{O} \quad \text{H}_2\text{O}_2}$$

41%

JOC (1967) 32 2669
      (1962) 27 666

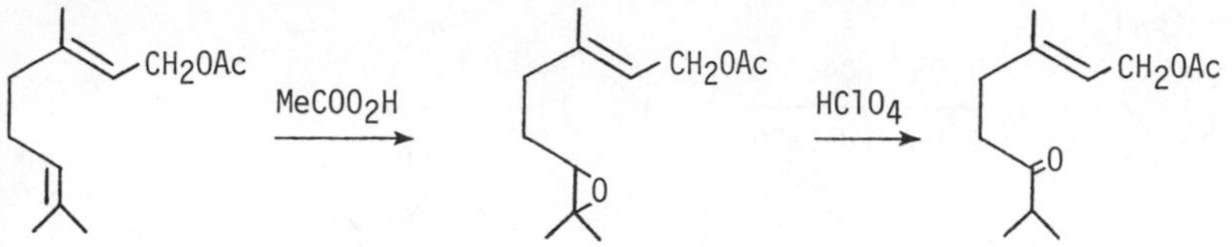

Tetr Lett (1964) 3481

Further examples of the epoxidation of olefins and the conversion of
epoxides into ketones are included in section 134 (Ethers and Epoxides
from Olefins) and section 174 (Ketones from Ethers and Epoxides)

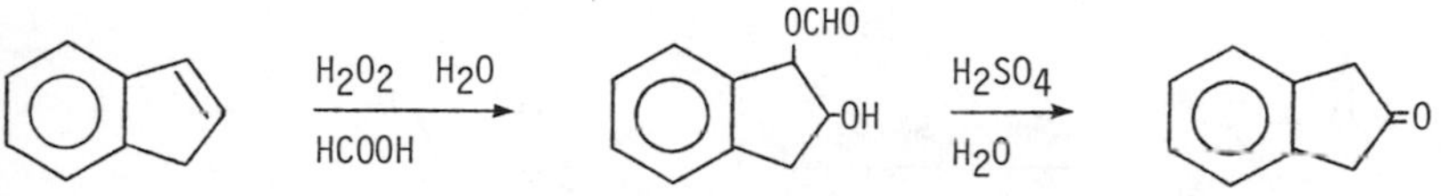

Org Synth (1961) 41 53

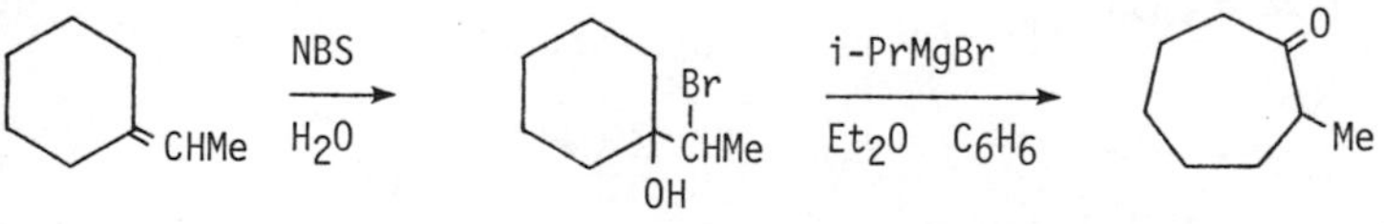

69-81%

JOC (1970) 35 2670

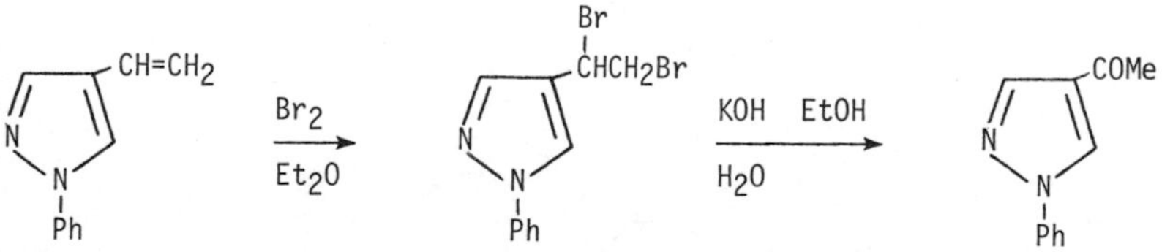

40%

JCS (1963) 3967

71%

$$\text{(indene with Ph, Ph)} \quad \xrightarrow[\text{2 } H_2 \text{ Pd-CaCO}_3]{\text{1 } O_3 \text{ HOAc}} \quad \text{(benzene with COPh, } CH_2COPh)$$

Ber (1954) **87** 993

Ozonides may also be reduced by the following reagents:

Zn-HOAc	JACS (1943) **65** 752
Ni	JACS (1941) **63** 3540
HCOOH	JACS (1948) **70** 4069
Tetracyanoethylene	Ber (1963) **96** 1564
	Bull Soc Chim Fr (1964) 729
Ph_3P	Angew (1956) **68** 473
$P(OMe)_3$-MeOH	JOC (1960) **25** 1031
NaI-HOAc	JOC (1968) **33** 1656
	Ber (1955) **88** 795
$PhNHNH_2$	Ber (1954) **87** 993

$$\text{(bicyclic with } =CH_2) \quad \xrightarrow[\substack{K_2CO_3 \; H_2O \\ \text{dioxane}}]{KMnO_4 \quad NaIO_4} \quad \text{(bicyclic with } =O)$$

JOC (1966) **31** 3028

$$\underset{\underset{Bu}{|}}{PhC{=}CHPh} \quad \xrightarrow[\text{dioxane } H_2O]{OsO_4 \quad NaIO_4} \quad \underset{71\%}{PhCOBu} \quad + \quad PhCHO$$

JACS (1966) **88** 476
(1957) **79** 6313
JCS **C** (1966) 655

$$\text{(cyclobutane} {=}CHPh) \quad \xrightarrow[\text{Pyr}]{OsO_4 \quad C_6H_6} \quad \text{(cyclobutane} \underset{OHOH}{-CHPh}) \quad \xrightarrow[\text{or NaIO}_4]{Pb(OAc)_4} \quad \text{(cyclobutane} {=}O)$$

JCS **C** (1966) 655

$$\xrightarrow[\text{CCl}_4]{RuO_4}$$

27%

Bull Soc Chim Fr (1964) 729
JCS (1962) 4745

Me
|
t-BuCH₂C=CH₂ →[K₂Cr₂O₇ H₂SO₄ H₂O] t-BuCH₂COMe   56%

JACS (1950) 72 3701

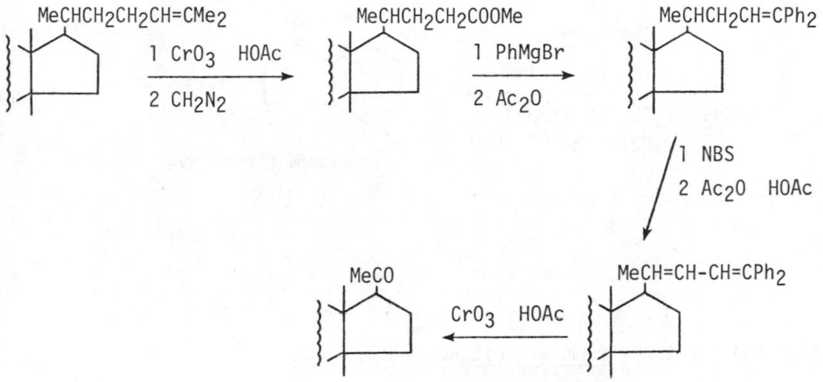

MeCHCH₂CH₂CH=CMe₂ →[1 CrO₃ HOAc / 2 CH₂N₂] MeCHCH₂CH₂COOMe →[1 PhMgBr / 2 Ac₂O] MeCHCH₂CH=CPh₂

↓ 1 NBS
  2 Ac₂O HOAc

MeCO ←[CrO₃ HOAc] MeCH=CH-CH=CPh₂

JACS (1970) 92 2059

PhC=CH₂   →[N-Nitrosopiperidine hν]   PhC=NOH --> PhCOMe
| |
Me   Me
                                          91%

JACS (1965) 87 4642

⬡=CH₂ →[N₂O / 300°] ⬡=O

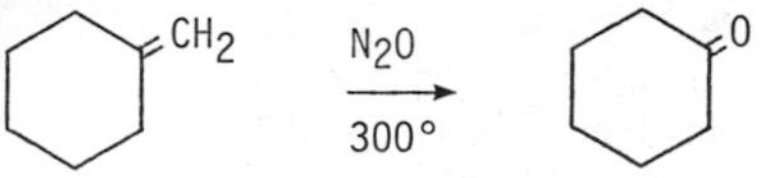

JCS (1951) 2999 3009

PhCH=CMe₂ →[NH₂NHNa Et₂O] Me₂C=NNH₂ --> Me₂CO
                             53%

Angew (1962) 74 650
(Internat Ed 1 456)

Ph$_2$C=CHEt $\xrightarrow{\text{NaN}_3 \quad \text{H}_2\text{SO}_4}$ PhCOCH$_2$Et   64%

JACS (1950) <u>72</u> 5777

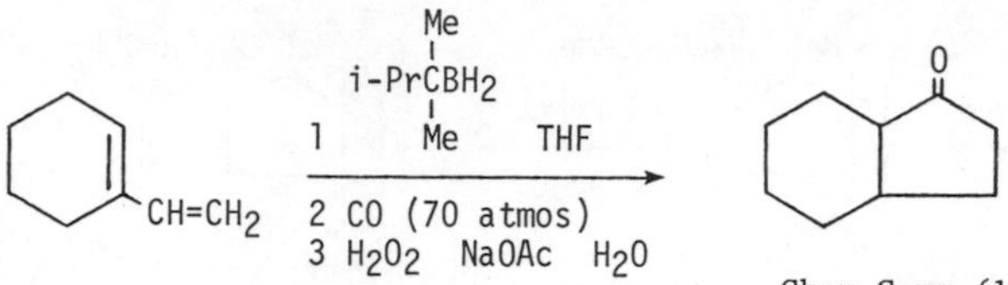

60%

Chem Comm (1968) 594

Section 180   Ketones from Miscellaneous Compounds

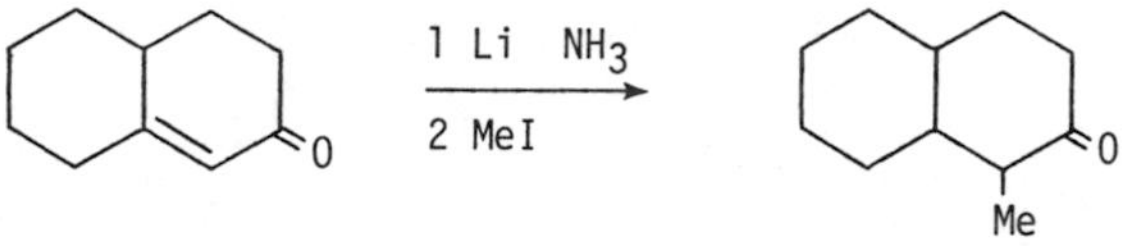

< 83%

JACS (1965) <u>87</u> 275
Tetrahedron (1964) <u>20</u> 357
JOC (1967) <u>32</u> 2851
Bull Soc Chim Fr (1969) 4356 4348

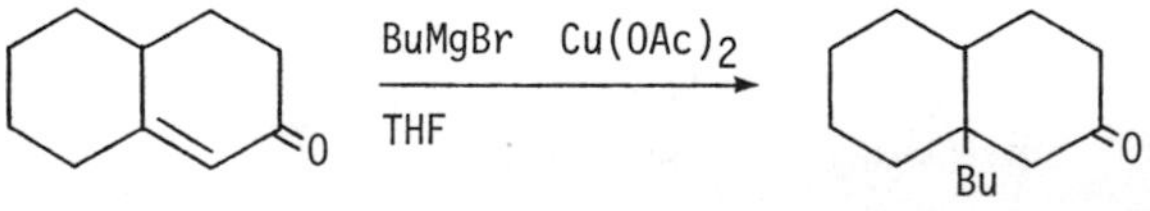

71%

JACS (1965) <u>87</u> 82
JOC (1966) <u>31</u> 3128
JCS (1943) 501

MeCH=CHCOMe $\xrightarrow[\text{Et}_2\text{O}]{\text{MeMgBr-Bu}_3\text{PCuI}}$ Me$_2$CHCH$_2$COMe   99%

JOC (1966) 31 3128

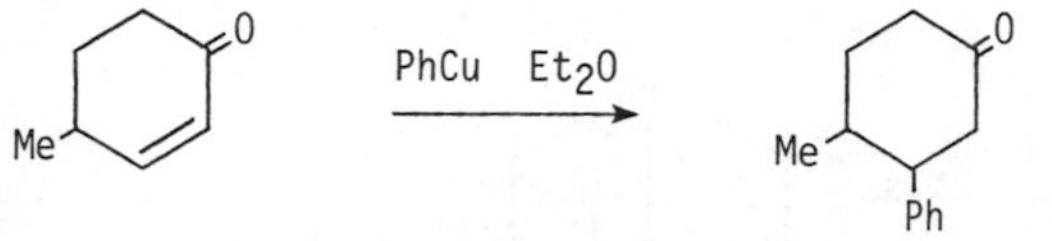

$\xrightarrow{\text{PhCu   Et}_2\text{O}}$

Tetr Lett (1970) 1579

$\xrightarrow[\text{HCl   EtOH}]{\text{H}_2\text{   Pd-C}}$ cis

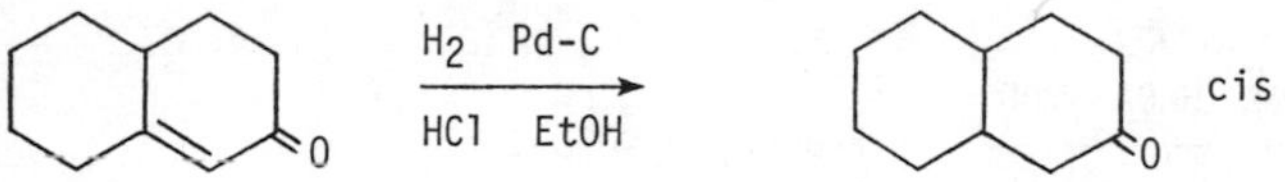

JOC (1958) 23 1853
Chem Ind (1966) 1796

For reduction of enones with H$_2$ and Rh see JOC (1968) 33 3695

PhCH=CHCOMe $\xrightarrow{\text{H}_2\text{   Ni   CH}_2\text{Cl}_2}$ PhCH$_2$CH$_2$COMe   93-96%

Bull Soc Chim Fr (1954) 522

Me$_2$C=CHCOMe $\xrightarrow{\text{Ni   H}_2\text{O}}$ Me$_2$CHCH$_2$COMe   95%

Bull Chem Soc Jap (1967) 40 1548

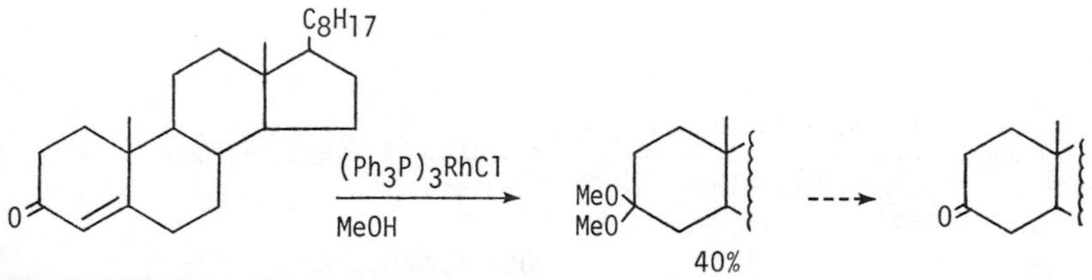

$\xrightarrow[\text{MeOH}]{\text{(Ph}_3\text{P)}_3\text{RhCl}}$

40%

Ber (1968) 101 1154

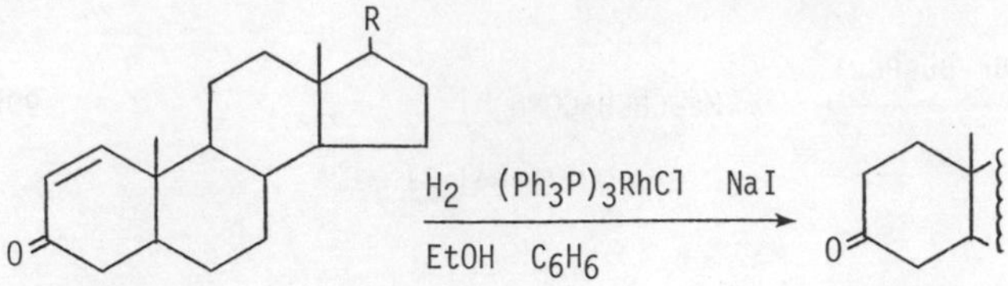

JACS (1966) **88** 4537

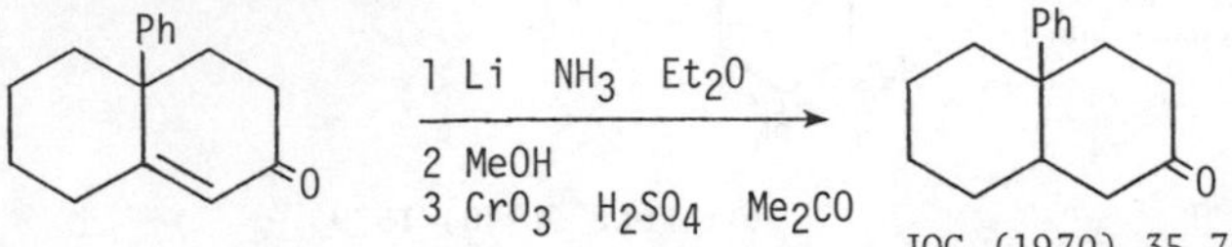

63%

JOC (1970) **35** 753
JACS (1965) **87** 275

$$t\text{-BuCH=CHCOBu-t} \xrightarrow{\text{Na} \quad \text{HMPA} \quad \text{THF}} t\text{-BuCH}_2\text{CH}_2\text{COBu-t}$$

JACS (1970) **92** 2783
Bull Soc Chim Fr (1968) 595

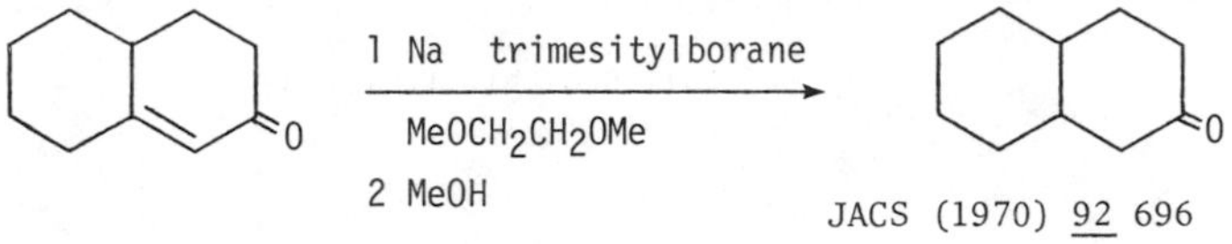

1 Na   trimesitylborane

MeOCH$_2$CH$_2$OMe

2 MeOH

85-95%

JACS (1970) **92** 696

$$\text{Me}_2\text{C=CHCOMe} \xrightarrow{\text{CrCl}_2 \quad \text{NH}_3 \quad \text{H}_2\text{O}} \text{Me}_2\text{CHCH}_2\text{COMe}$$

Angew (1968) **80** 271
(Internat Ed **7** 247)

$$\text{PhCH=CHCOPh} \xrightarrow[\substack{\text{dimethylpyridine-3,5-dicarboxylate} \\ 156°}]{\text{Diethyl 1,4-dihydro-2,6-}} \text{PhCH}_2\text{CH}_2\text{COPh}$$

JCS (1960) 3257

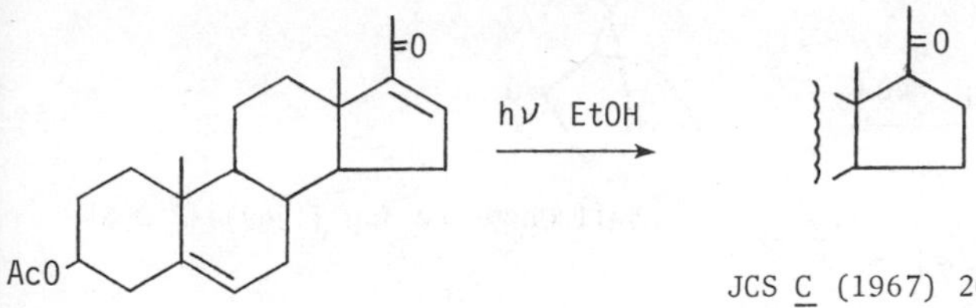

hν EtOH

40%

JCS C (1967) 2032

$Me_2C=CHCOMe$ $\xrightarrow{Ph_3SnH}$ Me_2CHCH_2COMe

Compt Rend (1965) 260 581

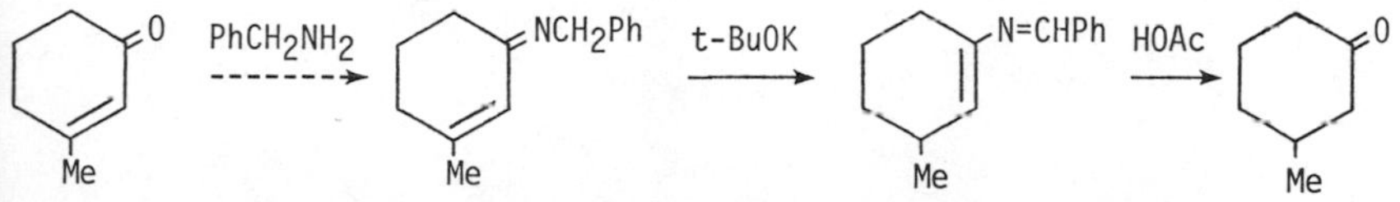

PhCH$_2$NH$_2$ t-BuOK   HOAc

40-70%

JACS (1967) 89 2794

Zn   HOAc

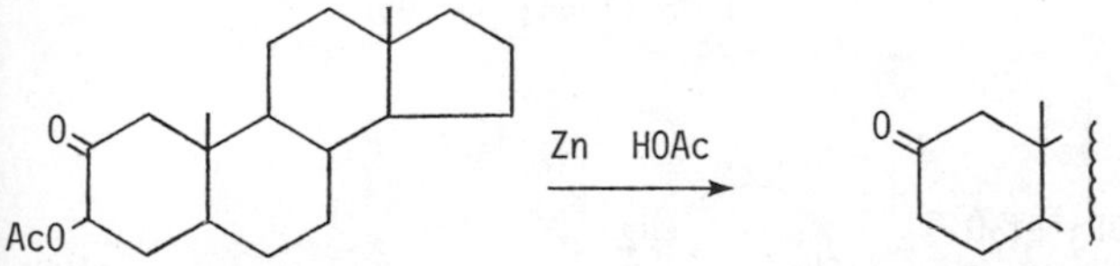

JCS C (1970) 244
JACS (1955) 77 4367

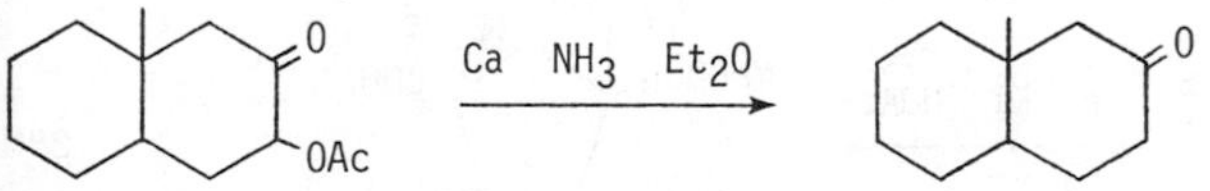

Ca   NH$_3$ Et$_2$O

80%

JOC (1969) 34 4188
JCS (1956) 4344

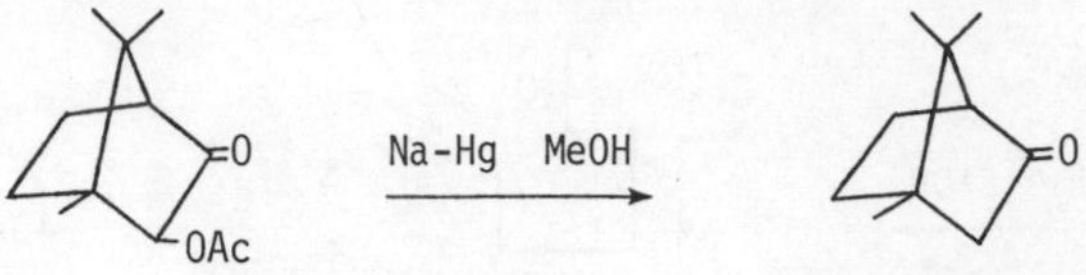

Na-Hg   MeOH

95%

Bull Chem Soc Jap (1969) 42 2068

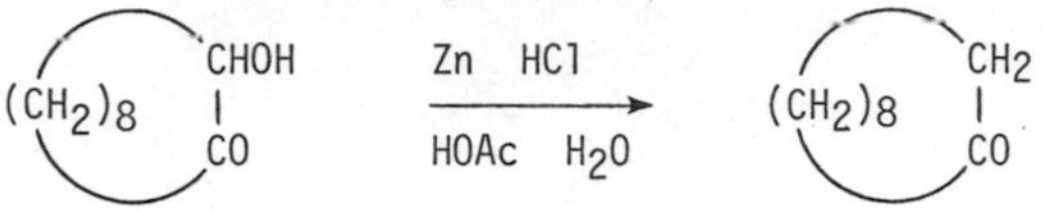

$(CH_2)_8$
CHOH
CO

Zn   HCl
———————
HOAc H_2O

$(CH_2)_8$
CH_2
CO

75-78%

Org Synth (1963) Coll Vol 4 218
JACS (1964) 86 3068

OH
|
$EtCOCEt_2$

HI   HOAc H_2O
————————————→

$EtCOCHEt_2$

80%

JACS (1964) 86 3068
JOC (1938) 3 456

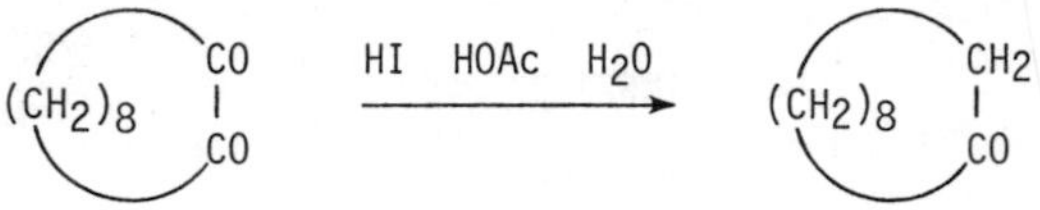

$(CH_2)_8$
CO
CO

HI   HOAc H_2O
————————————→

$(CH_2)_8$
CH_2
CO

80%

JACS (1964) 86 3068

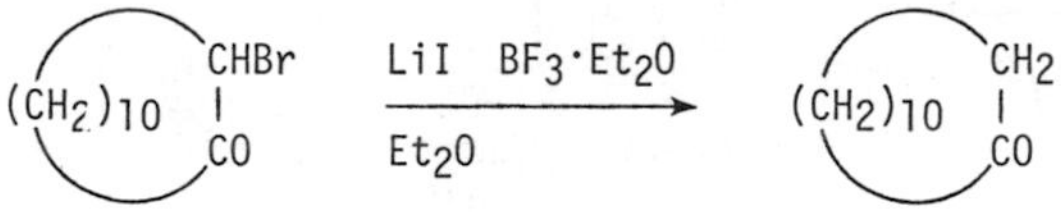

$(CH_2)_{10}$
CHBr
CO

LiI $BF_3 \cdot Et_2O$
————————————→
Et_2O

$(CH_2)_{10}$
CH_2
CO

100%

Tetr Lett (1971) 137

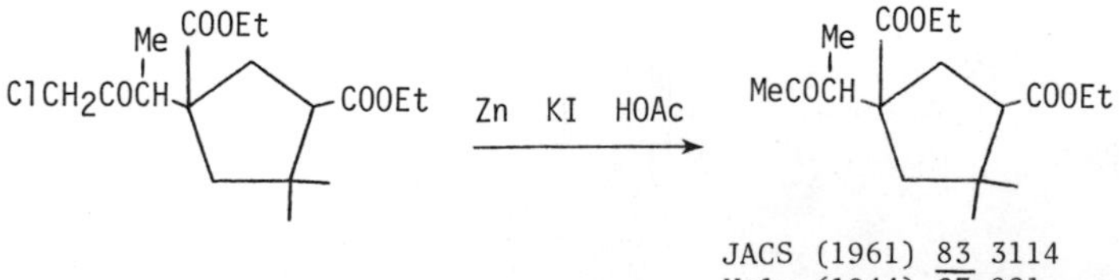

Me   COOEt
|
$ClCH_2COCH$ COOEt

Zn   KI   HOAc
————————————→

Me   COOEt
|
$MeCOCH$ COOEt

88%

JACS (1961) 83 3114
Helv (1944) 27 821
Tetr Lett (1968) 1575

PhCOCH$_2$Cl $\xrightarrow{\text{CrSO}_4 \quad \text{DMF} \quad \text{H}_2\text{O}}$ PhCOMe   79%

JACS (1963) 85 2768
     (1945) 67 1728
Angew (1968) 80 271
(Internat Ed 7 247)

$\xrightarrow[\text{MeOH} \quad \text{Me}_2\text{CO}]{\text{H}_2 \quad \text{Pd} \quad \text{KOAc}}$

JCS (1964) 5535

PhCOCH$_2$Cl $\xrightarrow{\text{Bu}_2\text{SnH}_2 \quad \text{Et}_2\text{O}}$ PhCOMe   95%

JOC (1963) 28 2165

$\xrightarrow[\text{or 2,6-dimethylphenol}]{\text{PhNMe}_2}$

70-76%

Chimia (1967) 21 464
JCS C (1969) 301

i-PrCH$_2$COCHPr-i $\xrightarrow[\text{2 EtOH}]{\text{1 Ph}_3\text{P} \quad \text{C}_6\text{H}_6}$ i-PrCH$_2$COCH$_2$Pr-i
$\quad$ |
$\quad$ Br

JCS (1962) 2337
Tetr Lett (1962) 471

PhCOCHMe $\xrightarrow{\text{Ph}_2\text{PH} \quad \text{CCl}_4}$ PhCOCH$_2$Me   100%
$\quad$ |
$\quad$ Br

JOC (1969) 34 2687

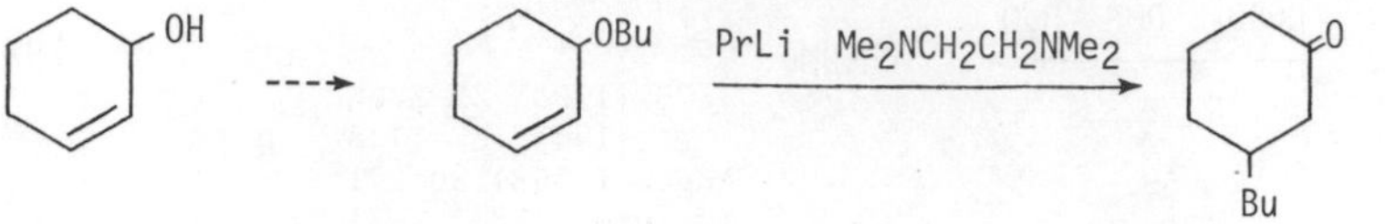
OH
OBu
PrLi $Me_2NCH_2CH_2NMe_2$
O
Bu

Tetr Lett (1969) 821

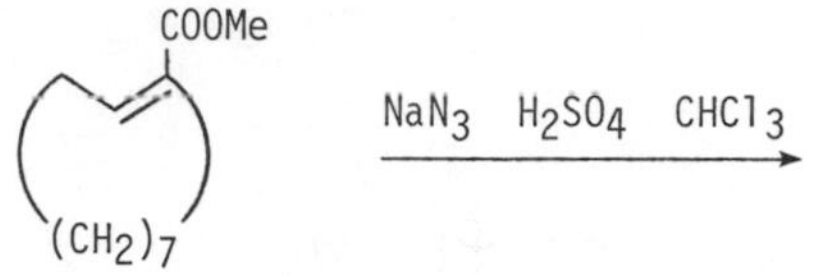
COOMe
$(CH_2)_7$
NaN_3 H_2SO_4 $CHCl_3$
O
$(CH_2)_6$

JOC (1968) 33 2157

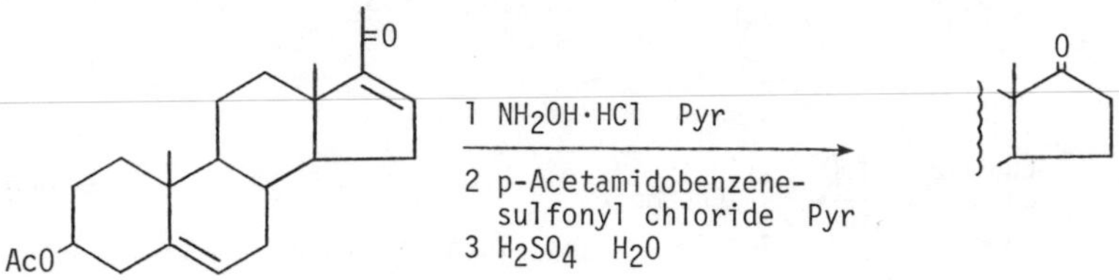
O
1 $NH_2OH\cdot HCl$ Pyr
2 p-Acetamidobenzene-sulfonyl chloride Pyr
3 H_2SO_4 H_2O
AcO
O

74%

JOC (1956) 21 520
Org React (1960) 11 1

$Me_2C(OH)CH(OAc)Bu$ —Zn 154°→ $Me_2CHCOBu$ 68%

Chem Comm (1968) 1639
JOC (1970) 35 660

$Me_2C{=}CHCH_2CH_2CH(OH)Me$ —Polyphosphoric acid→ $Me_2CHCH_2CH_2CH_2COMe$ 80%

Bull Soc Chim Fr (1963) 1799
JACS (1965) 87 2772

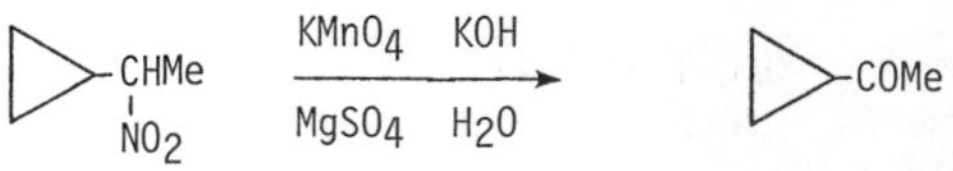
CHMe
NO_2
$KMnO_4$ KOH
$MgSO_4$ H_2O
COMe

77%

JOC (1962) 27 3699

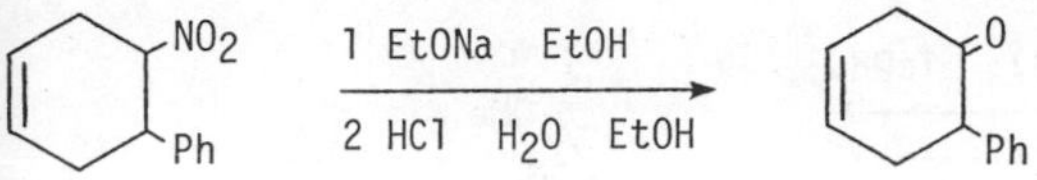

JOC (1952) 17 581
Chem Rev (1955) 55 137

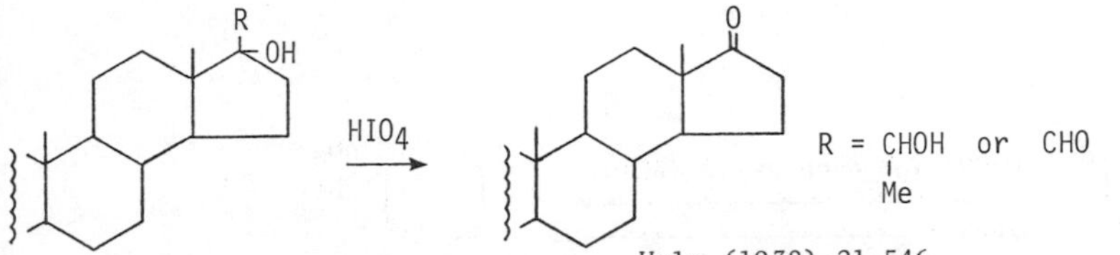

78%

R = CHOH or CHO
        |
        Me

Helv (1938) 21 546
     (1940) 23 170
     (1941) 24 945
Org React (1944) 2 341

Section 180A   Protection of Ketones
                 ooooooooooooooooooooo

Ketals are stable to the following reagents: NaOH, NaH, Na-NH$_3$, RMgX, NaBH$_4$,
LiAlH$_4$, H$_2$-Pd, CrO$_3$ and SOCl$_2$-Pyr

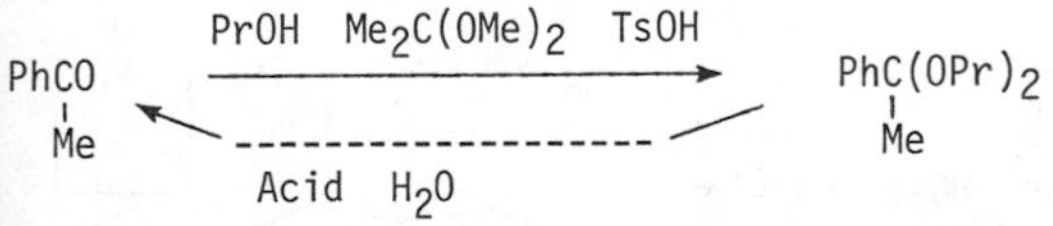

PrOH Me$_2$C(OMe)$_2$ TsOH

PhCO ← — — — — — — — — — — — — — → PhC(OPr)$_2$
  | |
  Me   Acid H$_2$O   Me

JOC (1960) 25 521 525

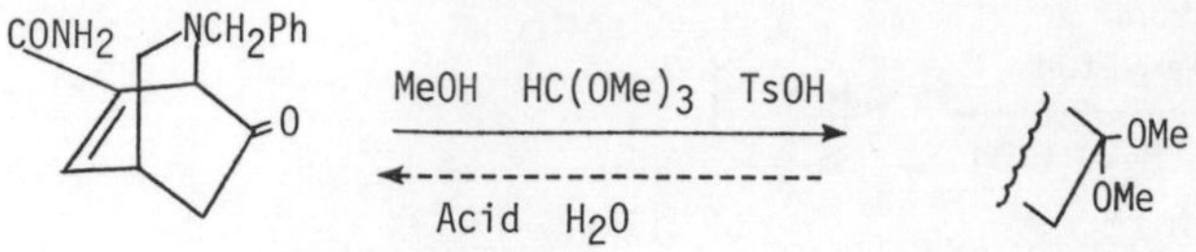
CONH2
NCH2Ph
O
MeOH HC(OMe)3 TsOH
Acid H2O
OMe
OMe

JACS (1968) 90 2448
Annalen (1962) 656 97

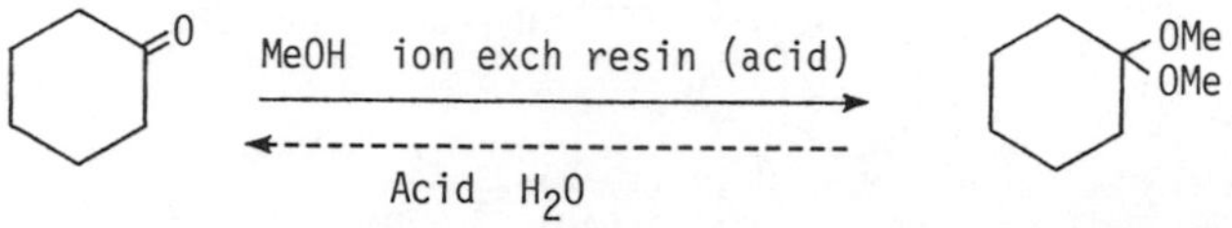
O
MeOH ion exch resin (acid)
Acid H2O
OMe
OMe

JOC (1959) 24 1731

The following catalysts may also be used for the preparation of ketals:

$(Ph_3P)_3RhCl$ . . Ber (1968) 101 1154
SeO_2 . . . . . JACS (1954) 76 6113

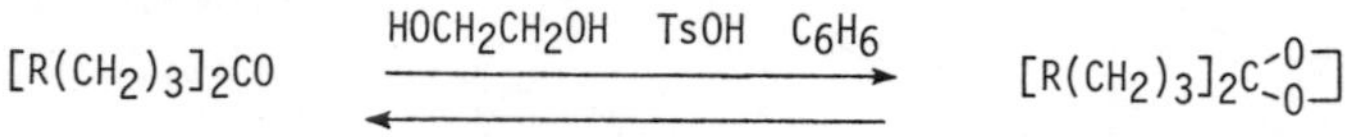
[R(CH2)3]2CO
HOCH2CH2OH TsOH C6H6
H3PO4 H2O
[R(CH2)3]2C O O

JCS (1962) 4722
JACS (1964) 86 2183

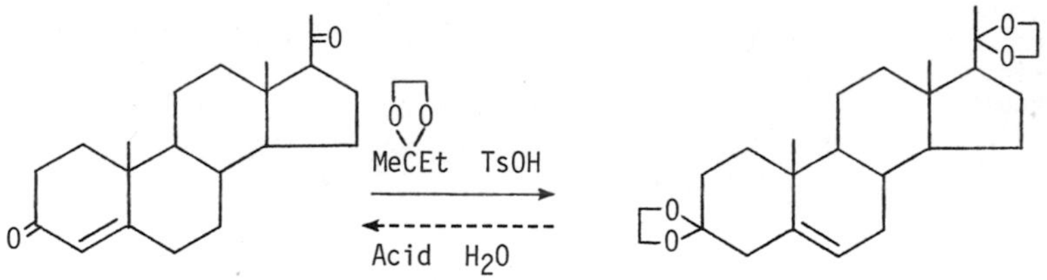
O
O
O O
MeCEt TsOH
Acid H2O
O
O
O
O

JACS (1954) 76 1359
(1953) 75 1716

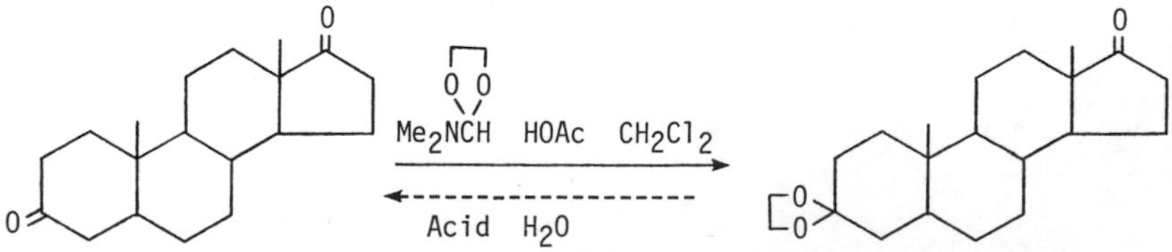
O
O
O O
Me2NCH HOAc CH2Cl2
Acid H2O
O
O
O

Steroids (1963) 1 45

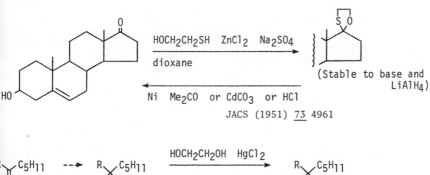

Me₂C(CH₂OH)₂ / TsOH ← Acid H₂O

$Me_2C(CH_2OH)_2$

TsOH

Acid H_2O

JACS (1968) 90 1253

1 ClCH₂CH₂OH

2 Et₃N

Acid H_2O

JOC (1968) 33 1280

$(PhCH_2)_2CO$

$HO(CH_2)_nSH$ TsOH C_6H_6

TsOH Me_2CO

$(PhCH_2)_2C{\overset{S}{\underset{O}{\bigg\langle}}}(CH_2)_n$

n=2 or 3

JACS (1953) 75 3704
(1954) 76 1945
For cleavage with Ni see JACS (1958) 80 4723

$HOCH_2CH_2SH$ $ZnCl_2$ Na_2SO_4

dioxane

Ni Me_2CO or $CdCO_3$ or HCl

JACS (1951) 73 4961

(Stable to base and
LiAlH₄)

C_5H_{11} --→ $R\underset{S\ S}{\bigvee}C_5H_{11}$

$HOCH_2CH_2OH$ $HgCl_2$

$R\underset{O\ O}{\bigvee}C_5H_{11}$

H_2SO_4 H_2O THF

JACS (1968) 90 3245

Thioketals are stable to the following reagents: NaOH, $LiAlH_4$, RMgX, CrO_3-Pyr

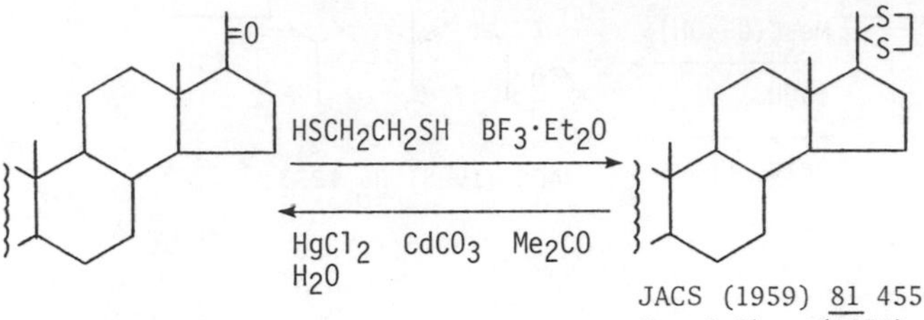
HSCH2CH2SH BF3·Et2O
HgCl2 CdCO3 Me2CO
H2O

JACS (1959) 81 4556
Can J Chem (1955) 33 716

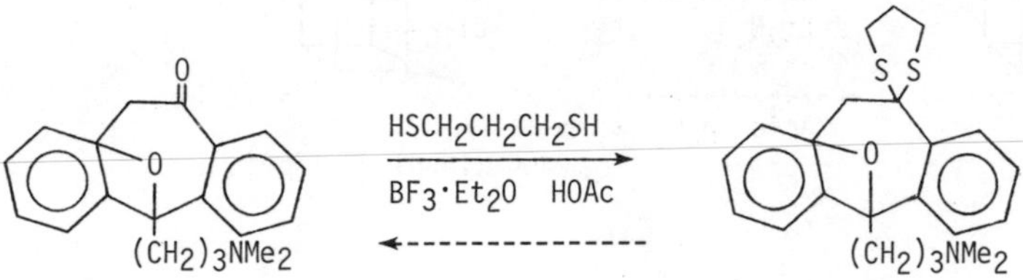
HSCH2CH2CH2SH
BF3·Et2O HOAc
(CH2)3NMe2
(CH2)3NMe2

J Med Chem (1970) 13 191

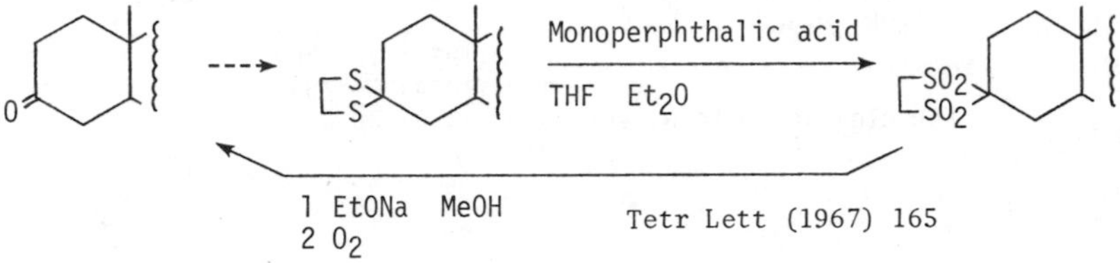
Monoperphthalic acid
THF Et2O
1 EtONa MeOH
2 O2

Tetr Lett (1967) 165

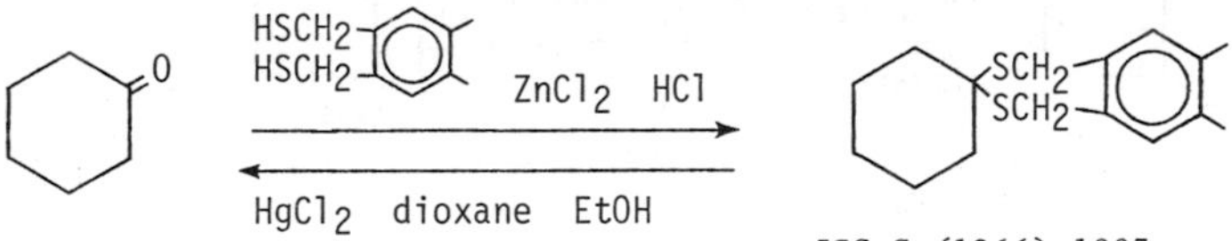
HSCH2
HSCH2
ZnCl2 HCl
HgCl2 dioxane EtOH

JCS C (1966) 1005

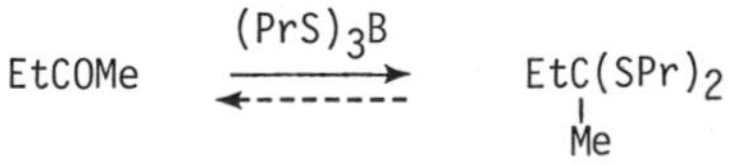
EtCOMe
(PrS)3B
EtC(SPr)2
Me

(Neutral conditions)

Can J Chem (1969) 47 859
(1965) 43 307

$$(RCOOCH_2)_2CO \underset{HgCl_2 \quad Me_2CO \quad H_2O}{\overset{EtSH \quad ZnCl_2}{\rightleftharpoons}} (RCOOCH_2)_2C(SEt)_2$$

Can J Chem (1955) 33 716

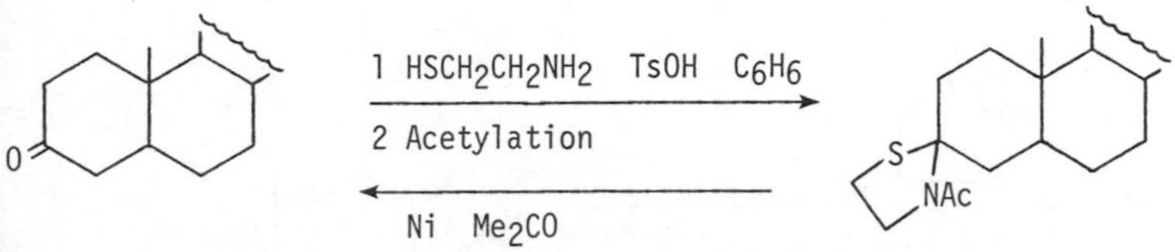

1 HSCH2CH2NH2 TsOH C6H6
2 Acetylation
Ni Me2CO
O
S
NAc

JOC (1962) 27 1112

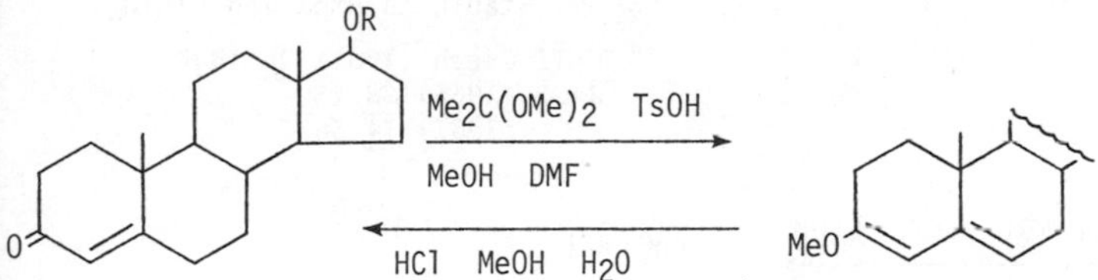

OR
Me2C(OMe)2 TsOH
MeOH DMF
HCl MeOH H2O
O
MeO

(Stable to base, RMgX and $LiAlH_4$)

JOC (1961) 26 3925

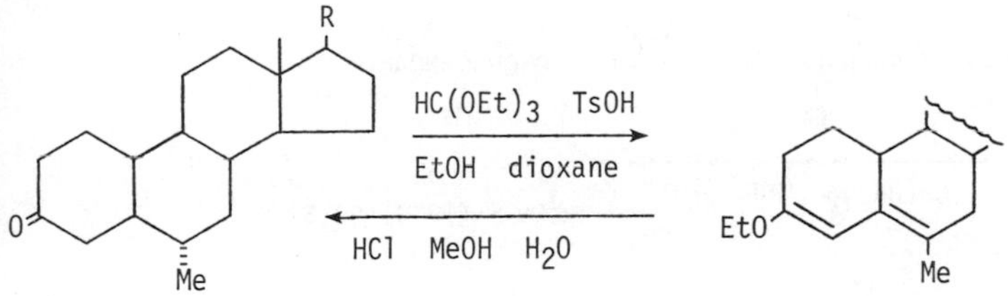

R
HC(OEt)3 TsOH
EtOH dioxane
HCl MeOH H2O
O
Me
EtO
Me

JACS (1959) 81 4566

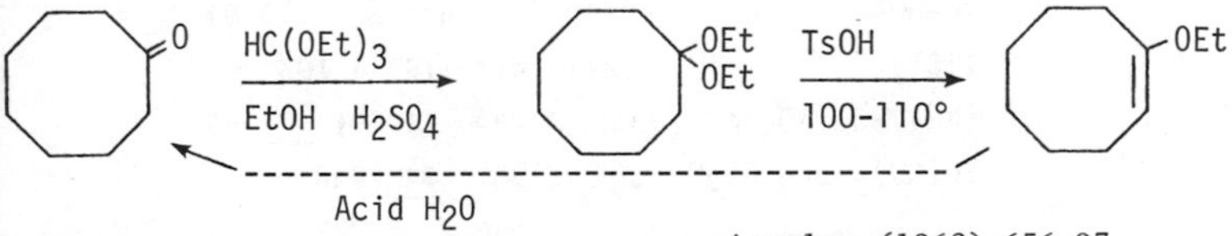

O
HC(OEt)3
EtOH H2SO4
OEt
OEt
TsOH
100-110°
OEt
Acid H2O

Annalen (1962) 656 97
Gazz (1962) 92 309
(Chem Abs 57 12572)

PhCH$_2$SH Pyr·HCl

PhCH$_2$S—

H$_2$O$_2$

PhCH$_2$SO—

Acid

(Stable to LiAlH$_4$)

LiAlH$_4$

(Stable to acid)

JACS (1951) $\underline{73}$ 1528

Pyrrolidine EtOH

O=

⁷CH$_2$

HOAc NaOAc MeOH

(Stable to RMgX and LiAlH$_4$)

Coll Czech (1961) $\underline{26}$ 1646
JACS (1956) $\underline{78}$ 430
(1952) $\underline{74}$ 3627

PhCO
|
Et

NH$_2$OH·HCl NaOH EtOH
⇄
Ni NaOH EtOH H$_2$O

PhC=NOH
|
Et

JCS $\underline{C}$ (1966) 531

PhCH$_2$CO
|
Me

⤍ PhCH$_2$C=NOH
|
Me

Ac$_2$O

PhCH$_2$C=NOAc
|
Me

Cr(OAc)$_2$ THF H$_2$O

JACS (1970) $\underline{92}$ 5276

Oximes may also be cleaved by the following reagents:

NaHSO$_3$	JOC (1966) $\underline{31}$ 3446
MeCOCH$_2$CH$_2$COOH-HCl	JACS (1959) $\underline{81}$ 4629
Zn-HOAc	J Indian Chem Soc (1969) $\underline{46}$ 44
TiCl$_3$	Tetr Lett (1971) 195
(NH$_4$)$_2$Ce(NO$_3$)$_6$	Can J Chem (1969) $\underline{47}$ 145
Fe(CO)$_5$-BF$_3$·Et$_2$O	JOC (1967) $\underline{32}$ 2938

For preparation of oximes of unreactive and acid or base sensitive ketones
(using Me$_2$SO-NH$_2$OH·HCl) see Chem Ind (1969) 240

(Stable to acid and base)

HOAc   H$_2$O

JOC (1962) $\underline{27}$ 914
For cleavage of methoximes with O$_3$ see JOC (1969) $\underline{34}$ 2961

Semicarbazones are stable to NaBH$_4$ and Oppenauer oxidation

Helv (1932) $\underline{15}$ 1220

Semicarbazones may also be cleaved by the following reagents:

H$_2$SO$_4$-H$_2$O	JCS (1946) 27
HCl-H$_2$O	JACS (1950) $\underline{72}$ 1751
MeCOCOOH	JCS (1953) 3864
	JACS (1955) $\underline{77}$ 1221
MeCOCH$_2$COMe-HCl	Annalen (1962) $\underline{656}$ 119
Ac$_2$O-Pyr	JOC (1956) $\underline{21}$ 795
(NH$_4$)$_2$Ce(NO$_3$)$_6$	Can J Chem (1969) $\underline{47}$ 145
MeCOO$_2$H	Ber (1961) $\underline{94}$ 712
NaNO$_2$-HOAc	Rec Trav Chim (1946) $\underline{65}$ 796
	JACS (1955) $\underline{77}$ 4781
	(1961) $\underline{83}$ 4249

PhC=NNMe$_2$ (Stable to base, B$_2$H$_6$, LiAlH$_4$ and CrO$_3$)

Chem Comm (1969) 445

For cleavage of dimethylhydrazones with O$_3$ see   JOC (1969) $\underline{34}$ 2961
"   "   "   "   " peracid see Ber (1961) $\underline{94}$ 712

Annalen (1962) $\underline{656}$ 119

Phenylhydrazones may also be cleaved by the following reagents:

MnO_2 JOC (1967) <u>32</u> 2252

$MeCOO_2H$ JOC (1967) <u>32</u> 2865

2,4-Dinitrophenylhydrazones are stable to CrO_3 and B_2H_6

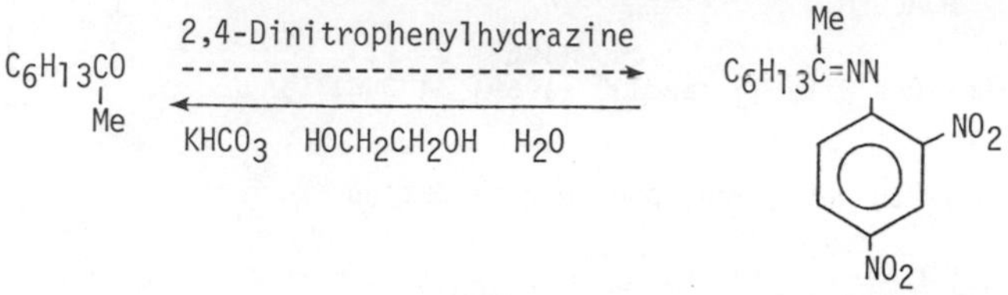

Aust J Chem (1968) <u>21</u> 271

2,4-Dinitrophenylhydrazones may also be cleaved by the following reagents:

$MeCOCH_2CH_2COOH-HCl$ JACS (1959) <u>81</u> 4629

$SnCl_2-HCl-Me_2CO$ Nature (1954) <u>173</u> 266

$CrCl_2-HCl$ JCS (1962) 4729

O_3 Chem Comm (1968) 433
JOC (1951) <u>16</u> 556

$CuCO_3 \cdot Cu(OH)_2-HCOOH$ Nature (1954) <u>173</u> 541

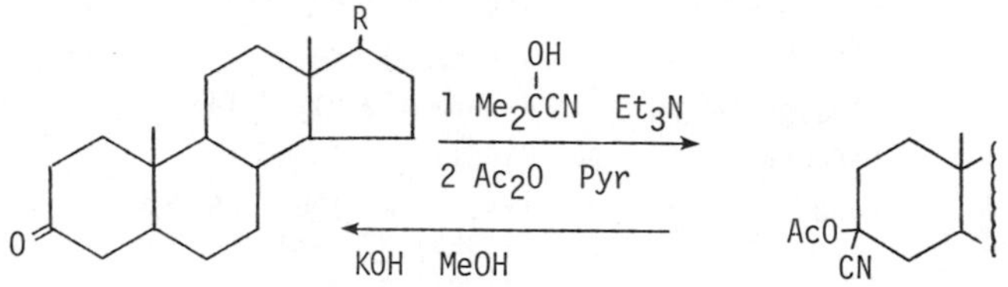

(Stable to acids, peracids, CrO_3

and Br_2)

Gazz (1961) <u>91</u> 1250
(Chem Abs <u>56</u> 10211)
Steroids (1967) <u>10</u> 411
JACS (1953) <u>75</u> 650

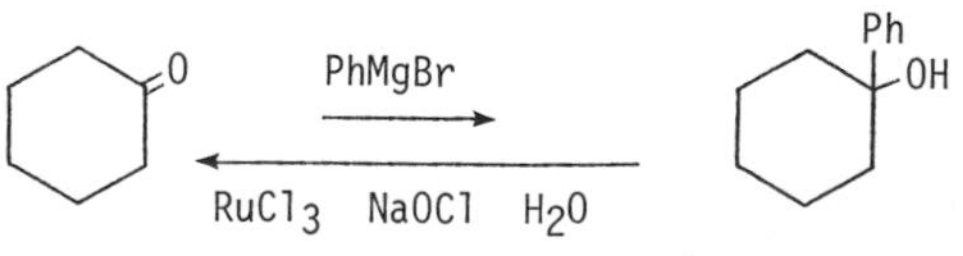

Chem Comm (1970) 1420

Chapter 13   PREPARATION
OF
NITRILES

Section 181   <u>Nitriles from Acetylenes</u>

$BuC\equiv CH$ $\xrightarrow[\substack{2\ MeLi\ \ Et_2O \\ 3\ (CN)_2}]{1\ (i\text{-}Bu)_2AlH}$ $BuCH=CHCN$ $\quad \dashrightarrow \quad$ $BuCH_2CH_2CN$

$\qquad\qquad\qquad\qquad\qquad\quad$ 87%

$\qquad\qquad\qquad\qquad\qquad\qquad\qquad\qquad$ JACS (1968) <u>90</u> 7139

$C_9H_{19}C\equiv CH$ $\xrightarrow[2\ ClCN]{1\ Mg\ \ EtMgBr}$ $C_9H_{19}C\equiv CCN$ $\quad \dashrightarrow \quad$ $C_9H_{19}CH_2CH_2CN$

$\qquad\qquad\qquad\qquad\qquad\qquad\qquad\qquad$ Annales de Chimie (1926) <u>5</u> 5

$MeC\equiv CH$ $\xrightarrow[NO]{COCl_2\ \ h\nu}$ $MeCN$

$\qquad\qquad\qquad\qquad\qquad\qquad$ JACS (1963) <u>85</u> 3506

Section 182   <u>Nitriles from Carboxylic Acids</u>

$\overset{\displaystyle CN}{\underset{\displaystyle (CH_2)_3COOH}{|}}$ $\xrightarrow{\text{Electrolysis   MeONa   MeOH}}$ $\overset{\displaystyle CN}{\underset{\displaystyle (CH_2)_6CN}{|}}$ $\qquad\qquad \sim 41\%$

$\qquad\qquad\qquad\qquad\qquad\qquad$ Z. Naturforsch (1947) 2b 185
$\qquad\qquad\qquad\qquad\qquad\qquad$ Advances in Org Chem (1960) <u>1</u> 1

i-PrCOOH $\xrightarrow{PhCH_2CN \quad H_2SO_4}$ i-PrCN 40%

JCS (1956) 1686

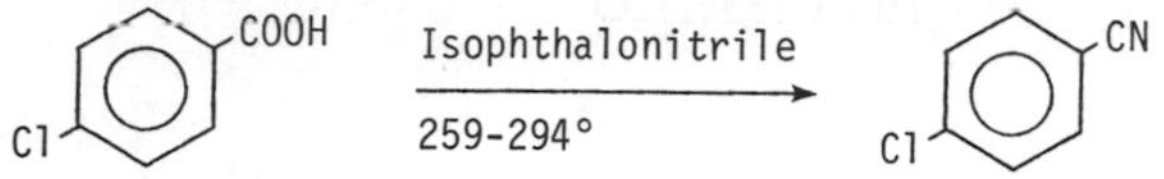
COOH
Cl
Isophthalonitrile
259-294°
CN
Cl

93%

JOC (1958) 23 1350

PhCH=CHCOOZn$_{1/2}$ $\xrightarrow{Pb(CNS)_2 \quad \Delta}$ PhCH=CHCN 27-36%

JACS (1916) 38 2120

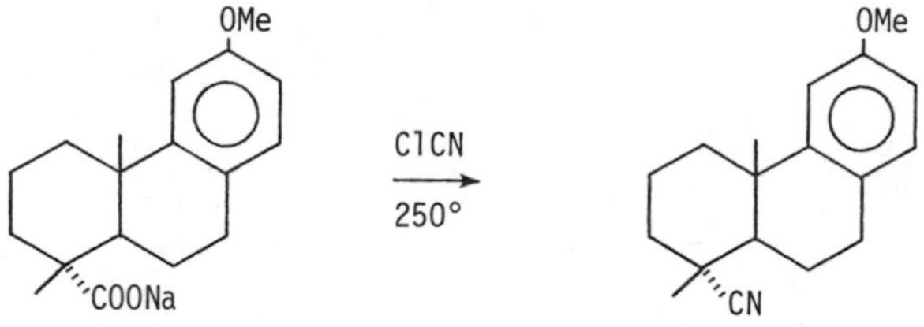
OMe
COONa
ClCN
250°
OMe
CN

26%

JCS (1961) 3185

t-BuCOOH $\xrightarrow[2\ DMF]{1\ ClSO_2NCO \quad pentane}$ t-BuCN 68%

Ber (1967) 100 2719
Tetr Lett (1968) 1631

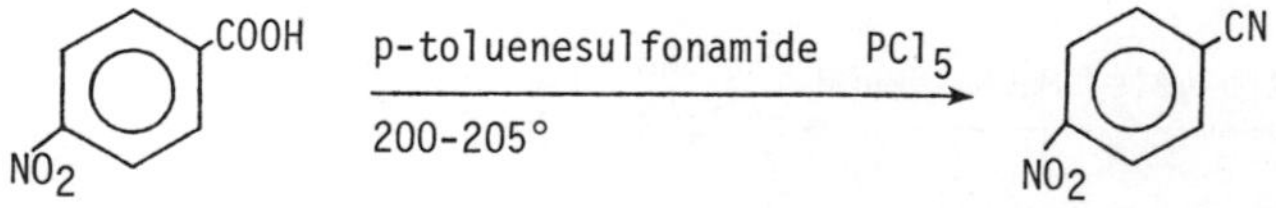
COOH
NO2
p-toluenesulfonamide PCl5
200-205°
CN
NO2

85-90%

Org Synth (1955) Coll Vol 3 646
JCS (1946) 763

JACS (1967) <u>89</u> 2338

$$\underset{\overset{|}{COOH}}{\underset{|}{(CH_2)_8}}{COOH} \xrightarrow[160\text{-}340°]{(NH_2)_2CO} \underset{\overset{|}{CN}}{\underset{|}{(CH_2)_8}}{CN} \qquad 46\text{-}49\%$$

Org Synth (1955) Coll Vol 3 768
(1963) Coll Vol 4 62

$$(t\text{-}Bu)_2CHCOOH \xrightarrow[2\ NaNH_2\ \ NH_3]{1\ SOCl_2} (t\text{-}Bu)_2CHCONH_2 \xrightarrow[C_6H_6]{SOCl_2} (t\text{-}Bu)_2CHCN \qquad 61\%$$

JACS (1960) <u>82</u> 2498

Further examples of the preparation of nitriles from amides are included in section 186 (Nitriles from Amides)

$$C_{14}H_{29}CH_2COOH \xrightarrow[2\ PhH\ \ AlCl_3]{1\ SOCl_2} C_{14}H_{29}CH_2COPh \xrightarrow[HCl]{i\text{-}PrCH_2CH_2ONO} \underset{NOH}{C_{14}H_{29}\overset{\|}{C}COPh}$$

$$\downarrow \begin{array}{c} TsCl \\ NaOH \end{array}$$

$$C_{14}H_{29}CN$$

JACS (1953) <u>75</u> 2347
JOC (1961) <u>26</u> 3507

Section 183   <u>Nitriles from Alcohols</u>
oooooooooooooooooooooo

$$Ph_3COH \xrightarrow[]{\underset{|}{CH_2COOH}\ \ ZnCl_2} Ph_3CCH_2CN \qquad 86\%$$

Annalen (1963) <u>661</u> 157

C$_6$H$_{13}$CHOH $\xrightarrow{\text{CCl}_4 \quad \text{Ph}_3\text{P}}$ C$_6$H$_{13}$CHCl $\xrightarrow{\text{NaCN} \quad \text{Me}_2\text{SO}}$ C$_6$H$_{13}$CHCN
| Me   Me   Me

JOC (1967) <u>32</u> 855

JACS (1970) <u>92</u> 336

Further examples of the preparation of nitriles from halides and sulfonates are included in section 190 (Nitriles from Halides and Sulfonates)

PhCH$_2$OH $\xrightarrow[\text{MeONa} \quad \text{MeOH}]{\text{NH}_3 \quad \text{O}_2 \quad \text{CuCl}_2}$ PhCN   30%

Rec Trav Chim (1963) <u>82</u> 757

Section 184   Nitriles from Aldehydes

PhCHO $\xrightarrow[\text{H}_2\text{O}]{\text{TsCl} \quad \text{KCN}}$ PhCHCN $\xrightarrow{\text{PhH} \quad \text{AlCl}_3}$ PhCHCN   57%
| OTs   Ph

JOC (1954) <u>19</u> 1699

PhCHO $\xrightarrow[\text{MeOCH}_2\text{CH}_2\text{OMe}]{\text{(EtO)}_2\text{OPCH}_2\text{CN} \quad \text{NaH}}$ PhCH=CHCN $\dashrightarrow$ PhCH$_2$CH$_2$CN
                                                         66%

JACS (1961) <u>83</u> 1733

$$PrCHO \xrightarrow[\text{HOAc}]{\overset{\overset{\displaystyle COOEt}{|}}{CH_2CN} \ H_2 \ Pd} \underset{PrCH_2\overset{\overset{\displaystyle COOEt}{|}}{CHCN}}{} \xrightarrow[2 \ \Delta]{1 \ Base} PrCH_2CH_2CN$$

Org Synth (1955) Coll Vol 3 385
JACS (1959) 81 5397

$$\xrightarrow[\text{PhCOCl}]{KCN} \qquad \xrightarrow[\text{Pd-BaSO}_4]{H_2}$$

Arch Pharm (1957) 290 218

$$PhCHO \dashrightarrow \underset{OH}{Ph\overset{|}{C}HCN} \xrightarrow[CHCl_3]{SOCl_2} \underset{Cl}{Ph\overset{|}{C}HCN} \xrightarrow[HOAc]{Zn} PhCH_2CN \qquad \sim 56\%$$

J Soc Chem Ind (1935) 54 98T

$$\xrightarrow[\substack{2 \ NaOH \\ 3 \ NH_2OH \cdot HCl \\ EtONa}]{\substack{1 \ \text{thiazolidine} \ NaOAc}} \qquad \xrightarrow{Ac_2O}$$

CH$_2$CCOOH with NOH

JACS (1940) 62 1512
JCS (1946) 958
Org React (1942) 1 210

$$PhCHO \xrightarrow[\substack{NaOAc \ \ Ac_2O \\ 2 \ NaOH \ \ H_2O \\ 3 \ HCl \ \ H_2O \\ 4 \ NH_2OH \cdot HCl \\ NaOH \ \ H_2O}]{1 \ AcNHCH_2COOH} \underset{NOH}{PhCH_2\overset{\|}{C}COOH} \xrightarrow{Ac_2O} PhCH_2CN \qquad 43\text{-}60\%$$

JACS (1942) 64 885
Org React (1946) 3 198

$$\xrightarrow[\text{Et}_2O]{NH_3 \ \ nickel \ peroxide}$$

50%

Chem Comm (1966) 17

$$C_6H_{13}CHO \xrightarrow{\quad NH_3 \quad Pb(OAc)_4 \quad C_6H_6 \quad} C_6H_{13}CN \qquad 59\%$$

Chem Ind (1965) 988

$$C_6H_{13}CHO \xrightarrow[\text{MeONa \quad MeOH}]{\quad NH_3 \quad O_2 \quad CuCl_2 \quad} C_6H_{13}CN \qquad 63\%$$

Rec Trav Chim (1963) 82 757

$$PhCHO \xrightarrow[\text{2 } I_2]{\text{1 } NH_3 \quad MeONa \quad MeOH} PhCN \qquad 50\%$$

Bull Chem Soc Jap (1966) 39 854

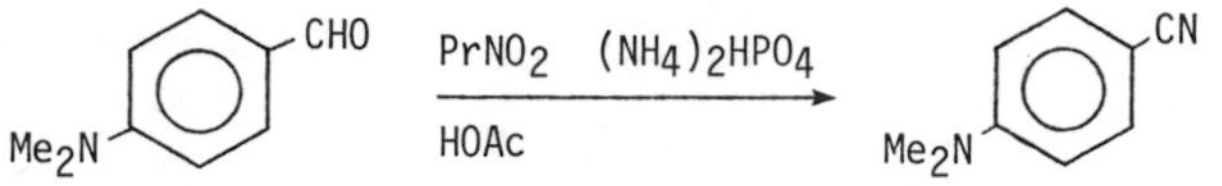

$$\xrightarrow[\text{HOAc}]{\quad PrNO_2 \quad (NH_4)_2HPO_4 \quad} \qquad 77\%$$

JACS (1961) 83 2203

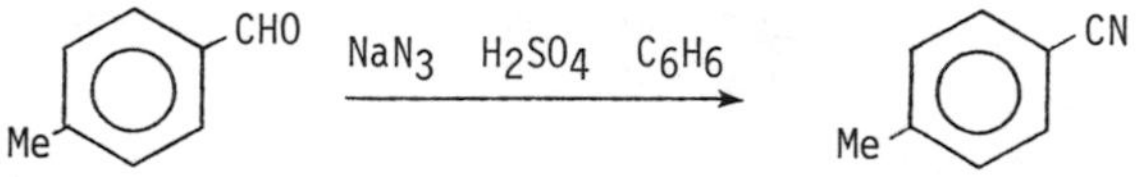

$$\xrightarrow{\quad NaN_3 \quad H_2SO_4 \quad C_6H_6 \quad} \qquad 44\%$$

JACS (1952) 74 1168
Org React (1946) 3 307

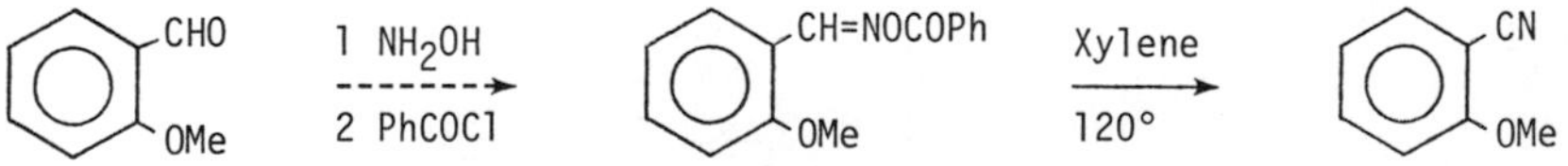

$$\xrightarrow[\text{2 } PhCOCl]{\text{1 } NH_2OH} \qquad \xrightarrow[120°]{Xylene}$$

JCS (1950) 1243

PhCH=CHCHO $\xrightarrow[\text{Pyr} \quad C_6H_6]{\text{CF}_3\text{COONHCOCF}_3}$ PhCH=CHCN   88%

JACS (1959) 81 6340

$\xrightarrow[\text{HOAc}]{\text{NH}_2\text{OH·HCl} \quad \text{NaOAc}}$ 67%

Chem Ind (1961) 1873
JCS (1965) 1564

$\xrightarrow[\text{H}_2\text{O} \quad \text{EtOH}]{\text{NH}_2\text{OH·HCl} \quad \text{NaOH}}$ ... $\xrightarrow{\text{Ac}_2\text{O}}$ 70-76%

Org Synth (1943) Coll Vol 2 622

The following reagents may also be used for dehydration of oximes to nitriles:

HCl-EtOH	Can J Chem (1967) 45 1014
MsCl-Pyr-collidine	Tetr Lett (1965) 2497
PhNCO-Et$_3$N	JOC (1961) 26 782
PhN=CCl$_2$	Bull Chem Soc Jap (1962) 35 1104
(PhO)$_2$POH-CCl$_4$-Et$_3$N	JOC (1969) 34 2805
p-chlorophenyl chlorothionoformate	Chem Comm (1970) 1014

BuCHO $\xrightarrow[C_6H_6]{\text{NH}_2\text{NMe}_2}$ BuCH=NNMe$_2$ $\xrightarrow[\substack{\text{2 MeONa} \quad \text{MeOH} \\ \text{or H}_2\text{O}_2 \quad \text{H}_2\text{O}}]{\text{1 MeI}}$ BuCN   42-47%

JOC (1962) 27 4372
    (1966) 31 4100

PhCH$_2$CHO $\xrightarrow{\text{Ph}_2\text{NNH}_2}$ PhCH$_2$CH=NNPh$_2$ $\xrightarrow{\text{O}_2 \quad h\nu}$ PhCH$_2$CN   41%

Tetr Lett (1970) 2085

PhCHO $\xrightarrow[\text{2 190-200°}]{\text{1 NH}_2\text{N} \diagup \text{TsOH C}_6\text{H}_6}$ PhCN   69%

Z Chem (1964) 4 304

Some of the methods listed in section 192 (Nitriles from Ketones) may
also be used for the preparation of nitriles from aldehydes

Section 185   Nitriles from Alkyls

This section lists examples of the replacement of alkyl groups by nitriles.
For the replacement of hydrogen by nitrile, RH → RCN, see section 191
(Nitriles from Hydrides)

PhEt $\xrightarrow[\text{389°}]{\text{O}_2 \quad \text{NH}_3 \quad \text{Cr}_2\text{O}_3}$ PhCN   8%

Tetrahedron (1968) 24 6277

Section 186   Nitriles from Amides

RhCl(PPh$_3$)$_3$ / 250°   25% (R=Et)
71% (R=H)

Tetr Lett (1970) 1963

PhCH$_2$CONH$_2$ --→ PhCH$_2$CONHBr $\xrightarrow[\text{C}_6\text{H}_6]{\text{Ph}_3\text{P}}$ PhCH$_2$CN   60%

JCS (1960) 2976

PhCONH₂ $\xrightarrow{\text{Ph}_3\text{P} \quad \text{CCl}_4 \quad \text{THF}}$ PhCN　　　　　　　82%

Tetr Lett (1970) 4383

JACS (1949) __71__ 2650

The following reagents may also be used for the conversion of amides into nitriles:

SOCl₂	Org Synth (1963) Coll Vol 4 436
SOCl₂-DMF	Chem Ind (1964) 752
TsCl-Pyr	JACS (1955) __77__ 1701
COCl₂-Pyr	JCS (1954) 3730
PhSO₃H (235°)	JCS (1946) 763
P₂O₅	Org Synth (1963) Coll Vol 4 144
POCl₃-NaHSO₃	JOC (1957) __22__ 1142
AlCl₃	JACS (1940) __62__ 1432
Et₃SiH-ZnCl₂	Compt Rend (1962) __254__ 2357
Catechyl phosphorus trichloride	Ber (1963) __96__ 1387
Me₂NCHF₂	Coll Czech (1963) __28__ 2047
DCC-Pyr	JOC (1961) __26__ 3356
NaBH₄-Pyr	Chem Pharm Bull (1969) __17__ 98
	JOC (1967) __32__ 846
BuLi	JOC (1967) __32__ 3640
	(1966) __31__ 3873

Section 187　　__Nitriles from Amines__
　　　　　　　　○○○○○○○○○○○○○○○○○○○○○○○

 $\xrightarrow[\text{H}_2\text{O}]{\text{NaNO}_2 \quad \text{HCl}}$ $\xrightarrow[\text{C}_6\text{H}_6]{\text{CuCN}}$ 64-70%

Org Synth (1932) Coll Vol 1 514
Rec Trav Chim (1961) __80__ 1075

Org Prep and Procedures (1969) $\underline{1}$ 221

Tetr Lett (1966) 5087
JOC (1958) $\underline{23}$ 1221

$$C_7H_{15}CH_2NH_2 \xrightarrow[C_6H_6]{\text{Nickel peroxide}} C_7H_{15}CN \qquad 96\%$$

Chem Pharm Bull (1963) $\underline{11}$ 296

$$PhCH_2NH_2 \xrightarrow{Pb(OAc)_4 \quad C_6H_6} PhCN \qquad 59\%$$

Tetrahedron (1967) $\underline{23}$ 721

$$BuCH_2NH_2 \xrightarrow{AgO \quad H_2O} BuCN \qquad 87\%$$

Tetr Lett (1968) 5685

$$PrCH_2NH_2 \xrightarrow[\text{2 CsF \quad MeCN}]{\text{1 Cl}_2 \quad NaHCO_3 \quad H_2O} PrCN \qquad >81\%$$

JOC (1968) $\underline{33}$ 1008

$C_5H_{11}CH_2NH_2$ $\xrightarrow{\quad IF_5 \quad Pyr \quad CH_2Cl_2 \quad}$ $C_5H_{11}CN$ 36%

JOC (1961) 26 2531

Section 188   Nitriles from Esters

The reaction ROTs $\longrightarrow$ RCN is included in section 190 (Nitriles from Halides and Sulfonates)

$EtCH_2COOMe$ $\xrightarrow{\quad MeONO \quad MeONa \quad DMF \quad}$ $EtCN$ 18%

J Prakt Chem (1969) 311 370

$PhCH=CHCH_2CH(COOEt)_2$ $\xrightarrow[\text{NaOEt}]{C_5H_{11}ONO}$ $PhCH=CHCH_2\underset{\overset{\|}{NOH}}{C}COOH$ $\xrightarrow{Ac_2O}$ $PhCH=CHCH_2CN$

JCS (1950) 926

Section 189   Nitriles from Ethers

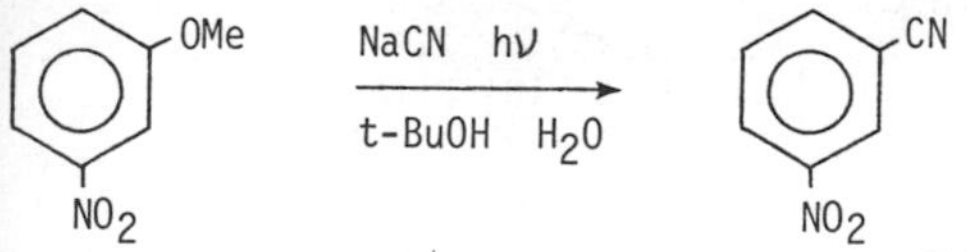

30%

JACS (1966) 88 2884

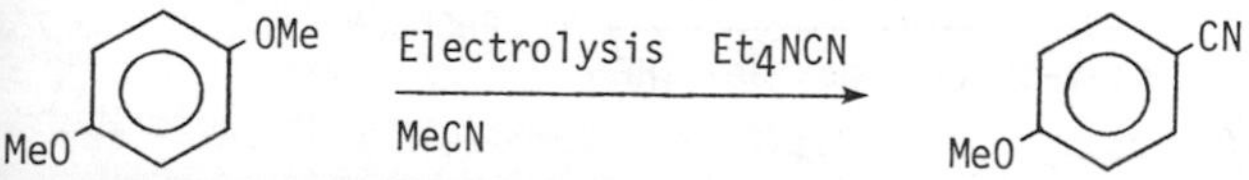

45%

JACS (1969) 4181

$$ArOMe \xrightarrow{KCN} MeCN$$

Ber (1933) 66B 1623

Section 190   Nitriles from Halides and Sulfonates
                ○○

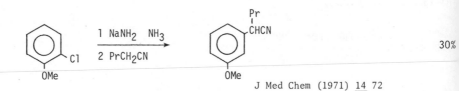

30%

J Med Chem (1971) 14 72

$$EtI \xrightarrow[EtONa]{\overset{COOEt}{\underset{CH_2CN}{|}}} Et_2\overset{COOEt}{\underset{|}{C}}CN \xrightarrow{NaOH \quad H_2O} Et_2\overset{COOH}{\underset{|}{C}}CN \xrightarrow{\Delta} Et_2CHCN$$

Δ

JACS (1959) 81 5397
     (1938) 60 131
Org React (1957) 9 107

$$EtBr \xrightarrow{CH_3CN \quad NaNH_2 \quad NH_3} EtCH_2CN$$

58%

JACS (1945) 67 2152
Ber (1968) 101 3113
Org React (1957) 9 107

Further examples of the alkylation of nitriles with halides are included
in section 193 (Nitriles from Nitriles)

$$PhBr \dashrightarrow PhB \xrightarrow[\text{di-t-butylphenoxide THF}]{ClCH_2CN \quad \text{potassium 2,6-}} PhCH_2CN$$

<75%

JACS (1969) 91 6854

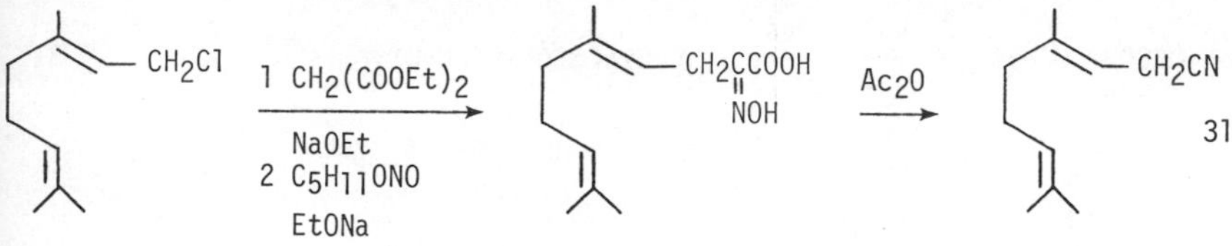

JCS (1950) 926

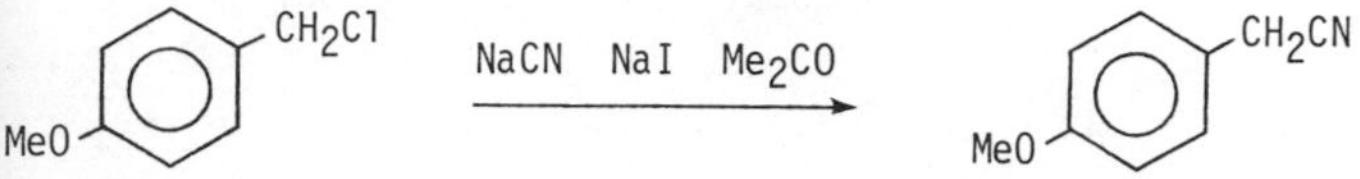

~75%

Org Synth (1963) Coll Vol 4 576

BuBr $\xrightarrow[\text{}]{\text{NaCN   Me}_2\text{SO}}$ BuCN   92%

JOC (1960) <u>25</u> 257 877

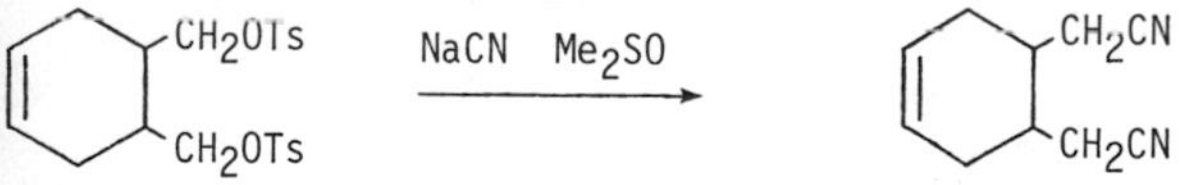

92%

Tetr Lett (1964) 2273
JACS (1970) <u>92</u> 336
JOC (1958) <u>23</u> 797

BuBr $\xrightarrow[\text{H}_2\text{O}]{\text{NaCN   PhCH}_2\text{NMe}_3^+ \text{ CN}^-}$ BuCN   69%

Chem Pharm Bull (1962) <u>10</u> 427

PhCH$_2$Br $\xrightarrow[\text{C}_6\text{H}_6]{\text{Ion exch resin (CN form)}}$ PhCH$_2$CN   72%

JOC (1963) <u>28</u> 698

C₇H₁₅C≡CCH=CHCH₂Br $\xrightarrow[\text{xylene}]{\text{CuCN NaI}}$ C₇H₁₅C≡CCH=CHCH₂CN   31%

JACS (1953) 75 3430
Ber (1954) 87 712

RBr $\xrightarrow[\text{2 ClCN or (CN)}_2]{\text{1 Mg Et}_2\text{O}}$ RCN   63% (R=PhCH₂CH₂)
                                                                          80% (R=Ph)

Annales de Chimie (1915) 4 28
Compt Rend (1911) 152 388

PhCH=CHBr $\xrightarrow[\text{MeOH}]{\text{K}_4\text{Ni}_2\text{(CN)}_6 \quad \text{KCN}}$ PhCH=CHCN   78%

JACS (1969) 91 1233
JOC (1969) 34 3626

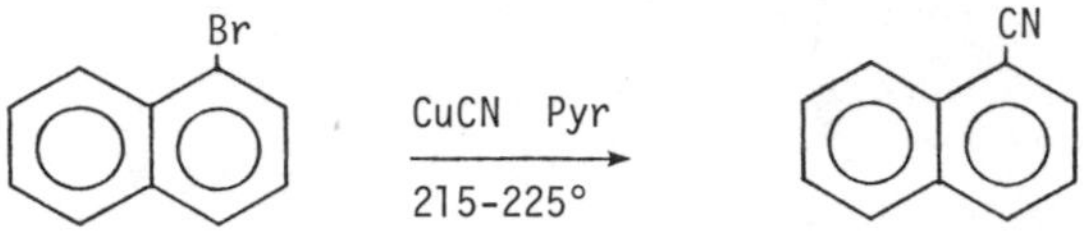

$\xrightarrow[\text{215-225°}]{\text{CuCN Pyr}}$ 82-90%

Org Synth (1955) Coll Vol 3 631 212
JACS (1966) 88 3318
JOC (1952) 17 298

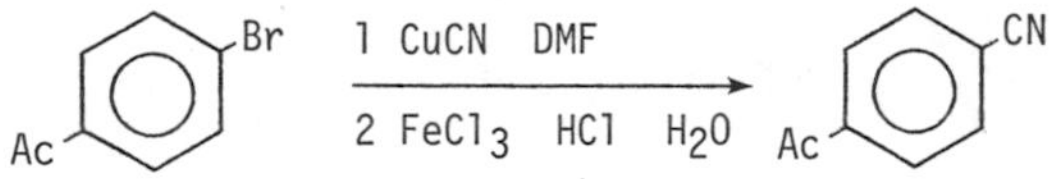

$\xrightarrow[\text{2 FeCl}_3 \quad \text{HCl} \quad \text{H}_2\text{O}]{\text{1 CuCN DMF}}$ 88%

JOC (1961) 26 2522

The following solvents may also be used in the above reaction:

N-Methylpyrrolidone JOC (1961) 26 2525
HMPA   JOC (1969) 34 3626
Me₂SO   Proc Chem Soc (1962) 113

PhCH₂Br $\xrightarrow[\text{2 NaNH}_2 \quad \text{NH}_3]{\text{1 Ph}_3\text{P}}$ PhCH=PPh₃ $\xrightarrow{\text{NO}}$ PhCN   24%

Chem Comm (1969) 166
Annalen (1969) 721 34
Bull Chem Soc Jap (1967) 40 2983

Section 191   Nitriles from Hydrides (RH)

This section lists examples of the replacement of hydrogen by CN,
RH ⟶ RCN. For the replacement of alkyl groups by CN, RR ⟶ RCN see
section 185(Nitriles from Alkyls)

$$\underset{\underset{Me}{|}}{\overset{\overset{Me}{|}}{i\text{-PrCH}}} \quad \xrightarrow{\text{ClCN \quad benzoyl peroxide}} \quad \underset{\underset{Me}{|}}{\overset{\overset{Me}{|}}{i\text{-PrCCN}}}$$

JACS (1969) 91 3028
Ber (1963) 96 670

$$MeCH=CH_2 \quad \xrightarrow{\text{Pd(CN)}_2 \quad DMF} \quad MeCH=CHCN \qquad\qquad 20\%$$

JACS (1966) 88 4105

$$C_5H_{11}C\equiv CH \quad \xrightarrow[\text{2 ClCN}]{\text{1 EtMgBr \quad Et}_2\text{O}} \quad C_5H_{11}C\equiv CCN \qquad\qquad 67\%$$

Bull Soc Chim Fr (1915) 17 228

~60%

JOC (1954) 19 1699
    (1949) 14 839

< 35%

JACS (1921) 43 898
JCS (1954) 678
Chem Met Eng (1921) 24 638
(Chem Abs 15 1703)

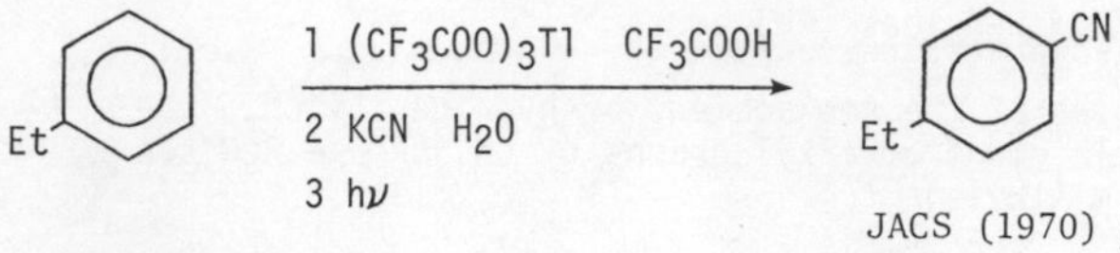

Et—⟨benzene⟩ → 1 (CF₃COO)₃Tl   CF₃COOH
                 2 KCN   H₂O
                 3 hν
→ Et—⟨benzene⟩—CN   80%

JACS (1970) 92 3520

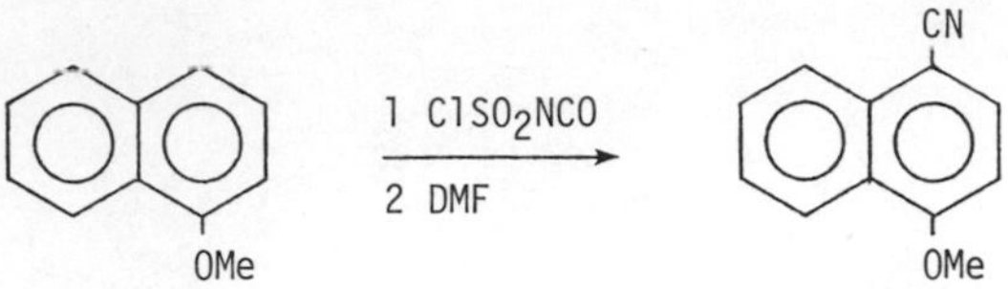

⟨naphthalene⟩—OMe → 1 ClSO₂NCO
                      2 DMF
→ ⟨naphthalene with CN⟩—OMe   81%

Ber (1967) 100 2719

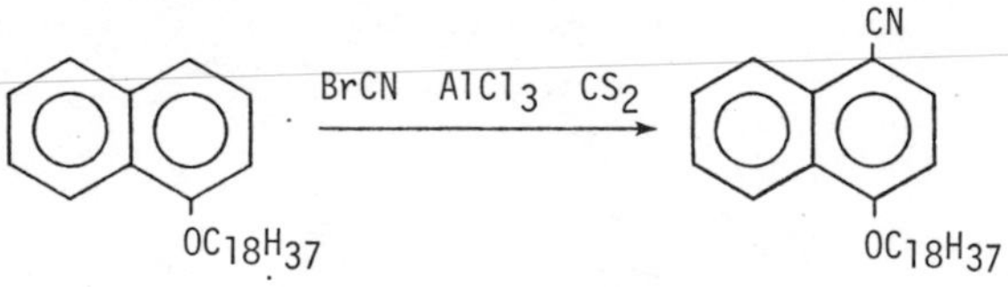

⟨naphthalene⟩—OC₁₈H₃₇ → BrCN   AlCl₃   CS₂
→ ⟨naphthalene with CN⟩—OC₁₈H₃₇   80%

JCS (1954) 678

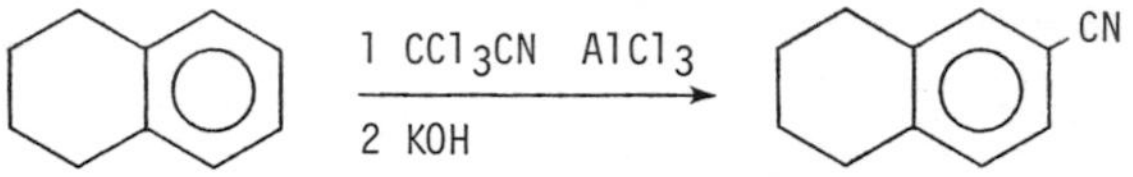

⟨tetralin⟩ → 1 CCl₃CN   AlCl₃
               2 KOH
→ ⟨tetralin⟩—CN   65%

Ber (1933) 66B 339

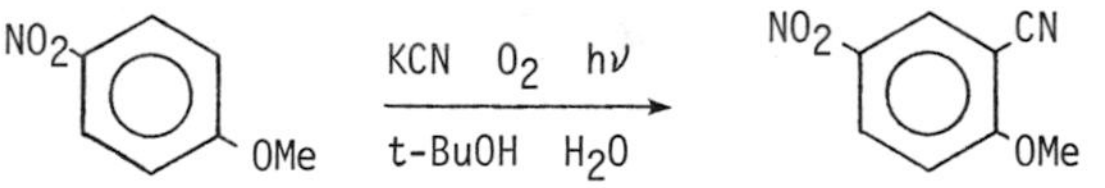

NO₂—⟨benzene⟩—OMe → KCN   O₂   hν
                      t-BuOH   H₂O
→ NO₂—⟨benzene⟩—CN, OMe   53%

JACS (1966) 88 2884

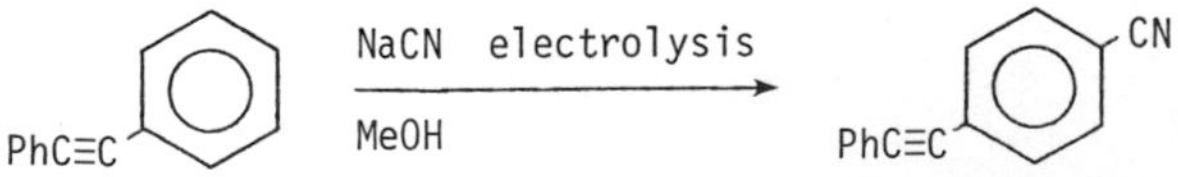

PhC≡C—⟨benzene⟩ → NaCN   electrolysis
                    MeOH
→ PhC≡C—⟨benzene⟩—CN   35%

Chem Comm (1970) 711
JACS (1969) 91 4181

Section 192   Nitriles from Ketones

EtCO—Me
$\xrightarrow[\text{HOAc}\quad C_6H_6]{\begin{array}{c}\text{COOEt}\\|\\\text{CH}_2\text{CN}\quad\text{alanine}\end{array}}$
$\begin{array}{c}\text{COOEt}\\|\\\text{EtC}=\text{CCN}\\|\\\text{Me}\end{array}$
$\xrightarrow[\text{Et}_2\text{O}]{\text{PhCH}_2\text{MgCl}}$
$\begin{array}{c}\text{PhCH}_2\\|\\\text{EtC}\!-\!\text{CHCN}\\|\qquad|\\\text{Me}\quad\text{COOEt}\end{array}$

$\Big\downarrow$ KOH

$\begin{array}{c}\text{PhCH}_2\\|\\\text{EtCCH}_2\text{CN}\\|\\\text{Me}\end{array}$

~70%

Org Synth (1963) Coll Vol 4 93
JCS C (1969) 121

$\xrightarrow[\text{2 165-175}°]{\begin{array}{c}\text{COOH}\\|\\1\ \text{CH}_2\text{CN}\quad\text{NH}_4\text{OAc}\end{array}}$

CH$_2$CN (cyclohexene)   - - →   CH$_2$CN (cyclohexane)

76-91%

Org Synth (1963) Coll Vol 4 234

EtCO—Me
$\xrightarrow[\text{HOAc}\quad\text{EtOH}]{\begin{array}{c}\text{COOEt}\\|\\\text{CH}_2\text{CN}\quad H_2\quad\text{Pd-C}\quad\text{NH}_4\text{OAc}\end{array}}$
$\begin{array}{c}\text{COOEt}\\|\\\text{EtCHCHCN}\\|\\\text{Me}\end{array}$
$\xrightarrow{\Delta}$
$\begin{array}{c}\text{EtCHCH}_2\text{CN}\\|\\\text{Me}\end{array}$ 64%

JACS (1959) 81 5397

$\xrightarrow[\text{NaOEt}\quad\text{DMF}]{(\text{EtO})_2\text{OPCH}_2\text{CN}}$

84%

JOC (1965) 30 505

Ph$_2$CO
$\xrightarrow[\text{2 H}_3\text{PO}_4]{\text{1 LiCH}_2\text{CN}\quad\text{THF}\quad\text{hexane}}$
Ph$_2$C=CHCN   - - →   Ph$_2$CHCH$_2$CN

58%

JOC (1968) 33 3402

$$\text{MsO-steroid} \xrightarrow[\text{2 POCl}_3 \quad \text{Pyr}]{\text{1 KCN} \quad \text{HOAc}} \text{(unsaturated nitrile)} \xrightarrow[\text{EtOAc} \quad \text{HOAc}]{\text{H}_2 \quad \text{Pd-C}} \text{(saturated nitrile)} \quad 35\%$$

JCS C (1968) 2283

$$C_{14}H_{29}CH_2COPh \xrightarrow[\text{HCl} \quad H_2O \quad \text{dioxane}]{\text{i-PrCH}_2CH_2ONO} C_{14}H_{29}\underset{\underset{\text{NOH}}{\|}}{C}COPh \xrightarrow[\text{NaOH} \quad H_2O]{\text{TsCl}} C_{14}H_{29}CN \quad >63\%$$

JACS (1953) 75 2347
JOC (1961) 26 3507

Some of the methods listed in section 184 (Nitriles from Aldehydes) may also be applied to the preparation of nitriles from ketones

Section 193   Nitriles from Nitriles

Review: The Alkylation of Esters and Nitriles   Org React (1957) 9 107

$$BuCH_2CN \xrightarrow[\text{toluene}]{BuBr \quad NaNH_2} Bu_3CCN \quad 80\%$$

JACS (1948) 70 3091

$$PhCH_2CN \xrightarrow{Br_2} PhCHCN \xrightarrow{PhH \quad AlCl_3} Ph_2CHCN \quad 74\%$$
$$\qquad\qquad\quad \underset{Br}{|}$$

JOC (1949) 14 839

$$PhCH_2CN \xrightarrow[\text{Me}_2SO \quad H_2O]{\text{MeBr} \quad \text{NaOH}} PhCHCN$$

$$\underset{Me}{|}$$

79%

JOC (1969) $\underline{34}$ 226

$$PhCH_2CN \xrightarrow[\text{2 Cyclohexyl bromide}]{\text{1 NaNH}_2 \quad NH_3} PhCHCN$$

65-77%

Org Synth (1955) Coll Vol 3 219
JCS (1946) 25
JOC (1958) $\underline{23}$ 1346

$$PhCH_2CN \xrightarrow[\text{2 CO}_2]{\text{1 BuLi}} \underset{COOH}{\overset{|}{PhCHCN}} \xrightarrow[\substack{\text{2 BuBr} \\ \text{3 } \Delta \text{ (decarbox)}}]{\text{1 NaNH}_2 \quad NH_3} \underset{Bu}{\overset{|}{PhCHCN}}$$

< 70%

JOC (1966) $\underline{31}$ 3873

For further examples of the alkylation of nitriles with halides see
section 190 (Nitriles from Halides and Sulfonates)

Section 194   Nitriles from Olefins
              °°°°°°°°°°°°°°°°°°°°°°°°°°°°

$$EtCH=CH_2 \xrightarrow{B_2H_6} (EtCH_2CH_2)_3B \xrightarrow[\text{t-butylphenoxide} \quad THF]{Cl_2CHCN \quad \text{potassium 2,6-di-}} (EtCH_2CH_2)_2CHCN$$

~85%

JACS (1970) $\underline{92}$ 5790

$$EtCH=CH_2 \xrightarrow{B_2H_6} (EtCH_2CH_2)_3B \xrightarrow[\text{t-butylphenoxide} \quad THF]{ClCH_2CN \quad \text{potassium 2,6-di-}} Et(CH_2)_3CN$$

JACS (1969) $\underline{91}$ 6854

BuCH=CH$_2$ $\xrightarrow[\text{THF}]{\text{B}_2\text{H}_6}$ (BuCH$_2$CH$_2$)$_3$B $\xrightarrow[\text{2 KOH H}_2\text{O}]{\text{1 N}_2\text{CHCN}}$ Bu(CH$_2$)$_3$CN   95%

JACS (1968) 90 6891

C$_6$H$_1$$_3$CH=CH$_2$ $\xrightarrow{\text{CH}_3\text{CN dibenzoyl peroxide}}$ C$_6$H$_1$$_3$(CH$_2$)$_3$CN   17%

JCS (1965) 1918 1932

$\xrightarrow[\text{(PhO)}_3\text{P}]{\text{HCN Pd[P(OPh)}_3]_4}$ CN   83%

Chem Comm (1969) 112

COOMe
|
(CH$_2$)$_3$CH=CH$_2$ $\xrightarrow{\text{HCN (28 atmos) Co}_2\text{(CO)}_8}$

COOMe
|
(CH$_2$)$_3$CHMe
|
CN   ~19%

JACS (1954) 76 5364

PhCH=CH$_2$ $\xrightarrow{\text{NNO}}$ PhCCH$_2$N (piperidine)
||
NOH
$\xrightarrow[\text{Et}_3\text{N}]{\text{TsCl}}$ PhCN

JACS (1965) 87 4642

Section 195   Nitriles from Miscellaneous Compounds
○○○

MeCH=CHCN $\xrightarrow{\text{Bu}_3\text{SnH MeOH}}$ MeCH$_2$CH$_2$CN   90%

Tetr Lett (1967) 4805
(1969)   489

$$\xrightarrow[\text{EtOAc \quad HOAc}]{\text{H}_2 \quad \text{Pd-C}}$$

JCS C (1968) 2283

92%

$$\underset{\overset{|}{\text{Cl}}}{\text{PhCHCN}} \xrightarrow{\text{Zn \quad HOAc}} \text{PhCH}_2\text{CN}$$

J Soc Chem Ind (1935) 54 98T

$$\xrightarrow[\text{EtOH}]{\text{H}_2 \quad \text{Pd-BaSO}_4}$$

Arch Pharm (1957) 290 218

51%

$$\text{C}_7\text{H}_{15}\text{CH}_2\text{NO}_2 \xrightarrow[\text{2 \quad Br}_2 \quad \text{CCl}_4]{\text{1 \quad Base}} \underset{\overset{|}{\text{Br}}}{\text{C}_7\text{H}_{15}\text{CHNO}_2} \xrightarrow[\text{C}_6\text{H}_6]{\text{Ph}_3\text{P}} \text{C}_7\text{H}_{15}\text{CN}$$

50%

JCS (1960) 2976

$$\text{PhC}\equiv\text{CCONH}_2 \xrightarrow{\text{NaOCl \quad H}_2\text{O}} \text{PhCH}_2\text{CN}$$

Rec Trav Chim (1920) 39 704
                (1927) 46 268

$$\text{PhCH}_2\text{COCOOH} \xrightarrow{\text{NH}_2\text{OH·HCl \quad H}_2\text{O}} \text{PhCH}_2\text{CN}$$

90%

Can J Chem (1961) 39 1340
JCS (1950) 926

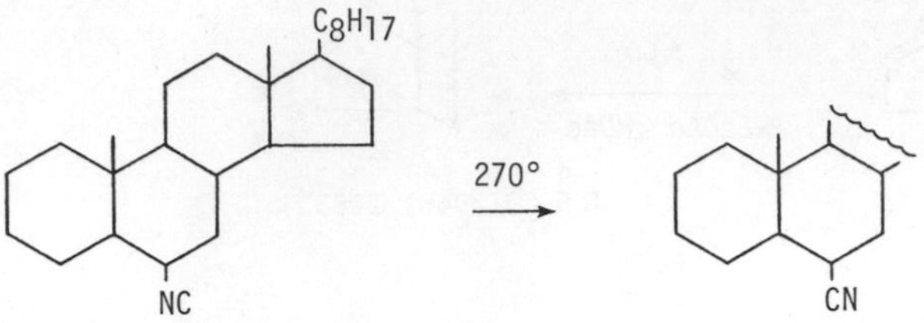

Tetr Lett (1966) 5087

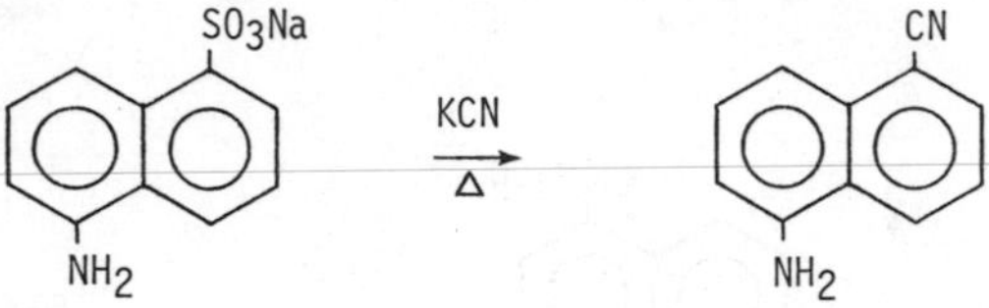

JCS (1954) 678

Chapter 14 PREPARATION OF OLEFINS

Section 196 Olefins from Acetylenes

$$BuC\equiv CH \xrightarrow[\text{2 BuLi}]{\text{1 Dicyclohexylborane}} BuCH_2CHLi_2 \xrightarrow{(C_5H_{11})_2CO} BuCH_2CH{=}C(C_5H_{11})_2 \quad 30\text{-}35\%$$

Tetr Lett (1966) 4315

$$BuC\equiv CH \xrightarrow[\text{2 NaOH}\ \ I_2\ \ \text{THF}\ \ H_2O]{\text{1 } (i\text{-Pr}\overset{Me}{\overset{|}{C}}H)_2BH\ \ \text{THF}} BuCH{=}CH\overset{Me}{\overset{|}{C}}HPr\text{-}i \ (\text{cis}) \quad 75\%$$

JACS (1967) 89 3652

$$BuC\equiv CH \xrightarrow[\text{THF}]{(i\text{-Bu})_2AlH} BuCH_2CH[Al(Bu\text{-}i)_2]_2 \xrightarrow[\text{2 HCHO}]{\text{1 MeONa}} BuCH_2CH{=}CH_2$$

Tetr Lett (1966) 6021

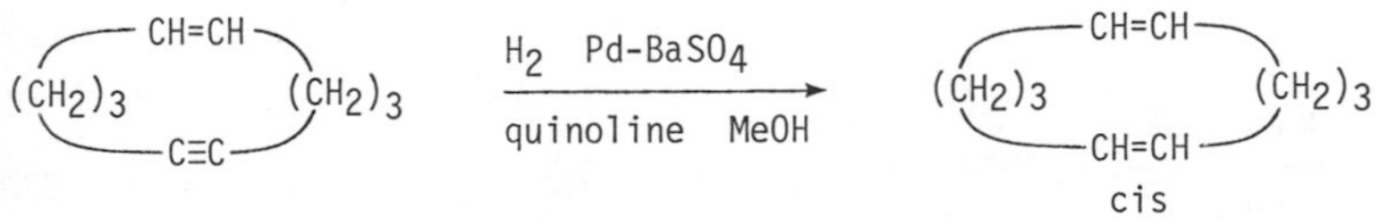

JACS (1956) 78 2518

$$\text{MeC(CH}_2)_3\text{C}\equiv\text{CEt} \xrightarrow[\text{NaBH}_4 \quad \text{EtOH}]{\text{H}_2 \quad \text{Ni(OAc)}_2} \text{MeC(CH}_2)_3\text{CH=CHEt}$$
$$\underset{\text{O O}}{}$$

80%

cis

Tetrahedron (1969) 25 5149

$$\text{BuC}\equiv\text{CEt} \xrightarrow[\text{LiCl} \quad \text{MeNH}_2]{\text{Electrolysis}} \text{BuCH=CHEt}$$

trans

58%

JOC (1968) 33 2727

$$\xrightarrow{\text{Na} \quad \text{NH}_3}$$

trans

<86%

JACS (1952) 74 3643
(1963) 85 622

$$\text{PrC}\equiv\text{C(CH}_2)_4\text{C}\equiv\text{CH} \xrightarrow[\text{NH}_3 \quad \text{Et}_2\text{O}]{\text{NaNH}_2 \quad \text{Na}} \text{PrCH=CH(CH}_2)_4\text{C}\equiv\text{CH}$$

trans

75%

JCS (1955) 3558

$$\xrightarrow{\text{Li} \quad \text{MeNH}_2}$$

trans

Chem Comm (1968) 634

$$\text{PhC}\equiv\text{CPh} \xrightarrow[\text{2 MeOH}]{\text{1 Li} \quad \text{THF}} \text{PhCH=CHPh}$$

cis

JOC (1970) 35 1702

$$\underset{\substack{\text{C≡CPh} \\ \text{COOH}}}{\bigcirc} \quad \xrightarrow{\text{CrSO}_4 \quad \text{DMF} \quad \text{H}_2\text{O}} \quad \underset{\substack{\text{CH=CHPh} \\ \text{COOH}}}{\bigcirc} \quad \text{trans}$$

85%

JACS (1964) __86__ 4358

$$\text{EtC≡CEt} \quad \xrightarrow[\text{2 HOAc}]{\text{1 NaBH}_4 \quad \text{BF}_3 \cdot \text{Et}_2\text{O} \quad \text{diglyme}} \quad \underset{\text{cis}}{\text{EtCH=CHEt}}$$

68%

JACS (1959) __81__ 1512

$$\underset{\substack{\text{OH} \\ \text{C≡CH}}}{\bigcirc} \quad \xrightarrow{(\text{i-Bu})_2\text{AlH} \quad \text{C}_6\text{H}_6} \quad \underset{\substack{\text{OH} \\ \text{CH=CH}_2}}{\bigcirc}$$

48%

JOC (1959) __24__ 627
    (1963) __28__ 1254
Ber (1956) __89__ 444

$$\text{EtC≡CEt} \quad \begin{cases} \xrightarrow[\text{2 MeLi}]{\text{1 (i-Bu)}_2\text{AlH}} \quad \underset{\text{cis}}{\text{EtCH=CHEt}} & \text{90\%} \\[2em] \xrightarrow[\text{2 100-130°}]{\text{1 (i-Bu)}_2\text{AlH-MeLi}} \quad \underset{\text{trans}}{\text{EtCH=CHEt}} & \text{88\%} \end{cases}$$

JACS (1967) __89__ 5085 5086

$$\text{EtC≡CEt} \quad \xrightarrow[\text{117-138°}]{\text{LiAlH}_4 \quad \text{diglyme}} \quad \underset{\text{trans}}{\text{EtCH=CHEt}}$$

97%

Tetrahedron (1967) __23__ 4509

$$\text{C}_{10}\text{H}_{21}\text{C≡C(CH}_2)_5\text{COOH} \quad \xrightarrow{\text{HI}} \quad \underset{\substack{| \\ \text{I}}}{\text{C}_{10}\text{H}_{21}\text{CH=C(CH}_2)_5\text{COOH}} \quad \xrightarrow[\text{HOAc}]{\text{Zn}} \quad \text{C}_{10}\text{H}_{21}\text{CH=CH(CH}_2)_5\text{COOH}$$

Compt Rend (1916) __162__ 944

$$\text{MeC} \equiv \text{CMe} \quad \xrightarrow[\text{2 HCl \quad H}_2\text{O}]{\text{1 B}_2\text{Cl}_4} \quad \text{MeCH=CHMe}$$

JACS (1967) <u>89</u> 4217 .

Section 197   Olefins from Carboxylic Acids, Acid Halides and Anhydrides

The decarboxylation of $\alpha\beta$- and $\beta\gamma$-unsaturated acids, RCH=CHCOOH $\longrightarrow$ RCH=CH$_2$ and RCH=CHCH$_2$COOH $\longrightarrow$ RCH$_2$CH=CH$_2$, is included in section 152 (Hydrides from Carboxylic Acids)

$$\begin{array}{c}\text{COOMe}\\ | \\ \text{(CH}_2\text{)}_4\text{COOH}\end{array} \quad \xrightarrow[\text{electrolysis \quad MeOH}]{\text{CH}_2\text{=CH(CH}_2\text{)}_8\text{COOH}} \quad \begin{array}{c}\text{COOMe}\\ | \\ \text{(CH}_2\text{)}_{12}\text{CH=CH}_2\end{array}$$

JCS (1953) 2393
(1954) 448 4219

For further examples of the Kolbe electrolysis see section 62 (Alkyls from Carboxylic Acids)

$$\text{BuCH}_2\text{CH}_2\text{COOH} \quad \xrightarrow{\text{Pb(OAc)}_4 \quad \text{Cu(OAc)}_2 \cdot \text{H}_2\text{O} \quad \text{Pyr}} \quad \text{BuCH=CH}_2 \qquad 72\%$$

Tetrahedron (1968) <u>24</u> 2215
JCS <u>C</u> (1969) 1047

Tetr Lett (1968) 2471   60%

$$\text{PhCH}_2\text{CH}_2\text{COCl} \quad \xrightarrow[\text{or RhCl(PPh}_3\text{)}_3 \quad \text{toluene}]{\text{PdCl}_2 \quad 200°} \quad \text{PhCH=CH}_2 \qquad 53\text{-}71\%$$

JACS (1968) <u>90</u> 94 99

COOH
|
Bu_2CCH_2Pr $\xrightarrow[\text{220-240°}]{\text{CuCN}}$ $Bu_2C=CHPr$ 73%

JACS (1950) $\underline{72}$ 2792

$C_{13}H_{27}CH_2CH_2COOH$ $\xrightarrow{\text{SeO}_2 \ \Delta}$ $C_{13}H_{27}CH=CH_2$

J Chem Soc Jap (1938) $\underline{59}$ 262 271

$\xrightarrow[\text{Et}_3\text{N Pyr H}_2\text{O}]{\text{Electrolysis}}$ 35%

Tetr Lett (1968) 5117 5123

$\xrightarrow{\text{Pb(OAc)}_4 \ \text{O}_2 \ \text{Pyr}}$ 76%

JACS (1968) $\underline{90}$ 113
Helv (1958) $\underline{41}$ 1191

$\xrightarrow{\text{PbO}_2 \ 170\text{-}190°}$ 30-37%

Helv (1958) $\underline{41}$ 1191
JACS (1952) $\underline{74}$ 4370

$\xrightarrow{\text{t-BuOOH}}$ $\xrightarrow[\text{or } h\nu]{\Delta}$ 38%

Chem Comm (1969) 98

Section 198   Olefins from Alcohols
                ○○○○○○○○○○○○○○○○○○○○○○○○

BuOH --> BuOCH$_2$CH=CH$_2$ $\xrightarrow{\text{PrLi \quad pentane}}$ BuCH=CH$_2$ ~73%

Tetr Lett (1969) 821

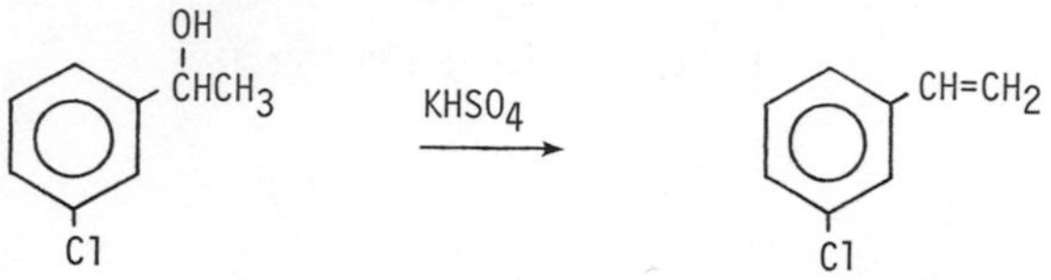

KHSO$_4$ 80-82%

Org Synth (1955) Coll Vol 3 204
          (1943) Coll Vol 2 606
          (1941) Coll Vol 1 430

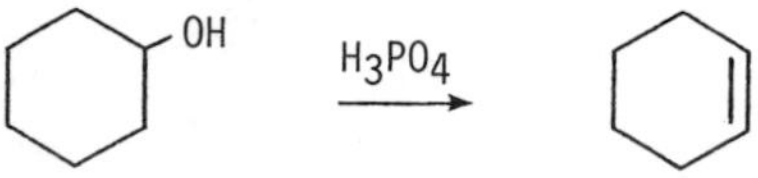

H$_3$PO$_4$ 79-84%

Org Synth (1943) Coll Vol 2 151

OH
Me$_2$CCH$_2$Me $\xrightarrow{\text{I}_2}$ Me$_2$C=CHMe

JACS (1915) 37 1748

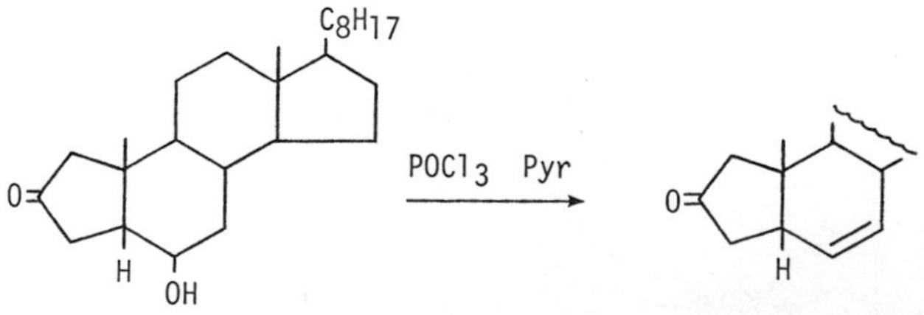

POCl$_3$ Pyr   83%

JACS (1961) 83 5003
Chem Pharm Bull (1961) 9 854
JCS C (1967) 1115

$(t\text{-}Bu)_2CCH_3$ $\xrightarrow{\text{SOCl}_2 \quad \text{Pyr}}$ $(t\text{-}Bu)_2C=CH_2$ ~70%
$\quad\quad$ |
$\quad\quad$ OH

JACS (1960) 82 2498

$\xrightarrow{\text{Me}_2\text{SO} \quad 160°}$

88%

JOC (1964) 29 123

$\xrightarrow[\text{Pyr}]{\text{MsCl}}$ $\quad$ $\xrightarrow[\substack{\text{Me}_2\text{SO} \\ \text{C}_6\text{H}_6}]{\text{t-BuOK}}$

57%

Steroids (1964) 4 55
JOC (1964) 29 742
(1961) 26 2883

Further examples of the preparation of olefins from sulfonates are included
in section 205 (Olefins from Halides and Sulfonates)

$\xrightarrow[\text{2 NaH \quad THF}]{\text{1 Et}_3\text{NSO}_2\text{NCOOMe}}$

55%

JACS (1970) 92 5224
JOC (1970) 35 2594

$\xrightarrow[\text{Pyr \quad Et}_2\text{O}]{\text{MeOSOCl}}$ $\quad$ $\xrightarrow{215\text{-}275°}$

JACS (1954) 76 1213

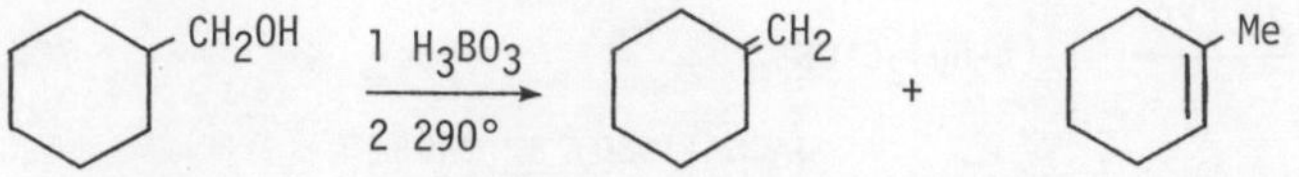

JOC (1961) 26 4193
Chem Rev (1960) 60 431

$C_{16}H_{33}CH_2CH_2OH$ $\xrightarrow[\text{(vapor phase)}]{Ac_2O\ (600°)}$ $C_{16}H_{33}CH=CH_2$ 68%

Chem Ind (1965) 681

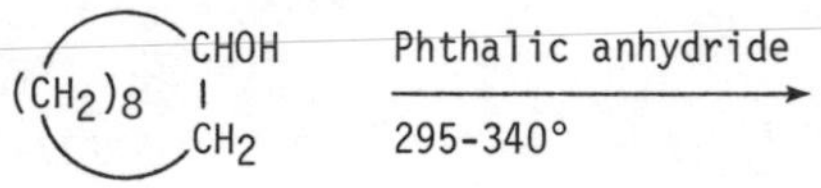

JACS (1952) 74 3636

For examples of the preparation of olefins from esters, by pyrolysis and by other methods, see section 203 (Olefins from Esters)

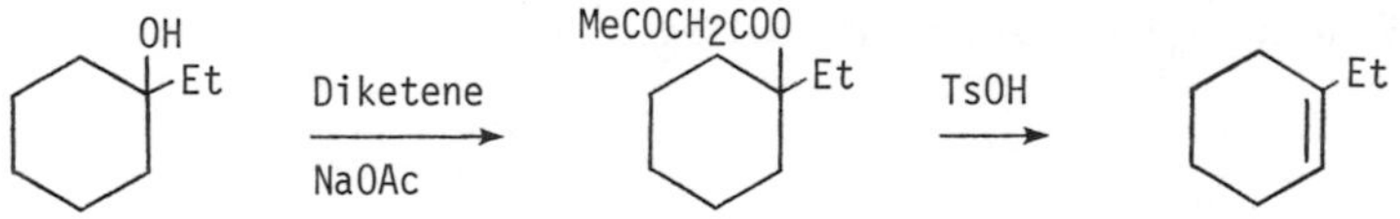

Arkiv Kemi (1961) 17 401

$C_6H_{13}CH_2CH_2OH$ $\xrightarrow[\text{2 } Bu_3P]{\text{1 } BrCH_2COCl\quad Pyr}$ $C_6H_{13}CH_2CH_2OCOCH_2\overset{+}{P}Bu_3\ \overset{-}{Br}$ $\xrightarrow{170°}$ $C_6H_{13}CH=CH_2$

~72%

JOC (1964) 29 1003
JACS (1961) 83 3336

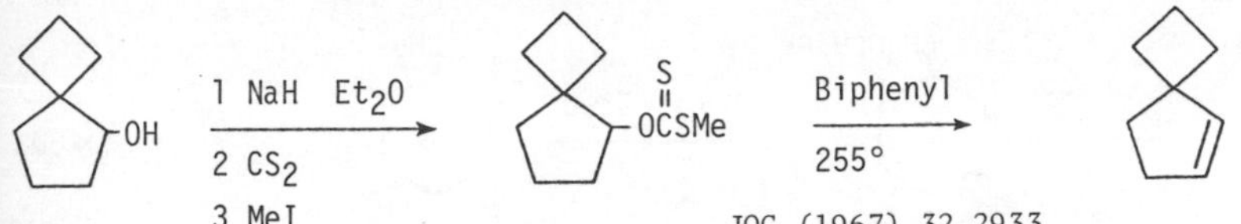

JOC (1967) 32 2933
    (1968) 33 2214
Can J Chem (1970) 48 970
Org React (1962) 12 1

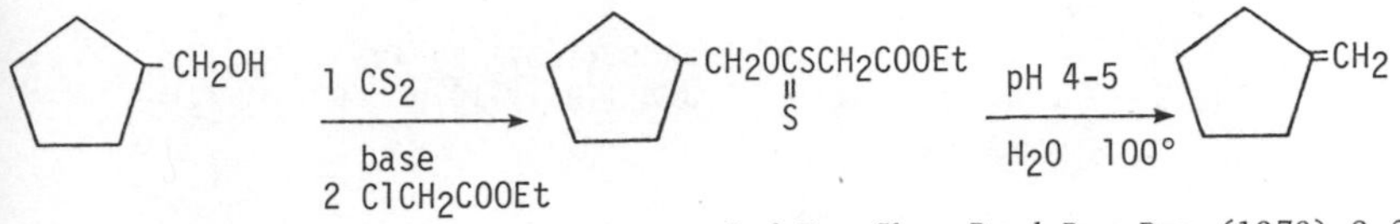

Ind Eng Chem Prod Res Dev (1970) 9 230

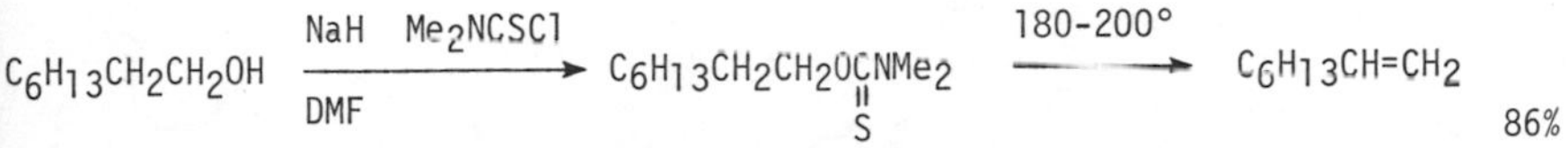

$C_6H_{13}CH_2CH_2OH$ $\xrightarrow[\text{DMF}]{\text{NaH Me}_2\text{NCSCl}}$ $C_6H_{13}CH_2CH_2OCNMe_2$ $\xrightarrow{180-200°}$ $C_6H_{13}CH=CH_2$

86%

JOC (1969) 34 3604

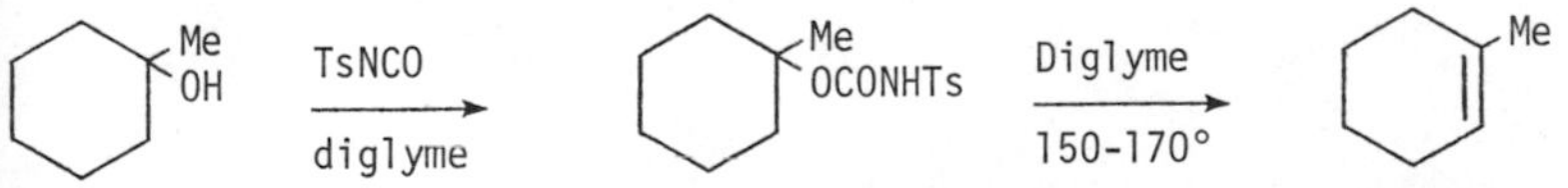

Chem Comm (1970) 606

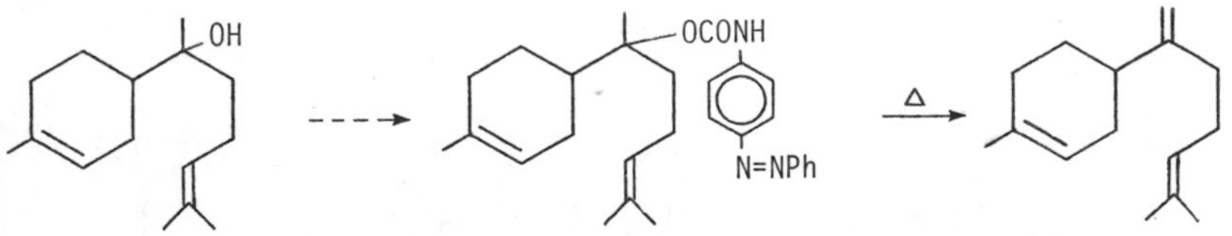

Aust J Chem (1954) 7 298
JACS (1953) 75 2118

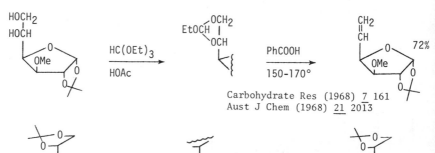

JCS (1958) 843

< 60%

JACS (1937) 59 403
JCS (1951) 1079

Chem Comm (1968) 1593
Tetr Lett (1968) 3655
JACS (1965) 87 934
     (1963) 85 2677

Carbohydrate Res (1968) 7 161
Aust J Chem (1968) 21 2013

Tetr Lett (1970) 5223

For the preparation of olefins from diols via bis sulfonates see section
205 (Olefins from Halides and Sulfonates)

Section 199   Olefins from Aldehydes

Examples of the decarbonylation of $\alpha\beta$-unsaturated aldehydes, RCH=CHCHO $\longrightarrow$
RCH=CH$_2$, are included in section 154 (Hydrides from Aldehydes)

PhCH=CHCHO $\xrightarrow{\text{PhCH}_2\text{COOH} \quad \text{Ac}_2\text{O} \quad \text{PbO}}$ PhCH=CHCH=CHPh   23-25%

Org Synth (1943) Coll Vol 2 229

PhCH$_2$CHO $\xrightarrow{\text{KOH} \quad \text{EtOH}}$ PhCH=CHCH$_2$Ph   91%

JOC (1966) 31 396

Izv (1960) 1030
(Chem Abs 54 24627)

EtCH$_2$CHO $\xrightarrow[\text{MeOH}]{\text{Br}_2}$ EtCH-CHOMe $\xrightarrow{\text{EtMgBr}}$ EtCH-CHOMe $\xrightarrow[\text{PrOH}]{\text{Zn}}$ EtCH=CHEt   ~20%
$\quad\quad\quad\quad\quad\quad$ Br Br $\quad\quad\quad\quad\quad\quad$ Br Et

JACS (1932) 54 751

JACS (1946) 68 1085

Reviews: The Wittig Reaction   Org React (1965) 14 270
                                        Quart Rev (1963) 17 406

New Reactions of Alkylidenephosphoranes and their Preparative Uses
                                        Angew (1965) 77 850
                                        (Internat Ed 4 830)

$$PhCHO \xrightarrow[\text{2 } 65°]{\text{1 } Ph_3P=CH_2 \quad Et_2O} PhCH=CH_2$$

67%

Ber (1954) <u>87</u> 1318

$$PhCHO \xrightarrow{Ph_3P=CHBu\text{-}t \quad Et_2O} PhCH=CHBu\text{-}t$$

84%

JACS (1965) <u>87</u> 4156

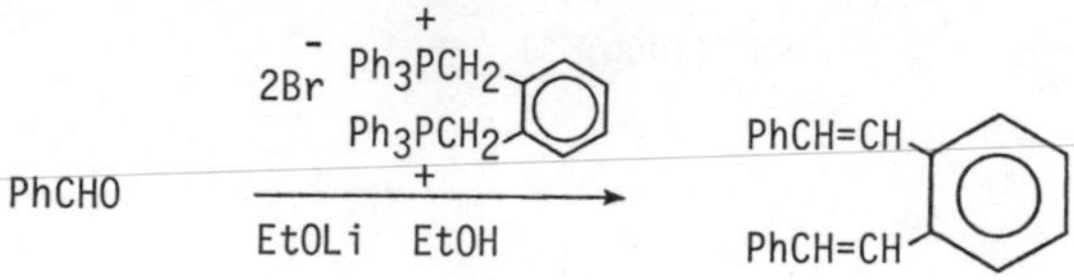

$$PhCHO \xrightarrow[\text{EtOLi EtOH}]{2Br^- \quad Ph_3PCH_2 \diagup \quad Ph_3PCH_2 \diagdown} $$

84%

Org React (1965) <u>14</u> 270

$$\xrightarrow[\text{MeOH MeCN}]{Ph_3PCH_2Pr^+ \quad F^-}$$

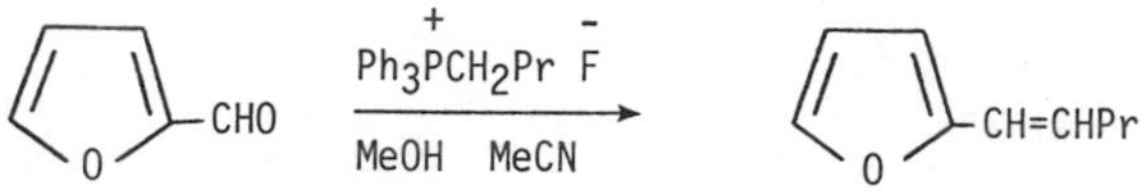

41%

Ber (1970) <u>103</u> 2077

$$PhCH=CHCHO \xrightarrow[\text{DMF MeONa}]{PhCH_2Br\text{-}(Me_2N)_3P} PhCH=CHCH=CHPh$$

89%

Annalen (1965) <u>682</u> 58

$$PhCHO \xrightarrow{Ph_2PO(CH_2Ph) \quad t\text{-}BuOK \quad C_6H_6} PhCH=CHPh$$

70%

Ber (1959) <u>92</u> 2499

$$PhCHO \xrightarrow[150°]{\overset{+ \quad -}{Ph_3PCH_2Et} \; Br \quad \overset{O}{\overset{/\backslash}{CH_2-CH_2}}} PhCH=CHEt \qquad 74\%$$

Angew (1968) 80 535
(Internat Ed 7 536)

$$PhCHO \xrightarrow[\text{2 Crystallization}]{1 \; Li\overset{Me}{\overset{|}{C}}HPO(NMe_2)_2} \overset{OH \; Me}{\underset{\text{diastereomer A}}{Ph\overset{|}{C}H-\overset{|}{C}HPO(NMe_2)_2}}$$

diastereomer A $\xrightarrow{\Delta}$ cis

diastereomer B $\xrightarrow{\Delta}$ trans

MnO₂ ↙   ↗ NaBH₄

$$\underset{\underset{Me}{|}}{PhCOCHPO(NMe_2)_2}$$

JACS (1966) 88 5653 5652
(1968) 90 6816
JOC (1969) 34 3053

$$PrCHO \xrightarrow{Ph_3P=CHPr} \underset{\underset{+ \quad -}{Ph_3P \; O}}{Pr\overset{|}{C}H\overset{|}{C}HPr} \xrightarrow[\substack{2 \; HCl \\ 3 \; t\text{-BuOK}}]{1 \; PhLi} \underset{\text{trans}}{PrCH=CHPr} \qquad 72\%$$

Angew (1966) 78 115
(Internat Ed 5 126)
JACS (1970) 92 226

$$\text{(cyclohexene)}-CHO \xrightarrow[\text{2 65°}]{1 \; Me_2\overset{Li}{\overset{|}{C}}PS(OMe)_2 \; THF} \text{(cyclohexene)}-CH=CMe_2 \qquad 68\%$$

JACS (1966) 88 5654

$$C_{11}H_{23}CHO \xrightarrow[\text{2 Toluene reflux}]{1 \; Li CH_2SO\overset{Li}{N}-\bigcirc-Me \; THF} C_{11}H_{23}CH=CH_2 \qquad 56\%$$

JACS (1966) 88 5656
(1968) 90 5548

Further examples of the Wittig Reaction and a list of bases which may be
used for the generation of Wittig reagents are included in section 207
(Olefins from Ketones)

32%

JACS (1970) 92 6635

$C_6H_{13}CHO$ $\xrightarrow[\text{2 BuLi}]{\text{1 Ph}_3\text{P=CHMe   THF}}$ $\underset{\underset{OLi \quad Me}{|\qquad|}}{C_6H_{13}CH-C=PPh_3}$ $\xrightarrow{\text{MeI}}$ $C_6H_{13}CH=CMe_2$ <50%

JACS (1970) 92 226

$C_6H_{13}CHO$ $\xrightarrow[\substack{\text{2 Hg(OAc)}_2 \\ \text{3 LiI   I}_2}]{\text{1 Ph}_3\text{P=CHMe}}$ $\underset{H}{\overset{C_6H_{13}}{>}}C=C\underset{Me}{\overset{I}{<}}$ $\xrightarrow{R_2CuLi}$ $\underset{H}{\overset{C_6H_{13}}{>}}C=C\underset{Me}{\overset{R}{<}}$ (R=alkyl)

Tetr Lett (1970) 447

PhCHO $\xrightarrow[\text{HMPA}]{\text{Me}_3\text{SiCH}_2\text{Ph   BuLi}}$ PhCH=CHPh   >50%

Tetr Lett (1970) 1137

$C_{11}H_{23}CHO$ $\xrightarrow{\text{CH}_2\text{I}_2 \quad \text{Mg-Hg} \quad \text{Et}_2\text{O}}$ $C_{11}H_{23}CH=CH_2$ 65%

Tetrahedron (1970) 26 1281
J Organometallic Chem (1968) 12 263

PhCHO $\xrightarrow{\text{C}_5\text{H}_{11}\text{CHLi}_2 \quad \text{THF}}$ PhCH=CHC$_5$H$_{11}$ 45-50%

Tetr Lett (1966) 4315

PhCH=CHCHO $\xrightarrow{\text{LiCH}_2\text{NC} \quad \text{THF}}$ PhCH=CHCH=CH$_2$ 28%

Angew (1968) 80 842
(Internat Ed 7 805)

Tetr Lett (1969) 4001

MeCH$_2$CHO --→ MeCH$_2$CH(OEt)$_2$ $\xrightarrow{\text{TsOH}}$ MeCH=CHOEt $\xrightarrow{(i\text{-Bu})_2\text{AlH}_4}$ MeCH=CH$_2$ 44%

JOC (1966) 31 329

Some of the methods included in section 207 (Olefins from Ketones) may
also be applied to the preparation of olefins from aldehydes

Section 200   Olefins from Alkyls, Methylenes and Aryls
ooooooooooooooooooooooooooooooooooooooooooo

PhCH$_2$CH$_2$Ph $\xrightarrow[\text{260°}]{\text{RhCl(PPh}_3)_3}$ PhCH=CHPh   37%

Tetr Lett (1970) 1825

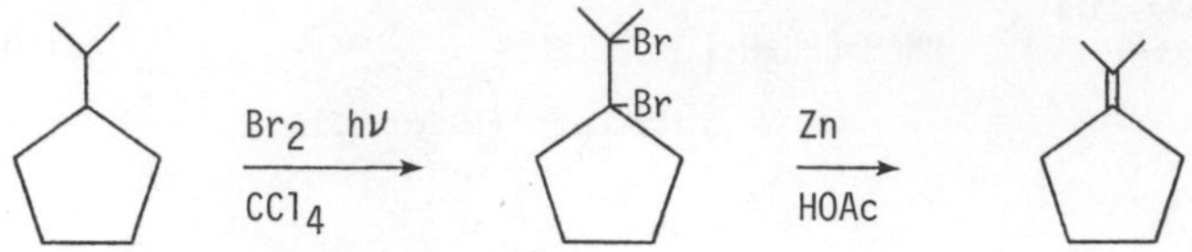

Rec Trav Chim (1964) **83** 67

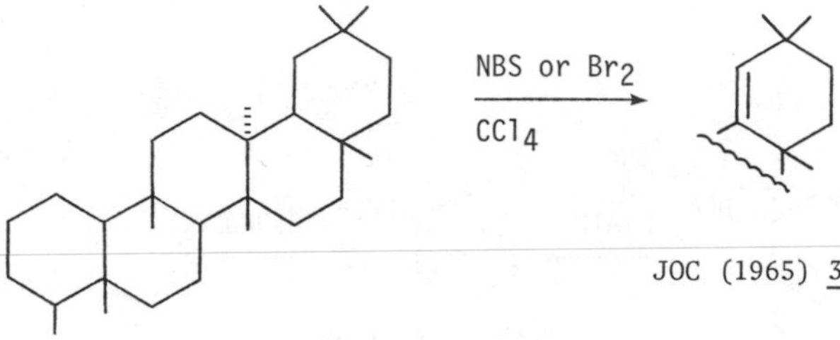

JOC (1965) **30** 2479

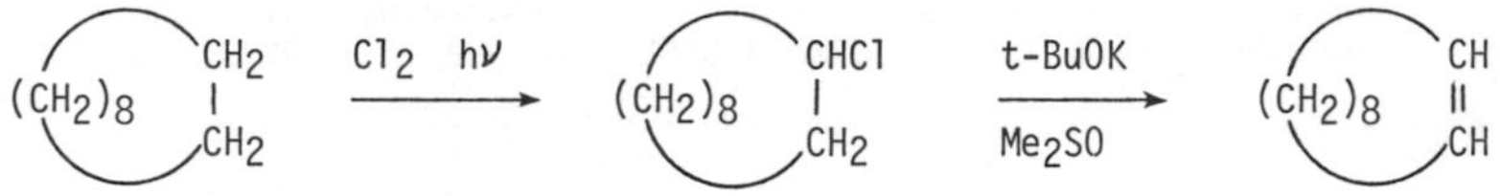

JOC (1967) **32** 510

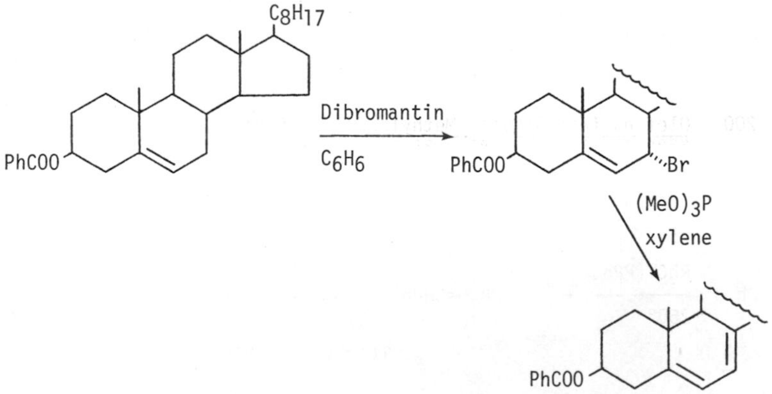

Helv (1958) **41** 70

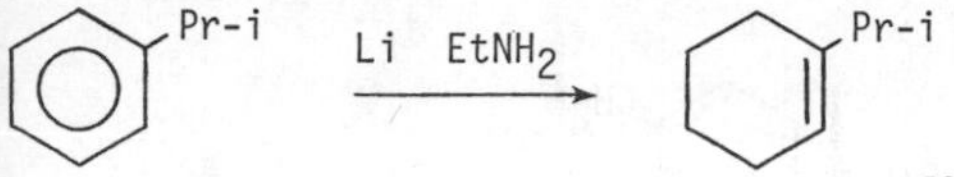

JOC (1963) 28 1094

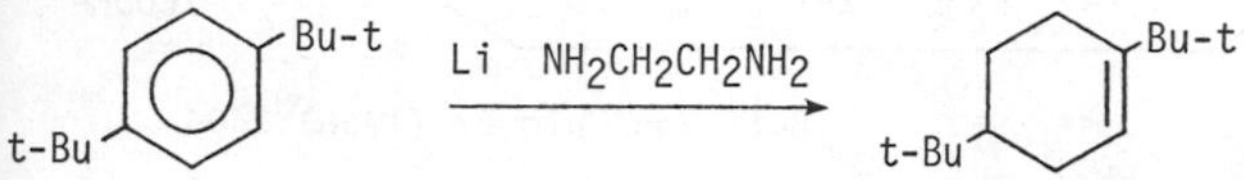

69%

JOC (1966) 31 965
Aust J Chem (1964) 17 55

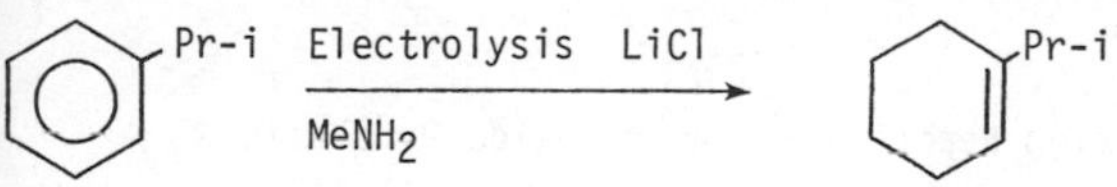

44%

JACS (1964) 86 5272

<81%

Section 201   Olefins from Amides

t-BuCH₂CMe₂   510°
      | ─────────────→   t-BuCH=CMe₂   +   t-BuCH₂C=CH₂
     NHAc   (vapor phase) |
                                                         Me

JOC (1958) 23 996
JACS (1959) 81 651
      (1958) 80 4588

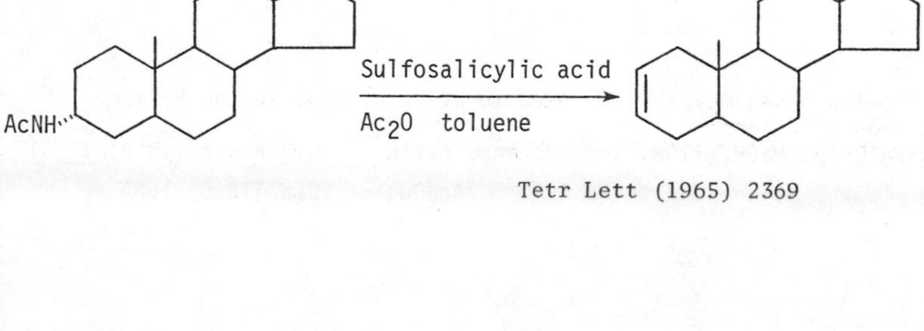

Sulfosalicylic acid
Ac₂O   toluene

Tetr Lett (1965) 2369

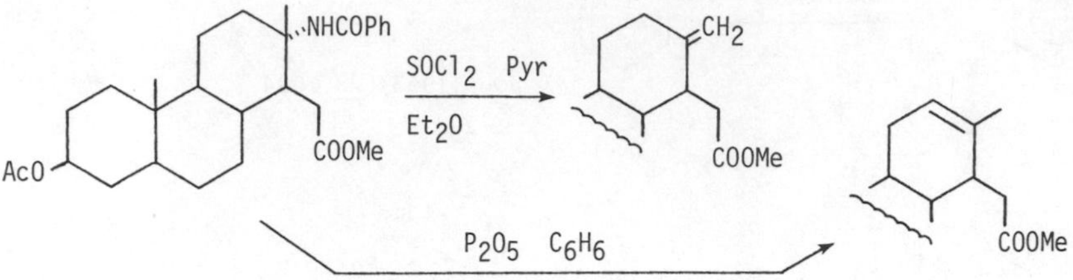

Bull Soc Chim Fr (1966) 2404

Section 202   Olefins from Amines

Review: The Hofmann Elimination Reaction and Amine Oxide Pyrolysis
Org React (1960) 11 317

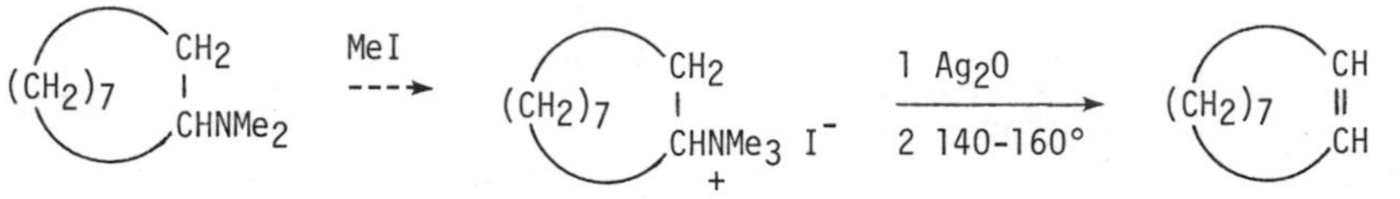

JACS (1952) 74 3643
JCS (1955) 4016

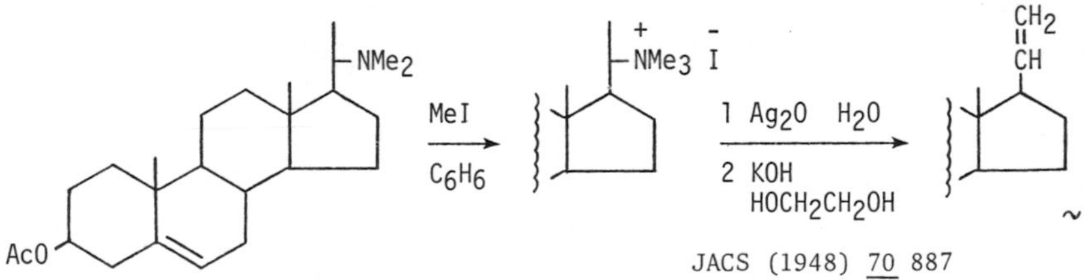

JACS (1948) 70 887

The following bases may also be used in place of Ag_2O in the Hofmann degradation of amines:
Ag_2SO_4-$Ba(OH)_2$, NaOH, TlOH, ion exchange resin   Org React (1960) 11 317

Ag_2CO_3 JCS (1935) 1685

$BuCH_2CH_2NH_2$ $\xrightarrow[\text{H}_2\text{O}]{\text{Me}_2\text{SO}_4 \quad \text{NaOH}}$ $BuCH_2CH_2\overset{+}{N}Me_3 \ (SO_4)^{-}_{1/2}$ $\xrightarrow[\text{reflux}]{\text{H}_2\text{SO}_4 \quad \text{H}_2\text{O}}$ $BuCH=CH_2$ 60%

Org React (1960) <u>11</u> 317

$(CH_2)_6 \overset{\displaystyle CH_2}{\underset{\displaystyle \overset{+}{C}HNMe_3 \ Br^{-}}{|}}$ $\xrightarrow{\text{(A) PhLi} \atop \text{or (B) KNH}_2}$ $(CH_2)_6 \overset{\displaystyle CH}{\underset{\displaystyle CH}{\|}}$ (A) cis

          (B) trans

66-68%

Annalen (1958) <u>612</u> 102

$\xrightarrow[\text{MeOH}]{\text{H}_2\text{O}_2 \quad \text{H}_2\text{O}}$ $\xrightarrow{160°}$ 79-88%

Org Synth (1963) Coll Vol 4 612

$\underset{\displaystyle \overset{|}{\underset{\displaystyle \overset{+}{\underset{\displaystyle -O}{N}Me_2}}{PhCHCHCH_3}}}{\overset{\displaystyle Me}{}}$ $\xrightarrow[\text{or H}_2\text{O} \quad 132\text{-}138°]{\text{Me}_2\text{SO} \quad \text{THF} \quad 25°}$ $\underset{\displaystyle}{\overset{\displaystyle Me}{PhC=CHMe}}$ + $\underset{\displaystyle}{\overset{\displaystyle Me}{PhCHCH=CH_2}}$

JACS (1962) <u>84</u> 1734

$\xrightarrow[\text{Ac}_2\text{O}]{\text{Sulfosalicylic acid}}$ > 70%

Tetr Lett (1965) 2369

$BuCH_2CH_2NH_2$ $\xrightarrow[\text{2 NaH \quad TsCl}]{\text{1 TsCl \quad DMF}}$ $BuCH_2CH_2NTs_2$ $\xrightarrow{\text{KI \quad DMF}}$ $BuCH=CH_2$ 31%

JACS (1969) <u>91</u> 2384

PhCH$_2$CH$_2$NH$_2$ $\xrightarrow[]{}$ PhCH$_2$CH$_2$N$\underset{SO_2}{\overset{CO}{\diagup}}$ $\xrightarrow[\Delta]{KOH}$ PhCH=CH$_2$ < 65%

(with COCl / SO$_2$Cl reagent)

Tetr Lett (1967) 3027
JACS (1969) 91 2384

Me$_2$CHCH$_2$NH$_2$ $\xrightarrow{C_8H_{17}ONO \quad CHCl_3 \quad HOAc}$ Me$_2$C=CH$_2$ 27%

JACS (1969) 91 1790

$\xrightarrow{BuONO}$

Tetr Lett (1969) 4001

Section 203   Olefins from Esters

For the preparation of olefins from sulfonic esters see section 205
(Olefins from Halides and Sulfonates)

HC≡C(CH$_2$)$_7$CH$_2$COOEt $\xrightarrow[Et_2O]{PhMgBr}$ HC≡C(CH$_2$)$_7$CH$_2\underset{OH}{C}$Ph$_2$ $\xrightarrow{220-230°}$ HC≡C(CH$_2$)$_7$CH=CPh$_2$
92%

JCS (1957) 1622

C$_{15}$H$_{31}$COOMe $\xrightarrow[Et_2O]{EtMgBr}$ C$_{15}$H$_{31}\underset{OH}{C}$Et$_2$ $\xrightarrow{HCOOH}$ C$_{15}$H$_{31}\underset{Et}{C}$=CHMe   96%

JACS (1945) 67 2239

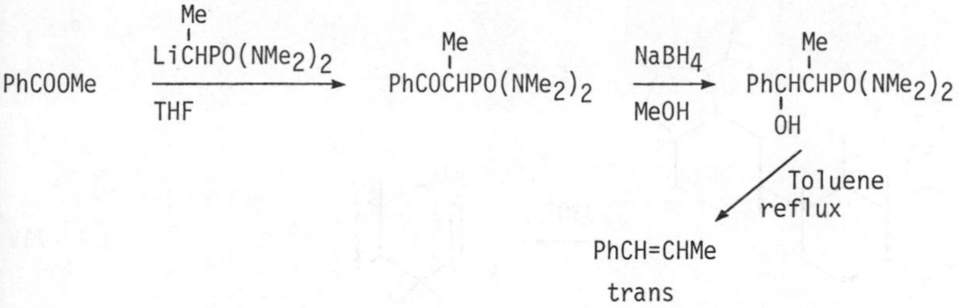
Me
LiCHPO(NMe2)2
PhCOOMe
THF
Me
PhCOCHPO(NMe2)2
NaBH4
MeOH
Me
PhCHCHPO(NMe2)2
OH
Toluene
reflux
PhCH=CHMe
trans

JACS (1966) 88 5653

Review: Pyrolytic Cis Eliminations Chem Rev (1960) 60 431

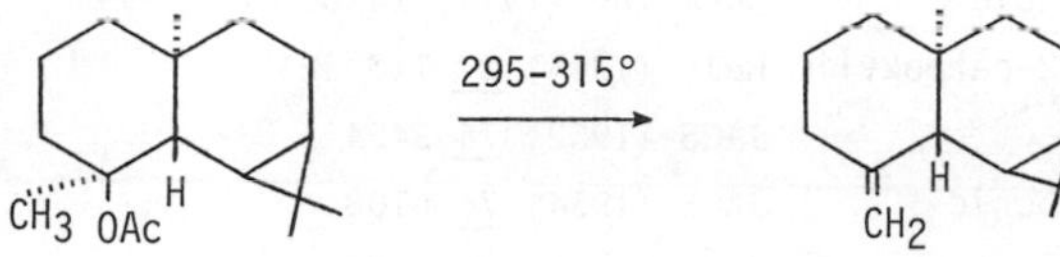
295-315°
CH3 OAc
H
H
CH2

JACS (1959) 81 1968

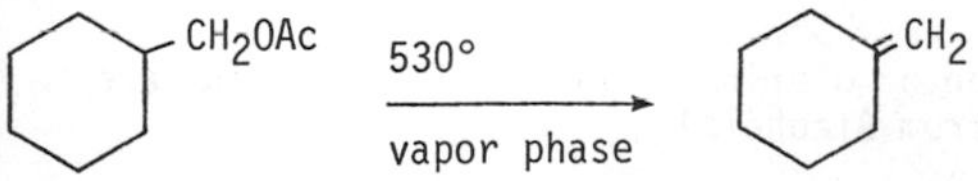
CH2OAc
530°
vapor phase
CH2

JACS (1959) 81 647 651

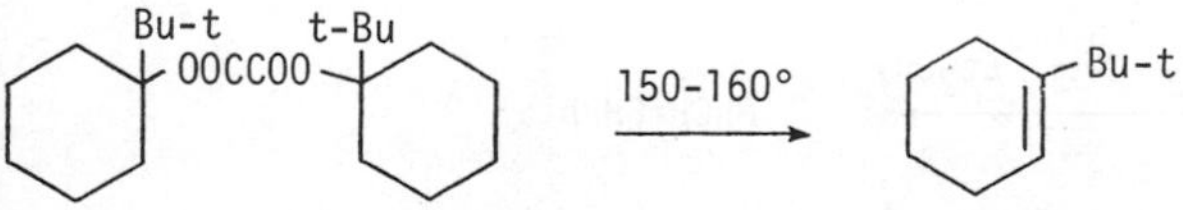
Bu-t t-Bu
OOCCOO
150-160°
Bu-t

<80%

JACS (1969) 91 3324
JOC (1965) 30 689

PhCOO $\xrightarrow{330°}$

71%

JCS (1965) 4379

Esters of the following acids have also been used for the preparation of olefins by liquid-phase pyrolysis:

Stearic	JACS (1948) 70 2690
	JCS (1957) 1998
d-Camphoric	JACS (1952) 74 3944
2,4,6-Triethylbenzoic	JACS (1953) 75 6011
2-Naphthoic	JACS (1954) 76 5692
3,5-Dinitrobenzoic	Chem Ind (1954) 1426
Anthraquinone-2-carboxylic	Helv (1944) 27 713 821
Ethyl carbonic	JACS (1952) 74 5454
2-Naphthyl carbonic	JACS (1954) 76 6108
Phenylcarbamic	JACS (1953) 75 2118
	Can J Chem (1953) 31 688

Further examples of the preparation of olefins by pyrolysis of esters are included in section 198 (Olefins from Alcohols)

PhCH$_2$CHPh
COO

Et   Et

Et

$\xrightarrow{\text{NaNH}_2 \quad \text{NH}_3 \quad \text{Et}_2\text{O}}$

PhCH=CHPh

JACS (1953) 75 6011

Section 204   Olefins from Ethers and Epoxides
oooooooooooooooooooooooooooooooooooo

H_3PO_4

triethylene glycol

Ber (1967) 100 1764

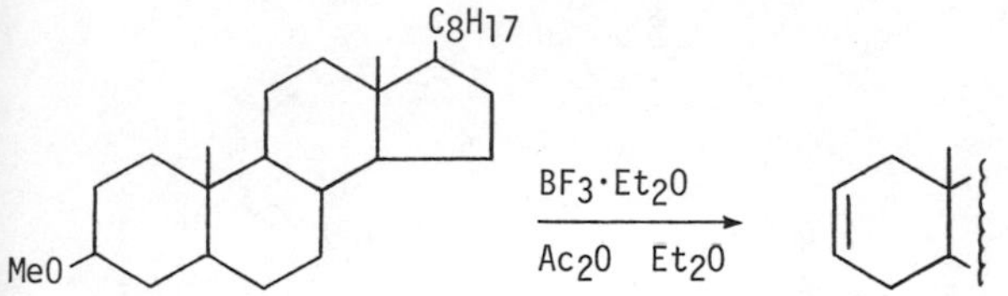

82%

C_8H_{17}

$BF_3 \cdot Et_2O$

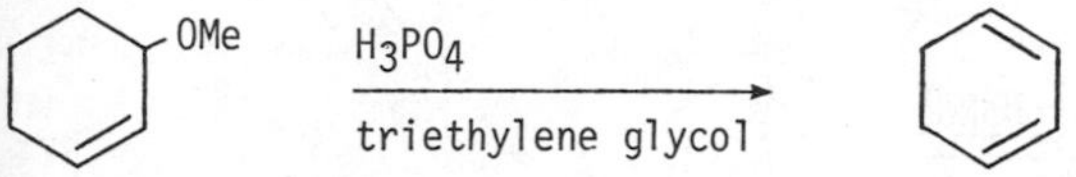

Ac_2O Et_2O

MeO

Tetr Lett (1964) 759

24%

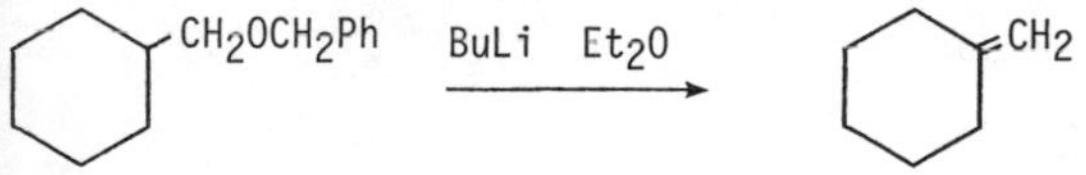

CH_2OCH_2Ph BuLi Et_2O

$=CH_2$

JCS (1958) 843

~60%

OMe
|
$C_6H_{13}CHCH_3$ i-PrLi   pentane

$C_6H_{13}CH=CH_2$

JACS (1951) 73 5708 1263

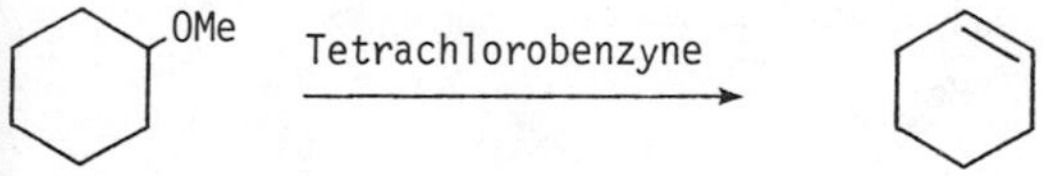

OMe

Tetrachlorobenzyne

Tetr Lett (1968) 4455

EtCH₂CH₂OBu $\xrightarrow[\text{140-150°}]{\text{N}_2\text{CHCOOEt}}$ EtCH=CH₂   6%

Rec Trav Chim (1955) <u>74</u> 143

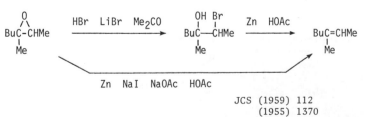

Tetrahedron (1968) <u>24</u> 3503

Chem Comm (1970) 144   80%

Chem Comm (1970) 1450

$\underset{\underset{\text{Me}}{|}}{\text{BuC-CHMe}}$ (epoxide) $\xrightarrow{\text{HBr LiBr Me}_2\text{CO}}$ $\underset{\underset{\text{Me}}{|}}{\overset{\text{OH Br}}{\text{BuC—CHMe}}}$ $\xrightarrow{\text{Zn HOAc}}$ $\underset{\underset{\text{Me}}{|}}{\text{BuC=CHMe}}$

Zn NaI NaOAc HOAc

JCS (1959) 112
(1955) 1370

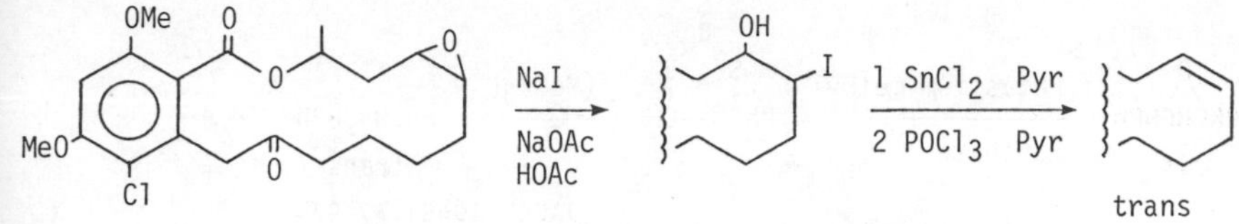

Aust J Chem (1966) 19 1265
JCS (1959) 112

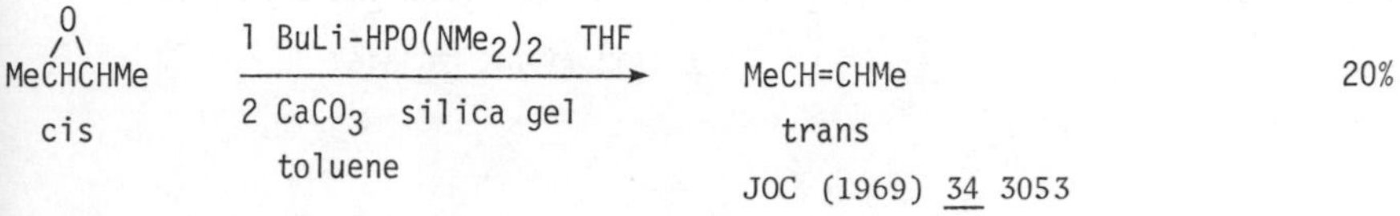

MeCH=CHMe   20%

trans

JOC (1969) 34 3053

PhCH=CH₂

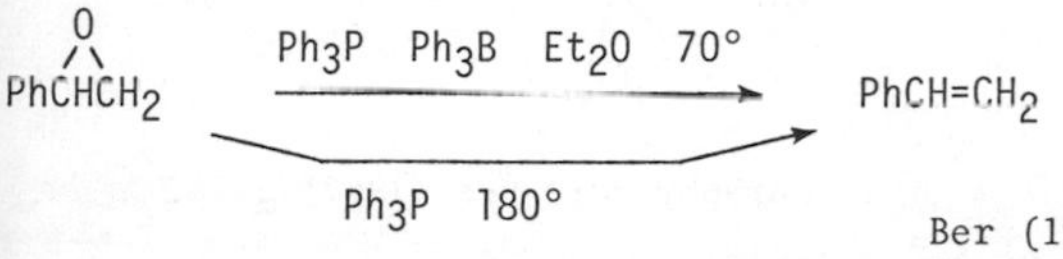

Ber (1955) 88 1654
Chem Ind (1959) 330

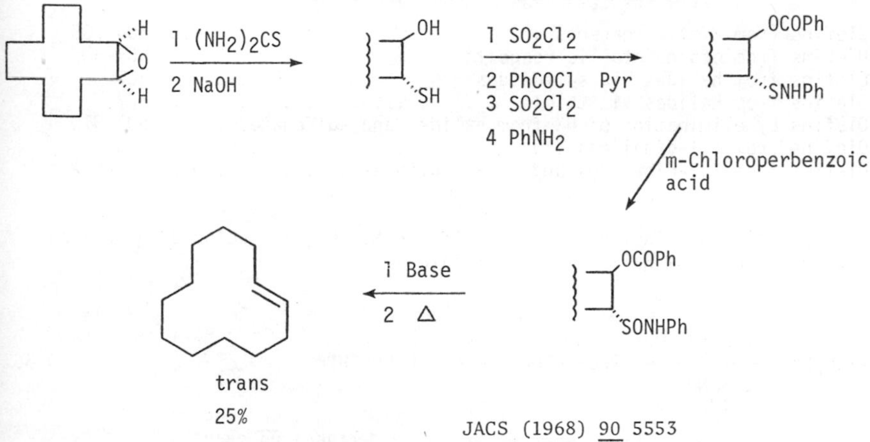

trans
25%

JACS (1968) 90 5553

$$\underset{cis}{\underset{PhCHCHPh}{\overset{O}{\overset{/\backslash}{}}}} \xrightarrow[\text{xanthate}]{\text{Potassium methyl}} \underset{PhCH-CHPh}{\overset{S}{\overset{\|}{\underset{S}{\overset{|}{}}\underset{}{\overset{}{}}S}}} \xrightarrow{(MeO)_3P} \underset{trans}{PhCH=CHPh}$$

JACS (1965) 87 934

$$\underset{PhOCH_2CHCH_2}{\overset{O}{\overset{/\backslash}{}}} \xrightarrow{KSCN \ H_2O} \underset{PhOCH_2CHCH_2}{\overset{S}{\overset{/\backslash}{}}} \xrightarrow{(EtO)_3P} PhOCH_2CH=CH_2$$

JOC (1961) 26 3467

$$\xrightarrow{KSeCN \quad MeOH}$$

64%

Carbohydrate Res (1967) 5 282

Section 205   Olefins from Halides and Sulfonates

The reduction of vinyl halides, $R_2C=CX \longrightarrow R_2C=CH$, is included in section 160 (Hydrides from Halides and sulfonates)

$$PrCH_2Br \xrightarrow[\substack{2 \ NaNH_2 \\ THF}]{1 \ Ph_3P} PrCH=PPh_3 \xrightarrow{O_2} PrCH=CHPr$$

< 66%

Ber (1963) 96 1899
    (1966) 99 2848
    (1961) 94 1987
Annalen (1969) 721 34

$$Ph(CH_2)_2CH_2Br \xrightarrow[\text{2 NaIO}_4]{\text{1 Ph}_3\text{P}} Ph(CH_2)_2CH_2PPh_3 \overset{+}{} \overset{-}{IO_4} \xrightarrow[\text{NH}_3]{\text{NaNH}_2} Ph(CH_2)_2CH=CH(CH_2)_2Ph$$

< 33%

Ber (1969) <u>102</u> 2259

Org React (1965) <u>14</u> 270

$$CH_2=CHCH_2Br \xrightarrow[\text{C}_6\text{H}_6]{\text{Ph}_3\text{P}} CH_2=CHCH_2PPh_3 \overset{+}{} \overset{-}{Br} \xrightarrow[\text{2 PhCHO}]{\text{1 PhLi   Et}_2\text{O}} CH_2=CHCH=CHPh \qquad 53\%$$

Ber (1954) <u>87</u> 1318
Annalen (1965) <u>682</u> 58

Further examples of the Wittig reaction are included in section 199
(Olefins from Aldehydes) and section 207 (Olefins from Ketones)

JACS (1946) <u>68</u> 1109

$$PhBr \xrightarrow[\text{2 CH}_3\text{COOEt}]{\text{1 Mg   Et}_2\text{O}} Ph_2\overset{\underset{|}{OH}}{C}CH_3 \xrightarrow{\Delta} Ph_2C=CH_2$$

Org Synth (1932) Coll Vol 1 226

$$O=\text{(cyclohexanone with Br at 4-position)} \xrightarrow{(CH_2=\overset{Me}{\underset{|}{C}})_2CuLi} O=\text{(cyclohexanone with } \overset{|}{\underset{|}{C}}=CH_2 \text{, Me groups)}$$

J Indian Chem Soc (1968) <u>45</u> 1026

$$RBr \xrightarrow[\text{DMF}]{[Me-\text{(Ni complex)}\overset{Br}{\diagdown}]_2} RCH_2\overset{|}{\underset{Me}{C}}=CH_2$$

90% (R= Me)
70% (R= CH$_2$=CH)
92% (R= PhCH$_2$CH$_2$CH$_2$)

JACS (1967) <u>89</u> 2755

$$\text{(cyclopentyl)}-Br \xrightarrow[\text{2 BrCH}_2\text{CH=CH}_2]{1 \text{ Mg} \quad Et_2O} \text{(cyclopentyl)}-CH_2CH=CH_2 \qquad 70\%$$

JACS (1946) <u>68</u> 1101
JOC (1970) <u>35</u> 22

Further examples of the coupling of halides are included in section 70 (Alkyls and Aryls from Halides and Sulfonates)

$$BuCl \xrightarrow[\text{2 BrCH}_2\text{CHBr}]{1 \text{ Mg} \quad Et_2O} \underset{OMe}{BuCHCH_2Br} \xrightarrow{Zn} BuCH=CH_2$$

(with OMe on BrCH$_2$CHBr)

JOC (1952) <u>17</u> 807

$$BuBr \xrightarrow[\text{2 MeCHCOMe}]{1 \text{ Li} \quad Et_2O} \underset{Me}{Bu\overset{OH}{\underset{|}{C}}-\overset{Cl}{\underset{|}{C}}HMe} \xrightarrow[\text{H}_2\text{O}]{NaOH} \underset{Me}{Bu\overset{O}{\overset{\diagup\diagdown}{C}}-CHMe} \xrightarrow[\text{(steps)}]{1 \text{ NaI} \quad NaOAc} \underset{Me \quad trans}{BuC=CHMe}$$

(BuBr with Cl on MeCHCOMe)

1 NaI   NaOAc
EtCOOH
2 SnCl$_2$ Pyr
3 POCl$_3$ Pyr

Zn   NaI   NaOAc   HOAc → cis + trans

JCS (1959) 112

$PrCH_2Br$ $\dashrightarrow$ $PrCH_2MgBr$ $\xrightarrow{\text{CF}_2\text{Br}_2 \ \ \text{Et}_2\text{O}}$ $PrCH=CHCH_2Pr$ 72%

Ber (1962) 95 1958

60%

Compt Rend (1965) 260 4535

$C_5H_{11}CH_2Br$ $\dashrightarrow$ $C_5H_{11}CH_2Li$ $\xrightarrow[\text{Et}_2\text{O}]{}$ $C_5H_{11}CH=CHCH_2C_5H_{11}$ 33%

trans

Tetr Lett (1964) 2547

$C_{15}H_{31}CH_2X$ $\xrightarrow{\text{NaH-CH}_3\text{SOMe}}$ $C_{15}H_{31}CH_2CH_2SOMe$ $\xrightarrow[\text{reflux}]{\text{Me}_2\text{SO}}$ $C_{15}H_{31}CH=CH_2$ <85%

X=Br or OTs

Chem Comm (1965) 29

$C_5H_{11}CH_2Br$ $\dashrightarrow$ $C_5H_{11}CH_2SH$ $\xrightarrow[\substack{\text{2 m-Chloroper-}\\ \text{benzoic acid}}]{\text{1 HCHO \ \ HCl}}$ $C_5H_{11}CH_2SO_2CH_2Cl$

$\downarrow \substack{\text{NaOH} \\ \text{H}_2\text{O}}$

$C_5H_{11}CH=CH_2$ <53%

JACS (1964) 86 4383

$\xrightarrow{\text{Collidine}}$

JOC (1961) 26 2883

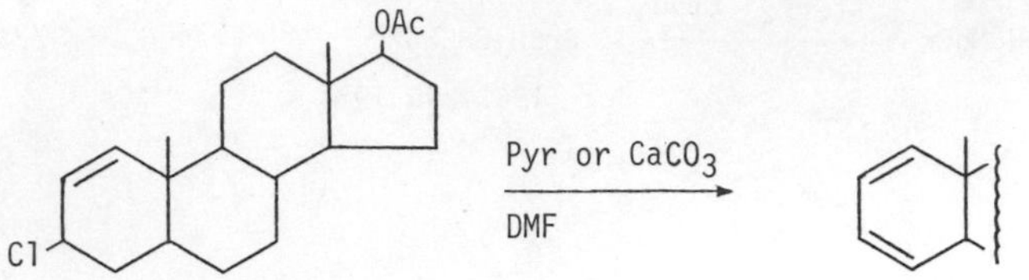

JOC (1963) 28 1976   80%

$BuCH_2CH_2Br$ $\xrightarrow[\text{}]{\text{HMPA   180-210°}}$ $BuCH=CH_2$ 61%

Chem Comm (1971) 113

$C_3H_7\overset{\underset{|}{Br}}{C}HCH_2Et$ $\xrightarrow[Me_2SO]{\text{1,5-Diazabicyclo[5.4.0]undec-5-ene}}$ $C_3H_7CH=CHEt$ 91%

Angew (1967) 79 53
(Internat Ed 6 76)
Ber (1966) 99 2012

$C_5H_{11}CH_2CH_2Br$ $\xrightarrow{\text{LiF   Li}_2\text{CO}_3\text{   HMPA}}$ $C_5H_{11}CH=CH_2$ ~51%

Bull Soc Chim Fr (1967) 2455
                    (1968) 283
JOC (1970) 35 76 1023

$Ph_3CCH_2CH_2Cl$ $\xrightarrow{\text{EtONa   Me}_2\text{SO}}$ $Ph_3CCH=CH_2$ 90%

JCS C (1967) 1115

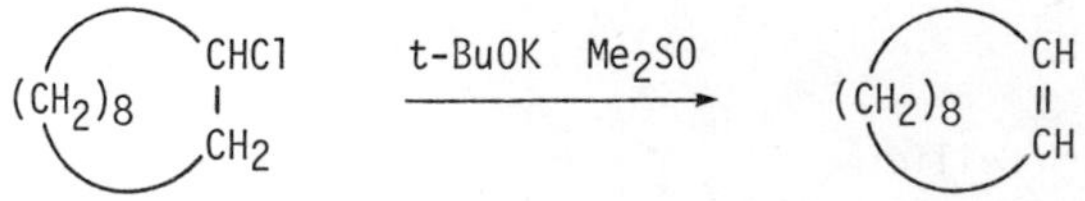

JOC (1967) 32 510
    (1965) 30 2054
Tetr Lett (1967) 2273

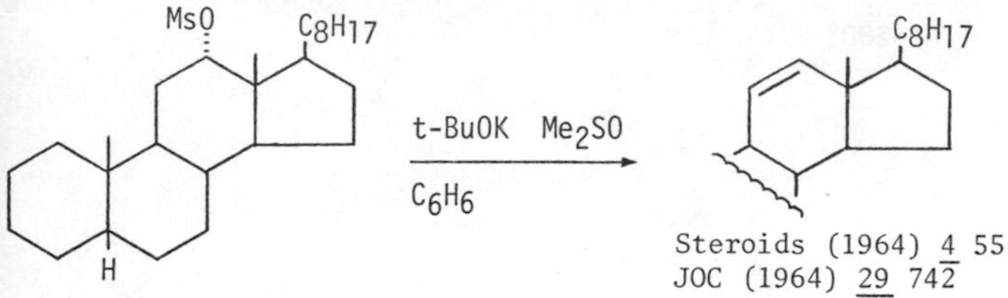

t-BuOK Me$_2$SO

C$_6$H$_6$

Steroids (1964) 4 55
JOC (1964) 29 742

65%

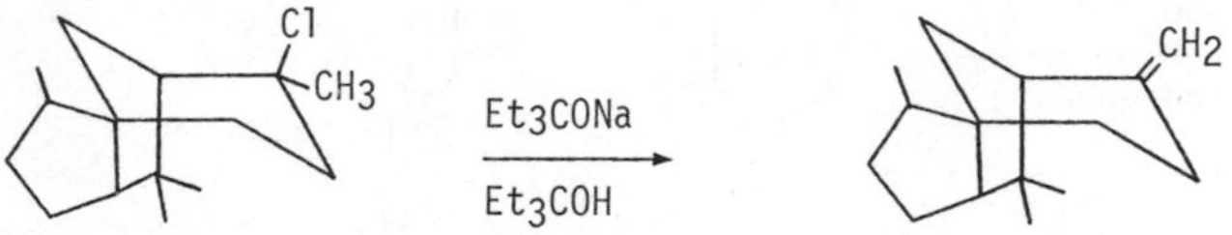

Et$_3$CONa

Et$_3$COH

JOC (1970) 35 196
Chem Comm (1968) 305

52%

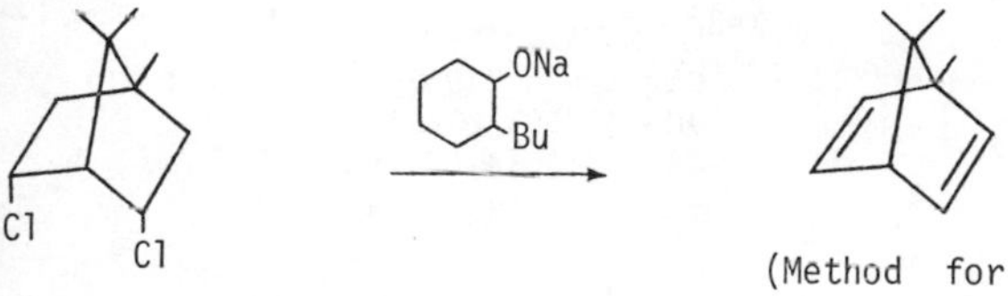

ONa

Bu

(Method for unreactive halides)

Annalen (1962) 652 96

94%

The following bases/solvents may also be used for the elimination of HX
from halides and sulfonates:

Lithium dicyclohexylamide	JOC (1967) 32 510
NaNH$_2$-HMPA	Bull Soc Chim Fr (1966) 1293
Potassium t-amylate	Helv (1967) 50 2111
Me$_2$SO	JACS (1959) 81 5428
	Tetr Lett (1968) 4191
(MeO)$_3$P-xylene	Helv (1958) 41 70
C$_{17}$H$_{35}$COOAg-C$_6$H$_6$	Ber (1942) 75B 660
Ag$_2$O-C$_6$H$_6$	Helv (1951) 34 1176

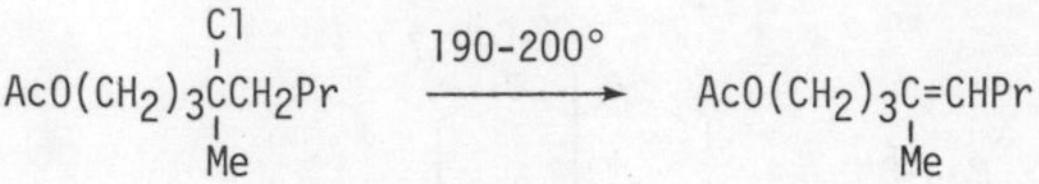

$$\underset{\underset{Me}{|}}{AcO(CH_2)_3\overset{\overset{Cl}{|}}{C}CH_2Pr} \xrightarrow{190\text{-}200°} \underset{Me}{AcO(CH_2)_3C=CHPr} \qquad\qquad 80\%$$

JOC (1948) 13 239

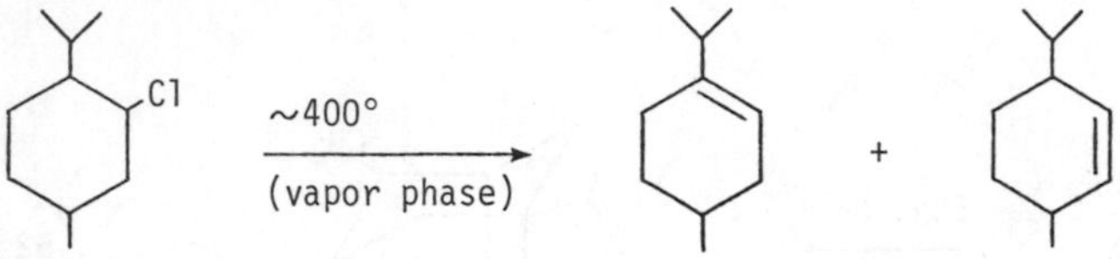

JCS (1952) 453
Chem Rev (1960) 60 431

$$Ph_2CX_2 \xrightarrow[\substack{\text{or NaI} \\ \text{or Fe(CO)}_5}]{Cu} Ph_2C=CPh_2$$

X=Br or Cl

Org Synth (1963) Coll Vol 4 914
JCS (1959) 678
JACS (1961) 83 1623

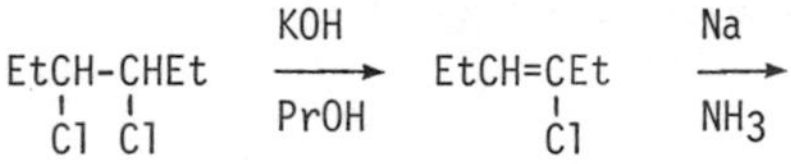

$$\underset{\underset{Cl\ \ Cl}{|\ \ \ |}}{EtCH\text{-}CHEt} \xrightarrow[PrOH]{KOH} \underset{\underset{Cl}{|}}{EtCH=CEt} \xrightarrow[NH_3]{Na} EtCH=CHEt$$

JACS (1951) 73 3329

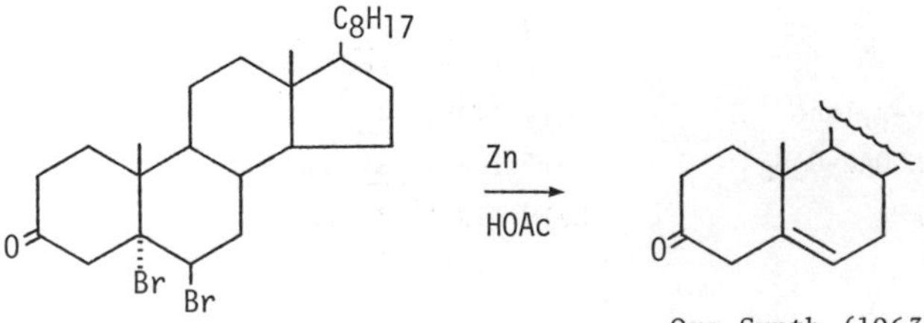

Org Synth (1963) Coll Vol 4 195
JOC (1970) 35 1733
JCS (1961) 4547

The following reagents may also be used for the conversion of 1,2-dihalides to olefins:

NaI, KI	JOC (1965) 30 1658
	(1959) 24 143
	Annalen (1967) 705 76
CrX$_2$ (X=Cl, OAc etc.)	JACS (1968) 90 1582
	(1967) 89 6547
	(1964) 86 4603
	JOC (1959) 24 1621 1629
	Tetrahedron (1968) 24 3503
Bu$_3$SnH	JOC (1963) 28 2165
	Acc Chem Res (1968) 1 299
Na-NH$_3$	JACS (1952) 74 4590
Sodium dihydronaphthylide	Chem Comm (1969) 78
Phenanthrene disodium	Angew (1964) 76 432
PhSNa, EtSNa	Rev Chim (Bucharest) (1962) 7 1379
	(Chem Abs 61 4208)
	JOC (1959) 24 143
Na$_2$Se	JOC (1966) 31 4292
(NH$_2$)$_2$CS	Chem Ind (1966) 1418
KSCN	Chem Ind (1966) 1418
Potassium cyclohexylphosphide	Ber (1961) 94 2664
(EtO)$_3$P	JOC (1970) 35 3181
NaH-Me$_2$SO	Chem Ind (1965) 345
LiAlH$_4$	Can J Chem (1964) 42 1294
PhLi	Annalen (1967) 705 76
PrMgBr	Bull Soc Chim Fr (1946) 604

NaI   Me$_2$CO   73%

JCS (1950) 598
JACS (1952) 74 4894

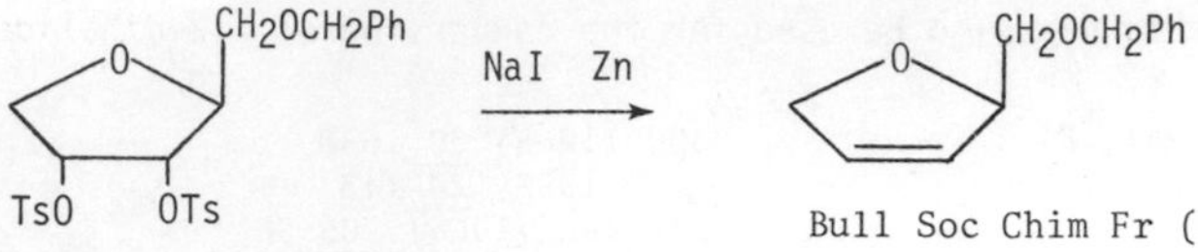

CH2OCH2Ph
NaI Zn
TsO OTs

82%

Bull Soc Chim Fr (1967) 4111

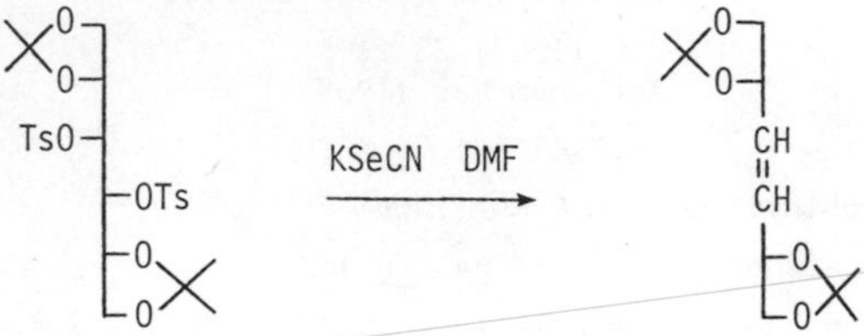

TsO
OTs
KSeCN DMF
CH
CH

45%

Carbohydrate Res (1967) 5 282

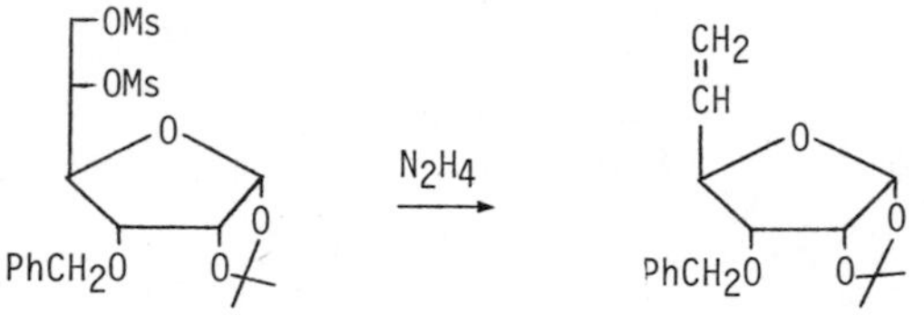

OMs
OMs
N2H4
PhCH2O
CH2
CH

17%

Ber (1969) 102 820

Section 206 Olefins from Hydrides (RH)

This section lists examples of the replacement of hydrogen by olefinic groups. For the dehydrogenation of alkyl groups, e.g. $RCH_2CH_2R \longrightarrow RCH{=}CHR$, see section 200 (Olefins from Alkyls, Methylenes and Aryls)

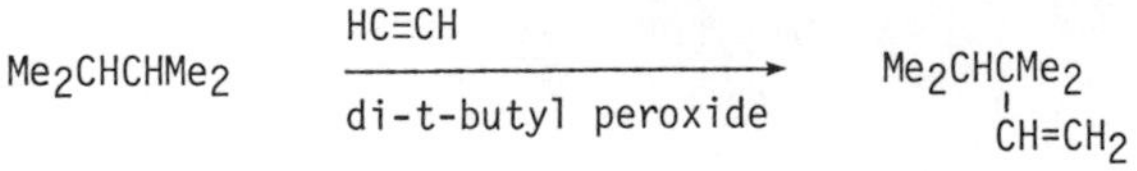

Me2CHCHMe2
HC≡CH
di-t-butyl peroxide
Me2CHCMe2
CH=CH2

JOC (1965) 30 3814

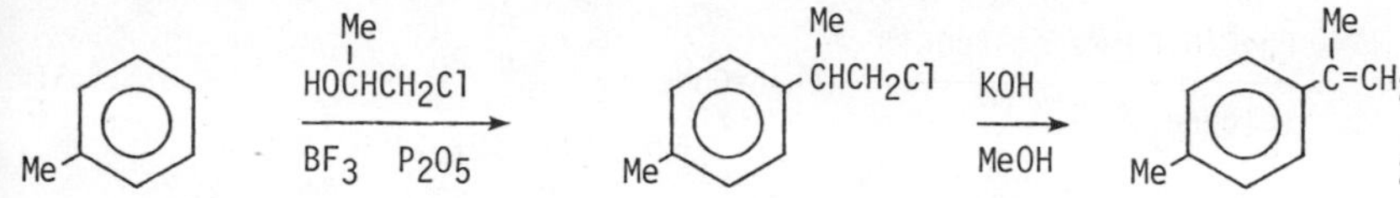

JACS (1948) 70 1772

Section 207   Olefins from Ketones

Reviews: The Wittig Reaction   Org React (1965) 14 270
                                       Quart Rev (1963) 17 406

New Reactions of Alkylidenephosphoranes and their Preparative Uses

Angew (1965) 77 850
(Internat Ed 4 830)

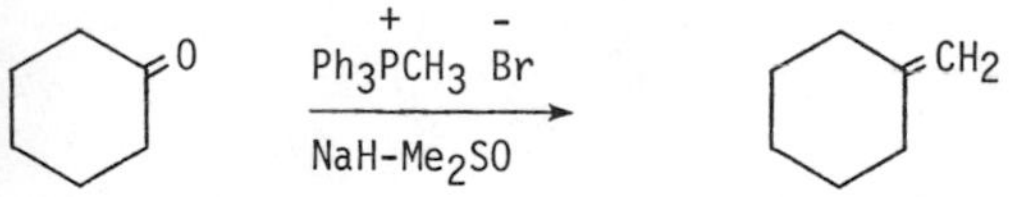

86%

Org React (1965) 14 270
Ber (1954) 87 1318

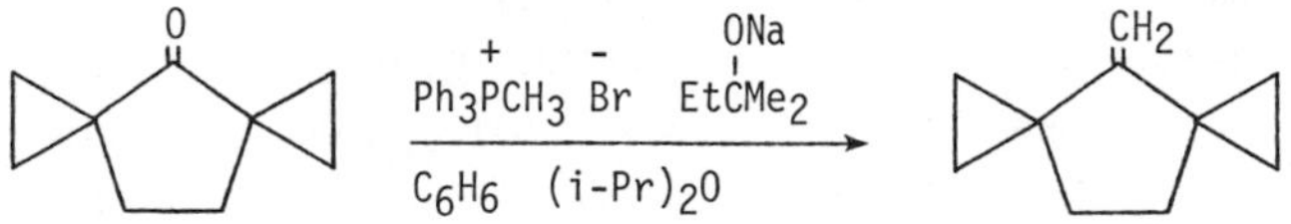

90%

Bull Soc Chim Fr (1967) 1936

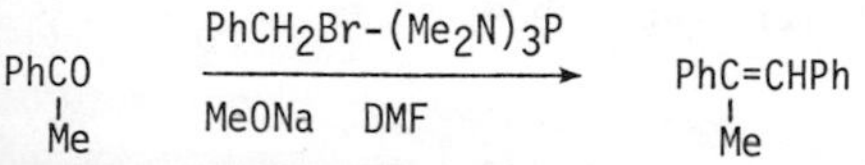

$PhCO$
$\underset{Me}{|}$
$\xrightarrow[\text{MeONa \quad DMF}]{PhCH_2Br-(Me_2N)_3P}$
$PhC{=}CHPh$
$\underset{Me}{|}$
35%

Annalen (1965) 682 58

Ph$_2$CO $\xrightarrow[\text{toluene}]{\text{Ph}_2\text{P(O)CH}_2\text{Me} \quad \text{t-BuOK}}$ Ph$_2$C=CHMe   51%

Ber (1959) $\underline{92}$ 2499

Ph$_2$CO $\xrightarrow[\text{2 Silica gel} \quad \text{C}_6\text{H}_6 \quad \text{reflux}]{\overset{\text{Me}}{\overset{|}{\text{1 LiCHPO(NMe}_2)_2}} \quad \text{THF}}$ Ph$_2$C=CHMe   87%

JACS (1966) $\underline{88}$ 5652 5653
JOC (1969) $\underline{34}$ 3053

JACS (1966) $\underline{88}$ 5656
    (1968) $\underline{90}$ 5548

Ph$_2$CO $\xrightarrow{\overset{\text{Me}}{\overset{|}{\text{LiCHPS(OMe)}_2}} \quad \text{THF}}$ Ph$_2$C=CHMe   93%

JACS (1966) $\underline{88}$ 5654

The following reagents/solvents may also be used for the generation of Wittig reagents:

NaH-Me$_2$SO, NaNH$_2$-NH$_3$, PhLi, BuLi,

EtOLi-EtOH, EtONa-EtOH	Org React (1965) $\underline{14}$ 270
Na	Chem Abs (1940) $\underline{34}$ 392
t-BuOK-Me$_2$SO	Ber (1965) $\underline{98}$ 604
PhLi-t-BuOK-t-BuOH	Angew (1964) $\underline{76}$ 683 (Internat Ed $\underline{3}$ 636)
Ethylene oxide	Angew (1968) $\underline{80}$ 535 (Internat Ed $\underline{7}$ 536)
Diazabicyclo[3.4.0]non-5-ene	Ber (1966) $\underline{99}$ 2012
Electrolysis (non basic)	JACS (1968) $\underline{90}$ 2728 Tetr Lett (1969) 3523

Further examples of the Wittig reaction are included in section 199 (Olefins from Aldehydes)

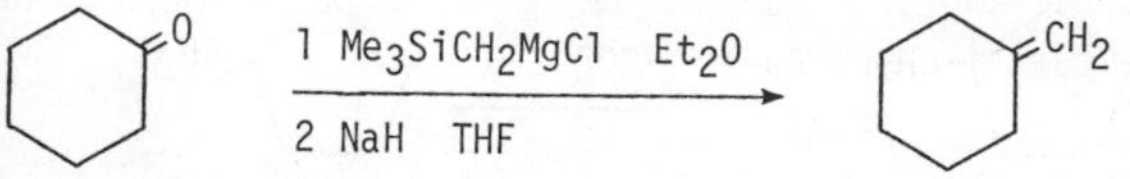

> 50%

Tetr Lett (1970) 1137
JOC (1968) 33 780

$(C_5H_{11})_2CO \xrightarrow{C_5H_{11}CHLi_2 \quad THF} (C_5H_{11})_2C{=}CHC_5H_{11}$ 30-35%

Tetr Lett (1966) 4315

$(C_5H_{11})_2CO \xrightarrow{CH_2I_2 \quad Mg \quad Et_2O} (C_5H_{11})_2C{=}CH_2$ 30%

Tetrahedron (1970) 26 1281

$Ph_2CO \xrightarrow{PhCH(Li)NC \quad THF} Ph_2C{=}CHPh$ 74%

Angew (1968) 80 842
(Internat Ed 7 805)

$Cl(CH_2)_2CH_2C(Me)O \xrightarrow[2\ Ac_2O]{1\ C_5H_{11}MgBr \quad Et_2O} Cl(CH_2)_2CH_2C(OAc)(Me)C_5H_{11} \xrightarrow[C_6H_6]{TsOH} Cl(CH_2)_2CH{=}C(Me)C_5H_{11}$

JOC (1948) 13 239 75%
JACS (1949) 71 819

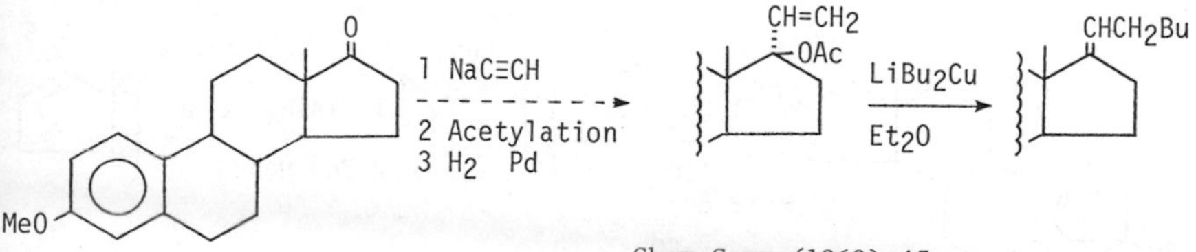

Chem Comm (1969) 43
JACS (1970) 92 735

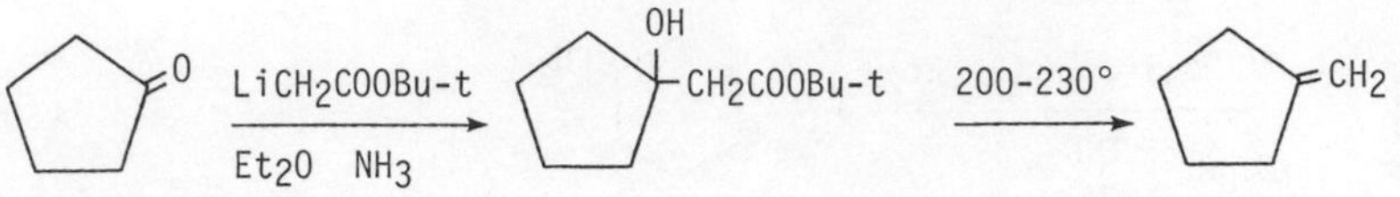

Bull Soc Chim Fr (1960) 1196

$Ph_2CO \xrightarrow{(i-PrO)_3P} Ph_2C=CPh_2$ low yield

JOC (1964) 29 2567

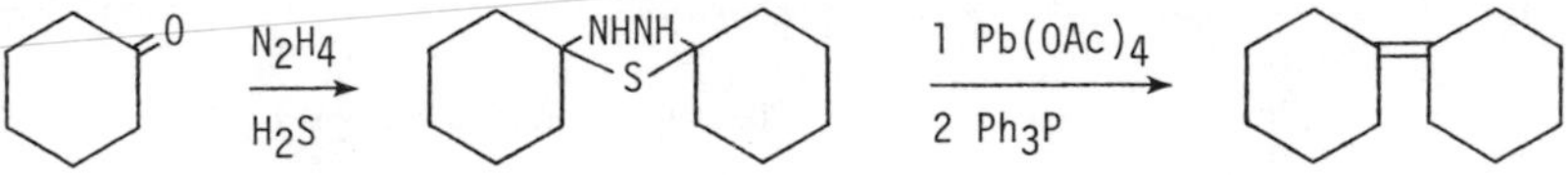

73%

Chem Comm (1970) 1226 1225

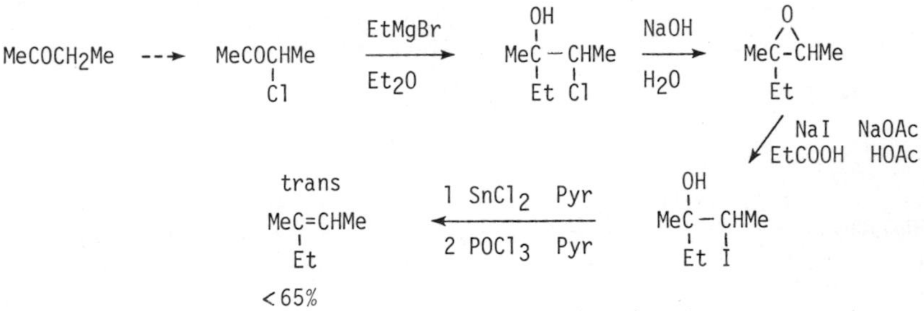

JCS (1959) 112

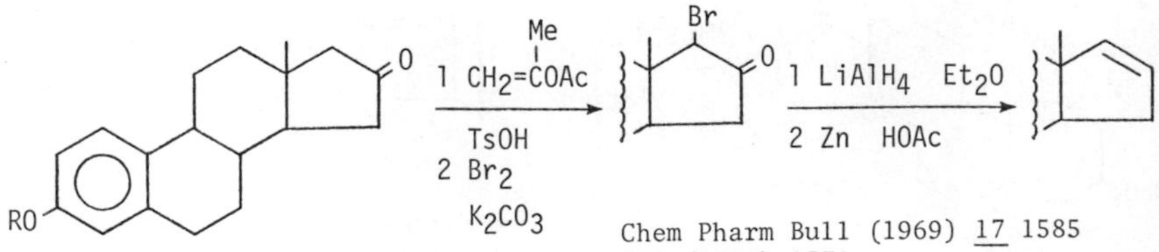

Chem Pharm Bull (1969) 17 1585
JCS (1955) 1370

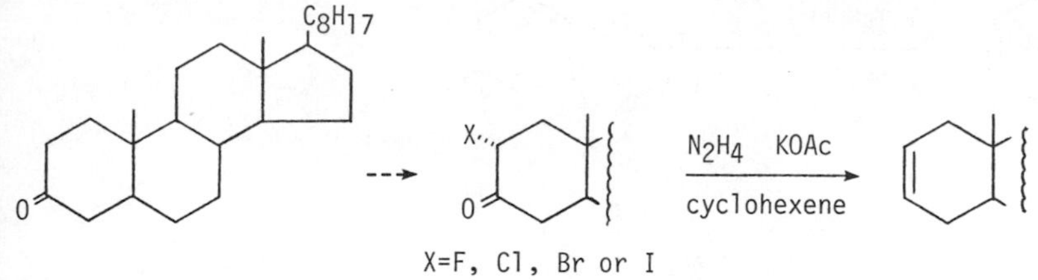

X=F, Cl, Br or I

JOC (1964) 29 958

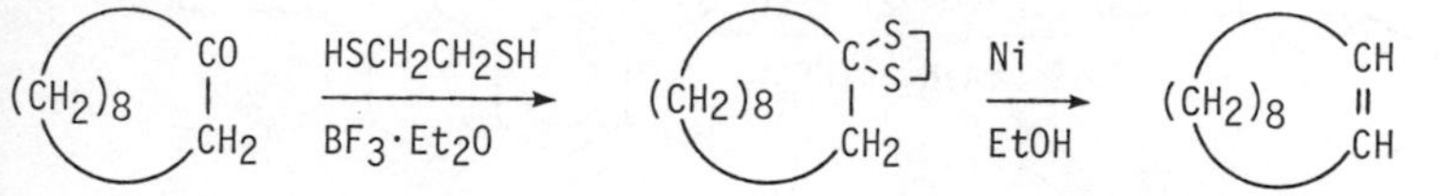

55%

Tetrahedron (1969) 25 2823
JOC (1963) 28 1443

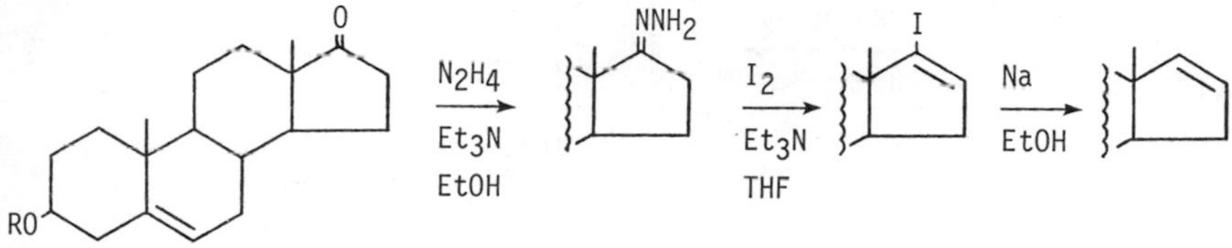

JCS (1962) 470

~100%

Annalen (1966) 691 41

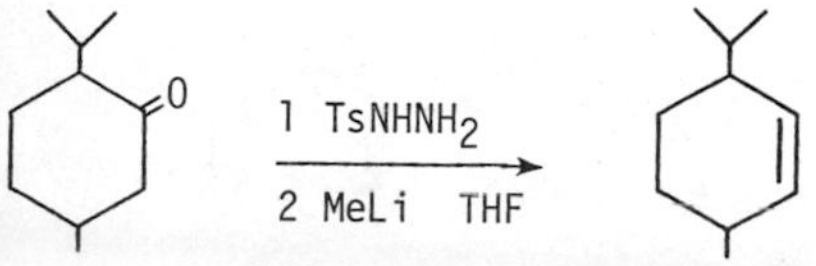

~79%

Synthesis (1970) 595
JACS (1967) 89 5734

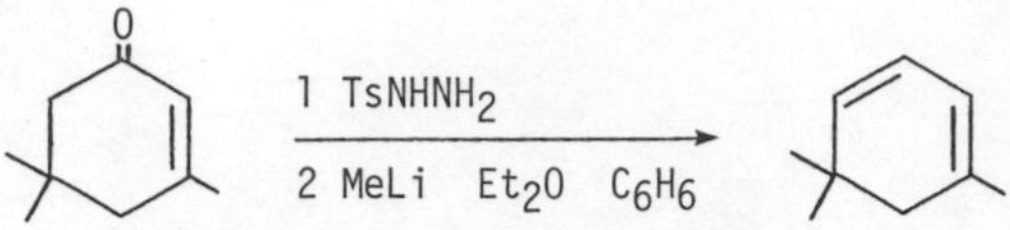
O
1 TsNHNH2
2 MeLi Et2O C6H6

~100%

JACS (1968) 90 4762

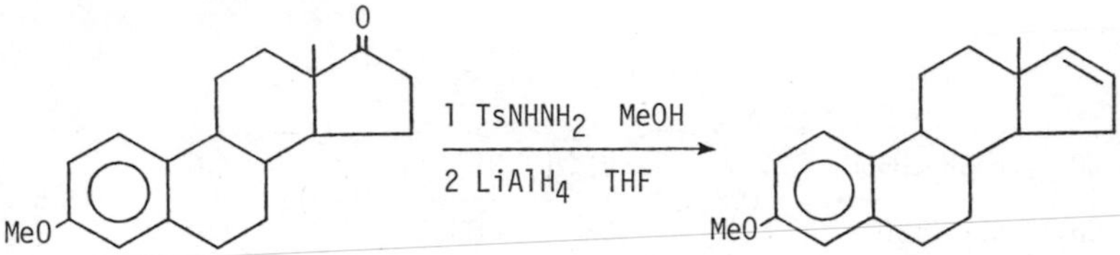
O
1 TsNHNH2 MeOH
2 LiAlH4 THF
MeO
MeO

70%

Tetrahedron (1963) 19 1127

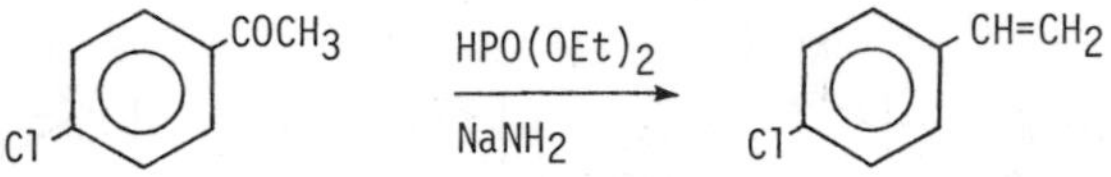
COCH3
Cl
HPO(OEt)2
NaNH2
CH=CH2
Cl

60%

Angew (1963) 75 138
(Internat Ed 2 98)

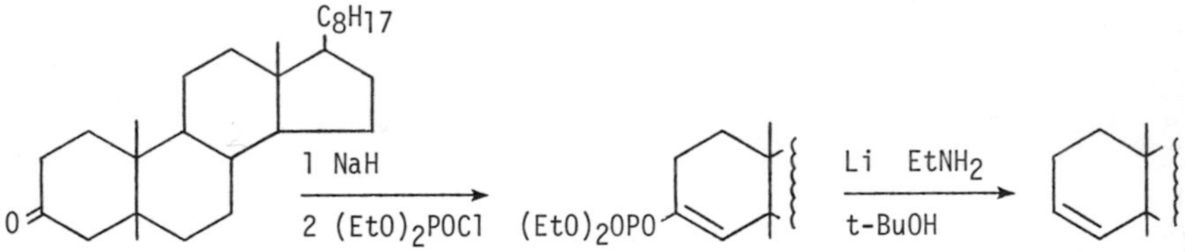
C8H17
O
1 NaH
2 (EtO)2POCl
(EtO)2OPO
Li EtNH2
t-BuOH

81%

Tetr Lett (1969) 2145
Chem Comm (1969) 112

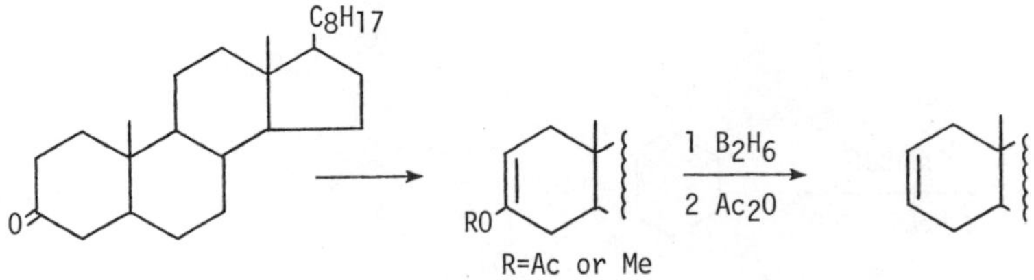
C8H17
O
RO
1 B2H6
2 Ac2O
R=Ac or Me

Gazz (1962) 92 309
(Chem Abs 57 12572)

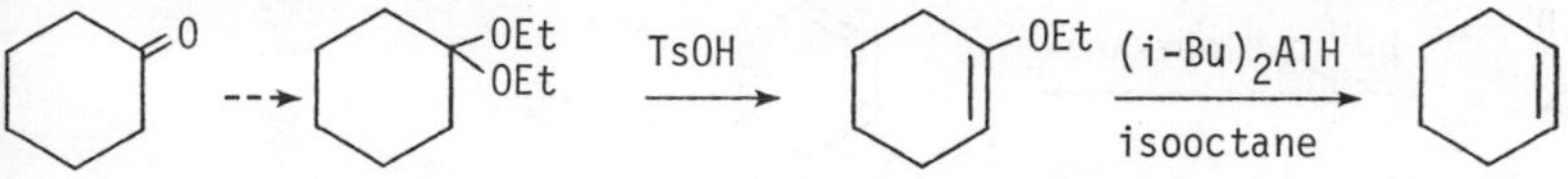

JOC (1966) 31 329

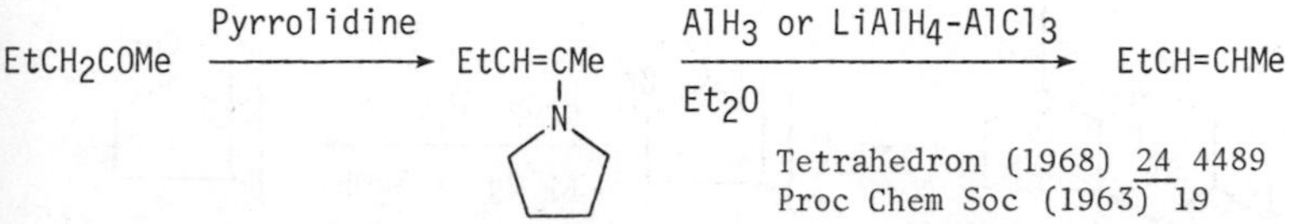

Tetrahedron (1968) 24 4489
Proc Chem Soc (1963) 19

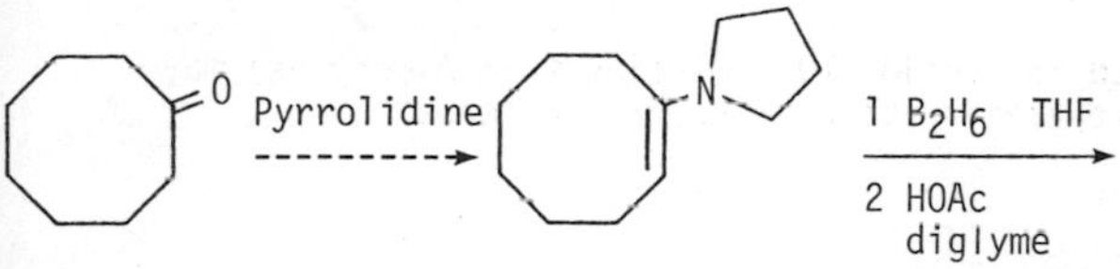

Tetr Lett (1964) 2039

H₂ catalyst

p-Nitroper-
benzoic
acid

110°

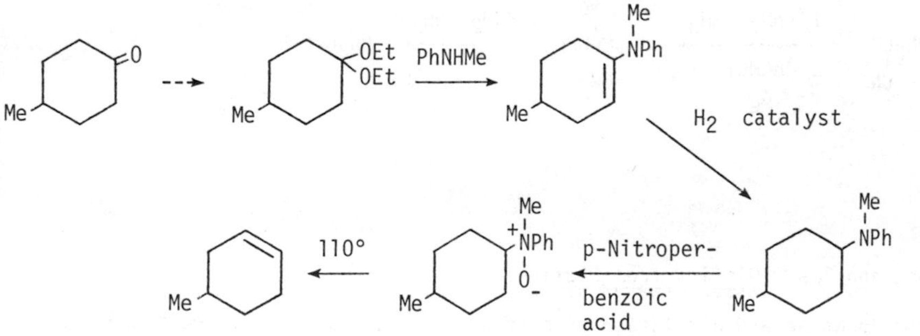

Tetr Lett (1964) 3853
Compt Rend (1967) C 264 710

$$PrCOCH_3 \xrightarrow[550° (1 \text{ sec})]{MnO_2} PrCH=CH_2 \qquad 62\%$$

Chem Comm (1969) 461

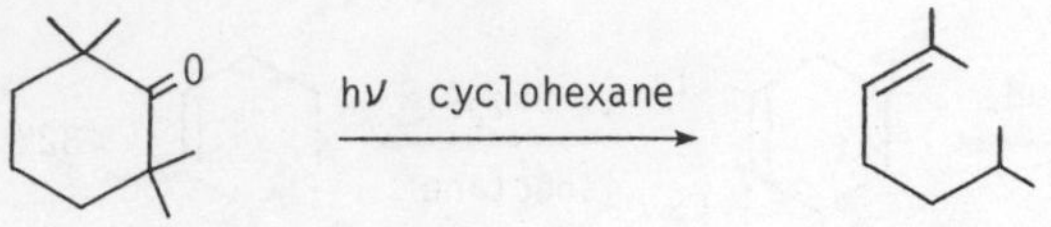

JOC (1966) 31 1393

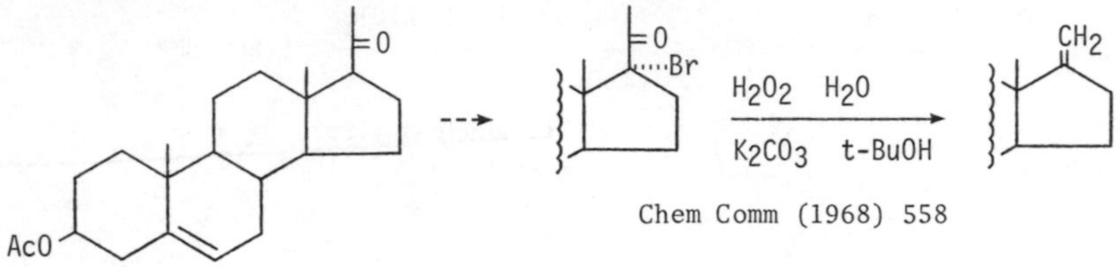

Chem Comm (1968) 558

<65%

Some of the methods listed in section 199 (Olefins from Aldehydes) may also be applied to the preparation of olefins from ketones

Section 208 Olefins from Nitriles

$Ph_2CH(CN)$ —(1 KNH_2 NH_3; 2 PhCHCl(Me))→ $Ph_2C(CN)CH(Me)Ph$ —(KNH_2 NH_3)→ $Ph_2C{=}C(Me)Ph$ 71%

JACS (1956) 78 82
(1960) 82 1786

Section 209 Olefins from Olefins

$BuCH{=}CH_2$ —(1 B_2H_6; 2 $PhHgCCl_2Br$)→ $BuCH_2CH_2CH{=}CHCH_2Bu$

JACS (1966) 88 1834

$$\underset{(CH_2)_6}{(CH=CH)} \xrightarrow[\text{t-BuOH}]{\text{CHBr}_3 \quad \text{t-BuOK}} \underset{(CH_2)_6}{\overset{CBr_2}{(CHCH)}} \xrightarrow{Mg \quad Et_2O} \underset{(CH_2)_6}{(CH=C=CH)}$$

$$\downarrow H_2 \quad Pd\text{-}C$$

$$15\% \quad \underset{(CH_2)_6}{(CH_2CH=CH)}$$

JOC (1961) 26 3518

$$PhCH_2CH=CH_2 \xrightarrow{NaH\text{-}Me_2SO} \underset{Me}{PhCH_2CHCH_2SOMe} \xrightarrow{\Delta} \underset{Me}{PhCH_2C=CH_2}$$

JOC (1964) 29 2699

$$MeCH=CHEt \underset{\longleftarrow}{\xrightarrow{WCl_6 \quad EtAlCl_2 \quad EtOH}} MeCH=CHMe \quad + \quad EtCH=CHEt$$

Tetr Lett (1967) 3327
JACS (1968) 90 4133

$$C_6H_{13}CH=CH_2 \xrightarrow[\text{Me}_3Al_2Cl_3 \quad PhCl]{(Ph_3P)_2Cl_2(NO)_2Mo} C_6H_{13}CH=CHC_6H_{13} \qquad 37\%$$

JACS (1970) 92 528

1 NaBH$_4$ BF$_3$
diglyme
2 C$_8$H$_{17}$CH=CH$_2$

62%

JACS (1967) 89 567 561
(1960) 82 2074

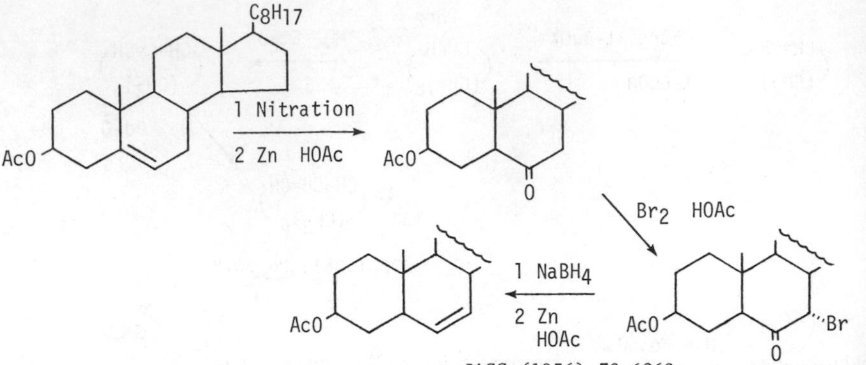

JACS (1956) 78 6269
JOC (1948) 13 424

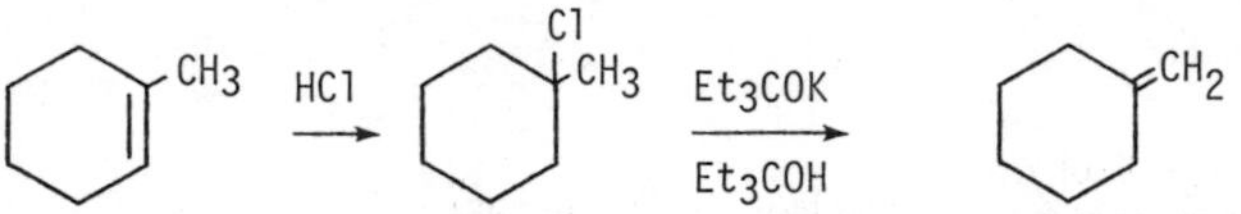

~62%

Chem Comm (1968) 305
Helv (1967) 50 2111

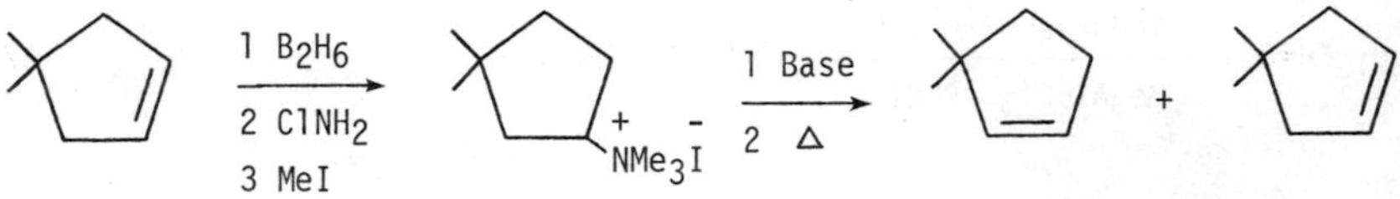

Tetr Lett (1968) 2253

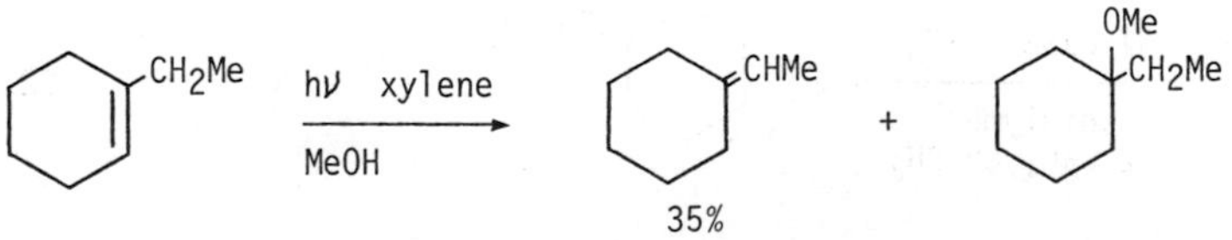

35%

JACS (1967) 89 5199

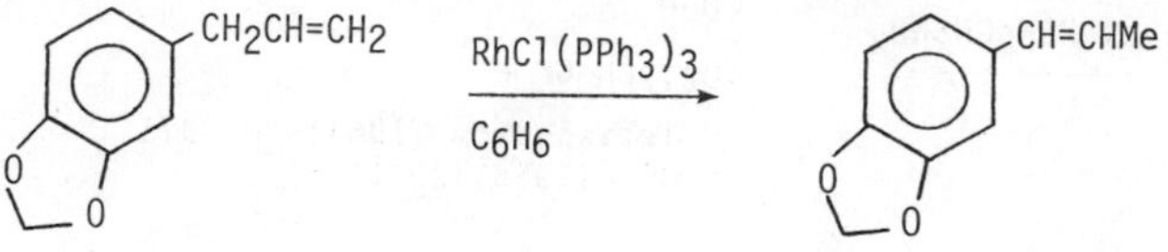

RhCl(PPh$_3$)$_3$

C$_6$H$_6$

60%

Tetr Lett (1968) 3797

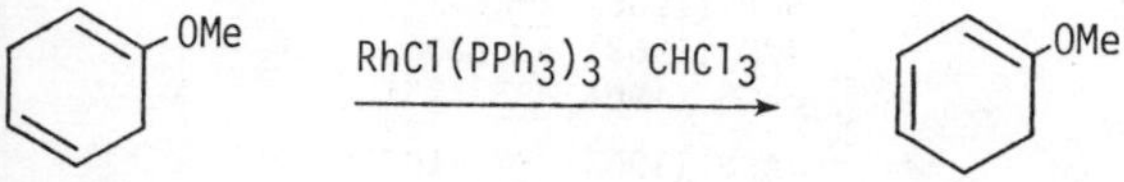

RhCl(PPh$_3$)$_3$　CHCl$_3$

80%

Tetr Lett (1968) 3797

$$EtCH_2CH_2CH=CH_2 \xrightleftharpoons{\text{RhCl}_3\cdot 3H_2O \quad EtOH} EtCH_2CH=CHCH_3 \quad + \quad EtCH=CHCH_2CH_3$$

JACS (1964) 86 1776
Helv (1967) 50 2445

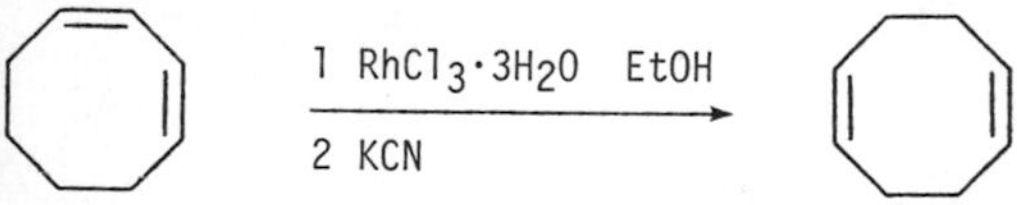

1 RhCl$_3$·3H$_2$O　EtOH

2 KCN

JACS (1964) 86 2516

$$PrCH_2CH_2CH_2CH=CH_2 \xrightarrow[\text{pet ether}]{Fe_3(CO)_{12} \quad h\nu} PrCH_2CH_2CH=CHCH_3 \quad + \quad PrCH_2CH=CHCH_2CH_3$$

$$+ \quad PrCH=CHCH_2CH_2CH_3$$

Proc Chem Soc (1964) 408

Review:　The Isomerization of Olefins.
　　　　　Part I.　Base Catalyzed Isomerization of Olefins

Synthesis (1969) 97

COOH
|
$(CH_2)_7CH_2CH=CH_2$ $\xrightarrow{\text{LiNHCH}_2\text{CH}_2\text{NH}_2}$ COOH
|
$(CH_2)_7CH=CHMe$

Tetrahedron (1964) <u>20</u> 2911
JOC (1958) <u>23</u> 1136

The following bases may also be used for the migration of double bonds:

LiNHEt   JACS (1964) <u>86</u> 5281

t-BuOK-Me$_2$SO   Ber (1966) <u>99</u> 1737
                          JOC (1968) <u>33</u> 221
                          JACS (1961) <u>83</u> 3731

Na-alumina   JACS (1965) <u>87</u> 4107
                          JCS <u>C</u> (1967) 2149
                                (1966) 260

Na-PhCH$_2$Na   Tetr Lett (1964) 467

$(CH_2)_7$ CH‖CH trans $\xrightarrow{\text{2-Naphthalenesulfonic acid}}$ $(CH_2)_7$ CH‖CH cis

JACS (1952) <u>74</u> 3643

MeCH=C(Me)—CH=CH—CH=CCHMe(OH)   cis $\xrightarrow[\text{EtOH}]{\text{I}_2 \; h\nu}$ MeCH=C(Me)—CH=CH—CH=CCHMe(OH)   trans

JACS (1954) <u>76</u> 5719

$(CH_2)_5$—CH=CH—$(CH_2)_5$ CH=CH cis $\xrightarrow{\text{I}_2 \; CCl_4}$ $(CH_2)_5$—CH=CH—$(CH_2)_5$ CH=CH trans

JCS <u>C</u> (1966) 260

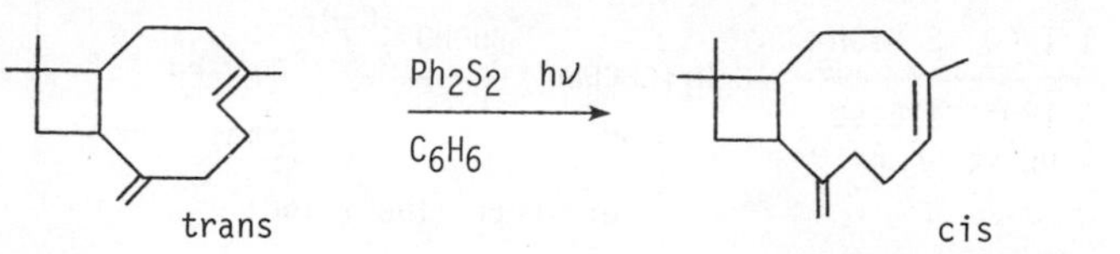

Ph_2S_2 hν
C_6H_6
trans
cis

Helv (1968) 51 548
JACS (1967) 89 2758
JCS C (1966) 260

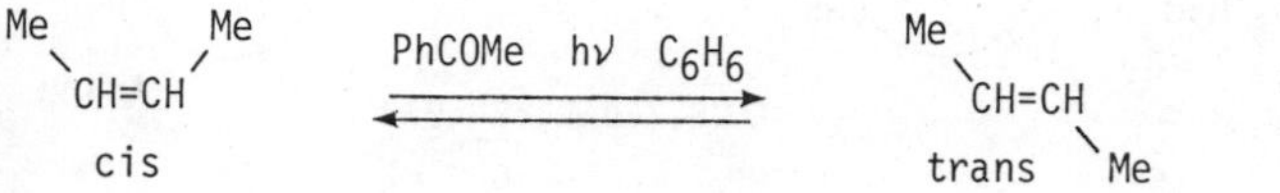

Me
Me
CH=CH
cis
PhCOMe hν C_6H_6
Me
CH=CH
trans
Me

JCS C (1966) 260

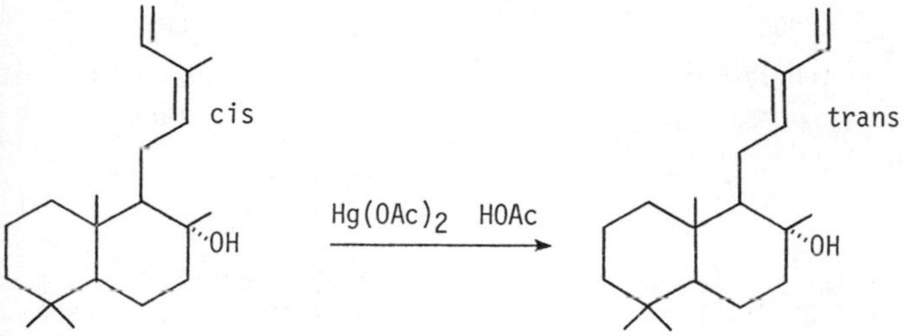

cis
OH
$Hg(OAc)_2$ HOAc
trans
OH

JCS C (1967) 2514

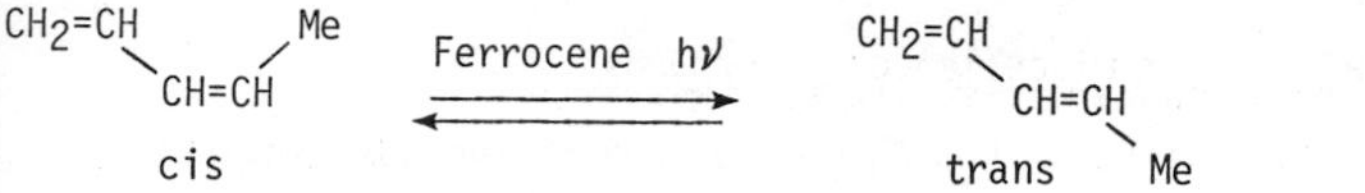

CH_2=CH
Me
CH=CH
cis
Ferrocene hν
CH_2=CH
CH=CH
trans
Me

JACS (1965) 87 1626

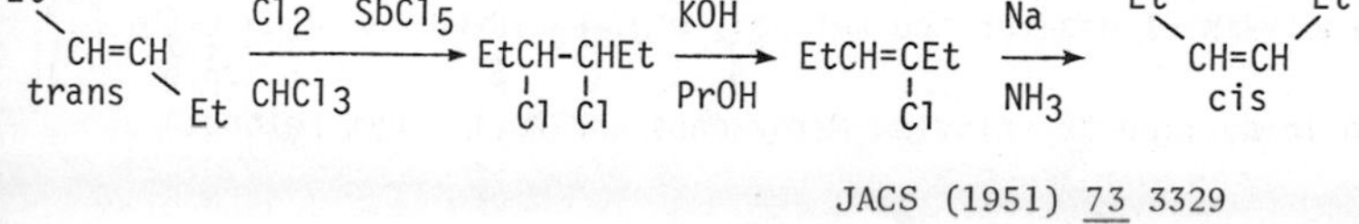

Et
CH=CH
trans
Et
Cl_2 $SbCl_5$
$CHCl_3$
EtCH-CHEt
Cl Cl
KOH
PrOH
EtCH=CEt
Cl
Na
NH_3
Et
Et
CH=CH
cis

JACS (1951) 73 3329

$$C_5H_{11} \diagdown_{CH=CH} \diagup^{Me} \quad \xrightarrow[\substack{3 \ 180° \quad 4 \ Base \\ 5 \ H_2SO_4 \quad 6 \ Base}]{1 \ INCO \quad 2 \ MeOH} \quad C_5H_{11}CHCHMe \xrightarrow{BuONO} \quad C_5H_{11} \diagdown_{CH=CH} \diagup^{} \quad 41\%$$

cis   NH   trans   Me

Tetr Lett (1969) 4001

$$(CH_2)_6 \diagdown^{CH}_{CH} \xrightarrow[\substack{2 \ HBr \quad H_2O \\ 3 \ H_2S}]{1 \ (SCN)_2} (CH_2)_6 \diagdown^{CHS}_{CHS} CS \xrightarrow[135°]{(C_8H_{17}O)_3P} (CH_2)_6 \diagdown^{CH}_{CH}$$

cis   JACS (1965) 87 934   trans

$$\begin{array}{c} \diagup^{C_6H_{13}} \\ CH \\ \| \\ CH \\ \diagdown_{(CH_2)_7COOR} \end{array} \xrightarrow[HCOOH]{H_2O_2} \begin{array}{c} C_6H_{13} \\ | \\ CHOH \\ | \\ CHOH \\ | \\ (CH_2)_7COOR \end{array} \xrightarrow[HOAc]{HBr \ H_2SO_4} \begin{array}{c} C_6H_{13} \\ | \\ CHBr \\ | \\ CHBr \\ | \\ (CH_2)_7COOR \end{array} \xrightarrow[EtOH]{Zn} \begin{array}{c} \diagup^{C_6H_{13}} \\ CH \\ \| \\ CH \\ \diagdown_{(CH_2)_7COOR} \end{array}$$

cis   trans

JCS (1954) 4219
(1953) 2393
(1951) 1079

Further examples of the conversion of diols and dihalides into olefins are
included in section 198 (Olefins from Alcohols) and section 205 (Olefins
from Halides and Sulfonates)

Section 210   Olefins from Miscellaneous Compounds
                ○○○

The reduction of vinyl halides, $R_2C=CR \longrightarrow R_2C=CHR$ is included in section
                                        |
                                        Cl
160 (Hydrides from Halides and Sulfonates)

The hydrogenolysis of unsaturated ketones, e.g. $R_2C=C-CO \longrightarrow R_2CH-C=CH$
                                                        | | | |
                                                        R R   R R
is included in section 72 (Alkyls, Methylenes and Aryls from Ketones)

$C_{15}H_{31}CH_2CH_2SOMe$ $\xrightarrow[\text{reflux}]{Me_2SO}$ $C_{15}H_{31}CH=CH_2$ 85%

Chem Comm (1965) 29
JOC (1964) 29 2699

$(Me_2CH)_2SO_2$ $\xrightarrow{\text{t-BuOK } Me_2SO}$ $MeCH=CH_2$ 100%

Chem Ind (1963) 1243
JOC (1967) 32 102

$\underset{\underset{Me}{|}}{PhCS}$ $\xrightarrow{\text{Ni xylene}}$ $\underset{\underset{Me}{|}\ \underset{Me}{|}}{PhC=CPh}$ 18%

JACS (1944) 66 1136

95%

JOC (1958) 23 1767
    (1961) 26 3467
Chem Pharm Bull (1960) 8 621

$PhCH_2CH_2SO_2Cl$ $\xrightarrow[Et_2O]{Et_3N \quad CH_2N_2}$ $\underset{SO_2}{PhCH_2\overset{\displaystyle CH-CH_2}{\diagdown\ \diagup}}$ $\xrightarrow{80°}$ $PhCH_2CH=CH_2$ 96%

Angew (1965) 77 41
(Internat Ed 4 70)

$\underset{\underset{OMe}{|}}{BuCHCH_2Br}$ $\xrightarrow{\text{Zn PrOH}}$ $BuCH=CH_2$ 78%

JACS (1932) 54 751

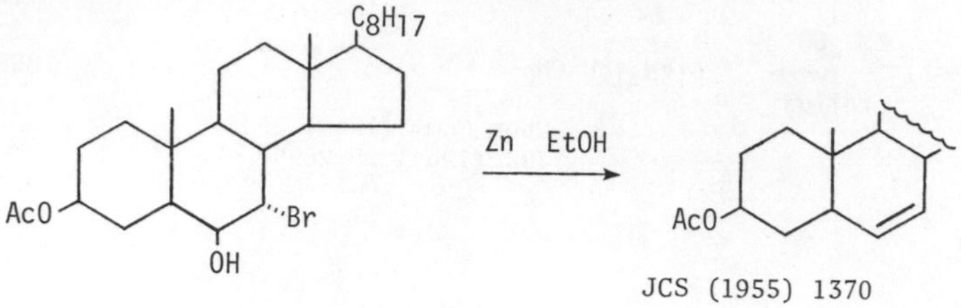

Zn EtOH

JCS (1955) 1370

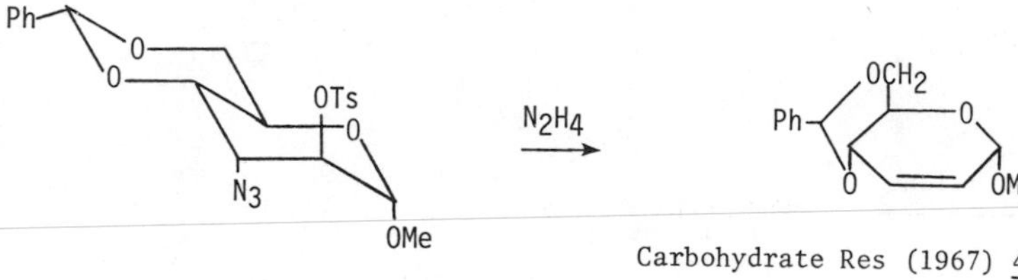

N_2H_4

72%

Carbohydrate Res (1967) 4 465

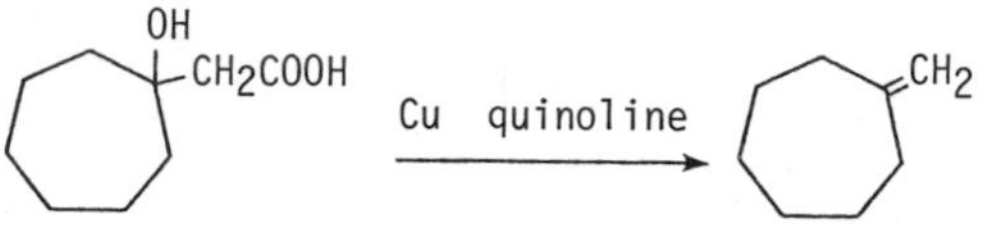

Cu quinoline

60%

Bull Soc Chim Fr (1960) 1196

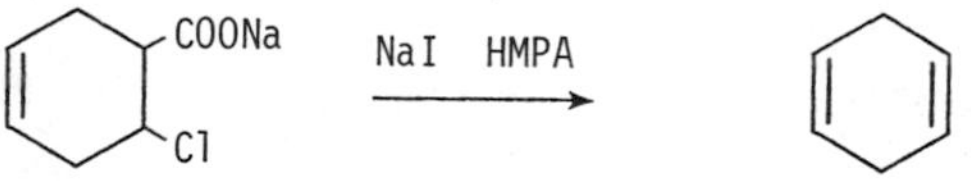

NaI HMPA

70%

JOC (1968) 33 4540

BuCOCCOOEt → 1 NaBH₄ MeOH → BuCH-CCOOH → 1 MsCl Na₂CO₃ → BuCH=CBu

$$\text{BuCOCCOOEt} \xrightarrow[\text{2 Hydrolysis}]{\text{1 NaBH}_4 \; \text{MeOH}} \text{BuCH-CCOOH} \xrightarrow[\text{2 Collidine}]{\text{1 MsCl} \; \text{Na}_2\text{CO}_3} \text{BuCH=CBu}$$

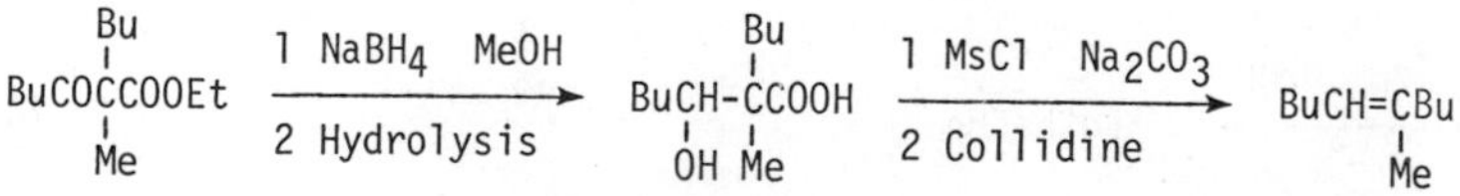

diastereomer A ⟶ trans
diastereomer B ⟶ cis

Tetr Lett (1968) 4569

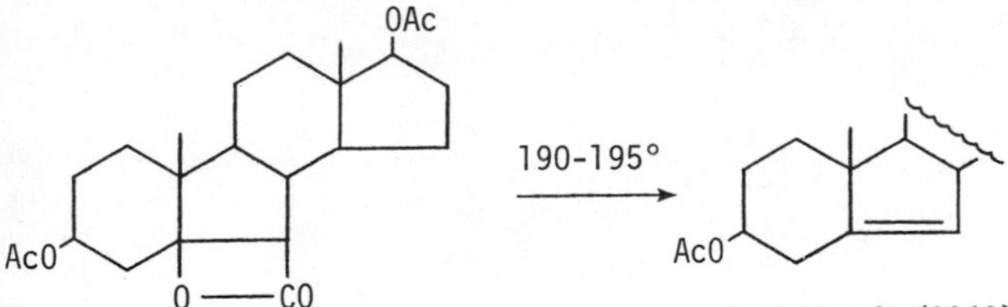

190-195°

91%

Coll Czech (1960) 25 2341

$C_5H_{11}CH_2CH_2CH_2COCOOH$ $\xrightarrow{h\nu \quad C_6H_6}$ $C_5H_{11}CH=CH_2$ 88%

JACS (1968) 90 1840

PhCH=CHCH₂CEt₂ $\xrightarrow{\Delta}$ PhCH₂CH=CH₂
       |
       OH   JCS (1965) 7242

1 Mg   THF
─────────
2 100°

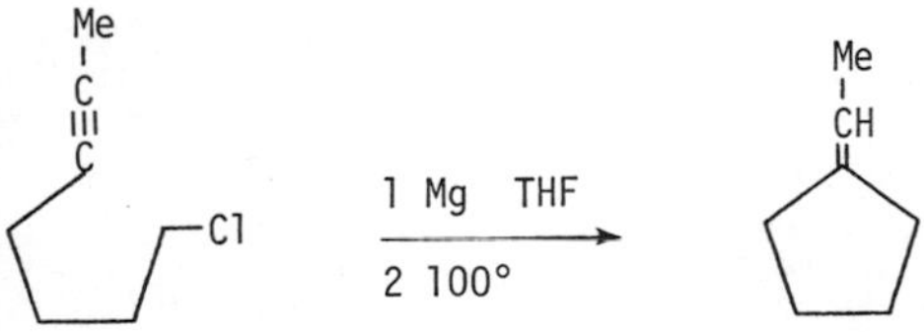

~90%

Tetr Lett (1968) 1457